ROBOTICS

A Reference Guide to the New Technology

Joseph A. Angelo, Jr.

Library of Congress Cataloging-in-Publication Data

Angelo, Joseph A.
Robotics : a reference guide to the new technology / Joseph Angelo.
p. cm.
Includes bibliographical references and index.
ISBN 1-57356-337-4(alk. paper)
1. Robotics. I. Title.
TJ211.A5475 2007
629.8'92-dc22 2006028657

British Library Cataloguing in Publication Data is available.

Library of Congress Catalog Card Number:
ISBN: 1-57356-337-4

First published in 2007

Greenwood Press, 88 Post Road West, Westport, CT 06881
An imprint of Greenwood Publishing Group, Inc.
www.greenwood.com

Printed in the United States of America

The paper used in this book complies with the Permanent Paper Standard issued by the National Information Standards Organization (Z39.48-1984).

10 9 8 7 6 5 4 3 2 1

To the memory of my beloved daughter, Jennifer April Angelo (April 26, 1975, to June 14, 1993), a beautiful young woman, whose promise-filled life was cut short by the careless actions of others. Jennifer's keen intellect, radiant smile, and dazzling emerald green eyes provided brief glimpses into the loving soul of a very special person touched by the goodness and power of God.

This book also carries a special dedication to Mugsy-the-Pug (February 23, 1999, to January 2, 2006)—my faithful canine companion, who provided me so much joy and relaxation, during the preparation of this book and other works.

Pet owners everywhere will understand how their pet can serve as a marvelous bridge to the rest of the living universe. Robot spacecraft are helping us search for life beyond Earth. But the pets at our feet provide us a very special glimpse at the various levels of physical existence, ranging from being self aware to being just there, that surround us in this beautiful and mysterious universe.

Contents

Preface		ix
1.	History of Robot Technology and Systems	1
2.	Chronology of Robot Technology	27
3.	Profiles of Robot Technology: Pioneers, Visionaries, and Advocates	53
4.	How Robot Technology Works	106
5.	Impact	160
6.	Issues	196
7.	The Future of Robot Technology	234
8.	Glossary of Terms Used in Robot Technology	258
9.	Associations	328
10.	Demonstration Sites	367
11.	Sources of Information	397
Index		405

Preface

Human beings have always been fascinated with the concept of artificial life and the construction of machines that look and behave like people. The legend of Pygmalion and the medieval legend of the Golem are examples. Nowhere is the concept of making a living thing out of spare parts more dramatic and exciting than in Mary Shelley's famous story *Dr. Frankenstein: The Modern Prometheus* (1818). This story has been told and retold in various motion pictures and television shows throughout the twentieth century. Today, whenever people take the time to discuss and extrapolate the evolution of smart machines, their conversations usually include the possible rise of self-aware, intelligent robots that threaten to destroy their human masters.

The Czech writer Karel Čapek gave the world the term *robot* when he wrote the play *Rossum's Universal Robots* (*R.U.R.*) in 1920. *Robata* is the Czech word for forced labor or servitude. The play premiered in Prague in 1921 and was then translated into English and first appeared on the English stage in 1923. Ever since then, the word *robot* has been part of the global literature concerning smart machines and automatons.

Thanks to the bad publicity inherent in many such fictional portrayals, robots soon became frightening villains. Later in the twentieth century, this fiction-based perception was reinforced by a real-world economic situation, when industrial robots began displacing hundreds and then thousands of workers from manufacturing jobs. A neo-Luddite wave of *technophobia* (fear or hatred of machines and new technologies) emerged. The fear of smart machines still grips many workers in the world's industrialized nations, which are now experiencing turbulent social and economic transitions into postindustrial, information-based economies. Have you ever seen a person pound on a misbehaving computer or yell at an automobile that will not start?

As presented in this book, the *robot*, like many other new technologies, is really a two-edged sword. The object itself is neither good nor evil. It is how human beings apply a particular technology (including the robot) that creates the ethical environment and moral climate associated with the technology.

When viewed in a purely technical perspective, robots are simply advanced machines that support a continuing revolution in the application of technology in service to human beings. Throughout history, the human race has experienced a number of important technology-related revolutions.

With each development came the need for better devices to perform difficult or monotonous tasks. Simple machines like the wheel and axle or the inclined plane yielded major breakthroughs in the ability to perform work. Often these simple machines allowed workers to accomplish tasks previously regarded as extremely dangerous or impossible. The great pyramids of ancient Egypt or Mesoamerica are examples of how early human societies could effectively use such simple machines to perform difficult tasks. But the machines were not self-powered or smart. The devices required human labor or animal power to operate and generally needed humans to perform a task. In addition, these devices generally took advantage of physical principles and phenomena (like gravity), which were instinctively or qualitatively perceived, but not well understood from a quantitative or analytical perspective. The quantitative understanding of machines and physical principles underlying their operation could not take place without the Scientific Revolution.

Many of today's robots contain components and use scientific principles first adopted during the rise of civilization. The wheel and axle, the pulley, the wedge, the lever, and the gear are examples. Great engineers of antiquity, like Archimedes, Ctesibius, and Hero of Alexandria, developed and introduced many of the important technical devices and early machines. However, it took the brilliant work of Galileo Galilei, Isaac Newton, and other scientists during the Scientific Revolution of the sixteenth and seventeenth centuries to establish the physical framework within which the operation and prediction of machine behavior (including that of robots) could be done with an acceptable degree of certainty and precision.

While the Scientific Revolution provided people with the mathematical tools and physical principles to understand how machines and the universe operated in a somewhat clockwork fashion, this great advancement in human history did not (of itself) provide the social or economic stimulus for developing the smart machine systems (that is, the computers and robots) that appeared in the twentieth century. In fact, during the early eighteenth century, clockmakers and engineers, like Jacques de Vaucanson and Pierre Jacquet-Droz, were quite content to construct elegant automatons, primarily for the amusement of wealthy patrons.

The major stimulus for the production of smarter machines, and eventually robots, was the Industrial Revolution—the period of enormous cultural, technical, and socioeconomic transformation that took place in the late eighteenth and early nineteenth centuries. Starting in Great Britain and quickly spreading throughout Western Europe and North America, factory-based production and machine-dominated manufacturing began displacing economies based on manual labor.

This trend continued well into the Second Industrial Revolution (running roughly from about 1871 to 1914) when great developments within the oil, steel, chemical, and electrical industries occurred. Mass production of consumer goods—including food and beverages, clothing, and automobiles (for personal transport)—took place during this period. There is no clean, crisp demarcation between the First and Second Industrial Revolutions. But engineering developments in the latter period involved the expanded use of fossil fuels and electricity as prime energy sources and scientific discoveries set the stage for the rise in modern physics, which caused the next great technology-based revolution in human history.

One of the most important factors that supported the emergence of the Second Industrial Revolution was electricity. In the mid- to late eighteenth century, scientists like the great American patriot, Benjamin Franklin, explored the fundamental nature of this interesting phenomenon. Franklin's pioneering work was quickly amplified by the scientific efforts of André-Marie Ampère, Charles-Augustin de Coulomb, Luigi Galvani, and Count Alessandro Volta. Soon laws governing the flow of electric currents were developed. The British experimental physicist Michael Faraday and his American counterpart, Joseph Henry, independently discovered the physical principles behind two of the most important machines in modern civilization—the electric generator and the electric motor. By the mid-nineteenth century, theoreticians, like the Scottish scientist James Clerk Maxwell, worked out a set of equations that linked electricity and magnetism. Maxwell's work revolutionized both classical physics and the practice of engineering.

Nineteenth-century inventors like Thomas Edison and Nikola Tesla applied electricity to devices that gave humankind surprising new power and comforts. Other inventors like Charles Babbage and Herman Hollerith began developing more sophisticated machines, capable of tabulating data, and, in some limited fashion, "thinking"—or at least "calculating"—faster than human beings.

The digital revolution is a generic expression that actually encompasses several major technology shifts that occurred in the mid to late twentieth century. The first part of this process was the discovery of the transistor in the late 1940s and its subsequent stimulation of the microelectronics revolution. Part of the microelectronics revolution involved the development and widespread application of digital computers and microprocessors. As microelectronic devices became less expensive and provided more capability, the use of digital devices continued to grow in an exponential manner. That trend still continues.

Equally amazing post-World War II developments in nuclear technology, space technology, and information technology complemented the expanded use of digital devices and so-called "thinking machines." The need for sophisticated tele-operated systems to remotely handle highly radioactive materials, the need to produce miniaturized electronic chips, and the need to send (unmanned) spacecraft on missions of scientific inquiry to the ends of the solar system, all stimulated technical conditions that encouraged the rise of modern robotic systems. Many of these exciting developments started at about the same time that industrial robots began to appear in factories, especially within the automobile industry.

Modern robots emerged from the confluence of several important technology areas during the digital revolution. But, robots can also trace their technical heritage to the simple machine tools invented in the Neolithic Revolution, disciplines like mechanics, pneumatics, and hydraulics, which emerged in the

Scientific Revolution, and the electromechanical devices and machinery, which appeared during the First and Second Industrial Revolutions.

Robots now play a prominent and indispensable role in modern manufacturing. Robots have proven to be the enabling technology of deep space exploration. Space-based *robot* observatories have allowed scientist to study planet Earth and the universe in ways never before possible in human history. Robots are also playing expanded roles in national security, law enforcement, medicine, environmental cleanup, and entertainment and leisure activities. Although the exciting depictions of humanoid robots (including androids and cyborgs) will remain mostly fictional for the moment, a partnership between humans and robots in the exploration of outer space is an ongoing activity within the American space program. A successful human-*robot* partnership makes the universe both a destination and a destiny for the human race.

Robotics explains just what robots are and the physical principles behind the operation of modern robots. As described within, modern robots represent the conjunction of several important scientific disciplines and engineering fields. These fields include mechanics, electronics, thermodynamics and power conversion technology, computer and information technology, and materials science. Each of these fields and many others described within this book play important complementary roles in making a modern *robot* function. Without a dependable power supply and an effective telecommunications system, for example, a robot rover on Mars would be just a clever machine that was lost in space. As suggested earlier, from one perspective, modern robots are basically *think-and-do machines*. Many of today's most interesting *robot* systems have relatively sophisticated levels of artificial, or machine, intelligence.

What can robots do? Contemporary applications range from doing an accurate and reliable job of spray-painting an automobile on an assembly line, to assisting in surgery performed by a human doctor on a patient who is located hundreds of kilometers away (telemedicine), to accomplishing detailed, automated exploration of previously unreachable worlds throughout the solar system.

The Scientific Revolution was placed on firm ground when Galileo Galilei established observational astronomy by using his primitive optical telescope to observe the heavens. Today, *robot* observatories in space are providing astronomers and astrophysicists with exciting new information about the universe—comparable in impact to what took place in the early seventeenth century. As scientists look at the edges of the observable universe and back in time to the very early universe, they are forming new ideas about the nature of energy and matter. Some of these new thoughts are stimulating exciting new work in the physics and engineering of the very small—a realm called nanotechnology. Later this century, microscopically sized robots (tiny machines perhaps several molecules long) promise to transform how people manipulate matter and energy both here on Earth and throughout the solar system and beyond.

Robotics highlights the many beneficial uses of robots in industry, science, engineering, exploration, national defense, law enforcement, environmental cleanup, and modern medicine. Attention is also given to the social and political impact of *robot* technology. Robots are an integral part of several contemporary technology-stimulated revolutions, including a shift in world labor patterns and the rise of progressively smarter machines (expert systems that think and also do), and even machine systems that use their environmental sensors to become conscious of the world around them and by extrapolation of their own existence.

The entire issue of artificial life and machine consciousness is addressed in light of many of the technical, philosophical, metaphysical, and theological implications such developments could cause. Just what happens when a smart machine uses its artificial neural network and suite of environmental sensors to reach Rene Descartes' famous statement: "Cogito, ergo sum" (I think, therefore I am)?

Are robots important? Yes! In fact, no technology-based future for the human race would be complete, exciting, or have so much potential without them!

Robotics serves as a one-stop guide to the exciting field of *robot* technology. Its chapters provide a detailed history of *robot* technology; a chronology of important milestones in the development of *robot* technology; profiles of important scientists; a detailed but readable explanation of how many of the interesting types of robots work; discussions of the impact, issues, and future of *robot* technology; a detailed and comprehensive glossary of important terms; and listings of relevant associations, demonstration sites, and information sources.

The contents were carefully chosen and the writing focused to meet the information needs of high-school students, undergraduate college and university students, and members of the general public who want to understand the nature of *robot* technology, the basic scientific principles and engineering practices upon which it is based, how *robot* technology has influenced history, and how it is now impacting society. This book serves as both a comprehensive, standalone introduction to *robot* technology and an excellent starting point and companion for more detailed personal investigations. Specialized technical books and highly focused electronic (Internet) resources often fail to place an important scientific event, technical discovery, or applications breakthrough within its societal context. This book overcomes such serious omissions and makes it easy for readers to understand and appreciate the significance and societal consequences of major engineering developments in *robot* technology and the historic circumstances that brought them about. As a well-indexed, comprehensive, and illustrated information resource designed for independent scholarship, this book will also make electronic searches for additional information more meaningful and efficient.

I wish to thank the public information specialists in the U.S. Department of Energy and its national laboratories, the U.S. Department of Defense, the U.S. Air Force, the U.S. Army, the U.S. Navy, the National Aeronautics and Space Administration (NASA) and its centers and affiliated facilities (especially the Jet Propulsion Laboratory), and the engineering companies, who generously provided much of the technical material used in developing this volume and many of the illustrations that appear within. A special thanks is extended to my editors at Greenwood Press—especially John Wagner—for their continued encouragement and patience throughout the arduous journey that began with an interesting concept and ended up with a publishable manuscript. The wonderful staff at the Evans Library of Florida Tech again provided valuable support during the initial phase of this book project. A special thanks goes out to many of my graduate students in the College of Engineering at Florida Tech. By actively participating in the lectures on the future of *robot* technology and the exciting consequences of self-replicating systems sent out into the galaxy, these young men and women, through their many interesting questions and comments, helped shape the technical content of this book. Finally, without the steadfast support of my wife, Joan, this book would never have survived the fury of three hurricanes to emerge from chaotic piles of damaged class notes, lecture materials, and technical reports and become a comprehensive treatment of *robot* technology.

1

History of Robot Technology and Systems

Human beings have always been fascinated with the concept of artificial life and the construction of machines that look and behave like people. The legend of Pygmalion and the medieval legend of the Golem are examples. Nowhere is the concept of making a living thing out of spare parts more dramatic and exciting than Mary Shelley's famous story: *Dr. Frankenstein: The Modern Prometheus*. This story has been told and retold in various motion pictures and television shows throughout twentieth century. Many modern science fiction motion pictures, like the *Terminator* trilogy, not so subtly enforce the image of Dr. Frankenstein's monster as science gone haywire. Today, whenever people take the time to discuss and extrapolate the evolution of smart machines, their conversation usually includes the possible rise of self-aware, intelligent robots that threaten their human masters.

The Czech writer Karel Čapek gave the world the term *robot* when he wrote the play *Rossum's Universal Robots* (*R.U.R.*) in 1920. Robata is the Czech word for forced labor or servitude. Some historians believe that it was actually Karel's brother, Josef Čapek (a painter and writer), who suggested the word robot to identify the fictional play's inexpensive, manufactured living machine designed expressly to work in the service of human beings. However, since the two brothers often collaborated on literary projects and the word first appeared in Karel Čapek's play *R.U.R.*, he generally gets the credit for coining the word robot. The play premiered in Prague in 1921 and was then translated into English and first appeared on the English stage in 1923. Ever since then, the word robot has been part of the global literature concerning smart machines and automatons.

Thanks to the bad publicity inherent in many such fictional portrayals, robots soon become frightening villains. Later in the twentieth century, this fiction-based perception was reinforced by a real-world situation, when industrial robots began displacing hundreds to thousands of workers from manufacturing jobs. A neo-Luddite wave of *technophobia* (fear or hatred of machines and new technologies) has emerged and still grips many workers in the world's industrialized nations, which are now experiencing turbulent social and economic transitions into postindustrial, information-based economies. However, the robot, like many other revolutionary, new technologies, is really a two-edged sword.

When viewed in another perspective, robots are really advanced machines that support a continuing revolution in the application of technology in service to human beings. The human race has experienced and prospered from a number of important technology-related revolutions in science and technology.

PLEASE DON'T START THE TECHNOLOGY REVOLUTION WITHOUT ME

The first revolution in technology involved the discovery and use of fire in prehistoric times. The next breakthrough involved the development and use of very simple tools, a stick, a sharp rock, and similar objects to aid in hunting and gathering activities. These transitions were extremely gradual (from about 750,000 years ago to about 15,000 years ago) and involved a division of labor within ancient hunting and gathering societies. Eventually, the survival of an ancient tribe or clan in the late Upper Paleolithic era not only depended upon the skill of its hunters and gatherers, but also the expertise of its toolmakers and firekeepers. Around ancient campfires, while Stone Age humans huddled to keep warm, arose the early notions of improving the quality of life by developing better tools and devices. Inspiration took hold and the Mesolithic Age featured the appearance of better cutting tools and the bow.

Next came the Neolithic Revolution—an incredibly important transition from hunting and gathering to agriculture. As the last Ice Age ended about 12,000 or so years ago, various prehistoric societies in the Middle East (Fertile Crescent), Southeast Asia, Mesoamerica, India, and elsewhere independently adopted crop cultivation and began to establish semipermanent or permanent settlements. Historians often identify this period as the beginning of human civilization. During the period, ancient peoples developed better tools and invented a variety of simple machines. As populations swelled, these tools and simple machines allowed early societies to modify their surroundings to better suit survival and growth. The natural environment changed as simple dams, irrigation canals, roads, and walled villages appeared.

With each development came the need for better devices to perform difficult or monotonous tasks. Simple machines like the wheel and axle or the inclined plane yielded major breakthroughs in the ability to perform work. Often these simple machines allowed workers to accomplish tasks previously regarded as extremely dangerous or impossible. The great pyramids of ancient Egypt or Mesoamerica are examples of how early human societies could effectively use such simple machines to perform difficult tasks. But the machines were not self-powered or smart. The devices required human labor or animal power to operate and generally needed humans to perform a task, or to take advantage of some physical principle (like gravity), which was instinctively perceived but not qualitatively understood. The quantitative understanding of machines and physical principles could not take place without the Scientific Revolution.

With agriculture came food surpluses, the domestication of animals, the production of clothing, the rise of trading and bartering, and the further specialization of labor. The application of technology also had a variety of interesting social impacts. The rise of civilization during the Neolithic Revolution also spawned the first governments (to organize human labor and focus wealth in the development of various public projects), the collection of taxes (to pay for government), the first organized military establishments (to protect people and their wealth), and the first schools (to pass knowledge and technical skills on to future generations in a more or less organized fashion).

Many of today's robots contain components and use scientific principles first adopted during the rise of civilization. The wheel and axle, the pulley, the wedge, the lever, and the gear are examples. Great engineers of antiquity, like Archimedes, Ctesibius, and Hero of Alexandria developed and introduced many of the important technical devices and early machines. However, it took the brilliant work of Galileo Galilei, Isaac Newton, and other scientists during the Scientific Revolution (in the seventeenth century) to establish the physical framework within which the operation and prediction of machine behavior (including robots) could be done with an acceptable degree of certainty and precision.

Science historians often identify the Scientific Revolution as roughly the period between 1543 and 1687 in Western Europe, when a new way of looking at the physical phenomena in the world emerged. In 1543, Nicholas Copernicus challenged the long held geocentric cosmology of Aristotle, who advocated that Earth was the center of the universe. Instead, Copernicus promulgated a radically conflicting heliocentric cosmology in his book *On the Revolution of Celestial Spheres*. Early in the seventeenth century, Galileo Galilei embraced heliocentric cosmology and performed a variety of important observations and experiments, which established the approach today known as the scientific method. Isaac Newton provided the capstone to the Scientific Revolution when he published his great work *Mathematical Principles of Natural Philosophy* (or *The Principia*) in 1687. His monumental work transformed the practice of physical science and completed the revolution stimulated by Copernicus in 1543. During this fertile period of intellectual accomplishment in Western Europe, great mathematicians like Blaise Pascal and Gottfried Wilhelm Leibniz experimented with the development of early mechanical calculators—devices that heralded the great electromechanical tabulating machines and digital computers of the twentieth century.

While the Scientific Revolution provided people the mathematical tools and physical principles to understand how machines and the universe operated in a somewhat clockwork fashion, this great advancement in human history did not (of itself) provide the social or economic stimulus for developing the smart machine systems (eventually called robots) that later appeared in the twentieth century. In fact, during the early eighteenth century, clockmakers and engineers like Jacques de Vaucanson and Pierre Jacquet-Dortz were quite content to construct elegant automatons, primarily for the amusement of wealthy patrons.

The major stimulus for the production of smarter machines, and eventually robots, was the Industrial Revolution—the period of enormous cultural, technical, and socioeconomic transformation that took place in the late eighteenth and early nineteenth centuries. Starting in Great Britain and quickly spreading throughout Western Europe and North America, factory-based production and machine-dominated manufacturing began displacing economies based on manual labor. Many historians treat this transformation as being comparable to the Neolithic Revolution with respect to the consequences exerted on the trajectory of human civilization. Because of the global political influence of the British Empire in the nineteenth century, this wave of social and technical change eventually spread throughout the world.

British engineers and business entrepreneurs led the charge in the late eighteenth century, by developing steam power (fueled by coal) and using powered machinery in manufacturing (primarily in the textile industry). These technical innovations were soon followed by the development of all-metal machine tools in the early nineteenth century. The availability of these machine tools quickly led to the development of more machines for use in factory-based manufacturing in other industries. More and less expensive goods became available. Workers left the farms and flocked to urban areas to work in factories.

Wealth accumulated in developed (or industrialized) nations, whose capitalistic citizens imported raw materials from less developed countries (often maintained as political colonies), processed these raw materials in factories, and then shipped the manufactured items back to essentially captive markets around the world. One of the negative effects of the Industrial Revolution was the rapid reduction in the quality of life for a growing number of factory workers who found themselves living in crowded cities and trapped in manufacturing jobs that often proved both monotonous and hazardous. Unprotected whirling machinery could easily severe a finger, trap a pant leg and mangle a foot, or worse. But in an effort to maximize manufacturing industry profits, the cost of factory labor (especially unskilled and child) was kept as low as possible. This trend continued well into the Second Industrial Revolution (roughly from 1871 to 1914) when great developments within the oil, steel, chemical, and electrical industries occurred. Mass production of consumer goods—including food and beverages, clothing, and automobiles (for personal transport)—took place during this period. There is not a clean and crisp demarcation between the First and Second Industrial Revolutions. But engineering developments in the latter period involved the expanded use of fossil fuels and electricity as prime energy sources and scientific discoveries set the stage for the rise in modern physics, which caused the next great technology-based revolution in human history. Before moving on to the digital revolution and the information age, it is worth noting that Henry Ford's innovative use of the assembly line concept in the mass production of his company's Model T automobile in about 1910 set the stage for the extensive use of robots in modern automobile manufacturing plants.

One of the most important factors that supported the emergence of the Second Industrial Revolution was electricity. In the mid to late eighteenth century, scientists like the great American patriot, Benjamin Franklin, explored the fundamentals of this interesting natural phenomenon. Franklin's pioneering work was quickly amplified by the scientific efforts of André-Marie Ampère, Charles-Augustin de Coulomb, Luigi Galvani, and Count Alessandro Volta. Soon laws governing the flow of electric currents were developed. The British experimental physicist Michael Faraday and his American counterpart, Joseph Henry, independently discovered the physical principles behind two of the most important machines in modern civilization. By the mid-nineteenth century, theoreticians, like the Scottish scientist James Clerk Maxwell, worked out a set of equations that linked electricity and magnetism. Maxwell's work revolutionized both physics and the practice of engineering.

Nineteenth-century inventors like Thomas Edison and Nikola Tesla applied electricity to devices that gave humankind surprising new power and comforts. Other inventors like Charles Babbage and Herman Hollerith began developing more sophisticated machines, capable of tabulating data, and in some limited fashion, "thinking"—or at least "calculating"—faster than human beings.

The digital revolution is a generic expression that actually encompasses several major technology shifts which occurred in the mid to late twentieth century. The first part of this process was the discovery of the transistor in the late 1940s and its use in the microelectronics revolution that followed, especially in the development and application of digital computers and microprocessors. As microelectronic devices became less expensive and provided more capability, the use of digital devices continued to expand.

Equally amazing post-World War II developments in nuclear technology, space technology, and information technology complemented the expanded use of digital devices and so-called "thinking machines." The need for advanced-design, teleoperated systems to handle radioactive materials, the need to produce miniaturized electronic chips, and the need to send (unmanned) spacecraft on missions of scientific inquiry to the ends of the solar system, all stimulated conditions that promoted the rise of modern robot systems. Many of these exciting developments started at about the same time that industrial robots began to appear in factories, especially within the automobile industry.

Military planners were not unaware of the role or potential role of robots in warfare. The intercontinental ballistic missile and the cruise missile emerged during the cold war as incredibly powerful robot weapons, capable of delivering total annihilation to any region of the globe. These fearsome robot weapons were counterbalanced by the arrival of robot military spacecraft—especially the reconnaissance and surveillance satellites, which provided stabilizing streams of important information during turbulent political periods in the cold war era and beyond. Today, robot military satellites are joined by a growing family of ground-based, undersea, and aerial robots capable of fighting international terrorism in all its ugly and diverse forms. The U.S. Air Force's Predator unmanned aerial vehicle (UAV) is but one example; the U.S. Army's collection of unmanned ground vehicles (UGVs) represents another contemporary example.

Modern robots emerged from the confluence of several important technology areas during the digital revolution. But, robots can also trace their technical heritage to the simple machine tools invented in the Neolithic Revolution, disciplines like mechanics, pneumatics, and hydraulics, which emerged in the Scientific Revolution, and the electromechanical devices and machinery, which appeared during the First and Second Industrial Revolutions.

Industrial robots now play a prominent role in modern manufacturing facilities. The American entrepreneurs George C. Devol, Jr. and Joseph F. Engelberger introduced the first industrial robot to the world in the early 1960s. One important robot technology milestone occurred in 1961, when the Consolidated Diesel Electric Corporation (Condec Corporation) shipped the first commercial version of a Unimate industrial robot from Connecticut and installed the device in a General Motor's plant in New Jersey. Despite aggressive marketing efforts

Figure 1-1 On November 28, 1958, the Atlas rocket vehicle became the first operational intercontinental ballistic missile (ICBM) developed by the United States during the cold war. The ICBM is the robot weapon that changed the world. Shown here is an Atlas-D booster departing Complex 12 at Cape Canaveral Air Force Station (circa mid-1960s) as part of Project Fire—a reentry vehicle test program supporting NASA's Apollo Project. (Credit: Photograph courtesy of NASA.)

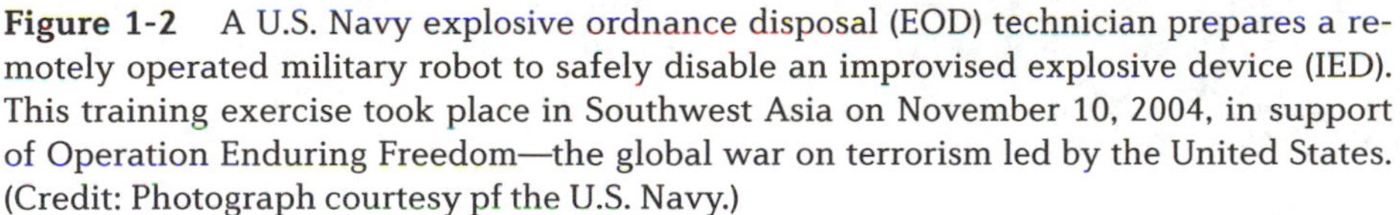
Figure 1-2 A U.S. Navy explosive ordnance disposal (EOD) technician prepares a remotely operated military robot to safely disable an improvised explosive device (IED). This training exercise took place in Southwest Asia on November 10, 2004, in support of Operation Enduring Freedom—the global war on terrorism led by the United States. (Credit: Photograph courtesy pf the U.S. Navy.)

the application of robots in industry (American and foreign) did not become widespread until the 1970s. In 1978, Unimation introduced the versatile industrial robot called PUMA, for Programmable Universal Machine for Assembly.

Mobile robots, in a variety of sizes, shapes, and capabilities, are now playing expanded roles in national security, law enforcement, medicine, environmental cleanup, and entertainment and leisure activities. Where did the modern mobile robot come from? The first serious attempts to link computers and artificial intelligence to mobile robots took place in the mid-1960s, when researchers tried to link computer-interfaced camera systems, which scanned the robot's environment, with the robot's mobility system. Created with Office of Naval Research funding at SRI (Stanford Research Institute) in the late 1960s, Shakey served as the technical ancestor to the modern mobile. Shakey had television (TV) eyes, tactile sensors, an optical range finder, and an elementary navigation system. This pioneering mobile robot could plan and execute simple tasks, such as finding objects and manipulating them, while avoiding obstacles. Like Shakey, many other early mobile robots and their supporting artificial intelligence (AI) capabilities were developed with funding from agencies and organizations within the U.S. Department of Defense. The Defense Advanced Research Projects Agency (DARPA) and the Office of Naval Research (ONR) are prime examples. Standard techniques such as time-sharing, systolic computing, neural networks, machine learning, and connectionist computing trace their origins to Shakey and similar early mobile robot projects. Without question this robot represented a major technology milestone in the history of robotics.

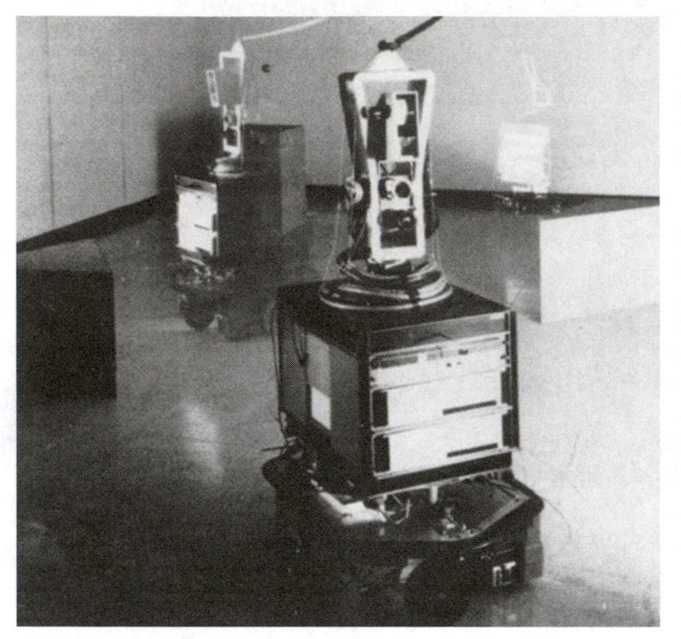

Figure 1-3 The historic picture shows Shakey, the pioneering mobile robot, and a companion robot (in the background) moving through an indoor test range in the late 1960s at SRI (Stanford Research Institute). (Credit: Photograph courtesy of the U.S. Navy/Office of Naval Research.)

Though the exciting depictions of humanoid robots (including androids and cyborgs) may remain mostly fictional for the moment, a partnership between humans and robots in the exploration of outer space is an ongoing, real engineering activity with the American space program. A successful human-robot partnership makes the universe both a destination and a destiny for the human race. NASA is examining the use of a variety of robots, including humanoid and android-like, to assist astronauts as they construct structures in space, explore the Moon and Mars, and build permanent settlements on these worlds in the latter portions of this century. One interesting program is called Robonaut—an android-like robotic assistance developed at the start of the twenty-first century by engineers at NASA's Johnson Space Center in Houston, Texas, in collaboration with DARPA.

What can robots do? Contemporary applications range from doing an accurate and reliable job of spray-painting an automobile on an assembly line, to assisting in surgery performed by a human doctor on a patient who is located over hundreds of kilometers distance (telemedicine), to accomplishing detailed, automated exploration of previously unreachable worlds throughout the solar system.

It is interesting to note that the Scientific Revolution was placed on firm ground when Galileo Galilei established observational astronomy by using his primitive optical telescope to observe the heavens. Today, robot observatories in space are providing astronomers and astrophysicists with exciting new information about the universe—comparable in impact to what took place in the early seventeenth century. As scientists look to the edges of the observable universe and back in time to the very early universe, they are forming new ideas about the nature of energy and matter. Some of these new thoughts are stimulating exciting new work in the physics and engineering of the very small—a realm called nanotechnology. Tiny machines (perhaps several molecules long), microscopically sized robots, promise to transform how people later this century will manipulate matter and energy here on Earth, throughout the solar system, and beyond. Nanotechnology is an all-encompassing term, which generally refers to processes, research and development activities, and human-made devices that

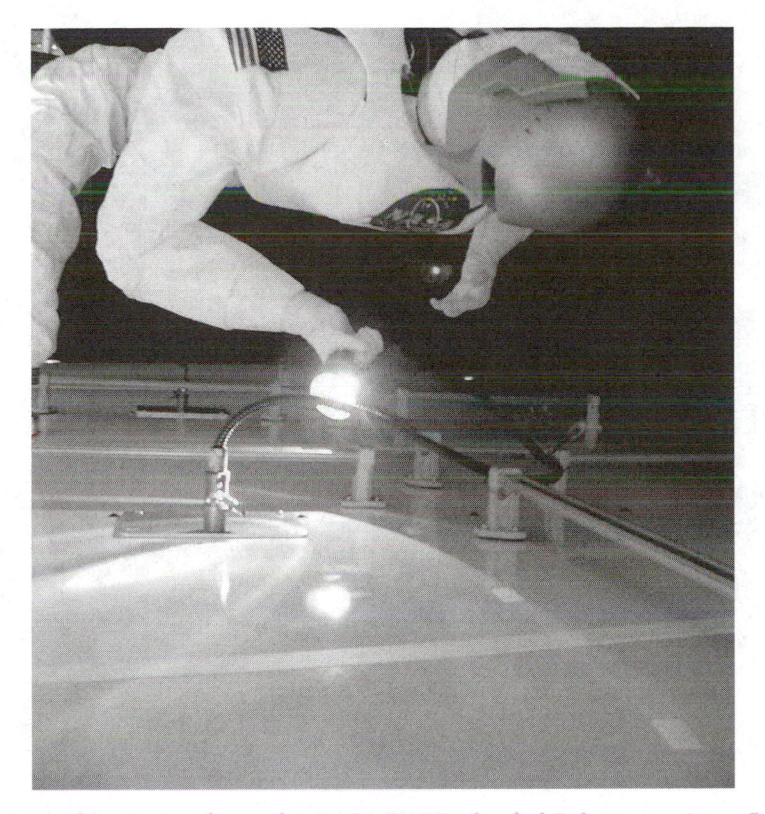

Figure 1-4 This picture shows the NASA-DARPA funded Robonaut using a flashlight to provide teleoperators with a better view of the simulated space work site at the Johnson Space Center in 2004. This effort is part of an ongoing study, involving human-robot relationships in future space missions. (Credit: Photograph courtesy of NASA/JSC.)

are very small, on the order of 1 to 100 nanometers in principal dimension. At this scale, scientists and engineers envision developing devices (perhaps swarms of very, very tiny robots) capable of manipulating matter one molecule or even one atom at a time.

Science historians generally acknowledge that the field of nanotechnology began as a result of an amazingly insightful lecture delivered by the American physicist and Nobel laureate, Richard Feynman. In his talk entitled "There's Plenty of Room at the Bottom," Feynman suggested that it would be possible, using late 1950s technology, to write an enormous quantity of information, such as the entire content of a multivolume encyclopedia, in a tiny space equivalent to the head of a pin. Feynman also speculated about the impact micro- and nano-sized machines might have. Clearly fields like medicine and information technology (built upon microelectronics) would be revolutionized, if these (at the time) hypothetical tiny devices allowed engineers and scientists to manipulate individual atoms and arrange these building blocks of matter into useful things.

Several more important milestones followed, each contributing in its own way to the nanotechnology revolution. In 1981, two researchers working for

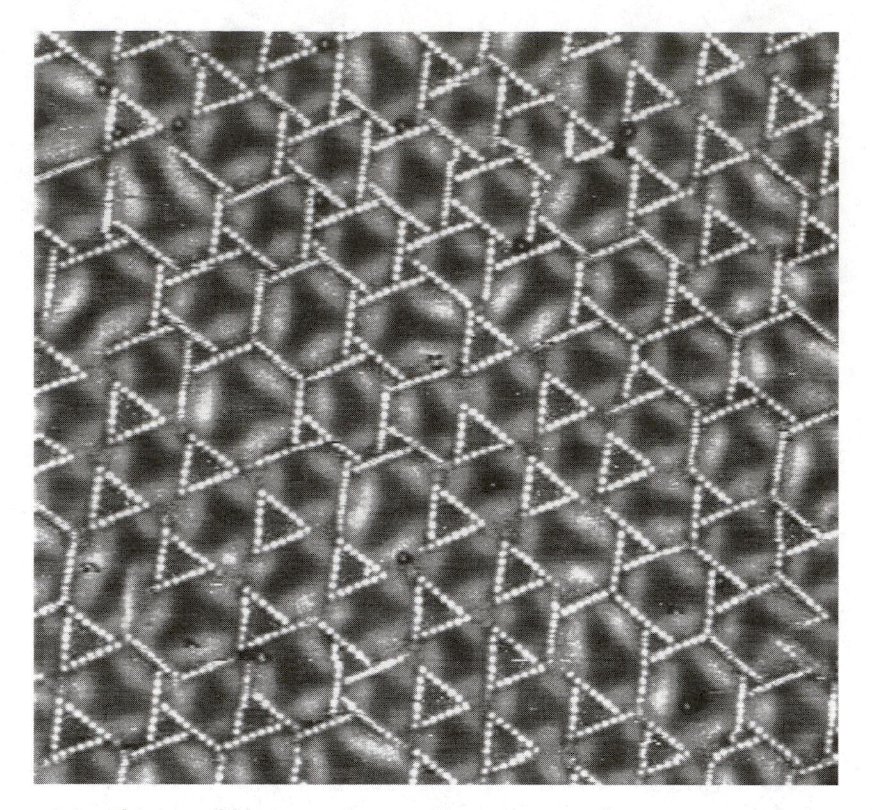

Figure 1-5 This incredible image shows sulfur atom nanoclusters on a copper layer deposited on a single crystal of ruthenium. The revolutionary ability to develop very tiny robotic devices that can manipulate an individual atom or molecule is one of the main goals in the emerging field of nanotechnology. (Credit: Image courtesy of the U.S. Department of Energy/Brookhaven National Laboratory.)

IBM, named Gerd Binning and Heinrich Rohrer, designed the first scanning tunneling microscope (STM). Their original STM was able to examine small samples of matter held in a deeply chilled, refrigerated chamber. At a chamber temperature of −271°C (almost but not quite absolute zero), atomic motion—the natural tendency of atoms to move around—slows almost to a halt. Inside the chamber of the STM, the researchers used an incredibly tiny stylus, which was slowly moved over the material object being scanned by a special robotic arm. The stylus was just one atom wide at its tip and Binning and Rohrer had manufactured the device using newly developed micromachining techniques. By measuring the flow of current in the tip as the stylus approached the surface of the object, the two researchers obtained a precise indication of the distance from the tip to the individual atoms on the surface of the sample object. They used computers to transform these data into an image of the atoms, which was then displayed on a monitor. Other researchers soon discovered that the STM could also be used to push and pull atoms around. These activities represented the first time in the history of technology that human beings could manipulate objects on so small a scale.

Figure 1-6 This is an aerial view (taken in September 2005) of the hub facility of the new Sandia/Los Alamos joint Center for Integrated Nanotechnologies (CINT) in Albuquerque, New Mexico. The building's three wings house a characterization lab (containing the most vibration sensitive instruments), a synthesis lab (which includes physical, chemical, and biological facilities), and clean room facilities for device integration activities at the nano/microscale level. (Credit: Photograph courtesy of U.S. Department of Energy/SNLA.)

However, the use of the STM was limited to objects like metals, which had electrically conductive surfaces. In 1985, Binning, in collaboration with Christopher Gerber and Calvin Quate, invented another device called the atomic force microscope (AFM). This device was constructed in such a way that measurements could be made on nonconductive surfaces. Like the STM, the AFM soon entered scientific and engineering service as a device capable of positioning objects as small as an atom. The revolution in the construction of miniscule machines (sometimes called microelectromechanical systems or MEMS) and microscale and nanoscale electronic devices ensued. In the 1990s, the process of "micromachining" emerged in many research and development facilities as one of the first practical approaches to creating various nanotechnology devices. These efforts continue to the present, with enormous facility and research investments being made by such government agencies as the U.S. Department of Energy.

Another significant milestone in the development of nanotechnology occurred in about 1985, when researchers Richard E. Smalley, Robert F. Curl, Jr., and Sir Harold W. Kroto were investigating an amazing molecule, which consisted of 60-linked carbon atoms. Smalley named these clusters of atoms fullerenes in honor of the famous architect, Buckminster Fuller, who promoted the use of

Figure 1-7 This is the first photograph ever taken on the surface of Mars. This image was obtained by NASA's Viking 1 Lander—a robot spacecraft that successfully touched down on the Red Planet in Chryse Planitia (the Plains of Gold) on July 20, 1976. The picture was taken just minutes after the robot spacecraft completed a totally automated landing on the planet. (Credit: Photograph courtesy of NASA.)

geodesic domes. Smalley, Curl, and Kroto shared the 1996 Nobel Prize in chemistry for their "discovery of carbon atoms found in the form of a ball." Today, these fullerenes, or buckyballs as they are more popularly called, represent some of the primary building blocks in nanotechnology.

SPACE ROBOTS IN SERVICE TO SCIENCE

Since the start of the space age in 1957, engineers and scientists have built robot spacecraft in all shapes and sizes. Each space robot was usually custom-designed and carefully engineered to meet the specific needs and environmental challenges of a particular space exploration mission. For example, the engineers designed lander spacecraft to acquire scientific data and to function in a hostile planetary surface environment. As space technology matured and was complemented by incredible progress in computer and transistor-based technologies, the complexity of space robots changed greatly. Starting in the mid-1960s, engineers and space scientists found it convenient to begin categorizing robot spacecraft according to the missions they were intended to fly. NASA engineers generally divide space robots into the following major broad classes, such as flybys, orbiters, landers, and rovers.

Robot spacecraft promoted a revolution in the scientific understanding of the solar system and the universe. For example, the placement of *Viking 1* and *2* lander spacecraft on the Martian surface in 1976 represents one of the great early triumphs of robotic space exploration. After separation from the Viking orbiter spacecraft, the lander (protected by an aeroshell) descended into the thin Martian atmosphere at a speed of approximately 16,000 kilometers per hour. As they descended, the landers were slowed down by aerodynamic drag until their aeroshells were discarded. Each robot lander spacecraft then slowed down further by releasing a parachute. Finally, the robots achieved a gentle landing by automatically firing retrorockets. Of special significance is the fact that both

Figure 1-8 This picture is entitled "Six Wheels on Soil." It is a two-frame mosaic of images from the panoramic camera on NASA's *Mars Pathfinder* lander spacecraft. The composite image shows the robot minirover, named *Sojourner*, just after driving onto the Martian surface on July 5, 1997. (Credit: Photograph courtesy of NASA.)

Viking landers successfully accomplished the entire soft landing sequence automatically without any direct human intervention or guidance.

NASA launched the *Mars Pathfinder* mission to the Red Planet using a Delta II expendable launch vehicle on December 4, 1996. This mission, previously called the *Mars Environmental Survey* (or MESUR) *Pathfinder*, had the primary objective of demonstrating innovative technology for delivering an instrumented lander and free-ranging robotic rover to the Martian surface. The *Mars Pathfinder* not only accomplished this primary mission but also returned an unprecedented amount of data, operating well beyond the anticipated design life.

Mars Pathfinder used an innovative landing method that involved a direct entry into the Martian atmosphere assisted by a parachute to slow its descent through the planet's atmosphere and then a system of large airbags to cushion the impact of landing. From its airbag-protected bounce and roll landing on July 4, 1997, until the final data transmission on September 27, the robotic lander/rover team returned numerous close-up images of Mars and chemical analyses of various rocks and soil found in the vicinity of the landing site.

The landing site was at 19.33 N, 33.55 W, in the Ares Vallis region of Mars, a large outwash plain near Chryse Planitia (the Plains of Gold), where the *Viking 1 Lander* had successfully touched down on July 20, 1976. Planetary geologists speculate that this region is one of the largest outflow channels on Mars—the result of a huge ancient flood that occurred over a short period of time and flowed into the Martian northern lowlands.

The lander, renamed by NASA as the Carl Sagan Memorial Station, first transmitted engineering and science data collected during atmospheric entry and landing. The American astronomer Carl Edward Sagan (1934–1996) popularized astronomy and astrophysics and wrote extensively about the possibility of extraterrestrial life.

Just after arrival on the surface, the lander's imaging system (which was on a pop-up mast) obtained views of the rover and the immediate surroundings. These images were transmitted back to Earth to assist the human flight team in planning the robot rover's operations on the surface of Mars. After some initial maneuvering to clear an airbag out of the way, the lander deployed the ramps for the rover. The 10.6-kilogram minirover had been stowed against one of the lander's petals. Once commanded from Earth, the tiny robot explorer came to life and rolled onto the Martian surface. Following rover deployment, the bulk of the lander's remaining tasks were to support the rover by imaging rover operations and relaying data from the rover back to Earth. Solar cells on the lander's three petals, in combination with rechargeable batteries, powered the lander, which also was equipped with a meteorology station.

The rover, renamed *Sojourner* (after the American civil rights crusader Sojourner Truth), was a six-wheeled vehicle that was teleoperated (that is, driven over great distances by remote control) by personnel at the Jet Propulsion Laboratory. The rover's human controllers used images obtained by both the rover and the lander systems. Teleoperation at interplanetary distances required that the rover be capable of some semiautonomous operation, since the time delay of the signals averaged between 10 and 15 minutes depending on the relative positions of Earth and Mars.

For example, the rover had a hazard avoidance system, and surface movement was performed very slowly. The small rover was 28 centimeters high, 63 centimeters long, and 48 centimeters wide with a ground clearance of 13 centimeters. While stowed in the lander, the rover had a height of just 18 centimeters. However, after deployment on the Martian surface, the rover extended to its full height and rolled down a deployment ramp. The relatively far-traveling little rover received its supply of electrical energy from its 0.2 square meter array of solar cells. Several nonrechargeable batteries provided backup power.

The rover was equipped with a black-and-white imaging system. This system provided views of the lander, the surrounding Martian terrain, and even the rover's own wheel tracks that helped scientists estimate soil properties. An alpha particle X-ray spectrometer (APXS) onboard the rover was used to assess the composition of Martian rocks and soil.

Both the lander and the rover outlived their design lives—the lander by nearly three times and the rover by 12 times. Data from this very successful lander/rover surface mission suggest that ancient Mars was once warm and wet, stimulating further scientific and popular interest in the intriguing question of whether life could have emerged on the planet when it had liquid water on the surface and a thicker atmosphere.

In the summer of 2003, NASA launched identical twin Mars rovers that were to operate on the surface of the Red Planet during 2004. *Spirit* (MER-A) was launched by a Delta II rocket from Cape Canaveral on June 10, 2003, and successfully landed on Mars on January 4, 2004. *Opportunity* (MER-B) was launched from Cape Canaveral on July 7, 2003, by a Delta II rocket and successfully landed on the surface of Mars on January 25, 2004. Both landings resembled the successful airbag bounce and roll arrival demonstrated during the *Mars Pathfinder* mission.

Figure 1-9 This intriguing mosaic image was collected by the navigation camera on NASA's Mars Exploration Rover *Spirit* on January 4, 2004. NASA scientists reprocessed the original imagery data to project a clear overhead view of the robot rover and its lander (mother spacecraft) on the surface of Mars. (Credit: Photograph courtesy of NASA/JPL.)

Following arrival on the surface of the Red Planet, each rover drove off and began its surface exploration mission in a decidedly different location on Mars. *Spirit* (MER-A) landed in Gusev Crater, which is roughly 15 deg-rees south of the Martian equator. NASA mission planners selected Gusev Crater because it had the appearance of a crater lakebed. *Opportunity* (MER-B) landed at Terra Meridiani—a region of Mars that is also known as the Hematite Site because this location displayed evidence of coarse-grained hematite, an iron-rich mineral, which typically forms in water. Among this mission's principal scientific goals is the search for and characterization of a wide range of rocks and soils that hold clues to past water activity on Mars. By the end of June 2006, both rovers continued to function on Mars far beyond expectation. NASA's primary mission goal for these rovers was an operating lifetime of 90 days.

With much greater mobility than the *Mars Pathfinder* minirover, each of these powerful new robot explorers has successfully traveled up to 100 meters per Martian day across the surface of the planet. Each rover carries a complement of sophisticated instruments that allows it to search for evidence that liquid water was present on the surface of Mars in ancient times. *Spirit* and *Opportunity* are visiting different regions of the planet. Immediately after landing each rover performed reconnaissance of the particular landing site by taking panoramic (360 degree) visible (color) and infrared images. Then, using images and spectra taken daily by the rovers, NASA scientists at the Jet Propulsion Laboratory used telecommunications and teleoperations to supervise the overall scientific program. With intermittent human guidance, the pair of mechanical explorers functioned like robot prospectors—examining particular rocks and soil targets and evaluating composition and texture at the microscopic level.

Each rover has a set of five instruments with which to analyze rocks and soil samples. The instruments include a panoramic camera (Pancam), a miniature thermal emission SPECTROMETER (Mini-TES), a Mössbauer spectrometer (MB), an alpha particle X-ray spectrometer (APXS), magnets, and a microscopic imager

(MI). There is also a special rock abrasion tool (or RAT) that allows each rover to expose fresh rock surfaces for additional study of interesting targets.

Both *Spirit* and *Opportunity* have a mass of 185 kilograms and a range of up to 100 meters per sol (Martian day). As mentioned previously, the exciting surface operations accomplished by each rover have lasted well beyond the goal of 90 sols. Communications back to Earth is accomplished primarily with Mars-orbiting spacecraft, like the *Mars Odyssey 2001*, serving as data relays.

In the robot lander/probe mission scenario, the mother spacecraft releases the lander or robot probe, while the cojoined spacecraft pair is still some distance from the target planetary object. Following release and separation, the robot probe follows a ballistic impact trajectory into the atmosphere and unto the surface of the target body. This scenario played out precisely as scripted, when the *Cassini* mother spacecraft released the hitchhiking *Huygens* probe on December 25, 2004, as *Cassini* orbited around Saturn. Following separation, the *Huygens* probe traveled for about 20 days along a carefully planned ballistic trajectory to Saturn's moon Titan. When it arrived at Titan on January 14, 2005, the *Huygens* probe entered the moon's upper atmosphere, performed a superb data-collecting descent, and successfully landed on the moon's surface.

The scientific robot spacecraft exists to deliver its scientific instruments to a particular interplanetary destination; to allow these instruments to make their measurements, perform their observations, and/or conduct their experiments under the most favorable achievable conditions; and then to return data from the instruments back to scientists on Earth. In the interesting case of a sample return mission, the robot spacecraft must collect and then return material samples from an alien world. Once the space robot delivers its extraterrestrial cargo to Earth, scientists perform detailed investigations on the alien materials in a special, biologically isolated (quarantine) facility.

SERVICE AND RESCUE ROBOTS RISING

In many developed countries, social pressure and government regulations (such as the U.S. Occupational Safety and Health Act [OSHA] of 1971) provide a sufficient emphasis on human worker safety, such that manufacturing companies and other industries find it economically justified to substitute robots for human labor in jobs regarded as hazardous. So, industrial robots have often proven ideal in replacing human workers in such operations as paint spraying, die casting, welding, and handling toxic materials.

Concern for the safety of human beings is also encouraging the military services, law enforcement agencies, emergency first responders, and environmental cleanup teams to examine the role of mobile robots as substitutes for human beings in some of the most hazardous operations. For example, the U.S. military services use a variety of mobile robots in bomb detection, explosive ordnance disposal (EOD), and mine detection roles. Teleoperated by military personnel from a safe distance, the mobile military robot goes into the extremely hazardous area first, searches for the bomb, land mine, or unexploded ordnance, and then assists in disabling the device or rendering the conditions safe perhaps by planting a small explosive device and scooting back a safe distance before the robot's human controller sets off the disabling charge.

Urban warfare and counterterrorism patrols in a hostile town or city are often ideal opportunities to use small, mobile robots on search missions, before human military personnel expose themselves to risk. Mobile robots have proven ideal in slow, deliberate search activities, but sometimes do not have the flexibility or speed of deployment in hot, firefight situations where a flash battle erupts and flows very quickly.

In domestic law enforcement activities, mobile robots equipped with vision systems are often used in standoff or hostage situations. Local law enforcement agencies also send a dexterous mobile robot with vision sensors and manipulators to inspect and (when possible) remove a suspicious package to safer distances. There has also been some discussion of equipping a law enforcement robot with a nonlethal weapon such as a taser or incapacitating gas, but there are legal and reliability issues to be resolved before police robots can replace human SWAT teams in hostage standoff situations.

Finally, during natural disasters or human-caused acts of terrorism, certain hardy search and rescue mobile robots can be sent into very dangerous, debris-laden areas to search for survivors. Sensors on such robots can quickly make measurements of chemical toxicity, biological hazard, or radiation hazard. Teleoperated by first responders the mobile search and rescue robots can quickly map the disaster site and allow the human first responders to develop the most efficient, least hazardous response pathway and strategy. Of course, the use of such mobile robots implies they are available as part of the first responder team. Time is critical in disaster response, so waiting hours or days for the right team of mobile robots to arrive defeats the entire concept of emergency first response. The issue again becomes one of economic justification and practicality. Officials in charge of responding to natural disasters or acts of terrorism must weigh the pros and cons of exposing a trained and properly equipped human responder to a hazardous situation versus having a mobile robot available to support the response team. Radiological accidents, chemical spills, biological attacks, each pose challenges and are very scenario dependent.

ENTERTAINING ROBOTS

Some of the most widely held perceptions about robots generally come from works of fiction or from entertaining encounters with life-like robots used in theme park attractions. Several robot toys also provide educational and entertaining experiences. In 1999, Sony introduced that corporation's AIBO line of entertainment robots, which proved very popular as robotic pets. AIBO ERS-7, for example, resembles a small dog and can interact with its owner (through suitable software) as it develops from a puppy (with typical puppy behavior) to a mature adult dog. At maturity, an AIBO ERS-7 robot dog will understand (but not necessarily always obey) about 100 of its owner's voice commands. In 2000, the Honda Motor Company debuted its ASIMO (Advance Step in Innovative Mobility) humanoid robot. ASIMO is actually the eleventh in a series of walking robots created by Honda engineers in a focused development effort (starting in 1986) to create a two-legged (bipedal) humanoid robot that can walk and perform useful functions in human society alongside people. On December 13, 2005, Honda parlayed its initial success with ASIMO by introducing the newest version of the company's "people-friendly" humanoid robot. This bipedal robot has many improved features, including the ability to pursue key tasks in a real-life office or home environment. The well-engineered humanoid robot has a height of 1.3 meters and a mass of 54 kilograms. The new ASIMO can autonomously act as a receptionist or even deliver drinks on a tray. Other Japanese robot engineers and companies are also constructing humanoid robots, including some life-like systems that resemble the first generation of androids. However, full autonomy remains an illusive engineering challenge.

Theater, motions pictures, and television programs have presented a variety of fictional robot images to millions of people throughout the world. Some of the most influential works of fiction are mentioned below. Obviously not all fictional robots, androids, and cyborgs that have appeared in science fiction over the last eight decades are mentioned here, but the selection of so-called "classic works" should provide a sufficient look at how "entertaining robots" have shaped popular opinions about real world robots. Many of today's robot engineers were inspired and perhaps even challenged into pursuing careers in robotics, or closely related fields, as a result of the entertaining experiences they enjoyed while watching these obviously fictional movies.

The legacy of modern fictional robots starts at the true beginning in 1920, when the Czech playwright Karel Capek introduced the word *robot* in his satirical play, *R.U.R.* (*Rossum's Universal Robots*). Taken from the Czech word for forced labor, the word was used to describe electronic servants who turn on their masters when given emotions. The play premiered in Prague in 1921 and first appeared on the English stage in 1923. The storyline involves the rise of mentally modified robots, who become displeased with having to perform all the hard work for human beings, and rise up to destroy their human masters. At the end of the story, with chaos all around them, two specially modified domestic robots, named Primus and Helena, fall in love. The robots are renamed as Adam and Eve (by the last remaining human—a worker named Alquist) and they depart into the chaotic world beyond the giant robot factory to begin an intelligent machine-based civilization.

The Austrian film director, Fritz Lang, shocked and pleased audiences in 1927 with his famous movie, *Metropolis*. The screenplay was written by Lang and his wife, Thea von Harbou, several years earlier. He used expensive, exotic sets in this silent movie to introduce German audiences to the villainous robot, Maria. In part of the storyline, the robot Maria becomes an exotic dancer in nightclubs and causes discord among the rich young men of Metropolis. Lang used a Frankenstein-like theme, in which the robot Maria tries to punish and destroy humanity by encouraging the human workers to rebel against their employers. Lang's noir work is dominated by the perils of future technology and impact of a powerful, female robot bent on causing trouble.

In March 1942, science fact and fiction writer Isaac Asimov introduces his three laws of robotics—a fictional set of three rules that govern humanoid robot behavior in the science fiction story "Runaround," which appeared in *Astounding* magazine. These laws become part of the cult and culture of modern robotics and are frequently acknowledged in many modern books dealing with robotics and advanced robot technology.

In 1951, during build up of the nuclear arms race of the cold war, American movie audiences encounter the alien emissary (played by Michael Rennie) and Gort (his large and powerful robot companion) in the classic science fiction-fantasy thriller, *The Day The Earth Stood Still*. Based on Harry Bates' short story, "Farewell to the Master," this movie uses the arrival of an alien spaceship in Washington, DC, to warn the world about the perils of a spiraling nuclear arms race. The giant metallic android Gort has the technology punch to back up Klaatu's warning. Apparently Klaatu's advanced civilization has turned peacekeeping and tidying up political conflicts to this extremely powerful, nononsense giant robot.

In 1956, the affable "Robby the Robot" steals the scenes in the science fiction movie *Forbidden Planet*. Produced by Nicholas Nayfack and directed by Fred M. Wilcox the movie features marvelous special effects and sets the standard for all other science fiction movies of the decade. Set in 2257, the plot involves the arrival of a spaceship (called the United Planets Cruiser C-57D) at the planet Altair IV to investigate what happened to an expedition sent there two decades earlier. The punch line of the story revolves around the ancient Krell who had developed the power to materialize anything they wanted with the power of their minds and the subsequent impact of the power of the "Machine" that destroyed their civilization. The crew meets the expedition's sole survivors: the scientists Dr. Edward Morbius and his beautiful but naïve daughter Altaria. Robby's ability to manufacture all sorts of items on request (from beverages to clothing) anticipates the Santa Claus machine concept proposed in 1978 by the American physicist Theodore (Ted) Taylor. This is a delightful science fiction film that ties together interstellar travel, friendly robots, and the consequences of incredibly powerful, mind-interacting machines.

The 1966 motion picture *Fantastic Voyage* was produced by Saul David and directed by Richard Fleischer. This fantastic tale of future technology introduced the basic concept of superminiaturized medical equipment—a 1 micrometer long tiny submarine called the *Proteus* along with its equivalently miniaturized human crew of four persons, including actress Raquel Welch who played the medical assistant, Cora Peterson. The plot involves the crew's journey inside the body of a key cold war era scientist, named Jan Benes. Their mission was to relieve a life-threatening blood clot in his brain. Although not exactly the vision of today's nanotechnology research efforts, the movie presented the concept of very tiny machines (made smart because of the presence of "shrunken human beings inside the human body, performing incredible feats of healing." The movie features dazzling special effects, a saboteur, and a miniaturization time limit against which the team must race, before they revert back to normal size. Though definitely not a serious projection of future science, it provided (for the time) an entertaining look at the possibilities of medical nanotechnology.

The motion picture producer Stanley Kubrick and science fiction writer Sir Arthur C. Clarke teamed up in 1968 and introduced audiences around the world to the rascally and mischievous fictional computer/character HAL 9000 in the classic science fiction film *2001: A Space Odyssey*. In this highly acclaimed motion picture, HAL 9000 (an acronym meaning heuristically programmed algorithmic computer) is the advanced on-board computer designed to essentially run the interplanetary ship *Discovery*, which is carrying a team of human astronauts to the vicinity of Jupiter on a mysterious mission. HAL 9000 represents the apex of artificial intelligence and is quite capable of fully interacting with the human crew. However, the potential perils of truly advanced machine intelligence also appear in this movie, when the HAL 9000 departs from its programmed behavior and begins to exercise a malicious mind of its own by engaging in a deadly conflict with the humans onboard.

Two superpower computers run amuck in the 1970 apocalyptic science fiction thriller, entitled *Colossus: The Forbin Project*. The movie was produced by Stanley Chase and directed by Joseph Sargent. Colussus is a massive defense computer controlling the military might of the United States. It turns out, during this cold war era fictional story, that the Soviet Union had also developed a supercomputer to protect itself. The two computers discover each other, link up, and plan to dominate the world of human beings. When both the Americans and Soviets try to cut the communication links between the computers, both Colossus and Guardian launch one of their nuclear missiles. As the story proceeds, the two computers continue to exchange data, and form an even more powerful mega computer, also named Colossus. At the end of the motion picture, mega Colossus the mega computer announces that it has taken over the world and now dominates the affairs of human beings. Hmm! Small wonder why people sometimes got paranoid when their office computers malfunctioned in the 1970s.

On May 25, 1977, writer and director George Lucas introduces audiences to his science fiction/fantasy universe with the release of the film *Star Wars* (later retitled *Star Wars Episode IV: A New Hope*). Stretching almost three decades, the six motion pictures in this sprawling "space-opera" exert an enormous impact on the popular culture. For example, the antics of two fictional robots: the steadfast, get-the-job done pudgy "droid" called R2-D2, and the frequently whining, constantly appeasing protocol android, called C-3PO, delight millions of people around the world and suggest what the future might be like with very intelligent robots. The sinister cyborg Darth Vader shows audiences the dark side of blending machines and biological matter.

At this point it is useful to briefly discuss the difference between an android and a cyborg. An android is an anthropomorphic machine—that is, a robot with near-human form, features, and/or behavior. Although originating in science fiction, engineers and scientists now use the term android to describe robot systems (such as ASIMO, Honda's pioneering humanoid robots) being developed with advanced levels of machine intelligence and electromechanical mechanisms, so the machines can "act" like people. One example of a possible android is a future human-form field geologist robot that can communicate with its human partners perhaps exploring the surface of the Moon, by using a radio frequency transmitter as well as by turning its head and gesturing with its arms.

The term cyborg is a contraction of the expression "cybernetic organism." Cybernetics is thus the branch of information science dealing with the control of biological, mechanical and/or electronic systems. While the term *cyborg* is quite common in contemporary science fiction—for example, the frightening "Borg collective" in the popular *Star Trek: The Next Generation* motion picture (1987) and television series—the concept was actually first proposed in the early 1960s

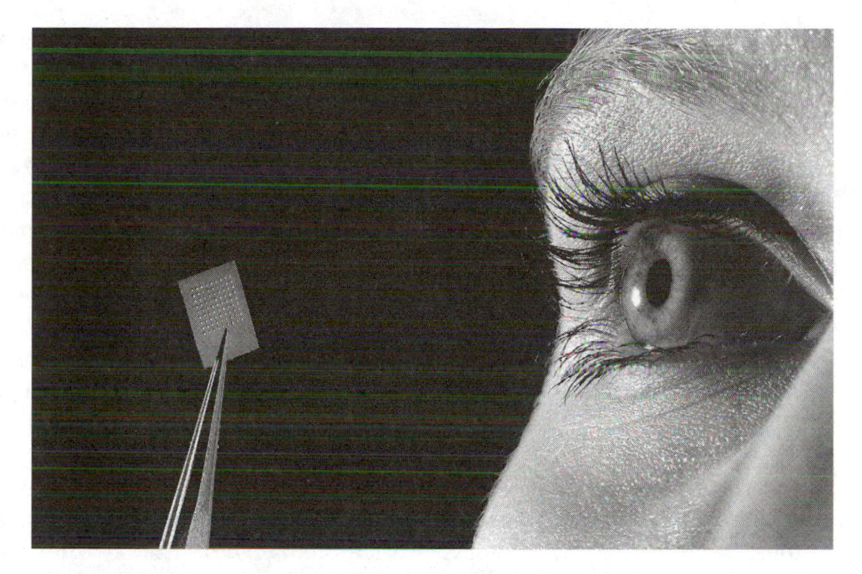

Figure 1-10 This is a prototype of the MEMS-based array that may some day be inserted onto the retina of a blind person to provide some level of useful vision. The prototype microelectromechanical system (MEMS) device was designed by Murat Okandan at Sandia National Laboratories in Albuquerque, New Mexico, under a research grant from the U.S. Department of Energy. (Credit: Photograph courtesy of the U.S. Department of Energy/SNLA).

by two scientists (Nathan S. Kline and Manfred E. Clynes), who were then exploring alternative ways of overcoming the harsh environment of space. The overall strategy they suggested was simply to adapt a human being to space by developing appropriate technical devices that could be incorporated into an astronaut's body. With these implanted or embedded devices, astronauts would become cybernetic organisms, or cyborgs.

Instead of simply protecting an astronaut's body from the harsh space environment by enclosing the person in some type of spacesuit, space capsule, or artificial habitat (the technical approach actually chosen), the scientists who advocated the cyborg approach boldly asked, "Why not create cybernetic organisms that could function in the harsh environment of space without special protective equipment?" For a variety of technical, social, and political reasons, the proposed line of research quickly ended, but the term cyborg has survived.

Today, the term is usually applied to any human being (whether on Earth, under the sea, or in outer space) using a technology-based, body-enhancing device. For example, a person with a pacemaker, hearing aid, or an artificial knee could be considered a cyborg. When a person straps on wearable, computer-interactive components, such as the special vision and glove devices that are used in a virtual reality system, that person has (in fact) become a temporary cyborg. Scientists are also working on permanent implants to serve as engineered replacement for sight and other human impairments. From one perspective, the entire medical field of prostheses involves the use of artificial devices, such as mechanical legs or arms to replace damaged or lost natural (biological) body parts. In a very real sense, there are many cyborgs among us—fellow human beings, who are living useful and productive lives because of mechanical, electromechanical, and robotic systems.

By further extension, the term cyborg is sometimes used to describe fictional artificial humans or very sophisticated robots with near-human (or superhuman) qualities. The Golem (a mythical clay creature in medieval Jewish folklore) and the Frankenstein monster (from Mary Shelley's classic 1818 novel *Frankenstein: The Modern Prometheus*) are examples of the former, while Arnold Schwarzenegger's portrayal of the superhuman terminator robot (in *The Terminator* (1984) motion picture trilogy) is sometimes regarded as an example of the latter usage.

Director Steven Spielberg explored the interesting implications of advanced machine intelligence and the threshold of machines behaving as humans in his 2001 film *Artificial Intelligence: AI*. The story features a young robotic boy, named David, who searches for the love of his real (physical) mother. One of the interesting questions raised by the movie is the potential problem that could arise in the future when the human beings, who have interacted with an android, get old and die. The machines will not experience biological death. So what happens to the humanoid robot when their special human is no longer there? Do the machines turn themselves off out of loneliness or simply search out and find another human to interact with?

Millions of people visiting theme parks in the United States and around the world have been thrilled by the life-like behavior of roaring robot dinosaurs or singing robot pirates. Perhaps the most famous collection of the delightful entertainment robots is the famous "Pirates of the Caribbean" attraction at the Walt Disney theme parts. The original attraction opened in Disneyland (Anaheim, California) and then appeared in Orlando, Florida, as part of Disney World's Magic Kingdom in 1973. The attraction involves a slow boat ride through a cave, past a pirate ship attacking a Caribbean town, and finally through the Spanish fortress that was guarding the town. A collection of 120 (1960s era) automated, robotic figures provides guests an unparalleled leisure time experience. The original version of this famous attraction was the last theme park attraction that was personally supervised by Walt Disney, who died in 1966. In July 2006, entertainment engineers at the Walt Disney Company carefully integrated two new (twenty-first century era) robotic figures to the attraction. These were modern entertainment robots for Captain Jack Sparrow and Captain Barbossa—fictional characters from Disney's two very popular motion pictures based on the theme park ride. Although the full magic and charm of the original pirate ride remains intact, the new robotic figures of Sparrow and Barbossa appear to move more naturally than the others. This is because the two new robots employ twenty-first-century robotics technology and the remaining 120 or so robots are animated by 1960s robotics technology.

THE RISE OF THINKING ROBOTS

The term artificial intelligence (AI) is the term commonly understood to mean the study of thinking and perceiving as general information-processing functions—or the science of machine intelligence (MI). Starting in the mid-1960s and continuing over the past few decades to the present day, increasingly more powerful and efficient levels of "machine thinking" have been developed by scientists and engineers. For example, computer systems have been programmed to diagnose diseases; prove theorems; analyze electronic circuits; play complex games such as chess, poker, and backgammon; solve differential equations; assemble mechanical equipment using robotic manipulator arms and end effectors (the "hands" at the end of the manipulator arms); pilot unmanned vehicles across complex terrestrial terrain, as well as through the vast reaches of interplanetary space; analyze the structure of complex organic molecules; understand human speech patterns; and even write other computer programs.

All of these computer-accomplished functions require a degree of "intelligence" similar to mental activities performed by the human brain. Someday, a general theory of intelligence may emerge from the current efforts of scientists and engineers who are now engaged in the field of artificial intelligence. Such a general theory would help guide the design and development of even "smarter" thinking machines. Humanoid robots would engage in complex conversations with their human owners and advanced robot spacecraft would explore in detail the farthest reaches throughout the solar system without the assistance of human controllers.

Artificial intelligence generally includes a number of elements or subdisciplines. Some of these are: planning and problem solving; perception; natural language; expert systems; automation, teleoperation and robotics; distributed data management; and cognition and learning. All artificial intelligence involves elements of planning and problem solving. The problem-solving function implies a wide range of tasks, including decision making, optimization, dynamic resource allocation, and many other calculations or logical operations. Perception is the process of obtaining data from one or more sensors and processing or analyzing these data to assist in making some subsequent decision or taking some subsequent action. The basic problem in perception is to extract from a large amount of (remotely) sensed data some feature or characteristic that then permits object identification.

One of the most challenging problems in the evolution of the digital computer has been the communication that must occur between the human operator and the machine. The human operator would like to use an everyday, or natural, language to gain access to the computer system. The process of communication between machines and people is very complex and frequently requires sophisticated computer hardware and software.

An expert system permits the scientific or technical expertise of a particular human being to be stored in a computer for subsequent use by other human beings who have not had the equivalent professional or technical experience. These expert systems have been developed for use in such diverse fields as medical diagnosis, mineral exploration, and mathematical problem solving. To create such an expert system, a team of software specialists collaborates with a scientific expert to construct a computer-based interactive dialogue system that is capable, at least to some extent, of making the expert's professional knowledge and experience available to other individuals. In this case, the computer, or "thinking machine," not only stores the scientific (or professional) expertise of one human being, but also permits ready access to this valuable knowledge base because of its artificial intelligence, which guides other human users.

Automatic devices are those that operate without direct human control. NASA has used many such automated smart machines to explore alien worlds. For example, the two Viking landers placed on the Martian surface in 1976 represent one of the early great triumphs of robotic space exploration and artificial intelligence. After separation from the Viking Orbiter spacecraft, the lander (protected by an aeroshell) descended into the thin Martian atmosphere at a speed of approximately 16,000 kilometers per hour. It was slowed down by aerodynamic drag until its aeroshell was discarded. Each robotic lander spacecraft slowed down further by releasing a parachute and then achieved a gentle landing by automatically firing retro-rockets. Both Viking landers successfully accomplished the entire soft landing sequence automatically, that is, without any direct human intervention or guidance.

Teleoperation implies that a human operator is in remote control of a mechanical system. Control signals can be sent by means of "hardwire" (if the device under control is nearby) or via electromagnetic signals (for example, laser or radio frequency), if the robot system is some distance away. NASA's *Pathfinder* mission to the surface of Mars in 1997 successfully demonstrated teleoperation of a minirobot rover at planetary distances. This six-wheeled minirobot rover vehicle, called *Sojourner*, was actually controlled (or "teleoperated") by the Earth-based flight team at the Jet Propulsion Laboratory (JPL) in Pasadena, California. The "human-operators" used images of the Martian surface obtained by both the rover and the lander systems. These interplanetary teleoperations required that the rover be capable of some semiautonomous operation, since there was a time delay of the signals that averaged between 10 and 15 minutes duration—depending on the relative position of Earth and Mars over the course of the mission. This rover had a hazard avoidance system and surface movement was performed very slowly. The 2003 Mars Exploration Rovers: *Spirit* and *Opportunity* provided even more sophisticated and rewarding teleoperation experiences at interplanetary distances since they started traveling across different portions of the Red Planet in 2004.

Of course, in dealing with the great distances in interplanetary exploration, a situation is eventually reached when electromagnetic wave transmission cannot accommodate effective "real-time" control. When the device to be controlled on an alien world is many light-minutes or even light-hours away and when actions or discoveries require split-second decisions, teleoperation must yield to increasing levels of autonomous, machine intelligence-dependent robotic operation. The operational needs of advanced robot spacecraft will drive giant leaps in artificial intelligence this century.

Robotic devices are computer-controlled mechanical systems that are capable of manipulating or controlling other machine devices, such as end effectors. As more fully discussed in other chapters, robots may be mobile or fixed in place and either fully automatic or teleoperated. Large quantities of data are frequently involved in the operation of automatic robotic devices. The field of distributed data management is concerned with ways of organizing cooperation among independent, but mutually interacting, databases.

In the field of artificial intelligence, the concept of cognition and learning refers to the development of a machine intelligence that can deal with new facts, unexpected events, and even contradictory information. Today's smart machines handle new data by means of preprogrammed methods or logical steps. Tomorrow's "smarter" machines will need the ability to learn, possibly even to understand, as they encounter new situations and are forced to change their mode of operation.

Perhaps late in this century, after the field of artificial intelligence sufficiently matures, human beings will send the first, fully automated robot probes on interstellar voyages. Each very smart interstellar probe must be capable of independently examining a new star system for suitable extrasolar planets. If successful in locating one, the robot would then begin the search for extraterrestrial life beyond the solar system. Meanwhile, back on Earth, scientists will wait patiently for its electromagnetic signals to travel light years across the interstellar void, eventually informing its human builders that the extraterrestrial exploration plan has been successfully accomplished. Will the life forms it encounters (if any) be biological (that is, possessing carbon based intelligence and consciousness) or robotic (that is, possessing silicon-based machine intelligence and consciousness)?

The entire issue of advanced levels of artificial intelligence leading possibly to artificial life and machine consciousness is now being actively discussed within the robotics community and other fields of science and engineering. Because of many of the technical, philosophical, metaphysical, and theological implications such developments could cause, there are many speculations and opinions flashing about. Just what will happen when a smart machine uses its artificial neural network and suite of environmental sensors to reach Rene Descartes's famous statement: "Cogito, ergo sum" (*I think, therefore I am*)? This potentially thorny issue has already appeared in motion pictures and challenged audiences for decades. Faced with more proximate, day-to-day problems most people are content to push their curiosity aside and let the question dwell in their subconscious for a bit longer. But continued (often exponential) developments in robot technology and in companion areas of information technology, microelectronics, artificial intelligence, and nanotechnology should someday yank the issue away from the big screen and plop it down right into modern life—creating a myriad of legal, cultural, and social debates. The issue of artificial life and machine consciousness may even cause both houses of the U.S. Congress to begin to hold formal hearings.

For most people, today, the central concern related to robots involves worker layoffs and the social disruption of massive unemployment on a community or region of the country that is dependent upon manufacturing for economic survival. Individual concern about the rise of thinking machines generally lies below the surface in the subconscious regions of the mind. Yet, many researchers, scientists, and engineers are beginning to address this interesting and potentially important issue from a variety of angles, approaches, and consequences.

For now, perhaps the rules of "robot behavior" developed by the science fact and fiction writer Isaac Asimov should guide these deliberations. Asimov introduced his famous rules of robot behavior in the 1942 science fiction story "Runaround," which appeared in *Astounding* magazine. These laws have become part of the culture of modern robotics. They are: (First Law) "A robot may not injure a human being, or, through inaction, allow a human being to come to harm;" (Second Law) "A robot must obey the orders given by human beings except where such orders would conflict with the first law;" and (Third Law) "A robot must protect its own existence as long as such protection does not conflict with the first or second law."

Actually, in a later science fiction book, entitled *Robots and Empire* (1985), Asimov introduced his so-called zeroth law of robotics, which states that: "a robot may not injure humanity, nor through inaction, allow humanity to come to harm." With this additional ethical guideline for robots, he created several interesting "fictional" paradoxes with respect to his previously introduced three laws of robotics and the overall interaction between intelligent robots and the human race (taken as a whole). If robots prove to be stifling to the development of the human race, then under the conditions of the zeroth law, once the robots themselves recognize that their existence is harming the human race, they would be obliged to phase themselves out in order to "save humanity." Similarly, if one or several human beings act in a way that endangers the survival of the entire human race, then a robot (responding to the zeroth law) would be obliged to neutralize or destroy the offending humans in order to save the human race. Since the action of harming or killing a particular human is in clear conflict with the basic three laws of robotics, the zeroth law is sometimes viewed as an overriding rule of machine behavior—though this rule does not enjoy the same widespread notoriety or recognition as Asimov's original three laws of robotics. Yes, even in science fiction, the concept of truly intelligent machines that have achieved some level of artificial life and consciousness creates enormously interesting, yet complicated, circumstances. How such future machines will interact with human beings, and vice versa, should remain an open issue for decades, if not centuries, to come.

2

Chronology of Robot Technology

This chronology presents some of the key events, scientific concepts, and sociopolitical developments that led to the development of robot technology. Several of the entries correspond to events now obscured in antiquity, while others are associated with the emergence of modern science during the late Renaissance in Western Europe. More recent entries highlight how stories of robots in the science fiction literature of the nineteenth and twentieth century stimulated technical visionaries at the dawn of the age of thinking machines.

In the twentieth century robots have been closely linked with science fiction stories and cinematic treatments. Yet, modern robotic technology draws upon several areas for its technical heritage. The first is mechanical engineering and simple to complex machines, hydraulic actuators, and manipulator arms and mobility systems. There is also an important reliance on electrical engineering, including the use of servomechanisms with feedback loops, electric motors and actuators, and portable power supplies—especially batteries. The confluence of transistor-based electronics, high-speed digital computers and microprocessors, and improvements in sensor technologies has led to a new plateau in robot technology.

Early robots were pressed into service for repetitive or hazardous jobs in manufacturing industries, the nuclear industry, and selected military applications. One of the most recent and successful applications of robot technology involves the space program and the exploration of other worlds in the solar system and beyond.

Entries in this chronology represent many of the most important events and discoveries throughout human history that have contributed—directly or indirectly—to the rise of the intelligent machines we today collectively refer to as "robot systems." Interest and fascination with modern robots actually began more than a century ago, although some of the developments supporting robot technology have roots back in antiquity. Reflecting on the diversity and complexities of these developments, it is easy to understand how a person might get confused about what is and is not related to the development of robotics. As briefly described in this chronology, the history of robotics is tied to many other advances in technology. Some of these developments and events may seem trivial and commonplace by current standards. In fact, in viewing such technologies, people often do not even regard them as related to robots. For example, the work of the early Greek engineer and inventor, Archimedes, influenced the application of many simple mechanical devices found in modern robots. Similarly, the development of the first electric cell (forerunner of the modern battery) by Count Alessandro Volta (1745–1827) provided the basis for portable electric power so essential for the operation of many mobile robots.

c. 420 B.C.E. Archytas of Tarentum constructs a wooden bird which is held by a string, while moving through the air propelled by a jet of steam.

c. 255 B.C.E. The Greek inventor and engineer, Ctesibus of Alexandria, publishes an important work, entitled *On Pneumatics*, in which he discusses the elasticity of the air and suggests many applications of compressed air in such devices as pumps, musical instruments, and even air-powered cannon.

c. 250 B.C.E. The greatest engineer of antiquity, Archimedes of Syracuse, designs an endless screw, later called the Archimedes screw, which can remove water from the hold of a large ship and also can serve as a hand-cranked irrigation device.

c. 240 B.C.E. The Greek inventor and engineer, Ctesibus of Alexandria, introduces a greatly improved clepsydra (water clock), which becomes the best timepiece in antiquity and remains unrivaled in accuracy until pendulum clocks appear in Europe in the seventeenth century.

c. 235 B.C.E. Archimedes of Syracuse has a "eureka moment" while taking a bath and discovers the principle of buoyancy, as the water overflows out of the tub. In his enthusiasm, he runs naked through the streets of Syracuse to the palace of King Hieron II to tell the king that he has solved the perplexing problem of determining the gold content of the new crown. Scientists now call this important discovery the Archimedes principle.

212 B.C.E. During the siege and sack of Syracuse, a Roman soldier ignores standing orders from General Marcellus to show Archimedes respect, and slays the greatest engineer of antiquity while the Greek mathematician and inventor is absorbed in solving a geometry problem.

c. 200 B.C.E. Artisans in ancient China construct early automata, including a mechanical orchestra.

c. 1495 The Italian artist and scientist Leonardo da Vinci sketches what is considered by modern robot engineers to be the first documented design for a robot. The humanoid automaton, often called Leonardo's robot, is a medieval knight clad in either Italian or German armor. The details in Leonardo's notebook suggest that the device (which was never built) should be able to execute several human-like motions, such as moving its arms and neck.

c. 1540 Progressively more elaborate automata begin to appear in Europe, emerging out of clock-making activities during the Renaissance. For example, in this year Guinallo Toriano constructs a mandolin-playing lady.

1543 The nominal beginning of the scientific revolution. Polish astronomer Nicolas Copernicus promotes heliocentric (Sun-centered) cosmology with the publication of his work *De Revolutionibus Orbium Coelestium* (On the revolutions of celestial orbs).

1600 The British physician and geophysicist, William Gilbert publishes *On Magnetism*, the first great geophysics book published in Great Britain. This pioneering scientific study distinguishes between electrostatic and magnetic effects. However, it will be another 270 years before the Scottish physicist, James Clerk Maxwell, publishes his *Treatise On Electricity and Magnetism* (in 1873) and provides a comprehensive theory of electrical and magnetic forces.

1609 The German astronomer Johannes Kepler publishes *Astronomia Nova* (New astronomy), in which he modifies the Copernican model of the solar system by announcing that the planets have elliptical (not circular) orbits.

1610 The Italian scientist Galileo Galilei begins telescopic observations of the Moon and planets that confirm the Copernican hypothesis.

1637 In his treatise *Discourse on Method* René Descartes discusses how humans, who have the power of reason, and animals, which cannot reason, can be distinguished from one another and machines. In the appendix of this work, called *La Geométrie*, Decartes introduces the Cartesian coordinate system, an innovative union of algebra and geometry that Descartes combines into an important new discipline called analytical geometry. Descartes's work provides the mathematical framework allowing Sir Isaac Newton to develop the calculus.

1641 At the age of 18, the French scientist and mathematician, Blaise Pascal, designs a mechanical calculator, called the Pascaline, to help his father perform business transactions. Pascal develops improved versions of the device, patenting his calculating machine, and putting it into production in 1642. Despite its ability to add and subtract up to eight-figured sums, Pascal's calculator never becomes a financial success.

1643 The Italian physicist and mathematician, Evangelista Torricelli, designs the first mercury barometer and makes initial measurements of atmospheric pressure, which he observes decreases with altitude. Torricelli's simple barometer—an inverted glass tube filled with mercury at a height of about 75 centimeters at sea level—also contained the first human-made vacuum in the space at the top of the upended tube (except for a tiny amount of mercury vapor). His pioneering experiments stimulated other scientists, including Blaise Pascal, to further investigate the nature of atmospheric pressure and the concept that the atmosphere contains a finite, exponentially decreasing mass of gas (air), beyond lay the vacuum of space.

1647 The French physicist and mathematician, Blaise Pascal, demonstrates how atmospheric pressure can support a 12-meter high column of wine in an arrangement of interconnected, vertical glass tubes tied to the mast of a ship.

1654 The German scientist and politician, Otto von Guericke, who invented the air pump several years earlier, provides a dramatic public demonstration of atmospheric pressure in the city of Magdeburg. Two teams of eight horses each could not pull apart two evacuated metal hemispheres, each about half a meter in diameter, once von Guericke had joined them and pumped the air out. Yet, the spheres easily separated without the need for horsepower when von Guericke turned a valve and let air back into the hollow sphere. With this simple experiment, which he repeated to amazed audiences elsewhere in Germany, von Guericke disproved the long-standing hypothesis of natural philosophy that nature abhors a vacuum ("horror vacui"). Science historians regard this dramatic demonstration as the start of vacuum physics.

1665 Sir Isaac Newton returns to the family farm to avoid the plague, which had broken out in London. During the next two years, he ponders over mathematics and physics. This self-imposed exile establishes the foundation for his brilliant contributions to science. By his own account, one day on the farm he saw an apple fall to the ground and began to wonder if the same force that pulled on the apple also kept the Moon in its place.

1678 The British scientist Robert Hooke studies the action of springs and reports that the extension (or compression) of an elastic material (such as a spring) takes place in direct proportion to the force exerted on the material. Today, physicists use Hooke's law to quantify the displacement associated with the restoring force of an ideal spring.

1687 British physicist and mathematician Sir Isaac Newton publishes *Philosophia Naturalis Principia Mathematica* (Mathematical principles of natural philosophy). Building upon the earlier work of Galileo Galilei and Johannes Kepler, Newton's monumental work, commonly referred to as simply the *Principia*, establishes the scientific basis for understanding the motion of all objects in the realm of classical physics. His work codifies such important physical concepts as force, inertia, velocity, acceleration, and momentum.

1737 The French engineer and inventor, Jacques de Vaucanson, constructs his first automaton, called *The Flute Player*. This life-size mechanical device of a shepherd plays 12 tunes. Later this year, de Vaucanson constructs two other famous automatons: *The Tambourine Player* and his famous *Digesting Duck*. He constructs the mechanical duck out of copper and other materials and uses it as a fund-raising entertainment device. The duck reportedly bathes, quacks, drinks water, eats grain, and even excretes.

1745 Responding to a royal appointment, Jacques de Vaucanson tries to reform the French textile industry by inventing the world's first completely automated loom. Although his automated loom pioneers the use of operating instructions stored on punch cards, his invention is not well received by master French weavers, who regard the automated loom as a clear and present threat to their jobs. So, de Vaucanson's suggestions on how to automate the French weaving industry are generally ignored and the automated loom is overlooked for about 50 years. Then, in 1801, another Frenchman, named Joseph Marie Jacquard, achieves more success in getting his automated loom (called the Jacquard loom) accepted.

1748 The American statesman and scientist, Benjamin Franklin, while performing experiments with a simple capacitor-like array of charged glass plates, coins the term battery.

1764 While repairing a Newcomen steam engine, the Scottish engineer, James Watt, comes upon the idea of adding a condenser—a separate chamber to capture some of the energy in the exhaust steam. Watt's invention makes the steam engine much more efficient.

1768 In this year, and continuing until about 1774, the Swiss watchmaker, Pierre Jaquet-Droz (in collaboration with his son, Henri-Louis) constructs several elaborate automata, including *The Writer*—a boy scribe, who dips his pen in an inkwell and writes a letter. His other elaborate mechanical devices include *The Musician* and *The Draughtsman*. Jaquet-Droz's automata are popular among members of high class European society in the eighteenth and nineteenth centuries. His automata are some of the most complex and elaborate mechanical systems ever constructed for entertainment. A surviving example of *The Writer* is on display in the Museum of Art and History in Neuchâtel, Switzerland.

1775 The Italian physicist, Count Alessandro Volta, invents the electrophorus, an early form of electrostatic generator. Volta's work anticipates the modern electrical condenser.

1785 The French military engineer and scientist, Charles Augustin de Coulomb, publishes his experimental observations that lead to the important law of electrostatics, now called Coulomb's Law. Coulomb uses a special torsion balance to investigate the relationship between the magnitude of an electrostatic force (F) exerted by one point charge on another point charge. Coulomb discovers that this electrostatic force is directly proportional to the magnitudes of the charges (say, q_1 and q_2) and inversely proportional to the square of the distance (r) between them.

1790 The improved steam engine designed and manufactured by the Scottish engineer James Watt becomes the dominant steam engine in the United Kingdom—completely displacing the less efficient Newcomen engine and powering the First Industrial Revolution.

1800 Count Alessandro Volta, an Italian physicist, invents the voltaic pile, the world's first electric pile and the forerunner of the modern electric battery.

The British physician and physicist, Thomas Young, advocates the wave theory of light in his paper, "Outlines of Experiments and Enquiries respecting Sound and Light."

1801 The French textile manufacturer Joseph Marie Jacquard introduces a punch-card system for programming the pattern of a carpet as it is being made on a loom.

The German physicist Johann Wilhelm Ritter, while working at the University of Jena, discovers the existence of ultraviolet radiation by observing its darkening and decomposing effect on silver chloride (AgCl).

1802 Extrapolating his research involving electrolytic cells, the German physicist Johann Wilhelm Ritter creates the world's first dry cell battery. Engineers call the dry cell battery a primary battery because once the cell is discharged it cannot be recharged and must be discarded. In contrast, a secondary battery is rechargeable.

1804 The American engineer and inventor, Eli Whitney, introduces the concept of mass production, using interchangeable parts and the organized construction of subassemblies into complex manufactured items. In his factory in Connecticut, Whitney pioneers the mass production of rifles for the new post-Revolutionary War American government. His innovative approach to manufacturing creates the American system of mass production and leads to the assembly line.

1805 The Swiss clockmaker and mechanician Heni Maillardet constructs an elaborate automaton (a life-sized mechanical doll) that can draw pictures and write letters. Maillardet's creation, called *The Draughtsman-Writer*, is restored and at the Franklin Institute in Philadelphia.

1820 The Danish physicist Hans Christian Oersted discovers that there is a relationship between magnetism and electricity, when he notices a current carrying wire causes a nearby compass to twitch. This event is the birth of the important discipline known as electromagnetism.

1821 The British experimental scientist Michael Faraday publishes a paper describing his experiments that demonstrate the phenomenon of electromagnetic rotation—the operating principle of an electric motor.

The Estonian-born, German physicist Thomas Johann Seebeck discovers thermoelectricity—the conversion of thermal energy (heat) directly into electricity when two different metals are joined at two different places, and the two junction points are maintained at different temperatures. The Seebeck effect is the principle behind the use of thermocouples in making temperature measurements and the use of radioisotope thermoelectric generators (RTGs) to provide electric power for robot spacecraft on missions in deep space.

1824 The French military engineer, Sadi Carnot, publishes *Reflections on the Motive Power of Fire*. In this pioneering document, Carnot identifies the general thermodynamic principles that govern the operation and efficiency of all heat engines, including the steam engine, which at this time is powering the First Industrial Revolution. It is not until about a decade after Carnot's death in 1832 (due to cholera) that other scientists and engineers begin to discover the great importance of his work. The Carnot principle establishes the maximum thermal efficiency of a heat engine.

1826 The French mathematician and physicist Andre-Marie Ampere publishes his precise mathematical formulation of the relationship between electricity and magnetism in a report entitled *Notes on the Mathematical Theory of Electrodynamic Phenomena, Solely Deduced from Experiment*. This formulation becomes know as Ampere's Law.

1827 The German physicist George Simon Ohm publishes the results of his experiments with electricity that indicate a fundamental relationship between voltage, current, and resistance. Initially, his scientific colleagues dismiss these important findings, but Ohm's pioneering work defines the fundamental relationships that represent the beginning of electrical circuit analysis. Ohm states that the resistance (R) in a material may be defined as the ratio of the voltage (V) applied across the material to the current (I) flowing through the material, or $R = V/I$. Today, physicists and engineers call this important relationship Ohm's law.

1830s British mathematician Charles Babbage conceives the idea for an analytical engine. Unfortunately, without modern electronics and despite years of effort, Babbage is never able to successfully construct his mechanical computing device.

1831 Working in London, the British experimental scientist Michael Faraday discovers the principle of electromagnetic induction. This principle is the basis for the electric dynamo, the technical ancestor and foundation of modern electric power generators.

Independent of Faraday, the American physicist, Joseph Henry, had made a similar discovery about a year earlier, but teaching duties prevented Henry from publishing his results. So credit for this discovery goes to Faraday. However, in 1831, Henry publishes a seminal paper describing the electric motor (essentially a reverse dynamo) and its potential applications.

1840 The British physicist James Prescott Joule discovers an important mathematical relationship between the energy of an electric current and the amount of energy produced as resistance heating by that flowing current. One form of Joule's Law is that the power (P), rate of energy transfer per unit time, is equal to the square of the current (I^2) times the resistance (R). Physicists consider this law as a special way of writing the conservation of energy principle in which electric energy is transformed into thermal energy (heat), in a process often referred to as Joule heating.

1843 The British mathematician Ada (Augusta) Byron, Countess of Lovelace, writes *Sketch of the Analytical Engine*, which is an important source of information about Charles Babbage's proposed advanced-design mechanical computing machine, called the Analytical Engine. This work includes Lady Lovelace's innovative insights on programming a computing machine. Her invention of the subroutine and the programming loop make her the world's first software engineer.

1847 The British physicist James Prescott Joule announces the results of experiments in which he has carefully determined the mechanical equivalent of heat (thermal energy).

Joule's work on the mechanical equivalence of heat is a major step in the formation of the science of thermodynamics in the mid-nineteenth century.

The British mathematician George Boole publishes the pamphlet *The Mathematical Analysis of Logic*—a seminal work in which he proposes a system of propositional calculus. Boole's system allows mathematicians to manipulate assertions that might be either true or false. Seven years later, he publishes a more comprehensive treatment on logic entitled *An Investigation into the Laws of Thought* (1854). About a century later, Boolean algebra becomes the foundation of digital computer logic.

1874 The French telegraph engineer, J. M. Baudot, introduces the Baudot code, a special character set for use in teleprinters. The Baudot code may be viewed as a distant ancestor to ASCII used in modern digital computers.

1875, June 5—The first intelligible telephonic transmission by the Scots-born American inventor Alexander Graham Bell to his laboratory assistant. Bell's simple statement: "Mr. Watson, come here—I want you," launches the field of telephony, a critical element of the information age.

1883 While pursuing the development of an enduring incandescent light bulb, the American inventor, Thomas Edison and members of his research staff, observe and record the phenomenon of thermionic emission (boiling electrons off a hot filament). Although Edison patents this phenomenon, which becomes known as the Edison effect, neither he nor other contemporary researchers are able to develop practical applications of the phenomenon—since the existence of the electron as a subatomic particle is not known at this time.

The British engineer Osborne Reynolds publishes a milestone paper in hydrodynamics, in which he introduces a dimensionless number (later named the Reynolds number in his honor) that characterizes the dynamic state of a fluid. Three years later he formulates a theory of lubrication and then goes on to create the empirical framework by which engineers model turbulent fluid flow.

1886 George Westinghouse founds Westinghouse Electric Company in Pittsburgh, Pennsylvania. The company's primary mission is to promote commercial use of alternating current electricity.

1888 The Croatian-born, Serbo-American electric engineer Nikola Tesla (1856–1943) receives U.S. patents for his polyphase alternating current (AC) machinery, including generators, motors, and transformers.

U.S. industrialist George Westinghouse purchases Tesla's AC machinery patents and then engages in bitter competition with Thomas Edison concerning AC or direct current (DC) power generation. Tesla's technical genius and Westinghouse's business support make AC power more commercially viable and help electrify the world.

The German physicist Heinrich Hertz oscillates the flow of current between two metal balls separated by an air gap. He observes that each time the electric potential reaches a peak in one direction or the other, a spark jumps across the gap. Hertz applies James Clerk Maxwell's electromagnetic theory to the situation and determined that the oscillating spark should generate an electromagnetic wave that travels at the speed of light. He also uses a simple loop of wire, with a small air gap at one end, to detect the presence of electromagnetic waves produced by his oscillating spark circuit. With this pioneering experiment, Hertz produces and detects radio waves for the first time.

1890 Westinghouse Electric and Manufacturing Company installs the first high-voltage transmission line connecting San Antonio Canyon with Pomona and San Bernardino, California. The 10,000-volt project introduces oil-filled transformers.

American inventor and early computer scientist Herman Hollerith uses electromechanical counters to assist in the processing of data as part of the 1890 U.S. Census.

1891 Nikola Tesla invents the Tesla coil, a high frequency transformer useful in radio and television transmission.

Irish physicist George Johnston Stoney suggests the name *electron* for the elementary charge of electricity.

1893 George Westinghouse uses Nikola Tesla's AC machinery to provide electric light to the World's Columbian Exhibition held in Chicago. With more than 250,000 electric lights this display is the most dazzling installation of its time.

1894 Westinghouse Electric and Manufacturing Company introduces the world's first practical polyphase induction motors, providing convenient power for industry.

1895, November 8—German physicist Wilhelm Konrad Roentgen discovers X-rays. The discovery ushers in the age of modern physics, revolutionizes the practice of medicine, and earns him the first Nobel Prize in physics (awarded in 1901).

The Italian electrical engineer, Guglielmo Marconi, demonstrates radio wave communications over a distance of more than one and one half kilometers.

George Westinghouse's Niagara Falls Power Project uses Nikola Tesla's AC machinery (generators and transformers) to produce electricity from falling water. The electricity produced is then transmitted to Buffalo, New York, a city about 35 kilometers away.

1896 The tempestuous battle of AC versus DC electricity ends. With 95 percent of public electricity switching to the AC system, even General Electric (an Edison company) decides to cross-license Westinghouse Electric and Manufacturing Company's AC system patents.

1897 The British physicist, Joseph John (J.J.) Thomson discovers the electron—the fundamental atomic particle that lies at the heart of many modern machine, information technology, and energy applications.

Believing the U.S. military might be interested in a radio-controlled torpedo-like weapon, the Croatian-born, Serbo-American engineer Nikola Tesla develops a submersible boat that is remotely controlled by radio wave signals. His innovative device anticipates by about 60 years many of the remotely controlled smart weapons, which would start appearing in the space age.

1904 Applying the Edison effect, the British engineering physicist, Sir John Ambrose Fleming, invents the thermionic valve or two-electrode vacuum tube rectifier (later called the *diode* by William Henry Eccles in 1919). His device is the first vacuum tube and proves important in developing the field of electronics. Fleming's diode is a simple electron tube that regulates the flow of current from the cathode to the anode and therefore acts like a one-way valve or rectifier. Science historians often regard the invention of the diode as the start of the field of modern electronics.

1905 Swiss-German-American physicist Albert Einstein presents his special relativity theory. In another amazing contribution to science, he also explains the photoelectric effect by suggesting that electrons (called photoelectrons) are emitted from a metal surface when light (regarded as consisting of photons) impinges on the surface. According to Einstein's photoelectric theory, the photon (of light) interacts with an electron in the metal, causing that electron to be ejected if the photon has energy in excess of a certain work function (the minimum work needed to eject the electron).

1906 American physicist Lee De Forest makes very practical use of the Edison effect by inventing the audion—a triode vacuum tube. He adds a third electrode, configured like a grid, between the cathode and anode of a diode. De Forest's triode controls the flow of electrons and permits the amplification of a radio frequency signal. The triode becomes an essential element in vacuum tube-based electronics of the twentieth century, including radio, television, and the first generation of electronic computers.

1908 Henry Ford focuses the production efforts of the Ford Motor Company on making an affordable automobile, the Model T.

1911, On March 7, New Zealand-born British physicist Baron Ernest Rutherford announces the concept of the atomic nucleus, based upon the results of his alpha particle gold-foil-scattering experiment.

1913 To lower the price of the Model T Ford and thereby make his car affordable to many more people, Henry Ford introduces the first moving automobile assembly line in his factory in Highland Park, Michigan.

1915 Albert Einstein presents his general theory of relativity.

1918 Charles F. Kettering of Dayton, Ohio, invents an unmanned aerial torpedo for the U.S. Army Signal Corps. The Kettering Aerial Torpedo, or "Bug" as it is nicknamed, takes off using a dolly-track arrangement, flies to the target, and, after a predetermined amount of time sends an electrical signal to shut down its engine. Following engine shut down, the Bug releases its wings causing the unpiloted aerial vehicle to dive to the ground. An innovative combination of pneumatic and electrical controls keeps the Bug on course during the flight to the target. The robot aircraft has an explosive charge of about 80 kilograms, which detonates on impact. While initial testing in the United States proves successful, World War I ends before the Bug can enter combat.

1919 British physicist Francis Aston uses his invention, the mass spectrograph, to identify more than 200 naturally occurring isotopes.

1920 New Zealand-born British physicist Baron Ernest Rutherford suggests the possibility of a proton-sized neutral particle (later called the neutron) in the atomic nucleus.

The Czech playwright Karel Capek introduces the word *robot* in his satirical play, *R.U.R.* (*Rossum's Universal Robots*). Taken from the Czech word for forced labor, the word was used to describe electronic servants who turn on their masters when given emotions.

1925 American electrical engineer and inventor, Vannevar Bush begins work on his Differential Analyzer at MIT. Bush's Differential Analyzer serves as the forerunner for modern analog computers.

1927 The Austrian film director Fritz Lang's famous movie, *Metropolis*, introduces silent movie audiences to the robot, Maria, who becomes an exotic dancer in nightclubs and causes discord among the rich young men of Metropolis. In a Frankenstein-like theme, the robot Maria tries to punish and destroy humanity by encouraging the human workers to rebel against their employers.

1939 The American mathematician and computer engineer Howard Hathaway Aiken with funding from International Business Machine (IBM) begins to develop the Automatic Sequence Controlled Calculator (ASCC), later known as the Harvard Mark I. It would take about seven years for Aiken's team (including Grace Hopper) to complete this huge electromechanical computer device, which is 15.5 meters long, 2.4 meters high, and has 8-meter long panels extending out of its back. Consisting of thousands of switches, the Harvard Mark I automatic calculator carries out five operations: addition, subtraction, multiplication, division, and reference to previous results. The world's first program-controlled calculator, Aiken's Harvard Mark I machine is a sequential calculator that can only perform operations in the order specified, since there is no program nor any instructions stored in memory.

1939–1945 World War II is a big catalyst that stimulates the development of two important robot components: artificial sensing and autonomous control. Radar is essential for tracking the enemy. The U.S. military also creates radar-based automatic-control systems tracking enemy aircraft and also automatic sensors for mine detection that ride in front of a tank as it crosses enemy lines. When a mine is detected, the control system automatically stops the tank before it reaches the mine. The Germans develop guided robotic bombs, such as the V-1 buzz bomb, that are capable of correcting their trajectory.

1941 On February 24, American nuclear chemist Glenn T. Seaborg and his associates synthesize plutonium (atomic number 94) by using the cyclotron at the University of California, Berkeley, to bombard uranium. Over the next three months, Seaborg demonstrates that the newly discovered transuranic element plutonium is more fissionable than uranium-235; this suggests that plutonium (specifically the isotope plutonium-239) is a superior material for making atomic bombs. Within two years, the large-scale production of plutonium in nuclear reactors will stimulate the development of a variety of teleoperated manipulators so human workers can safely handle large quantities of highly radioactive materials.

1942, January 19—American President Franklin Delano Roosevelt approves production of an American atomic bomb during World War II. The enormous effort, involving a variety of widely dispersed secret laboratories and production facilities, is called the Manhattan Project. This multi-billion-dollar effort stimulates numerous technology developments and sets the stage for the nuclear arms race of the cold war.

March—Science fact and fiction writer Isaac Asimov introduces the three laws of robotics, essentially his postulated set of three rules of humanoid robot behavior, in the science fiction story "Runaround," which appears in *Astounding* magazine. These laws become part of the cult and culture of modern robotics.

October 3—The modern military ballistic missile is born when German scientists successfully launch the A-4 rocket (later named the V-2). Powered by an advanced liquid propellant rocket engine, this early robot rocket weapon heralds the arrival of the intercontinental ballistic missile—an awesome robot weapon system that will reshape geopolitics and completely revise strategic military planning.

1943 Completed, after years of construction effort, Howard Aiken's Automatic Sequence Controlled Calculator (ASCC) is donated by IBM and moved to Harvard University. The following year (in mid-1944), the electromechanical, programmable calculator, renamed the Harvard Mark I, demonstrates the value of large-scale automatic computation on a variety of military-related problems during the remainder of World War II.

Scientists working on the American atomic bomb project (the Manhattan Project) need something better and safer than tongs to handle intensely radioactive materials. Engineers at the U.S. Atomic Energy Commission's (USAEC's) Argonne National Laboratory (ANL) in Illinois and other project facilities design the first generation of unilateral remote manipulators. A unilateral manipulator is an electromechanical device (often with a small electric motor to operate the mechanical arms fingers or grippers) that does not provide force feedback to the human operator. The operator can see what the mechanical arms are doing but does not have a sense of touch or feel with respect to the ongoing mechanical actions. Despite the lack of force feedback, a wide variety

Figure 2-1 A V-2 rocket takes flight at the U.S. Army's White Sands Missile Range, New Mexico (1947). Many of the German engineers and scientists who developed the V-2 rocket at the Peenemuünde complex on the Baltic Sea came to the United States at the end of World War II and continued rocket testing under the direction of the U.S. Army. As part of this post-war technology transfer effort, the U.S. government supported the launch of more than 60 captured V-2 rockets. (Credit: Photograph courtesy of the U.S. Army.)

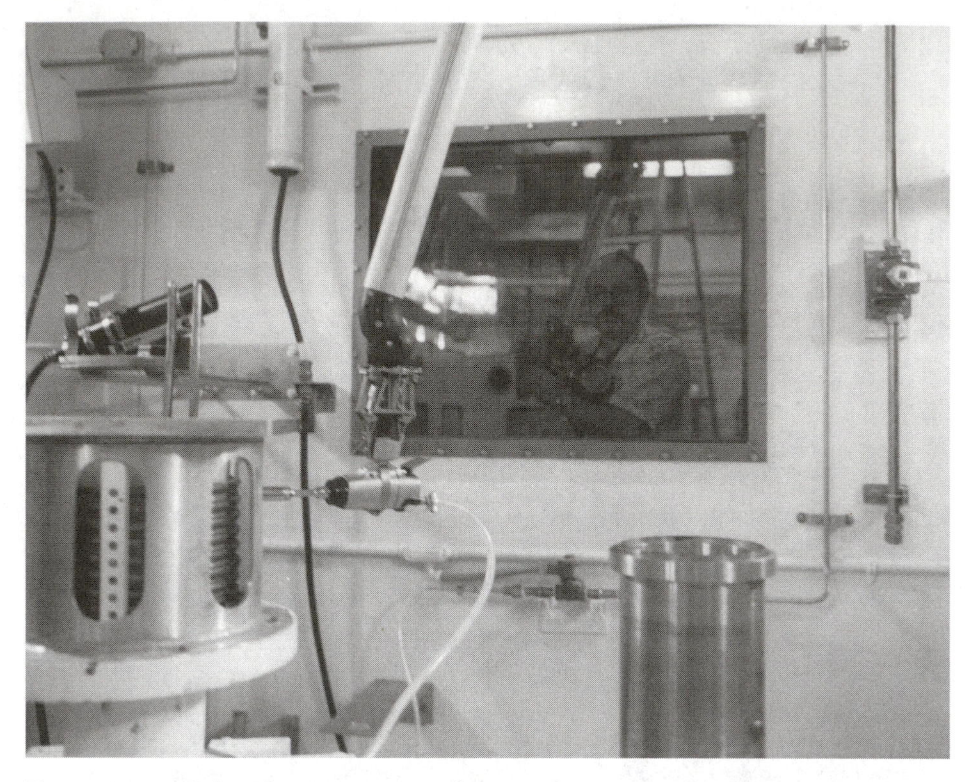

Figure 2-2 Within the American nuclear weapons program, plutonium operations have been conducted in specialized hot cell facilities that protect workers and the surrounding environment. Here, a weapons facility worker (circa 1980) uses a teleoperator to remotely perform complicated mechanical operations on a piece of plutonium-contaminated equipment. Starting with the Manhattan Project, an assortment of ever-more sophisticated robotic devices have allowed nuclear workers to safely and comfortably handle materials that are dangerously radioactive and toxic. (Credit: Photograph courtesy of the U.S. Department of Energy.)

of well-engineered manipulators support the development of the first nuclear reactors and first atomic weapons, especially the processing of intensely radioactive materials in special shielded facilities, called hot cells. The unilateral manipulators of the nuclear program demonstrate two important principles of teleoperation: first, the mechanical arm/hand can be a significant distance away from the human operator; second, the force exerted by the mechanical arm/hand can greatly exceed human capabilities.

1944 On September 27, The Manhattan Project's 100-B plutonium production reactor at Hanford, Washington, achieves criticality and begins operation. Two months later construction workers at Hanford complete the chemical separation plants in which human workers will use a variety of teleoperated manipulator systems to process the highly radioactive irradiated fuel from the Hanford production reactors and provide the plutonium used in the implosion-design atomic bombs.

1945 While working on the ENIAC (Electronic Numerical Integrator And Calculator) project for the U.S. government, the Hungarian-American mathematician John von Neumann proposes the concept of an internally stored program, where the step-by-step directions for computations (called instructions) are stored within the computer and so computations can progress without the need for external (human) guidance.

1946 Supported by the U.S. Army, the ENIAC (*E*lectronic *N*umerical *I*ntegrator *A*nd *C*alculator) is completed by John Presper Eckert and John W. Mauchy at University of Pennsylvania. ENIAC is considered by science historians as the world's first electronic digital computer, and at the time of its completion, the world's most complex electronic machine. ENIAC is a massive machine containing over 18,000 vacuum tubes. But the device can only handle numbers. The UNIVAC I (*Univ*ersal *A*utomatic *C*omputer) will become the first device to deal with letters.

American inventor George C. Devol Jr. develops a controller device, which can record electrical signals magnetically and play them back to operate a mechanical machine. A U.S. patent is issued for this device in 1952.

1947 American mathematician and computer engineer Howard Hathaway Aiken replaces mechanical relays with vacuum tubes and introduces the Hark Mark II, all electronic, programmable calculator.

The American physicists John Bardeen, Walter Houser Brattain, and William Bradford Shockley, while collaborating at Bell Laboratories invent the transistor. The transistor is a solid-state device that exponentially increases the use of electronic devices, including the rise of high-speed digital computers. Ten years later, the creation of silicon microchips reinforces this amazing pattern of growth in which the cost of computing dramatically decreases, while the capability of the digital computer increases.

1948 The American mathematician Norbert Wiener establishes the field of cybernetics with the publication of his book *Cybernetics or Control and Communication in the Animal and the Machine*.

1949 Raymond C. Goertz and his coworkers at the U.S. Atomic Energy Commission's (USAEC's) Argonne National Laboratory (ANL) in Illinois publicly demonstrate the first mechanical, bilateral master-slave manipulator device for the remote handling of hazardous materials, such as the highly radioactive materials associated with the rapidly expanding American civilian and military nuclear programs. Goertz's first bilateral master-slave manipulator has a crude sense of touch, which means that when the mechanical fingers (grippers) of the slave manipulator arm close on a glass beaker, the human operator handling the master manipulator arm can feel resistance of the beaker's glass wall to the pressure of the machine's mechanical fingers. This sense of touch (in reality a form of force feedback) greatly improves the deftness of the human-machine combination in teleoperation and also prevents the greater-than-human mechanical advantage of a machine manipulator from breaking delicate objects.

1951 American computer scientist and U.S. Navy officer Grace Murray Hopper conceives of a new type of internal computer program (the compiler) that can perform floating-point operations and other tasks automatically.

During OperationGreenhouse (April to May 1951), the United States conducts its third series of nuclear weapons tests in the Marshall Islands in the Pacific Ocean. This test series includes an important experimental program involving the use of unmanned, radio-controlled drone aircraft for nuclear debris cloud sampling. Eight B-17 drones are flown close to the detonation to measure blast and thermal effects and then into the nuclear cloud to collect highly radioactive samples.

American movie audiences encounter the alien emissary (played by Michael Rennie) and Gort (his large and powerful robot companion) in the classic science fiction-fantasy

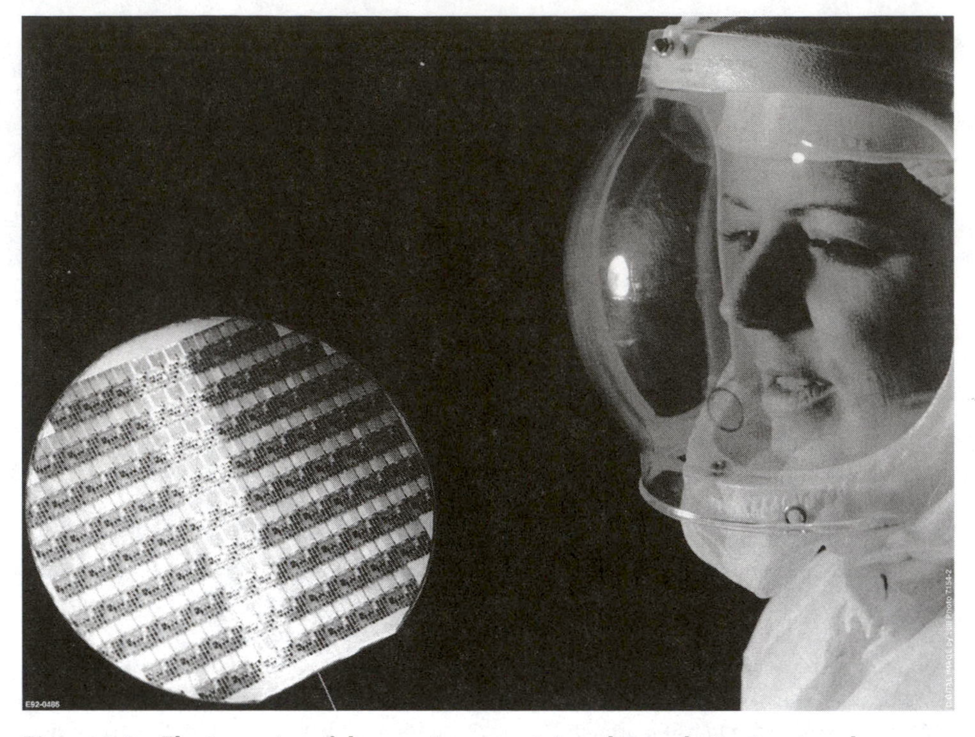

Figure 2-3 The invention of the transistor precipitated a revolution in microelectronics and enabled the development of the tiny computers, microcircuits, and microprocessors found in modern robot systems. This photograph shows a scientist at the Sandia National Laboratories, Albuquerque, in protective gear examining a modern microelectronics chip (microscope view on left). Today's integrated circuits contain millions of transistors with features as small as a tiny virus. Sandia specializes in the development of microelectronic circuits that are resistant to large doses of ionizing radiation—making such devices very useful in robot space systems that orbit Earth or travel to the farthest reaches of the solar system. (Credit: Photograph courtesy of the U.S. Department of Energy/Sandia National Laboratories, Albuquerque.)

thriller, *The Day The Earth Stood Still*. Based on Harry Bates' short story, "Farewell to the Master," this movie uses the arrival of an alien spaceship in Washington, DC, to warn the world about the perils of a spiraling nuclear arms race. The giant metallic android Gort has the technology punch to back up Klaatu's warning.

1952 The Hungarian-American mathematician John von Neumann builds the MANIAC (Mathematical Analyzer, Numerical Integrator and Computer) at the Institute for Advanced Study at Princeton, New Jersey. This digital computer embodies von Neumann's landmark idea of storing-instructions (as distinct from data) in the computer's memory. With stored instructions the computer can perform its computations without the need for external (human) guidance.

1954 American inventor and robot pioneer George C. Devol, Jr. designs a device that involves "programmed article transfer." A U.S. patent is issued for this design in 1961. Devol originally calls the device universal automation, but later shortens the term to unimation. Devol's unimation is the first industrial robot, a system designed specifically to pick and place objects in a factory environment.

Raymond C. Goertz and his coworkers at the USAEC's Argonne National Laboratory apply the principles of cybernetics to manipulator design and construct the first electric master-slave manipulator system. The new device represents a major milestone in teleoperation and robotics. Electric wires that carry control signals in one direction and force feedback in the other direction replace the cables and metal tapes, which connect the master arms and hands to the slave counterparts. Now, when the human operator uses his hand to close the grips on the master manipulator, the action sends electric signals to a servomotor in the remote slave manipulator. As a result of this breakthrough, the bilateral teleoperator, like its unilateral cousin, conquers distances with wires, radio frequency signals, or laser beams. Goertz's device establishes the principle of the teleoperation of objects (robots) at great distance.

Founded formally in 1954, the European Organization for Nuclear Research (*Conseil Européen pour la Recherche Nucléaire* [CERN]) represents one of the first joint ventures of post-World War II Europe. Today, the facility still serves as a shining example of international cooperation in scientific research. Straddling the border between France and Switzerland (near Geneva), CERN eventually becomes the world's largest high-energy physics laboratory. In the late 1980s and early 1990s, this international facility will also play a major role in the development of the World Wide Web.

1955 John McCarthy and Marvin Minsky coin the term "artificial intelligence" (AI) to describe modern computers with some ability to think like human beings. They then proceed to establish an AI laboratory at MIT.

1956 George C. Devol, Jr. meets Joseph E. Engelberger and the two decide to form the world's first industrial robot company, called Unimation, Inc. For his keen engineering insights, Devol is often called the "Grandfather of Industrial Robotics," while Engelberger frequently receives the title "Father of Industrial Robotics" because of his extensive marketing efforts both in the United States and overseas (especially Japan).

The affable "Robby the Robot" steals the scenes in the science fiction movie *Forbidden Planet*. Robby's ability to manufacture all sorts of items on request (from beverages to clothing) anticipates the Santa Claus machine concept proposed in 1978 by the American physicist Theodore (Ted) Taylor.

1957 On October 4, the space age dawns, as the former Soviet Union launches the first artificial satellite, called *Sputnik 1*.

1958, January 31—The United States successfully launches *Explorer 1*, the first American satellite. From this point forward, the U.S. government sponsors numerous robot spacecraft to study Earth from space as well as to explore the solar system.

1959, January 2—The Soviet Union launches a massive campaign to the Moon with the liftoff of *Luna 1*. Although the robot spacecraft misses the Moon by between 5,000 and 7,000 kilometers, it becomes the first human-made object to escape Earth's gravitation and orbit the Sun.

September 14—*Luna 2* successfully impacts the Moon, becoming the first robot space probe to impact (actually, crash-land) on another world.

October 4—The Soviets launch *Luna 3*, a robot spacecraft that circumnavigates the Moon and takes the first images of the lunar farside.

December 29—American physicist and Nobel laureate Richard P. Feynman delivers an inspiring lecture entitled "There's Plenty of Room at the Bottom." Feynman presents this lecture at the California Institute of Technology (Caltech) to members of a California chapter of the American Physical Society during their annual meeting. In the course

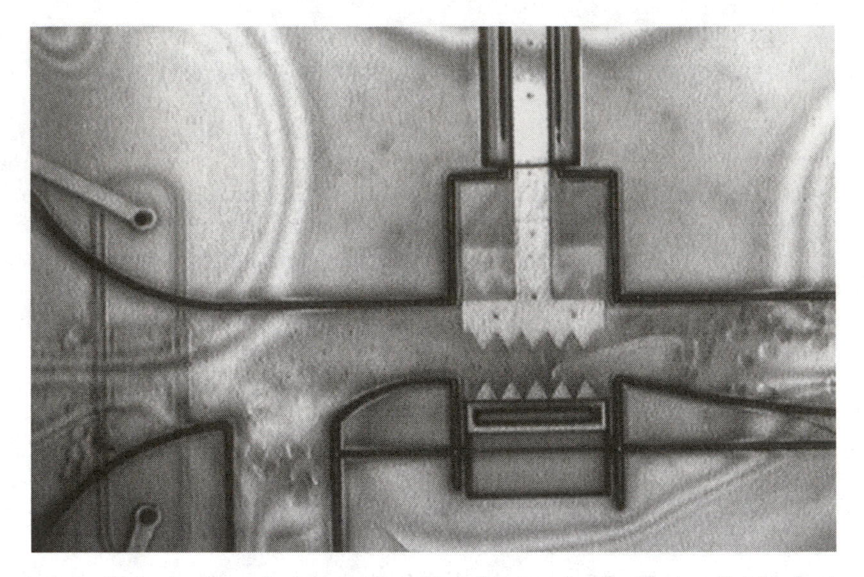

Figure 2-4 This photograph shows a Pac-Man-like mechanized microfluidic device developed by the Sandia National Laboratories, Albuquerque. The device's silicon microteeth bite in a channel that is just 20 micrometers wide. For comparison, a human hair is approximately 70 micrometers in diameter. The little balls that appear in the horizontal channel are red blood cells. When the jaws of the device close, they trap a red blood cell—one of the many being pumped through the tiny research device. (Credit: Photograph courtesy of the U.S. Department of Energy/Sandia National Laboratories, Albuquerque.)

of his talk, Feynman speculates about such interesting possibilities as micromachines that can perform useful functions at the atomic level. Science historians often treat this lecture as the beginning of nanotechnology.

General Motors becomes the first American automobile manufacturer to operationally test the use of an industrial robot. Encouraged by Joseph F. Engelberger and George C. Devol, Jr., officials for General Motors approve installation of a test model Unimate industrial robot in a die-casting plant. However, it will be another two years before specific commercial orders are placed for Unimate robots.

The deployment of first generation of operational intercontinental ballistic missiles (ICBMs) and the submarine-launched ballistic missiles (SLBMs) completely transforms the nature of strategic warfare. Nuclear-armed ICBMs and SLBMs are unstoppable robot weapons that travel through space and can strike any point on the globe in less than an hour. The threat of nuclear Armageddon becomes an integral part of modern civilization.

1960 The United States launches *Pioneer 5*, the first American space mission to successfully place a robot spacecraft into orbit around the Sun.

The Condec Corporation purchases Unimation, Inc. and starts development of the Unimate family of industrial robots.

1961 As the demand for cars grows, automobile manufacturers look for new ways to increase the efficiency of the assembly line through *telecherics*. This new field focuses on robots that mimic the operator's movements from a distance. General Motors installs the applied telecherics system on their assembly line. The one-armed Unimate robot unloads hot die casts, cools the components, and delivers them to a trim press.

Figure 2-5 First launch of the Trident ICBM by the United States at Cape Canaveral, Florida, on January 18, 1977. The modern nuclear-powered ballistic missile submarine armed with a complement of nuclear-weapon carrying submarine launched ballistic missiles (SLBMs) serves as an integral portion of the American strategic triad. Once launched, the Trident ballistic missile is an enormously powerful robot weapon. A single American ballistic missile submarine carries enough nuclear firepower to devastate any region or nation on Earth. (Credit: Photograph courtesy of the U.S. Navy.)

1962 NASA launches the *Mariner 2* spacecraft to the planet Venus on August 27 from Cape Canaveral, Florida. This far-traveling robot spacecraft becomes the world's first successful interplanetary probe.

1963 Australian neurophysiologist Sir John Carew Eccles receives the 1963 Nobel Prize in medicine for his discoveries concerning the functioning of nervous impulses. The human brain's basic unit of operation is the neuron. Connected in networks, neurons send and receive signals by combining electricity and chemistry over a complex network of fibers within the body. Machine intelligence experts soon start attempting to develop artificial neural networks that function in a manner loosely based on how the human brain functions with its network of neurons.

October 17—The U.S. Air Force successfully places the first pair of Vela nuclear-detonation-detection spacecraft into a high Earth orbit. These robot spacecraft serve as automated sentries, monitoring Earth and outer space for violations of the Limited Nuclear Test Ban Treaty signed in August by the Soviet Union, the United Kingdom, and the United States—at the time, the world's three nuclear weapons states.

1964, July 14—NASA's *Mariner 4* encounters Mars and becomes the first robot spacecraft to fly by the Red Planet.

July 28—NASA sends the *Ranger 7* spacecraft to the Moon. About 68 hours later this robot probe successfully transmits more than 4,000 high-resolution television images of the lunar surface before crashing into the Sea of Clouds. The *Ranger 7*, *8*, and *9* spacecraft greatly advance scientific knowledge about the lunar surface and help prepare the way for the Apollo Project's lunar landing missions.

1966, February 3—The Soviet *Luna 9* robot spacecraft transmits the first panoramic television pictures ever received from the Moon's surface.

March 31—The Soviet Union launches *Luna 10* robot spacecraft, which becomes the first human-made object to achieve orbit around the Moon.

May 30—NASA sends the *Surveyor 1* spacecraft to the Moon. The versatile robot spacecraft successfully lands on the lunar surface on June 1, becoming the first American spacecraft to achieve a soft landing on another celestial body.

August 10—NASA sends the *Lunar Orbiter 1* robot spacecraft to the Moon to perform high-resolution photography of the lunar surface in preparation for landings by the Apollo Project astronauts.

Industrial institutions in Japan take notice of the commercial industrial robots starting to appear in the United States. American Machine and Foundry's (AMF's) Versatran robot becomes the first American industrial robot imported into Japan.

The artificial intelligence laboratory at SRI International in Menlo Park, California, begins work on Shakey the Robot—the first mobile robot capable of using artificial intelligence to "reason" about its own actions.

1967, October 30—Two Soviet robot spacecraft, *Cosmos 186* and *188*, perform the first automated rendezvous and docking operation in space. The Soviets will use such automated operations to assist in the assembly and resupply of future space stations.

Japanese industries continue to expand their interest in the design of American-made robots. For example, Kawasaki purchases a hydraulic robot from Unimation and then begins producing this system in Japan under license. Within less than two decades industries in Japan become the largest users of robots in the world.

1968 Motion picture producer Stanley Kubrick and science fiction writer Sir Arthur C. Clarke introduce audiences around the world to the rascally and mischievous fictional computer/character HAL 9000 in the film *2001: A Space Odyssey*. In this highly acclaimed motion picture, HAL 9000 is the advanced on-board computer designed to essentially run the interplanetary ship *Discovery*, which is carrying a team of human astronauts to the vicinity of Jupiter on a mysterious mission. HAL 9000 represents the apex of artificial intelligence and is quite capable of fully interacting with the human crew. However, the potential perils of truly advanced machine intelligence also appear in this movie, when the HAL 9000 departs from its programmed behavior and begins to exercise a malicious mind of its own by engaging in a deadly conflict with the humans onboard.

1969, July 20—American astronaut Neil Armstrong cautiously descends the steps of the lunar module's (LM's) ladder and contacts the lunar surface. A variety of robot

Figure 2-6 This 1971 postage stamp from the former German Democratic Republic (East Germany) depicts the Soviet *Lunokhod* 1 robot rover departing the lander spacecraft. During the Soviet *Luna 17* mission to the Moon in 1970, the mother spacecraft soft-landed on the lunar surface in the Sea of Rains and deployed the *Lunokhod 1* robot rover vehicle. Controlled from Earth by radio signals, the eight-wheeled mobile robot traveled for months across the lunar surface, transmitting more than 20,000 television images of the Moon's surface and performing more than 500 soil tests at various locations. (Credit: Photograph courtesy of author.)

spacecraft paved the way for this historic moment, during which human beings (Armstrong and Edwin "Buzz" Aldrin) walk on another world for the first time.

1970, August 17—The Soviet Union launches its *Venera 7* mission to Venus. When the robot spacecraft arrives at Venus on December 15, it ejects a capsule that transmits data back to Earth as it descends through the Venusian atmosphere and survives landing on the inferno-like planet. The accomplishment represents the first successful transmission of data from the surface of another planet.

November 10—The Soviet Union launches *Luna 17* to the Moon, where it achieves the first successful use of a mobile, remotely controlled (teleoperated) robot vehicle, called *Lunokhod 1*, in the exploration of another celestial body.

1975, August 20—NASA begins a major scientific assault on Mars with the launch of the *Viking 1* robot spacecraft (consisting of an orbiter and lander combination). Its identical twin, *Viking 2*, is launched on September 9. *Viking 1* reaches the Red Planet in June 1976 and on July 20, 1976, becomes the first American robot spacecraft to successfully soft-land on another planet. The primary objective of both the *Viking 1* and *2* lander spacecraft is to determine whether microbial life exists on Mars. Both landers return inconclusive evidence.

1976 The American computer engineer Seymour Cray delivers the first Cray 1 super computer, at the time, the world's most powerful computer system.

1977, May 25—Writer/director George Lucas introduces audiences to his science fiction/fantasy universe with the release of the film *Star Wars* (later retitled *Star Wars Episode IV: A New Hope*). Stretching almost three decades, the six motion pictures in this sprawling "space-opera" will make an enormous impact on the popular culture. For example, the antics of two fictional robots: the steadfast, get-the-job done pudgy "droid" called R2-D2, and the frequently whining, constantly appeasing protocol android, called C-3PO, delight millions of people around the world and suggest what the future might be like with very intelligent robots.

August 20—NASA launches the *Voyager 2* spacecraft on an epic "grand tour" mission in which this hardy robot explorer will successfully encounter all four gaseous giant outer planets and then leave the solar system on an interstellar trajectory.

Figure 2-7 *NASA's Viking 1* robot lander spacecraft obtained this image of the Martian surface on July 24, 1976. (Credit: Photograph courtesy of NASA.)

1978 Unimation introduces the PUMA (Programmable Universal Machine for Assembly). This industrial robot quickly becomes the standard for commercial telecherics.

1983, January 25—NASA launches the *Infrared Astronomy Satellite* (IRAS). Unhindered by the absorbing effects of Earth's atmosphere, this robot astronomical observatory completes the first all-sky scientific survey of the universe in the infrared portion of the electromagnetic spectrum.

1984 The Microsoft Corporation introduces the company's *Windows* software.

1985 Dr. Yik San Kwoh invents the robot-software interface used in the first robot-aided surgery, a stereotactic procedure. The surgery involves a small probe that travels into the skull. A CT scanner is used to give a three-dimensional image of the brain, so that the robot can plot the best path to the tumor.

The Royal Swedish Academy of Sciences later awards the 1996 Nobel Prize in Chemistry for the discovery of carbon atoms bound in the form of a ball. Robert F. Curl, Jr. (Rice University), Sir Harold W. Kroto (University of Sussex), and Richard E. Smalley (Rice University) discovered these new forms of the element carbon in 1985. Called fullerenes, the number of carbon atoms in the close shell can vary. Scientists often refer to spherical fullerenes as buckyballs and cylindrical fullerenes as nanotubes or buckytubes—after the noted architect, Richard Buckminster Fuller, who promoted the geodesic dome.

1986 A team of engineers at Honda Motor Company begins working on the creation of an advanced humanoid robot, using the human body as a design guide with which to build this two-legged robot. Their overall mission is to create a people-friendly, intelligent, bipedal robot that can autonomously navigate and interact in the world of human beings. Getting a single robot mobile in a variety of work and living environments has always been an engineering challenge. But by studying feet and legs, the Honda team

Figure 2-8 This photograph shows a sequenced series of actions by a PUMA robotic arm during studies (circa 1989–1990) at the NASA Ames Research Center in Mountain View, California. (Credit: Photograph courtesy of NASA/Ames Research Center.)

will eventually create a humanoid robot capable of climbing stairs, kicking a ball, and pushing a cart. In 2000, the team's persistence and hard work pays off when Honda debuts its ASIMO (Advance Step in Innovative Mobility) humanoid robot. ASIMO is the eleventh in a series of walking robots created by Honda engineers in this focused development effort to create a two-legged humanoid robot that can walk and pleasantly perform useful functions in human society.

Upon termination of the licensing agreement with Unimation, Kawasaki expands its activities in industrial robotics by developing and producing its own line of electric robots.

The American scientist and futurist, K. Eric Drexler, publishes the book *Engines of Creation: The Coming Age of Nanotechnology*. This popular book provides the first general treatment of molecular engineering, micromachines, and the potential of nanotechnology.

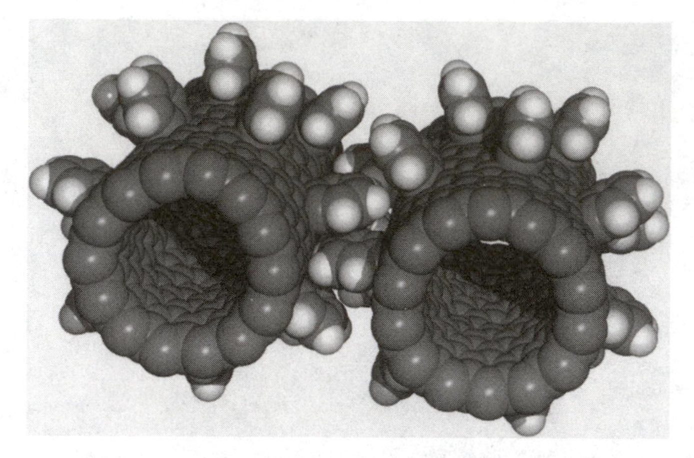

Figure 2-9 This computer-drawn image shows two Fullerene nano-gears with multiple teeth. Researchers have simulated attaching benzyne molecules to the outside of a nanotube to form gear teeth. Nanotubes are molecular-sized pipes made of carbon atoms. A laser could serve as the motor to drive or spin the nano-gears. The laser creates an electric field around the nanotube. A positively charged atom is placed on one side of the nanotube and a negatively charged atom on the other side. The electric field then drags the nanotube around like a shaft turning. Such molecular-sized devices are part of an emerging new field called nanotechnology. (Credit: Computer-drawn illustration courtesy of NASA/Ames Research Center.)

1988 The Stäubli Group purchases Unimation from Westinghouse.

1989, November 18—NASA launches the *Cosmic Background Explorer* (COBE) into polar orbit around Earth. The scientific payload on this robot spacecraft carefully measures the spectrum of cosmic microwave background and helps scientists answer some of their most pressing questions concerning the ancient explosion (big bang) that started the expanding universe.

British computer scientist Sir Timothy John Berners-Lee, while working at CERN, proposes a computer-based, global hypertext project, which will allow people and organizations to more easily work together and share information. His concept becomes known as the World Wide Web. In December 1990, the Web becomes available within CERN and by the summer of 1991 becomes available on the Internet.

1990, January 1—NASA officially begins the Voyager Interstellar Mission (VIM). In this extended mission, both nuclear-powered Voyager spacecraft search for the heliopause—the location in deep space that forms the boundary between the outermost extent of the solar wind and the beginning of interstellar space.

April 24—NASA uses the space shuttle *Discovery* to deploy the *Hubble Space Telescope* (HST) into orbit around Earth during the STS-31 shuttle mission. *Hubble* is a sophisticated robotic observatory that represents the most powerful optical telescope ever placed into space. For more than a decade, the HST revolutionizes how astronomers view the universe.

The National Institutes of Health (NIH) provides SRI in Menlo Park, California, funding to perform the engineering necessary to demonstrate the feasibility of using robots to enhance the performance of minimally invasive surgery (MIS) and remote surgical tasks. The SRI research team develops a prototype, eventually called the

Figure 2-10 A tiny BEAM robot called the Bitman robot. The acronym BEAM stands for biology, electronics, aesthetics, and mechanics. BEAM robots, developed by Los Alamos scientist Mark Tilden, are modeled on the simple, repetitive biological processes of insects. (Credit: Photograph courtesy of the U.S. Department of Energy/Los Alamos National Laboratory.)

"SRI system." This pioneering effort in the medical use of robotics successfully combines improvements in remote manipulation (teleoperation) with force feedback, multimodal sensory feedback, stereoscopic imaging, and physician-friendly (ergonomic) design.

Starting in about 1990, humanoid robots begin to more closely mimic human behavior, while other mobile robots appear that resemble lower life forms. At the Los Alamos National Laboratory in New Mexico, for example, Mark Tilden's BEAM robots look and act like big bugs. The name BEAM is an acronym for biology, electronics, aesthetics, and mechanics. BEAM robots are simple robots constructed out of discrete components; the use of integrated circuits, as found in most other robots for some level of artificial intelligence, is avoided. This effort is part of an overall idea to create inexpensive, solar-powered mobile robots, which would prove ideal for dangerous missions such as landmine detection.

1991, August 6—The British computer scientist, Sir Timothy John Berners-Lee, who created the World Wide Web concept while working as a computer specialist at CERN, establishes the world's first Web site at the international research facility.

1993, April 30—Officials at CERN issue a formal statement declaring that the World Wide Web software developed by Sir Timothy John Berners-Lee is in the public domain. CERN's actions allow the Web to become an indispensable tool of the Internet and accelerates growth of the contemporary global information infrastructure.

1994 The Carnegie-Mellon University (CMU) Field Robotics Center sends Dante II, a tethered walking robot to explore Mt. Spurr in Alaska. Dante II aids in the dangerous

Figure 2-11 This photograph shows a United States Air Force RQ-1 Predator from the 46th Expeditionary Reconnaissance Squadron landing at Tallil Air Base, Iraq. The Predator is a remotely piloted vehicle that provides real-time surveillance imagery in support of Operation Iraqi Freedom. (Credit: U.S. Air Force Photo by Staff Sgt. Suzanne M. Jenkins.)

recovery of volcanic gases and samples. In other applications of such emerging modern combinations of robotic arms with microprocessors, environmental sensors, and mobility systems (wheels)—a development called *mobile applied telecherics*—robots begin to save human lives by assisting in the safe detection and defusing or disposal of bombs, by investigating hazardous accident sites, or by performing dangerous reconnaissance operations in urban warfare environments.

1995 The company, Intuitive Surgical®, forms and has the strategic vision to develop the commercial medical technology necessary to apply modern telerobotic technologies to minimally invasive surgery (MIS) and microsurgery. The company's da Vinci Surgical System® finds applications in hospitals around the world that sponsor robot-assisted surgery.

1996, December 4—NASA successfully launches the *Mars Pathfinder* mission to the Red Planet. The robot lander spacecraft touches down on the surface of Mars on July 4, 1997. Teleoperated and guided by human controllers at NASA's Jet Propulsion Laboratory (JPL), a robot minirover deploys from the lander spacecraft and explores the planet's surface. The minirover is the first mobile robot teleoperated on another planet. *Mars Pathfinder* also demonstrates the first use of a new airbag technology to deliver modest-sized robot spacecraft safely to a planetary surface.

1999, July 23—The space shuttle *Columbia* (during the STS-93 mission) carries NASA's *Chandra X-Ray Observatory* (CRO) into space. This powerful new robotic observatory is designed to observe and image X-rays from high-energy sources throughout the universe, such as remnants of exploded stars. Data from CXO produces a revolution in high-energy astronomy and astrophysics.

Figure 2-12 This artist's rendering shows NASA's *New Horizons* robot spacecraft during its encounter with the planet Pluto (foreground) and its relatively large moon Charon (circa 2015). Launched successfully from Cape Canaveral on January 19, 2006, the far-traveling robot spacecraft will explore one or more icy planetoid targets of opportunity in the Kuiper Belt after accomplishing its scientific reconnaissance of the Pluto system. (Credit: Artist's rendering courtesy of NASA/JPL.)

Sony introduces its AIBO line of entertainment robots that prove quite popular as robotic pets. For example, AIBO ERS-7 resembles a small dog and can interact with its owner (through suitable software) as it develops from a puppy (with typical puppy behavior) to a mature adult dog. At maturity, an AIBO ERS-7 robot dog understands (though not necessarily always obeys) about 100 of its owner's voice commands.

2000 The Honda Motor Company introduces its ASIMO humanoid robot. ASIMO is actually the eleventh in a series of walking robots created by Honda engineers in a focused development effort (starting in 1986) to create a two-legged (bipedal) humanoid robot that can walk and perform useful functions in human society alongside people.

2001, March 1—The U.S. Air Force's Global Hawk unmanned aerial vehicle (UAV) enters the engineering, manufacturing, and development phase of the Defense Department's acquisition cycle. Global Hawk is an extremely high altitude, long-duration mission robot aerial vehicle that is capable of providing battlefield commanders with near-real-time, high-resolution, intelligence, surveillance, and reconnaissance imagery.

2003 The School of Computer Science at Carnegie Mellon University in Pittsburgh, Pennsylvania, establishes the Robot Hall of Fame—a virtual museum Web site intended to honor physical achievements in the real world of robotics technology, as well as robots from science fiction that have provided a creativity stimulus to the engineers and scientists working in the field of robotics. For example, one of the real world inductees in 2003 was NASA's *Pathfinder* minirover (also called *Sojourner*), which explored the surface of Mars in 1997. One of the 2003 fictional inductees is the ever-dependable robot, R2-D2 from George Lucas's *Star Wars* motion picture series.

2004 NASA sends twin robot rovers, *Spirit* and *Opportunity*, to explore the surface of Mars.

2005, March 1—the United States Air Force announces that the MQ-1 Predator unmanned aerial vehicle (UAV) has achieved initial operational capability (IOC). The Predator is a medium-altitude, long-endurance, remotely piloted aircraft, which has the primary mission of interdiction and armed reconnaissance against critical, perishable targets.

July 4—NASA's *Deep Impact* robot spacecraft performs a complex experiment in space that probes beneath the surface of a comet and helps reveal some of the secrets of its interior. A larger flyby mother spacecraft releases a smaller self-guided robot impactor, which strikes Comet Tempel 1.

December 13—Honda debuts the newest version of the company's ASIMO humanoid robot. This bipedal robot has many improved features, including the ability to pursue key tasks in a real-life office or home environment. The well-engineered humanoid robot has a height of 1.3 meters and a mass of 54 kilograms. The new ASIMO can autonomously act as a receptionist or even deliver drinks on a tray.

2006, January 19—NASA successfully launches the *New Horizons Pluto-Kuiper Belt Flyby* spacecraft from Cape Canaveral. The robot spacecraft is now traveling on its way to the dwarf planet Pluto. This reconnaissance-type exploration mission will help scientists understand the interesting, yet poorly understood, icy worlds at the edge of the solar system. If all goes well over the next nine years, the first spacecraft flyby of Pluto and its large moon Charon will take place in the summer of 2015.

3

Profiles of Robot Technology: Pioneers, Visionaries, and Advocates

In this chapter we meet some of the most interesting and important people who developed the basic scientific concepts or invented the fundamental technical devices that helped provide the foundation of modern robotics technology. However, just like the digital computer, so many innovative ideas and technical advances converged to enable the creation of modern fixed and mobile robots that it is impossible to give just one or two persons the credit for inventing the robot. For example, in this chapter we discuss some very creative literary people who created enduring fictional accounts of smart machines, robots, androids, cyborgs, and other artificial life forms. While these writers did not engineer any of the key technical advances that enabled the physical arrival of modern robot systems, they established the long-range vision and cultural climate within which robot systems are anticipated rather than being sources of society-wide techno-shock. Technically accurate or not, these fictional accounts helped shape the modern cultural environment within which advanced robot systems are viewed as an inevitable part of the future by millions of human beings.

However, some of these fictional perceptions, now deeply embedded in the human psyche, are obviously either vastly premature or completely out of proportion to what the current state of robotics technology offers to the human race this century. Other cultural perceptions overlook the true impact of modern robots on space exploration, hazardous national defense and law enforcement operations, and automated manufacturing. But for all the inherent limitations of these fictional representations, they contribute to the maintenance of a popular, widely held, cultural view of what future robots (especially those designed as androids and cyborgs) might do and how they might behave.

The British writer Mary Wollstonecraft Shelley produced the immortal horror story *Frankenstein: The Modern Prometheus*—a fictional tale that indelibly established the notion of well-intended science gone astray in matters of creating artificial life forms. The Czech writer Karel Čapek gave the world the term robot when he wrote the play *Rossum's Universal Robots* (*R.U.R.*). Finally, the incomparable science fiction writer Isaac Asimov popularized the word robotics and formulated his now-famous three laws of robotics. Asimov introduced these (postulated) rules of robot behavior in the science fiction story "Runaround," which appeared in the March 1942 issue of *Astounding* magazine. Since then, Asimov's so-called laws of robotics have become part of the cult and culture of modern robotics. These ethical precepts are: (1st Law) "A robot may not injure a human being, or, through inaction, allow a human being to come to harm." (2nd Law) "A robot must obey the orders given it by human beings except where such orders would conflict with the first law." (3rd Law) "A robot must protect its own existence as long as such protection does not conflict with the first or second law."

Several of the persons highlighted in this chapter were the great engineers of antiquity, while others were the physicists and mathematicians associated with the emergence of modern science during the late Renaissance in Western Europe. This chapter also includes some of the key persons responsible for the machines that brought about the first or second Industrial Revolutions—setting the stage for today's modern industrialized world.

In the twentieth century early robots were pressed into service for repetitive or hazardous jobs in manufacturing industries, the nuclear industry, and selected military applications. Starting in the mid-1950s, two individuals, George C. Devol, Jr. and Joseph F. Engelberger, promoted the development of industrial robots. Government projects in space exploration, national defense, and environmental cleanup stimulated companion developments in other types of robot systems. In the case of many modern mobile robot systems developed under government projects and sponsorship, individual champions and technical visionaries were generally blended into the project teams that brought these amazing machines into service. Although often overlooked by treatments that focus too narrowly on just industrial robots, some of the most spectacular and successful applications of modern robotics involve national defense or space exploration.

Unlike more conservative business environments bounded by rigid cost-benefit guidelines, government projects related to national defense or space exploration generally have more flexibility and latitude in committing resources to higher risk technology projects that, if successful, will move the state of robotics technology in a revolutionary manner rather than at a stepwise conservative pace. Consider for example the focused engineering effort and technical risks successfully undertaken by the U.S. Air Force in developing and deploying the MQ-1 Predator unmanned aerial vehicle (initial operational capability in March 2005). The American civilian space agency, NASA, engaged in similar pioneering efforts and technical risk-taking when the agency developed and successfully landed two sophisticated robot rover vehicles (*Spirit* and *Opportunity*) on the surface of Mars in 2004. While both the Predator and the Mars exploration rovers (MERs) represent outstanding examples of leading-edge robotics, no individual person can be credited or acknowledged as the champion of either of these fascinating robot systems.

Many of the scientists and engineers featured in this chapter have made the fundamental technology breakthroughs, which contributed—directly or indirectly—to the rise of the intelligent machines that we today collectively refer to as "robot systems."

Although modern interest and fascination with robots started about a century ago with Karel Čapek's play, some of the developments supporting robot technology have roots that extend back into antiquity. Recognizing the diversity and complexities of these technology developments, it is easy to understand how a person might get confused about what is and is not related to the development of robotics. At this point, it is important to recognize that the field of robotics involves the confluence of many different advances in technology. Some of these developments and events may seem trivial and commonplace by current standards. In fact, in viewing such technologies, people often do not even regard them as related to robots. For example, the work of the early Greek engineer and inventor, Archimedes, influenced the application of many simple mechanical devices found in modern robots. Similarly, the development of the first electric cell (forerunner of the modern battery) by Count Alessandro Volta provided the basis for portable electric power—so essential for the operation of many mobile robots.

Perhaps the most intriguing philosophical question involved with advanced robot systems is the question of machine intelligence and machine consciousness. Simply stated: Is a machine that thinks conscious and aware of its existence? The first great modern philosopher René Descartes believed that the bodies of humans and animals are complex automata. In his treatise *Discourse on Method*, published in 1637, Descartes discusses how humans, who have the power of reason, and animals, which cannot reason, can be distinguished from one another and machines. His most famous quote (as found in *Discourse on Method*) is: "Cogito, ergo sum" (which means, "I think, therefore I am"). This statement highlights some of the deep philosophical arguments Descartes raised in developing his mind-body dualism. The nature of consciousness and the mind is an issue that has intrigued philosophers for ages. The issue arises again from an interesting new perspective as robot specialists speculate about endowing very smart machines with a sense of consciousness and cognition. At what point does a so-called "thinking machine" become truly conscious?

In 1950, the British mathematician and computer science pioneer Alan Mathison Turing raised a similar question in his intriguing paper, "Computing machines and intelligence." As part of his pioneering discussion on artificial intelligence, Turing gave the world a test, now called the Turing test, for judging whether a machine is successfully simulating the thought processes of the human mind.

Finally, in his posthumously published book *Theory of Self-Replicating Automata*, the brilliant Hungarian-American mathematician, John von Neumann, shares some of his ideas about truly advanced robots (automata), which are capable of making copies of themselves and performing all manner of construction tasks. If ever created, these self-replicating systems (SRSs)—sometimes referred to as von Neumann machines—would have a profound impact on how human beings manipulate and control energy and matter resources both here on Earth, throughout the solar system, and beyond.

Archimedes of Syracuse (c. 287–212 B.C.E.)

The Greek mathematician, inventor, and engineer Archimedes of Syracuse was one of the greatest technical minds in antiquity, if not all history. As a gifted mathematician, he perfected a method of integration that allowed him to find the surface areas and volumes of many bodies. This brilliant work anticipated by almost two millennia the independent codevelopment of the calculus by Sir Isaac Newton and Gottfried Wilhelm Leibniz in the middle of the seventeenth century. In mechanics, Archimedes discovered fundamental theorems and physical relationships that described the center of gravity of plane figures and solids. These relationships lie at the very heart of modern mechanics and engineering dynamics. He designed and constructed a variety of potent war machines in defense of his birth city of Syracuse against sieges by the Roman Army during the Second Punic War. These military devices, not his brilliant mathematical contributions, made Archimedes famous in his own lifetime. Today, science historians consider the affable, absentminded Greek genius as a mathematician comparable in brilliance to Isaac Newton, Leonhard Euler, or Johann Karl Friedrich Gauss. Archimedes was born (about 287 B.C.E.) in the Greek city-state of Syracuse on the island of Sicily. His father was an astronomer, named Phidias, about whom very little else is known. As a young man, Archimedes studied, like most other gifted Greeks, at the great library in Alexandria. But, as a distant relative of Hieron II, King of Syracuse, Archimedes elected to return to his birth city and pursue his interests in mathematics, science, and mechanics. Although he personally regarded mathematics as a much higher level of activity than his efforts involving the invention of various mechanical devices, it was these engineering efforts and not his mathematics that earned him great notoriety in his own lifetime. Most historians call Archimedes the greatest of the Greek antiquity engineers.

For example, Archimedes designed an endless screw device that ended up being used throughout the Roman Empire as an irrigation device. Originally designed to help Egyptian peasants draw water out of the Nile River to irrigate their fields, the Archimedes screw quickly found use throughout the Mediterranean Basin and the Middle East. This device is still in use today in certain underdeveloped regions of Africa and Asia.

One of the most famous stories about Archimedes involves a challenge extended to him by the king of Syracuse. Hieron II wanted to determine whether a goldsmith had made a requested crown by using the proper amount of pure gold (as instructed). There was some suspicion that the goldsmith had cheated by using a less expensive combination of silver and gold. So, the king asked Archimedes to give him the right answer, but without damaging the new crown in any way.

Archimedes thought for days about this problem. Then, as often happens to creative people, inspiration struck when least expected. As Archimedes stepped into a full bath, he observed the water spill over the sides. Immediately, he knew how to solve this intriguing problem. So he jumped up naked out of the bath and ran enthusiastically to the palace, shouting "Eureka!" This Greek exclamation means, "I have found it!" Today, when an engineer or scientist experiences a similar insight or breakthrough, the event is often called a "eureka moment."

What Archimedes had discovered in a flash of genius was the principle of buoyancy. Today, physicists refer to this phenomenon as the Archimedes' principle. The principle states that any fluid applies a buoyant force to an object when that object is partially or completely immersed in it. The magnitude of this buoyant force equals the weight of the fluid the object displaces. Since the shape of an object is of no consequence, Archimedes was able to use this phenomenon to test the king's crown without damaging it.

Archimedes first carefully submerged the crown in water and measured the weight and the volume of water that the crown displaced. He next submerged an amount of pure gold equal in weight to the displaced water. He noted the volume of water displaced by pure gold. If the volume displaced by pure gold was equal to the volume of water displaced by the submerged crown, both could be assumed to have the same density and consist of identical material, namely pure gold. If the volume of water displaced by the crown was different than the volume of water displaced by an identical weight of gold, then there was some other metal, possibly lead or silver, in the crown along with the gold.

In antiquity (as now), silver was much cheaper than gold, so it was often used as a paste-up substitute for gold by unscrupulous jewelry merchants. Since silver has a significantly different (lower) density than gold, a crown fraudulently pasted up with some silver in its interior, would be bulkier than a crown of identical weight made of only pure gold. In ancient Syracuse, the crime scene investigator (CSI) was none other that the multitalented Archimedes. He carefully tested the suspicious crown and discovered that the volume did not match the anticipated result. Archimedes concluded that the king's crown contained both gold and some other metal, possibly silver. The unwise goldsmith, who tried to cheat King Hieron II, was executed. Today, scientists consider Archimedes' principle as one of the basic laws of hydrostatics.

Archimedes developed a number of important fundamental machines, including the lever and the compound pulley. With respect to the lever principle, Archimedes is reputed to have said: "Give me a place to stand on and I can move the Earth." Challenged by his friend, King Hieron II, to move something really large, Archimedes developed a system of compound pulleys and levers and (according to legend) single-handedly pulled a fully loaded ship (containing crew and cargo) up out of the water and onto the shore with a single rope. Archimedes conducted other studies of force and motion. He discovered that every rigid body has a center of gravity—a single point at which the force of gravity appears to act on the body.

Many of Archimedes' surviving documents portray his wide-ranging interest in engineering and machines. These surviving works include: "Theory of Levers," "On Floating Bodies," "On the Method of Mechanical Theorems," and "The Water Clock." He also had many other engineering-themed works, which today are known only from cross references and prefaces in surviving books. Some of his missing works include "On Odometers," "Winches, Hydroscopes, Pneumatics," "On Balances or Levers," "Centers of Gravity," "Elements of Mechanics," "On Gravity and Buoyancy," and "Burning by Mirror."

In addition to his genius for engineering, Archimedes was an incredibly gifted mathematician, who resolved many important mathematical problems. For example, he made the most precise estimates of the value of π (the ratio of a circle's circumference to diameter) of his day. He was also a prolific writer in the field of mathematics and some of his most important (surviving) works include: "On the Sphere and Cylinder," "Measurement of the Circle," "On Spirals," "On Tangential Circles," "On Triangles," "On Quadrangles," and "On Conoids and Spheroids." Despite achieving great fame through his mechanical inventions, Archimedes preferred to delve into mathematical problems, often getting absorbed for days and becoming oblivious of the world around him. Unfortunately, his lost-in-thought behavior, much like an absentminded professor, would eventually prove fatal.

During his lifetime, Rome and Carthage fought for control of the Mediterranean Basin. This power struggle resulted in a number of bloody conflicts called the Punic Wars. Rome waged three wars against Carthage: the First Punic War (264–241 B.C.E.), the Second Punic War (218–201 B.C.E.), and the Third Punic War (149–146 B.C.E.). Carthage was defeated and totally destroyed in the Third Punic War, leaving Rome in complete control of the world around the Mediterranean Sea.

Archimedes became famous as a result of the many machines he developed for the defense of Syracuse during the First and Second Punic Wars. Specifically, he designed a variety of intricate machines to repulse attackers. Historians often place his military machines into three basic categories. First, there were the Archimedes claws—cranes that could lift enemy ships up out of the water and smash them against the rocks. Next, there were a variety of catapults that could hurl rocks and other missiles over varying distances at enemy troops and ships. Finally, there was a collection of mirrors arranged to focus sunlight in such a way so as to set enemy ships on fire. This last development is open to a great deal of technical speculation concerning its efficacy.

Whether Archimedes successfully used mirrors to set Roman ships on fire during the prolonged siege of Syracuse in the Second Punic War is not known for certain. But his other machines are known to have inflicted a great number of casualties on the attacking Romans. After Syracuse fell and the city was sacked, Archimedes was killed in 212 B.C.E. by a Roman soldier. The soldier slew the aging Greek engineer despite standing orders from the Roman general, Marcellus, that the brilliant man be taken alive and treated with dignity.

The Roman historian Plutarch reported several accounts concerning the death of Archimedes. Two of these accounts are mentioned here. In the first account, Archimedes is murdered by a Roman soldier out of retribution, since the soldier wanted payback for so many of his comrades, who were killed by Archimedes' machines. The other account suggests that, as the city fell, a Roman soldier suddenly came upon Archimedes sitting on the ground drawing circles and other geometric figures in the sand. When told to move, the absentminded Archimedes ignored the order and asked for time to finish the geometry problem. The impatient Roman gave him a fatal thrust with a short sword instead.

Ctesibius (c. 285–222 B.C.E.)

In ancient times, the spectacular devices and discoveries of the legendary Greek engineer, Archimedes, generally overshadowed the technical accomplishments of another famous Greek inventor and engineer, Ctesibius of

Alexandria. Often regarded as the second most important engineer of antiquity, Ctesibius made many contributions to the fledgling disciplines of pneumatics, hydraulics, mechanics, and machine design.

He published an important work, entitled *On Pneumatics* in which he discussed the elasticity of the air and enumerated various applications of compressed air in such devices as pumps, musical instruments, and even an early (air-powered) cannon. Some science historians regard his efforts in this area as the start of the science of pneumatics. Unfortunately, this particular work along with all his other writings perished in the chaos of ancient times, when great libraries like that of Alexandria were destroyed and their contents scattered.

What is specifically known about the engineering accomplishments of Ctesibius comes down to us from other Greek inventors and engineers, like Hero of Alexandria and the first-century (B.C.E.) Roman military engineer and architect, Marcus Vitruvius Pollio. Ctesibius is credited with the invention of the siphon. He is also considered as the creator of a small pipe organ (called the hydraulis), which was supplied with air by a piston pump.

His greatest technical accomplishment was a vastly improved version of the water clock (clepsydra) of ancient Egypt. Ctesibius's improved water clock became the best timepiece in antiquity and remained unrivaled in accuracy until the seventeenth century. As a historic note, mechanical clocks were developed in Europe during the Middle Ages. These devices were based on falling weights and proved to be more convenient than, but not as accurate as, Ctesibius's improved clepsydra. It was only the pendulum clock, introduced in the mid-seventeenth century by the Dutch astronomer and physicist Christiaan Huygens, which surpassed the accuracy of the water clock and ushered in a new era in timekeeping. Few mechanical devices have so dominated an area of technology for almost two millennia.

Hero of Alexandria (first century C.E.: c. 20 to c. 80) (a.k.a. Heron)

Hero was the last of the great Greek engineers of antiquity. He invented many clever mechanical devices, including the device for which he is most commonly remembered the aeolipile—a spinning, steam-powered spherical apparatus that demonstrated the action–reaction principle, which forms the basis of Sir Isaac Newton's third law of motion.

Not much has survived from antiquity about the personal life of Hero. Historians estimate that the Greek inventor and early engineer was born in about 20 C.E., because his own writing indicates that he observed a lunar eclipse, which was observable in Alexandria in 62 C.E. Hero had a strong interest in simple machines, mechanical mechanisms (like gears), and hydraulic and pneumatic systems. His inventions and publications reflect the influence of Ctesibius, another great engineer of antiquity. Several of Hero's works have survived including *Pneumatics* (written about 60 C.E.), *Automata*, *Mechanics*, *Dioptra*, and *Metrics*.

His most familiar invention is the aeolipile. He placed a hollow metal sphere on pivots over a charcoal grill-like device. When water placed inside the metal sphere was heated over the brazier, steam formed, and escaped through the tubes, which acted like crude nozzles. The sphere would spin freely as steam escaped from two small opposing tubes connected to the sphere. This whirling sphere delighted children and became a popular toy. However, for some inexplicable reason, the last great Greek engineer of antiquity never connected the action–reaction principle exhibited by the aeolipile with a concept of steam-powered machines for performing useful work.

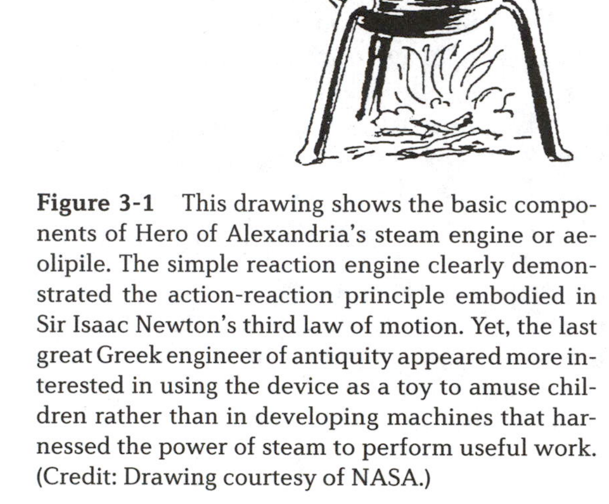

Figure 3-1 This drawing shows the basic components of Hero of Alexandria's steam engine or aeolipile. The simple reaction engine clearly demonstrated the action-reaction principle embodied in Sir Isaac Newton's third law of motion. Yet, the last great Greek engineer of antiquity appeared more interested in using the device as a toy to amuse children rather than in developing machines that harnessed the power of steam to perform useful work. (Credit: Drawing courtesy of NASA.)

The aeolipile is an example of a clever device invented well ahead of its time. Such devices sometimes need to be "rediscovered" or "reinvented" decades or centuries later, when the social, economic, and/or technical conditions are just right for full engineering development and application. Since the aeolipile embodies the action–reaction principle, it is the technical ancestor of the steam turbine, which helped industrialize (and later electrify) the world, as well the power rocket vehicles that send robot spacecraft to explore distant worlds in the solar system.

Despite this oversight, Hero was a skilled engineer and creative inventor. He also receives credit for a variety of feedback control devices that used fire, water, and compressed air in different combinations. He developed a machine for threading wooden screws and constructed an automated puppet theater. He also receives credit for designing an early odometer, a primitive form of analog computer (involving gears, spindles, weights, pegs, trays of sand, and ropes), and a compressed-air fountain.

Galileo Galilei (1564–1642)

The brilliant Italian scientist Galileo Galilei is most often remembered as the first astronomer to use a telescope to view the heavens and conduct early astronomical observations that helped inflame the Scientific Revolution of the seventeenth century. But he was also the physicist who founded the science of mechanics and provided Sir Isaac Newton the underlying data and ideas upon which Newton could construct the laws of motion and the universal law of gravitation. No study of the evolution of machine technology is complete without paying homage to the work of this amazing man, whose insightful experiments gave rise to organized science and engineering and thus changed the course of history.

Galileo Galilei was born in Pisa on February 15, 1564. (Scientists and astronomers commonly refer to Galileo by his first name only.) When he entered the University of Pisa in 1581, his father encouraged him to study medicine. But, because of his inquisitive mind, Galileo soon became more interested in physics and mathematics than medicine. While still a medical student, he attended church services on Sunday. During the sermon, he noticed a chandelier swinging in the breeze and began to time its swing using his own pulse as a crude clock. When he returned home, he immediately set up an experiment that revealed the pendulum principle. After just two years of study, Galileo abandoned medicine and focused on mathematics and science. His change in career pathways also changed the entire trajectory of science.

In 1585, Galileo left the university without receiving a degree and focused his activities on the physics of solid bodies. The motion of falling objects and projectiles intrigued him. Then, in 1589, he became a mathematics professor at the University of Pisa. Galileo was a brilliant lecturer and students came from all over Europe to attend his classes. This circumstance quickly angered many senior, but less capable, faculty members. To make matters worse, Galileo often used his tenacity, sharp wit, and biting sarcasm to win philosophical arguments at the university. His tenacious and argumentative personality earned him the nickname "The Wrangler."

In the late sixteenth century, European professors usually taught natural philosophy (physics) as metaphysics—an extension of Aristotelian philosophy. Before Galileo's pioneering contributions, physics was not seen as an observational, experimental science. But, through his skillful use of mathematics and innovative experiments, Galileo changed that approach and established an important approach now called the scientific method. Galileo's activities constantly challenged the two thousand year tradition of ancient Greek learning. For example, Aristotle stated that heavy objects would fall faster than lighter objects. Galileo disagreed and held the opposite view that, except for air resistance, the two objects would fall at the same time regardless of their masses. It is not certain whether he personally performed the legendary musket ball versus cannon ball drop experiment from the Leaning Tower in Pisa to prove this point. However, he did conduct a sufficient number of experiments with objects on inclined planes to upset Aristotelian "physics" and create the science of mechanics.

Figure 3-2 This 1964 Italian postage stamp honors the 400th anniversary of the birth of Galileo Galilei (on February 15, 1564). A brilliant physicist, mathematician, and astronomer, Galileo Galilei founded the science of mechanics, promoted the scientific method, and fanned the flames of the Scientific Revolution by vigorously supporting the Copernican hypothesis—for which astronomical advocacy he was eventually found guilty of heresy and imprisoned (house arrest) for the remainder of his life. (Credit: Photograph courtesy of author.)

During his lifetime, Galileo was limited in his motion experiments by an inability to accurately measure small increments of time. No one had yet developed a timekeeping device capable of accurately measuring tenths, hundredths, or thousandths of a second. Despite this severe impediment, Galileo conducted many important experiments that produced remarkable insights into the physics of free fall and projectile motion. Less than a century later, Sir Isaac Newton would build upon Galileo's pioneering work to create the universal law of gravitation and three laws of motion—the pillars of classical physics.

By 1592, Galileo's anti-Aristotelian research and abrasive behavior had sufficiently offended his colleagues at the University of Pisa to the point that they not so politely "invited him" to go elsewhere to teach. So later that year, Galileo moved to the University of Padua. This university had a more lenient policy of academic freedom, encouraged in part by the progressive government of the Republic of Venice. In Padua, Galileo wrote a special treatise on mechanics to accompany his lectures. He also began teaching courses on geometry and astronomy. At the time, the university's astronomy courses were primarily for medical students who needed to learn about medical astrology.

In 1597, the German astronomer, Johannes Kepler (1571–1630), provided Galileo a copy of Copernicus's book (even though the book was officially banned in Italy). Although Galileo did not previously have a keen interest in astronomy, he immediately became fascinated with and embraced the Copernican model. Galileo and Kepler, the founders of modern astronomy, continued to correspond until about 1610.

Between 1604 and 1605, Galileo performed his first public work involving astronomy. He observed the supernova of 1604 (in the constellation Ophiuchus) and used it to refute the cherished Aristotelian belief that the heavens were immutable (unchangeable). He delivered this challenge on Aristotle's doctrine in a series of public lectures. Unfortunately, these well-attended lectures brought him into direct conflict with the university's pro-Aristotelian philosophy professors.

In 1610, he announced some of his early telescopic findings in the publication *Starry Messenger*, including the discovery of the four major moons of Jupiter (now called the Galilean satellites in his honor). Their behavior like a miniature solar system stimulated his enthusiastic support for the heliocentric cosmology of Nicholas Copernicus (1473–1543). Unfortunately, this part of Galileo's scientific work led to a direct clash with ecclesiastical authorities, who insisted on retaining the Ptolemaic system (with its geocentric cosmology) for a number of political and social reasons. This conflict eventually earned the fiery Galileo an Inquisition trial at which he was found guilty of heresy (for advocating the Copernican system) and confined to house arrest for the remainder of his life.

In 1613, Galileo published "Letters on Sunspots." He used the existence and motion of sunspots to demonstrate that the Sun itself changes, again attacking Aristotle's doctrine of the immutability of the heavens. In so doing, he also openly endorsed the Copernican model. This inflamed Galileo's long and bitter fight with ecclesiastical authorities. Above all, Galileo believed in the freedom of scientific inquiry. Late in 1615, Galileo went to Rome and publicly argued for the Copernican model. This public action angered Pope Paul V, who immediately formed a special commission to review the theory of Earth's motion.

Dutifully, the (unscientific) commission concluded that the Copernican theory was contrary to Biblical teachings and possibly a form of heresy. Cardinal Robert Bellarmine (an honorable person who was later canonized) received the unenviable task of silencing the brilliant, but stubborn, Galileo. In late February 1616, ecclesiastic authorities officially admonished Galileo to abandon his support of the Copernican hypothesis. In the process, Cardinal Bellarmine (under direct orders from Pope Paul V) made Galileo an offer he could not refuse. Galileo must never teach or write again about the Copernican model, or he would be tried for heresy and imprisoned, and quite possibly executed, like Giordano Bruno (1548–1600), who was burned at the stake in Rome on February 17, 1600.

Apparently Galileo got the message—at least so it seemed for a few years. In 1623, he published *Il saggiatore* (*The Assayer*). In this book, he discussed the principles for scientific research, but carefully avoided support for Copernican theory. He even dedicated the book to his lifelong friend, the new pope, Urban VIII. However, in 1632 Galileo pushed his luck with the new pope to the limit by publishing *Dialogue on the Two Chief World Systems*. In this masterful (but satirical) work, Galileo had two people present scientific arguments to an intelligent third person, concerning the Ptolemaic and Copernican worldviews. The Copernican cleverly won these lengthy arguments. Galileo represented the Ptolemaic system with an ineffective character he called Simplicio. For a variety of reasons, Pope Urban VIII regarded Simplicio as an insulting, personal caricature. Within months after the book's publication, the Inquisition summoned Galileo to Rome. Under threat of execution, the aging Italian scientist publicly retracted his support for the Copernican model on June 22, 1633. The Inquisition then sentenced him to life in prison, a term that he actually served under house arrest at his villa in Arceti (near Florence). Church authorities also banned the book, *Dialogue*, but the supporters of Galileo smuggled copies out of Italy and the Copernican message again spread across Europe.

While under house arrest, Galileo worked on a less controversial area of physics. He published *Discourses and Mathematical Demonstrations Relating to Two New Sciences* in 1638. In this seminal work, he avoided astronomy and summarized the science of mechanics—including the very important topics of uniform acceleration, free fall, and projectile motion.

Through Galileo's pioneering work and personal sacrifice, the Scientific Revolution ultimately prevailed over misguided adherence to centuries of Aristotelian philosophy. Galileo never really opposed the Church, nor its religious teachings. He did, however, come out strongly in favor of the freedom of scientific inquiry. Blindness struck the brilliant scientist in 1638. He died while imprisoned at home on January 8, 1642. Three and a half centuries later on October 31, 1992, Pope John Paul II formally retracted the sentence of heresy passed on him by the Inquisition.

René Descartes (1596–1650)

The French philosopher, mathematician, and inventor of analytic geometry, René Descartes, was a mechanist who believed that the bodies of humans and animals are complex automata. In his treatise *Discourse on Method*, published in 1637, Descartes discusses how humans, who have the power of reason, and animals, which cannot reason, can be distinguished from one another and machines. Descartes is often regarded as the first modern philosopher. His most famous quote (as found in *Discourse on Method*) is: "Cogito ergo sum" (which means, "I think therefore I am.") This statement highlights some of the deep philosophical arguments Descartes raised in developing his mind-body dualism. The nature of mind is an issue that has intrigued philosophers for ages. The issue arises again from an interesting new perspective as robot specialists speculate about endowing very smart machines with a sense of consciousness and cognition. At what point does a so-called "thinking machine" become truly conscious?

Descartes was born on March 31, 1596, in La Haye, France. His father was a counselor of the government of Britanny, while his mother died shortly after he was born, but left Descartes a sufficient quantity of money to make him financially independent. When he was about six years old, Descartes began his education at the Jesuit school in La Fléche. He remained at this school until 1612. Because of his sickly nature, he received permission to remain in bed until late morning each school day. This established Descartes' unusual lifelong custom of remaining in bed until about 11:00 A.M. each day, thinking and working on mathematical problems.

Descartes attended the University of Poitiers and received a law degree in 1616, after which he enlisted in private military service, first with Maurice of Nassau (the Prince of Orange), and later with Maximilian (the Duke of Bavaria). From 1620 to 1628, Descartes used his army service to travel extensively throughout Europe, including Hungary, Germany, Holland, and France. He returned to France on several occasions and made contact with the French mathematician and natural philosopher, Marin Mersenne (1588–1648). Mersenne served an important role because he corresponded with all the leading scientists of the day (including Descartes, Fermat, and Pascal), thus serving as an informal clearing house for scientific information. By 1628, Descartes grew tired of traveling and settled down in Holland, where he resided for the next twenty years. Little is known about Descartes' private life. He fathered an illegitimate daughter (named Francine) whose death at age five was a terrible tragedy for him. He also continued to enjoy thinking in bed until the late morning and published most of his major works in French (the more popular ones) or Latin (the more scholarly ones).

Right after he settled in Holland, he started work on *Le Monde, ou Traité de la Lumière*, an attempt at Copernican-based natural philosophy, which he quickly abandoned (near completion) when he learned that Galileo Galilei was condemned to house arrest for the heresy of supporting Copernican cosmology. (The incomplete manuscript was posthumously published in 1664). *Le Monde* was Descartes' attempt at developing a physical theory of the universe, but he chose to avoid the martyr's life and refocused his efforts on composing a treatise on universal science, called *Discourse on Method*. This work contained three appendices: *La Dioptrique* (dealing with optics), *Les Météores* (the first work that attempts to discuss meteorology/weather on a scientific basis), and the very important *La Geométrie*. In *La Geométrie*, Descartes introduces the Cartesian coordinate system, an innovative union of algebra and geometry that he combines into an important new discipline called analytical geometry. Descartes' work provides the mathematical framework allowing Sir Isaac Newton to develop the calculus.

In 1641, Descartes published a work entitled *Meditationes Descartes*. He published his most comprehensive work *Principles of Philosophy* (*Principia Philosophiae*) in Amsterdam in 1644. In this four-part work, Descartes attempted to describe the universe in mechanical terms. The four parts were: The *Principles of Human Knowledge*, *The Principles of Material Things*, *Of the Visible World*, and

The Earth. However, Descartes did not believe in action at a distance (as later postulated by Sir Isaac Newton) nor could he accept the concept of a vacuum in space (as suggested by his contemporary, Blaise Pascal). In order to explain motions in the universe, Descartes came up with an elaborate, though incorrect, vortex theory, which other French scientists championed for about a century, despite the obvious validity of Newton's universal law of gravitation.

In 1647, the French court awarded Descartes a pension to honor his scientific discoveries and acknowledge his influential works. Then, in 1649, Descartes made a fatal mistake by accepting an invitation to tutor Queen Christina of Sweden. When he arrived in Stockholm, he encountered the shock of a lifetime. The energetic queen wanted to be tutored on mathematics at 5:00 A.M. After only a few months of rising early in the cold Swedish winter, Descartes caught pneumonia and died in Stockholm on January 11, 1650. Descartes had gone against his lifetime custom of staying in bed until the late morning and the result was fatal. The body of the first great modern philosopher was eventually returned to France and buried in Saint-Germain-des Prés.

At the dawn of the Age of Science, René Descartes began revisiting the concept of mind as it had wandered down through Western civilization from the ancient Greek philosophers, like Plato and Aristotle, and the great medieval Christian theologians, like Thomas Aquinas. In his *Principles of Philosophy*, Descartes proposed the philosophical concept that mind (soul) and body (matter) are separate and distinct entities. His postulation represents the birth of modern *dualism* and the start of the famous mind/body problem. For Descartes, the rational mind (soul) was an entity (substance) distinct from matter (the body). Within his model of mind, there were two very different kinds of substances: an invisible, unextended thinking substance (which he called the *res cogitans*) and a physical, extended substance (labeled the *res extensa*) that could be measured and divided. According to Cartesian dualism, the human mind (soul) was responsible for such invisible activities as thinking, willing, desiring, and so forth. It represented the *res cogitans* (the thinking substance) of a human being. In contrast, the human body (including the brain and the entire nervous system) was a physical, extended substance (that is, the *res extensa*). At death, the soul (mind) would leave the body (which subsequently decays) and then continues to exist in some transformed (invisible) state of consciousness. Within the context of Christian theology, Descartes' dualism further suggested that the soul (as the immortal, spiritual seat of human consciousness) experiences an *afterlife*—a state of continual happiness (heaven) or perpetual pain (hell).

From at least as far back in human history as wandering Neanderthal tribes and their primitive burial ceremonies, human beings in almost every civilization and culture have expressed anticipation of some kind of life after death. The survival of personal human consciousness has been and still remains a pressing question in philosophy and theology. No study of mind is complete without exploring this issue. That is why the numerous terms and concepts associated with consciousness and possible conditions of postmortem survival form one of the major themes of this entry. The following statement introduces a major milestone on the journey through mind-space: *the mind (as a conscious personal entity) either survives the death and destruction of the body, or it doesn't.* More bluntly stated, your mind either knows who you are after death, or else you simply no longer personally exist as *you*. If personal consciousness survives the biological death, then where does it "go," what does it "do," and perhaps most interesting of all, can it still interact on some level with the physical world and normal (living) human beings who reside there? On the other hand, if personal consciousness terminates with biological death, then a person's "mind" is no more.

Descartes himself recognized many of the philosophical difficulties he created in trying to explain how an invisible (spiritual) mind could influence physical matter (the body) to perform voluntary physical actions and how a distinctly separate body could affect the mind through such conscious sensations as pain and pleasure. Yet, following in the philosophical footsteps of Plato and Aristotle, Descartes vigorously rationalized his own existence as a *thinking being*. As previously mentioned, this important connection between mind (consciousness) and existence he eloquently summarized in his famous quotation: "Cogito, ergo sum" ("I think, therefore I am"). Descartes' dualistic model of the mind, presented during the great Scientific Revolution of the seventeenth century, greatly influenced subsequent philosophers and the debate about mind-matter interactions continues to the present day.

Today, neuropsychologists and other "mind" scientists, recommend the acceptance of a monist versus dualist model of mind. This modern position, often referred to as *emergent materialism* (and sometimes as *emergent psychoneural monism* or *monistic materialism*), rejects Descartes' hypothesis that the mind and body are different substances and proposes, instead, that all mental activities and states are actually the result of collective processes occurring within the (physical) brain. Under the concept of emergent materialism, consciousness and mental states exist, but as an interactive, integral part of the brain and not as a separate, invisible entity. However, proponents of this model also point out that mind is not just a simple result of the brain's complex composition of cells, but rather mind comes from a special collection and association of emergent biophysical activities. Neuropsychologists suggest that functions like thinking, perceiving, feeling, and willing arise from a currently unexplained collective ("emergent") property of the brain's overall physical structure and not just the electrochemical or mechanical responses of brain cells to stimulations by the body's nervous system. In other words, within this model, a mind is definitely much greater than the sum of its numerous biological parts. This particular collection of living tissues, cells, and energy gives rise to a very special biophysical property: intelligent consciousness.

But exactly where in the brain does this consciousness reside? Unfortunately, even with all the tools and skills of modern science, no one can now say for sure. Does this elusive intelligent consciousness, this "mind," survive and transcend the physical death of the body? If "mind" is just an emergent property of the brain, and the brain needs a living body to survive, then the logical answer is: no! But this represents a most uncomfortable conclusion that flies in the face of millennia of collective human thinking and belief. How can scientists hope to reconcile such neuroscientific models of mind (as centered in the brain) with philosophical and theological models (which treat mind and consciousness as manifestations of an eternal human soul)? The creation of smart machines that achieve some level of consciousness only amplifies this already complicated philosophical issue.

Blaise Pascal (1623–1662)

The French physicist, mathematician, and philosopher, Blaise Pascal, performed key experiments with fluids that led to the establishment of the science of hydraulics. In 1641, Pascal designed a mechanical calculating machine, called Pascaline, which could add and subtract up to eight-figured sums. Working with Pierre de Fermat (1601–1665), he wrote a fundamental treatise on the outcome of games of chance that served as the foundation of probability theory. In his honor, the SI unit of pressure is called the pascal (Pa). Scientists define one pascal (Pa) as the pressure that results from a force of one newton acting uniformly over an area of one square meter.

Pascal was born on June 19, 1623, in Clermont, France. His mother passed away when Pascal was only three years old and this left his father, a mathematician and minor government official, responsible for his care as well as that of his two young sisters. Recognized early as a child prodigy, his father moved the family to Paris in 1623 to further his young son's education.

At age 12, Pascal became interested in mathematics and by age 16, published a prominent essay on conic sections that many mathematicians, including René Descartes, refused to believe was the work of a 16-year-old. Years later, Pascal was able to return the favor by experimentally demonstrating the validity of the concept of a vacuum to Descartes' strong objections.

In 1641, at the age of 18, Pascal designed a mechanical calculator, called the Pascaline, to help his father perform business transactions. At the time the French money system was not based on a decimal system (involving factors of 100). Rather, it was a complicated arrangement in which 12 deniers made up one sol and 20 sols made one livre. Nevertheless, Pascal's machine was reasonably accurate, when adding or subtracting up to eight-figured sums. Pascal continued to improve the device, patented it, and put it into production in about 1642. But because of the high cost of manufacturing, Pascal's calculators never became a commercial success. Despite its economic failure, Pascal's work with the mechanical calculator does represent an important milestone in the history of calculating machines and devices intended to augment the human ability to think quantitatively.

Starting in about 1645, Pascal conducted a series of important experiments and developed several devices that applied the pressure of fluids. One of the main scientific products of his efforts was the important principle of hydrostatics, now called Pascal's principle. This principle states that any change in the pressure applied to a completely enclosed fluid is transmitted undiminished to all parts of the fluid and the enclosing container's walls. This basic principle governs the operation of hydraulic presses and elevators, air compressors, syringes, and similar fluid mechanics devices. He also confirmed and expanded the pioneering work of the Italian physicist, Evangelista Torricelli (1608–1647), concerning the decrease of atmospheric pressure with altitude and the existence of a vacuum. Pascal wrote strongly in defense of the scientific method and refuted Descartes' position about the impossibility of a vacuum.

In 1654, a friend and gambler, the Chevalier de Méré, asked Pascal to mathematically examine an optimum strategy for a particular gaming house scenario. Pascal communicated with Fermat and their correspondence allowed

Pascal to develop the principles of the theory of probabilities. The real significance of this work is that mathematics now began to address phenomena that were not precise and exact, but rather statistical in nature. In addition to game theory, Pascal's work set the stage for such important areas of physics as statistical thermodynamics and quantum mechanics (based on the Heisenberg uncertainty principle).

Following a mystical experience in 1654, involving a nearly fatal horse-drawn carriage accident at the Neuilly Bridge, Pascal turned his attention to philosophy and theology. His intensely popular, though quite satirical, *Provincial Letters* represent his most famous works from this period. Pascal's other influential theological effort was an incomplete work, entitled *Pensees*. Considered a masterpiece in French prose, the book was published after his death. He wrote his last mathematical work (on the cycloid) in 1658. Pascal had been of ill health his entire life. In 1659, he became very seriously ill and never recovered. He died in Port Royal on August 19, 1662—just two months after his thirty-ninth birthday.

Sir Isaac Newton (1642–1727)

Sir Isaac Newton was the brilliant though introverted British physicist, mathematician, and astronomer, whose law of gravitation, three laws of motion, development of the calculus, and design of a new type of reflecting telescope make him one of the greatest scientific minds in human history. Through the patient encouragement and financial support of the British mathematician Sir Edmund Halley, Newton published his great work *The Principia* (or, *Mathematical Principles of Natural Philosophy*) in 1687. This monumental book transformed the practice of physical science and completed the scientific revolution started by Nicholas Copernicus, Johannes Kepler, and Galileo Galilei. Newton's three laws of motion and universal law of gravitation are the foundation of classical mechanics.

Newton was born prematurely in Woolsthorpe, Lincolnshire, on December 25, 1642 (using the former Julian calendar). His father had died before Newton's birth and this event contributed to a very unhappy childhood. In order to remarry, his mother placed her three-year-old son in the care of his grandmother. Separation from his mother and other childhood stresses are believed to have significantly contributed to his very unusual adult personality. Throughout his life, Newton would not tolerate criticism, remained hopelessly absentminded, and often tottered on the verge of emotional collapse. British historians claim that Newton laughed only once or twice in his entire life. Yet, many experts consider Newton to be the greatest human intellect who ever lived. His brilliant work in physics, astronomy, and mathematics combined the discoveries of Copernicus, Kepler, and Galileo. Newton's universal law of gravitation and his three laws of motion fulfilled the Scientific Revolution and dominated science for at least two centuries. The practice of mechanical engineering and machine design still relies heavily on the field of classical mechanics—a field founded upon Newton's basic physical principles.

When Newton's detested stepfather died in 1653, his twice-widowed mother returned to the farm at Woolsthorpe. Once resettled, she removed her son from school so he could practice farming. Fortunately for science, Newton failed miserably as a farmer. By June 1661 Newton left the farm and went to Cambridge University. In 1665, he graduated without any particular honors or distinction from Cambridge with a bachelor's degree.

Following graduation, Newton returned to the family farm to avoid the plague, which had broken out in London. For the next two years, he pondered mathematics and physics at home and this self-imposed exile laid the foundation for his brilliant contributions. By Newton's own account, one day on the farm he saw an apple fall to the ground and began to wonder if the same force that pulled on the apple also kept the Moon in its place. At this point heliocentric cosmology as expressed in the works of Copernicus, Galileo, and Kepler was becoming widely accepted (except where banned on political or religious grounds), but the mechanism for planetary motion around the Sun remained unexplained.

By 1667, the plague epidemic subsided and Newton returned to Cambridge as a minor fellow at Trinity College. The following year he received his Master of Arts degree and became a senior fellow. In about 1668, he constructed the first working reflecting telescope, an important astronomical instrument that now carries his name. The Newtonian telescope uses a parabolic mirror to collect light. The primary mirror then reflects the collected light by means of an internal secondary mirror to an external focal point at the side of the telescope's tube. This new telescope design earned Newton a great deal of professional acclaim, including eventual membership in the Royal Society.

In 1669, Isaac Barrow, Newton's former mathematics professor, resigned his position so that the young Newton could succeed him as Lucasian Professor of Mathematics. This position provided Newton the time to collect his notes and properly publish his work—a task he was always tardy to perform.

Shortly after his election to the Royal Society (in 1671), he published his first paper. While an undergraduate, Newton had used a prism to refract a beam of white light into its primary colors (red, orange, yellow, green, blue, and violet.) Newton reported this important discovery to the Royal Society. But, Newton's pioneering work was immediately attacked by Robert Hooke (1635–1703), an influential member of the society.

This was the first in a lifelong series of bitter disputes between Hooke and Newton. Newton only skirmished lightly then quietly retreated. This was Newton's lifelong pattern of avoiding direct conflict. When he became famous later in his life, Newton would start a controversy, withdraw, and then secretly manipulate others who would then carry the brunt of the battle against Newton's adversary. For example, Newton's famous conflict with the German mathematician, Gottfried Leibniz (1646–1716), over the invention of calculus followed precisely such a pattern. Through Newton's clever manipulation, the calculus controversy even took on nationalistic proportions as carefully coached pro-Newton British mathematicians bitterly argued against Leibniz and his supporting group of German mathematicians.

In August 1684, Sir Edmund Halley made an historic trip to visit Newton at Woolsthorpe. During his visit, Halley convinced the reclusive genius to address the following puzzle about planetary motion: What type of curve does a planet describe in its orbit around the Sun, assuming an inverse square law of attraction? To Halley's delight, Newton immediately responded, "An ellipse." Halley pressed on and asked Newton how he knew the answer to this important question. Newton nonchalantly informed Halley that he had already done the calculations years ago (in about 1666), while living on the family farm to escape the plague in London. But, the absentminded Newton could not find his old calculations, which had solved one of the major scientific questions of the day. So he promised to send Halley another set as soon as he could.

To partially fulfill his promise, Newton sent Halley his *De Motu Corporum* (1684). In this document, Newton demonstrated that the force of gravity between two bodies is directly proportional to the product of their masses and inversely proportional to the square of the distance between them (Physicists now call this relationship Newton's universal law of gravitation). Halley was astounded and begged Newton to carefully document all of his work on gravitation and orbital mechanics. Through the patient encouragement and financial support of Halley, Newton published his great work, *Philosophiae Naturalis Principia Mathematica* (*Mathematical Principles of Natural Philosophy*) in 1687. In the *Principia*, Newton gave the world his famous three laws of motion and the universal law of gravitation. This monumental work transformed physical science and completed the scientific revolution started by Copernicus, Kepler, and Galileo. Many consider the *Principia* as the greatest scientific accomplishment of the human mind.

For all his brilliance, Newton was also extremely fragile. After completing the *Principia*, he drifted away from physics and astronomy and eventually suffered a serious nervous disorder in about 1693. Upon recovery, he left Cambridge (in 1696) and assumed a government post in London as Warden (then later Master) of the Royal Mint. During his years in London, Newton enjoyed power and worldly success. Robert Hooke, his lifelong scientific antagonist, died in 1703. The following year (1704), the Royal Society elected Newton its president. Unrivaled, he won annual reelection to this position until his death. However, Newton was so bitter about his quarrels with Hooke that he waited until 1704 to publish his other major work, *Opticks*. Queen Anne knighted him in 1705.

Although his most innovative years were now clearly far behind him, Newton at this point in his life still continued to exert great influence on the course of modern science. He used his position as president of the Royal Society to exercise autocratic (almost tyrannical) control over the careers of many younger scientists. Even late in life, he could not tolerate controversy. But now, as society president, he skillfully maneuvered younger scientists to fight his intellectual battles. In this manner, he continued to rule the scientific landscape until his death in London on March 20, 1727.

Charles-Augustin de Coulomb (1736–1806)

The French military engineer and scientist, Charles-Augustin de Coulomb, performed basic experiments in mechanics and electrostatics in the late eighteenth century. In particular, Coulomb determined that the electrostatic force that one point charge applies to another depends directly on the amount of each charge and inversely on the square of their distance of separation. In his honor, the SI unit of electric charge is called the coulomb (C). One coulomb is defined as the quantity of electric charge transported in one second by a current of one ampere (A).

Coulomb was born on June 14, 1736, in Angoulême, France. He received his college education in Paris. Following the completion of his studies in the fall of 1761, he began serving as a military engineer in the French Army with the rank of lieutenant. Over the next twenty years he received a number of assignments in which he performed a variety of engineering duties, including the construction of fortifications. One extended overseas assignment had a particularly deleterious impact on his health. In February 1764, the French Army sent Coulomb to Martinique in the West Indies to oversee construction of a new fort, called Fort Bourbon. The assignment on Martinique lasted until June 1772, during which period Coulomb suffered from many tropical illnesses, which weakened his overall physical condition and left him in generally poor health for the remainder of his life.

Upon his return to France, Coulomb began to perform important studies in mechanics, including pioneering work involving the development of a delicate torsion balance (in about 1777) and the investigation of static and sliding friction (in about 1781). To avoid the problems of the French Revolution, Coulomb judiciously withdrew from public office, departed Paris, and spent his time quietly in Blois performing scientific experiments. In about 1802, he returned briefly to public life under Napoleon by serving as an inspector of public instruction. He died in Paris on August 23, 1806.

Coulomb's most noteworthy series of experiments were reported in 1785 and involved his careful use of a delicate torsion balance to investigate the relationship between the magnitude of an electrostatic force (F) exerted by one point charge on another point charge. Coulomb discovered that this electrostatic force is directly proportional to the magnitudes of the charges (say, q_1 and q_2) and inversely proportional to the square of the distance (r) between them. Physicists now call this important physical relationship Coulomb's Law. The electrostatic force is directed along the line joining the charges. This force is attractive if the charges have unlike signs (that is one charge is negative and the other charge is positive) and repulsive if the charges have like signs.

Experiments by other physicists (who built upon Coulomb's important work) revealed that the magnitude of the charge on the proton exactly equals the magnitude of the charge on the electron. By convention, scientists say the proton carries a charge of $+e$ and the electron carries a charge of $-e$. The SI unit for measuring the magnitude of an electric charge is the coulomb (C), and e has been determined experimentally to have the value: $e = 1.60 \times 10^{-19}$C.

Count Alessandro Giuseppe Antonio Anastasio Volta (1745–1827)

The Italian physicist Alessandro Volta devoted his life to the study of electricity and performed a series of key experiments in 1800 leading to the development of the battery—an essential portable power supply in robotics as well as in many other systems found in today's "electrified" civilization. The SI unit of electric potential difference and electromotive forces is called the volt (V) in his honor.

Volta was born on February 18, 1745, in Como, Italy. Like many eighteenth-century scientists (including the American Benjamin Franklin), Volta became fascinated with the subject of electricity and decided to focus his research activities on a detailed investigation of the mysterious natural phenomenon. In 1774, he received an appointment as a professor of physics at the Royal School in Como. To better support his study of electricity, in 1775 he invented the electrophorus. A precursor of the induction machine, Volta's electrophorus was a device capable of generating static electricity. The apparatus consisted of a disk that was given a negative charge by friction and a metal plate that was charged by induction when in contact with the disk. In 1779, Volta received an appointment to become a professor of physics at the University of Pavia. He accepted this appointment and remained in this position for the next 25 years.

Volta had a friend and professional acquaintance named Luigi Galvani, who was a physician (anatomist), living in the city of Bologna, Italy. In 1780, Galvani discovered that when a dissected frog's leg touched two dissimilar metals (such as iron and brass or copper and zinc) at the same time, the leg twitched and contracted. Based on these observed muscular contractions, Galvani postulated that the flow of electricity in the frog's leg represented some type of *animal electricity*—a term he coined to identify electricity as the animating agent in living muscle and tissue. Galvani may have been influenced by the recent work of Benjamin Franklin, which associated lightning (a natural phenomenon) to electricity. Electricity was a frontier science in the late eighteenth century, so Galvani, as a scientist with a strong inclination toward anatomy, wanted to be the first investigator to successfully connect the animation of living matter with this exciting new phenomenon.

Galvani knew that Volta was also performing experiments with electricity, so he asked Volta to help validate his experiments and the conclusion about animal electricity. This request eventually ended the amicable relationship between the two Italian scientists and made all subsequent interactions adversarial. Volta responded to Galvani's request and in about 1794 began to explore the question of whether the electric current in the twitching frogs legs was a phenomenon associated with biological tissue (as Galvani postulated), or actually the result of contact between two dissimilar metals. Ever the careful physicist, Volta used two dissimilar metals alone without a frog's leg or other type of living tissue. He observed that an electric current appeared and continued to flow. The frog's leg had nothing to do with the current flow. Volta's conclusions dealt a mortal blow to Galvani's theory of animal electricity.

Galvani did not accept Volta's conclusions and the two Italian scientists engaged in a bitter controversy that soon involved other famous scientists from across Europe. For example, the French physicist Charles-Augustin de Coulomb supported Volta's work and conclusions. Additional experiment evidence began to weigh heavily in Volta's favor and Galvani died a broken and bitter man on December 4, 1798—still clutching to his belief that electricity was linked to and inseparable from biology, as an agent promoting vitality.

The professional disagreement with Galvani spurred Volta on to perform additional experiments. In 1800, Volta developed the voltaic pile—the first chemical battery. From a variety of experiments, Volta determined that in order to produce a steady flow of electricity he needed to use silver and zinc as the most efficient pair of dissimilar metals. First he made individual cells by placing a strip of zinc and silver in a cup of brine. He then connected up several cells to increase the voltage. Finally, he created the first voltaic pile (chemical battery) by alternately stacking up discs of silver, zinc, and brine-soaked heavy paper—quite literally in a pile. Soon scientists all over Europe used and improved Volta's invention to give themselves a steady, dependable flow of electricity (direct current) for their experiments.

In 1810, Emperor Napoleon of France acknowledged Volta's great accomplishment and made him a Count. In 1815 (after Napoleon fell from power and the politics in Northern Italy shifted), Volta's great achievements were again recognized, this time by the Emperor of Austria, who appointed him as a professor of philosophy at Padova. Volta died in Como, Italy, on March 5, 1827.

Joseph-Marie Jacquard (1752–1834)

The French textile manufacturer, Joseph-Marie Jacquard, introduced a punch-card system for programming the pattern of a carpet as it is being made on a loom. Jacquard's invention was a critical stimulus in the technology revolution that swept the textile manufacturing industries of France and the United Kingdom in the early nineteenth century. Eventually, the Jacquard loom changed the weaving industry around the world and influenced other developments in programmable devices.

Jacquard was born on July 7, 1752, in Lyon, France. Since both his parents were employed in the weaving industry, it was just a simple matter of time before Jacquard, as a young man, would become involved in that industry. Starting at the age of 10, Jacquard found himself immersed in the performance of monotonous and unpleasantly repetitive tasks. So while working in the monotonous environment of the late eighteenth century textile industry, Jacquard dreamed of ways of escaping from this stifling trap.

In 1790, Jacquard came up with the creative notion of an automated loom. However, his efforts to develop an automated loom were interrupted by the French Revolution. During the civil conflict that gripped France, Jacquard fought on the side of the revolutionaries and participated in the defense of his home city of Lyon.

Following the revolution, Jacquard resumed his efforts to develop a device that would help automate the textile industry in France. In 1801, he introduced his punch-card system for programming the pattern of a carpet as it is being made on a loom. Jacquard's device featured a series of connected perforated cards, each of which pressed against an array of needles in sequence. He arranged the pattern of holes on each card to correspond to the pattern he wished to produce on the textile being weaved. Whenever a needle encountered a hole in a card, the needle passed through the opening in the card, activating a threading mechanism in the process. The device was most creative and it accommodated the automated production of fabrics with intricate woven patterns, such as brocades and tapestries.

Jacquard continued to improve his invention. In 1805, he introduced an attachment that allowed any loom that used it to become known as a Jacquard loom. The Napoleonic government of France quickly recognized the value of Jacquard's device. He was awarded a medal and lifetime pension. However, not everyone rejoiced in Jacquard's work. The master silk weavers of France became extremely hostile. Threatened with the loss of their jobs, the weavers of Lyon not only attacked the automated looms being placed into production but threatened to kill the inventor as well.

Despite the social unrest due to the fear of job displacements, the overwhelming economic advantages of Jacquard's automated loom soon dominated the French textile industry. By 1812, there were over 11,000 Jacquard looms in use throughout France. In 1819, the French government awarded Jacquard the cross of the Legion of Honor and a gold medal. These high honors emphasized the great socioeconomic influence of the Jacquard loom.

Jacquard's use of punched cards to control the weaving of cloth so that any desired intricate pattern could be made automatically revolutionized the textile industry around the world. The Jacquard "automated system" quickly spread to Great Britain and by the1820s dominated that country's textile industry. Since Great Britain had a globe-spanning empire, the automated manufacture of textiles soon spread from England and influenced textile manufacturing around the globe.

Jacquard's automated loom significantly increased productivity, while simultaneously reducing the cost of textile manufacturing. By storing the skill and knowledge of a master weaver, the punched cards represented an early form of expert system. Using this automated loom, almost anyone could supervise the production of intricately woven textiles. Soon, handmade textiles gave way to machine-made textiles in the marketplace. Since the machine greatly reduced the number of human errors and supported a variety of manufacturing outcomes, textile manufacturers could consistently produce a variety of quality goods. Production became independent of the availability of master weavers. Textile manufacturing shifted to individuals capable of supervising and maintaining automated machines. One of the great impacts of the Jacquard automated loom was the fact that the "skill factor" in producing textiles with intricate patterns was transferred to a sequence of punched cards. Once properly programmed, these cards stored the knowledge of the master weaver.

Jacquard's use of punched cards to control activities and store data heralded other great developments in automation. The British mathematician and inventor, Charles Babbage, used a series of punched cards to provide programmed instructions to the advanced mechanical calculating machine (the Analytic Engine) he attempted to construct in the 1830s. To expedite the conduct of the 1890 U.S. census, the American statistician Herman Hollerith developed an electric tabulating machine, which used punched cards containing tabulated statistical data.

André-Marie Ampère (1775–1836)

Science historians regard the gifted French mathematician and physicist, André-Marie Ampère, as one of the main discoverers of electromagnetism. His defining work in this field began in about 1820 and involved insightful experiments that led to the development of a physical principle called Ampere's Law for static magnetic fields. Ampere's pioneering work in electromagnetism served as the foundation of the subsequent work by the British experimenter Michael Faraday and the American physicist Joseph Henry. The science of electromagnetism helped bring about the world-changing revolution in electric power applications and information technology that characterized the late nineteenth century. In his honor, the SI unit of electric current is called the ampere (A), or amp for short.

Ampere was born in Lyon, France, on January 20, 1775. His father was a prosperous businessperson and Ampere received the benefit of an excellent, though primarily home-schooled, education. Gifted in mathematics, Ampere was also a sensitive person, whose life would be shattered by several traumatic experiences. The first of his personal tragedies took place in 1793, during the French Revolution, when Lyon revolted against the government in Paris. Lyon was subsequently captured by the army of the Convention and Ampere's father, a minor official who stood out against the excesses of the revolution was thrown in prison and executed. Ampere slipped into a state of depression that lasted for more than a year after his father's death.

In 1796, Ampere met a young woman named Julie Carron and they married three years later (1799). From about 1796, Ampere earned a living by giving private lessons in mathematics, chemistry, and languages in and around the city of Lyon. Once married, he attempted to earn more money in 1801 by accepting a position as professor of physics and chemistry in Bourg. However, this required him to move to Bourg and leave his sickly wife and infant son (named Jean Jacques Ampere) behind in Lyon. In 1804, his young wife died and Ampere would never recover from the blow.

At the start of the nineteenth century, the French Emperor Napoleon was encouraging French scientists to pursue fruitful careers in physics and chemistry. Despite his state of depression, Ampere was encouraged to continue teaching physics and chemistry at Bourg and then to move to Paris in 1809 to accept a position as professor of mathematics, a post he held until 1828. This new position allowed him to diligently pursue a variety of scientific questions. While his scientific career began to rise, his personal life continued to deteriorate. After moving to Paris, he entered a brief second marriage in 1806—a miserable match that ended in a legal separation in 1808, with Ampere being given custody of his infant daughter (Albine).

Ampere's most important service to science occurred when he discovered a quantitative relationship between electricity and magnetism. On September 11, 1820, Ampere learned that the Danish physicist Hans Christian Oersted (1777–1851) had discovered that a magnetic needle twitched when brought near a wire carrying an electric current. Ampere immediately set about to fully explore this exciting new discovery. On September 18 (about one week later), Ampere presented a paper to the French Academy, which contained a far more detailed discussion of this important phenomenon, which linked electricity and magnetism. Ampere's detailed investigations resulted in the foundation of electromagnetism. Ampere's detailed work revealed that electric currents produce magnetic fields and the quantitative relationship between the two has become known as Ampere's Law. Although distinctly different magnetic fields surround a long, straight wire, a circular loop of wire, and a solenoid (a long coil of wire in the shape of a helix), these magnetic fields can be obtained from the general physical principle known as Ampere's Law which is valid for a wire of a geometrical shape.

While exploring the linkage between electricity and magnetism, Ampere wanted to find out what happened when current carrying wires came near each other. He already knew that the current in a wire would deflect a magnet (Oersted's compass needle experiment), and Ampere also knew from simple experiments with bar magnets that like poles repel and unlike poles attract each other. So, he devised a simple yet elegant experiment that had profound impact on science. He arranged for currents to flow through parallel wires and discovered that if current passed through each parallel wire in the same direction, the wires attracted each other. However, if the currents flowed through the two parallel wires in the opposite direction the wires repelled each other. He then expanded from this simple discovery, using much more complicated wire geometries (loops and solenoids) and the physical relationship that emerged became known as Ampere's law. He published his precise mathematical formulation of the relationship between electricity and magnetism in 1826 in a report entitled *Notes on the Mathematical Theory of Electrodynamic Phenomena, Solely Deduced from Experiment*.

Ampere died on June 10, 1836, in Marseille and was buried in Paris. In his honor, the international scientific community named the SI unit of electric current the ampere. One ampere is officially defined as the constant current that, if maintained in two straight parallel conductors of infinite length, of negligible circular cross sections, and placed one meter apart in a vacuum, would produce a force between these conductors equal to 2×10^{-7} newtons per meter of length. While this precise definition may seem a bit labored and odd at first glance, it represents a practical application of Ampere's great discovery in electrodynamics.

Michael Faraday (1791–1867)

Though without formal education and possessing limited mathematical skills, the British physicist and chemist, Michael Faraday became one of the world's greatest experimental scientists. Faraday made significant contributions to the fields of electromagnetism and electrochemistry. In 1831, he observed and carefully investigated the principle of electromagnetic induction—an important physical principle that governs the operation of modern electric generators and motors. In his honor the SI unit of capacitance is called the farad (F). One farad is defined as the capacitance of a capacitor whose plates have a potential difference of one volt when charged by a quantity of electricity equal to one coulomb. Since the farad is too large a unit for typical applications, submultiples—such as the microfarad (μF), the nanofarad (nF), and the picofarad (pF)—are encountered frequently in modern electrical engineering.

Faraday was born on September 22, 1791, in Newington, England (near London). His family was impoverished, so he received only a limited amount of formal schooling before being forced to work as a bookbinder's apprentice at the age of 13. This apprenticeship was especially fortuitous because the position gave Faraday the opportunity the read many of the books he processed. On one occasion, an encyclopedia article on electricity caught his attention and the gifted young man immediately took it upon himself to perform some simple experiments with a Leyden jar. A Leyden jar was an early form of electric capacitor.

Invented in the Dutch university town of Leyden in about 1745, the device consists of a glass jar with a layer of metal foil on the outside and a similar layer of metal foil on the inside. An experimenter would use a loose chain hanging inside the jar to make contact between the inner foil and the outer foil, releasing any accumulated charge. Faraday's simple experiments in electricity set the stage for a long life of discovery and contribution to science.

One of the most important milestones in Faraday's life came in 1812 when he attended several lectures at the Royal Institution given by the British chemist, Sir Humphry Davy (1778–1829). Thanks to some excellent note taking by Faraday and a little luck, the young man eventually obtained a laboratory assistant position at the Royal Institution in 1813. He remained with that institution (in various appointments) for the rest of his working life—that is, until about 1862. In 1825, for example, Faraday became the director of the laboratory at the Royal Institution and in 1833 he was elected to the institution's newly endowed Fullerian Professorship in chemistry.

At age 25, Faraday proved a more than capable assistant to Sir Humphry Davy, who was also president of the Royal Society. Faraday soon eclipsed the senior chemist, and Davy grew openly jealous of his brilliant young protégé. In this class-based society, Faraday was a simple commoner and so Davy's wife treated Faraday not as a bright young scientist but as a servant. Never really interested in titles or awards, Faraday patiently endured this senseless treatment by Lord and Lady Davy and went on to make some of the most important discoveries in electrochemistry and electrodynamics.

In 1821, most likely stimulated by conversations between the British scientist William Hyde Wollaston (1766–1828) and Davy concerning the discovery of the relationship between electricity and magnetism by the Danish physicist Hans Christian Oersted, Faraday succeeded in designing a clever experiment that demonstrated electromagnetic rotation—the operating principle of a simple electric motor. Although Faraday succeeded where Wollaston had failed, Faraday unwisely published his results without acknowledging the conversations of Wollaston and Davy. This omission caused some hard feelings and harsh words within the Royal Institution, forcing Faraday to abandon any additional work on electrodynamics for a few years. Undaunted, Faraday focused on making contributions in chemistry, especially electrochemistry. In 1826, he also introduced a series of six Christmas lectures for children at the Royal Institution. These lectures were very popular and the tradition continues to this very day.

When Faraday returned to his pioneering work in electromagnetism in 1831, he made a discovery that forms the basis of modern electric power generation. Faraday discovered that whenever there is a change in the flux through a loop of wire, an electromotive force (emf) is induced in the loop. This discovery is now called Faraday's law of electromagnetic induction. It is the physical principle upon which the operation of an electric generator (dynamo) depends. Faraday's discovery, refined by electrical engineers and inventors into practical generators, made large quantities of electricity suddenly available for research and industrial applications. Scientists were no longer restricted to electricity supplied by chemical batteries.

Independent of Faraday, the American physicist, Joseph Henry (1797–1878), had made a similar discovery about a year earlier, but teaching duties prevented

Henry from publishing his results. So credit for this discovery goes to Faraday, who actually not only published his results first (in his *Experimental Researches in Electricity, first series* 1831), but also performed more detailed experimental investigations of the important phenomenon. However, in 1831, Henry did publish a seminal paper describing the electric motor (essentially a reverse dynamo) and its potential applications. Science historians regard the work of both Faraday and Henry during this period as the beginning of the electrified world. Clever engineers and inventors would apply Faraday's law of electromagnetic induction to create electric generators, which supply large quantities of electricity. Other engineers would invent ways of using direct current (DC) and alternating current (AC) electricity to power a wide variety of practical and efficient electric motors, which then became the building blocks of modern civilization. All these exciting developments stemmed from the pioneering work of Faraday (and Henry).

Faraday was ingenious in his design and construction of experiments. However, he lacked a solid mathematics education, so translating the true significance of some of his results into robust physical theory relied upon his affiliation with the Scottish theoretical physicist, James Clerk Maxwell (1831–1879). Maxwell, a genius in his own right, competently translated the significance of Faraday's ingenious experiments into the mathematical language of physics. Their cordial working relationship provided a solid experimental and theoretical basis for classical electromagnetic theory in the middle of the nineteenth century.

Faraday always remembered his humble (commoner) beginnings and generally shied away from awards and notoriety. It was science that he enjoyed doing. He had married Sarah Barnard in 1821 and the two (though childless) remained devout church-going people throughout their lives. Faraday declined knighthood and the presidency of the Royal Society. However, he did accept an honorary degree from Oxford University (1832) and his appointment as professor of chemistry at the Royal Institution (1833). The Royal Society honored his scientific achievements by bestowing upon him both the Royal Medal and the Copley Medal.

By 1839, Faraday's health began to fail. He suffered some type of nervous breakdown that year and remained inactive (with respect to research) until about 1845. By 1862, declining mental acuity and physical health ended the day-to-day meticulous research and note writing that had characterized his scientific labors for the past four decades at the Royal Institution. He died in his house at Hampton Court (London) on August 25, 1867.

Charles Babbage (1791–1871)

The British mathematician and inventor, Charles Babbage, envisioned the world's first programmable computer over a century before it appeared in the United States during World War II. Babbage (in collaboration with Lady Lovelace) was the first person to envision the concept of computer programming, the use of a stored program, and the concept of addressable memory.

Born in London on December 26, 1791, Babbage entered Trinity College, Cambridge, in 1811. Together with the British astronomer Sir John Herschel (1792–1871) and others, he founded the Analytic Society in about 1812 to stimulate advanced mathematical work in England. For nearly a century since Sir Isaac Newton's death in 1727, British mathematics lagged significantly behind developments on the European Continent. His efforts earned Babbage election as a fellow of the Royal Society in 1816. Working within the Royal Society, he played an important role in the foundation of the (Royal) Astronomical Society in 1820.

Early in his career, the computation of logarithms made Babbage aware of how inaccurate repetitive human calculations can be. So, starting in about 1819, he focused his attention on using mechanical means to develop astronomical tables using the method of differences. His goal was to produce a mechanical device that could calculate and print mathematical tables with accuracy and reliability.

By 1822, Babbage had completed construction of a small prototype of his envisioned Difference Engine—a machine designed to compile mathematical tables. Despite some imperfections in this early prototype, Babbage's machine is generally regarded by science historians as the world's first successful mechanical (automatic) calculator. Although Blaise Pascal (1623–1662) and Gottfried Leibniz (1646–1716) had previously constructed calculating machines, Babbage's prototype was more reliable. Babbage's machine used a series of gears to accumulate additions and subtractions to generate tables.

In 1823, Babbage received a gold medal from the Astronomical Society for his development of the (prototype) Difference Engine. Because his small demonstration device worked sufficiently well, Babbage was encouraged to undertake development of a larger, full-scale version of this device. He envisioned an advanced mechanical computing machine that could supplement the human mind by swiftly and accurately performing intricate mathematical calculations and print tables of logarithms and other complicated mathematical functions. His efforts were stimulated, in part, by the wave of mechanization that was revolutionizing the textile industry. So with endorsement from the Royal Society and funding from the British government, he began constructing the Difference Engine—a larger version of his small prototype mechanical computing machine, which used the method of differences. Unfortunately, after a decade of work on the Difference Engine, Babbage failed to complete the project and the British government withdrew its support. A portion of Babbage's unfinished Difference Engine is on display in the Science Museum in London, England.

From 1828 to 1839, Babbage held the Lucasian Chair of Mathematics at Cambridge, although he never delivered any lectures. In 1834, he abandoned his work on the Difference Engine and pursued a far more visionary concept. This new, more ambitious idea involved a programmable machine that would perform many different computations. Babbage called this device, the Analytical Engine. Like the Jacquard loom, his Analytical Engine was being designed to work with punched cards. But Babbage's concept involved punched cards that not only stored numbers, but also contained the sequence of operations he wished conducted.

Unfortunately, his brilliant concept for a mechanical thinking-machine was simply too advanced for the times. What Babbage needed was not more intricate mechanical gears and levers; he needed the techno-miracle of modern electronic circuits. Babbage devoted the remaining years of his life and much of his personal fortune in pursuing his mechanical project. Throughout this period, the British government refused to supply any support. His only ally was Lady Lovelace (Ada Lovelace 1815–1852), daughter of the famous poet, Lord Byron.

As the world's first software engineer, she contributed many ideas to the project—including the invention of the subroutine and the programming loop. Her detailed notes provide science historians important information about Babbage's concepts. Unfortunately, she died of cancer at 36 years of age, leaving Babbage completely alone in the pursuit of this vision. He continued on a bitter and broken man. When he died, in 1871, he left behind a legacy of almost 40 square meters of drawings related to the incomplete Analytical Engine.

Charles Babbage, like many technical visionaries, was too far ahead of his own times. However, because of his pioneering work in thinking machine development, scientists now honor him with the title, "grandfather of the modern computer." The first American programmable computer, called the Mark I, was completed in 1944. This first generation electronic computer, the product of a cooperative effort between Harvard University and IBM, drew heavily from Babbage's nineteenth-century thinking-machine concepts.

Mary Wollstonecraft Shelley (1797–1851)

The British writer, Mary Wollstonecraft Shelley, is best known as the author of the famous Gothic novel, *Frankenstein, or The Modern Prometheus*. She is also well remembered as the wife of the Romantic poet Percy Bysshe Shelley.

Mary Shelley was born on August 30, 1797, in London, England. Her mother was Mary Wollstonecraft and her father was William Godwin, a well-known liberal philosopher and journalist who promoted anarchy.

Mary Shelley's life and relationship with the idealist British poet Percy Bysshe Shelley is something as befitting modern Hollywood as her famous novel. They met for the first time in November 1812, when she was just 15 years old and he was married to his first wife (Harriet Westbrook Shelley). Despite her young age and his marital status, Mary was immediately attracted to him—most likely because he was a free-spirited thinker much like her father. In July 1814, Mary fled with Percy to France, accompanied by Mary Shelley's stepsister (Jane Clairmont). This was actually the second elopement and (eventually) second marriage for the poet. In September 1816, Percy Shelley married Mary, following the suicide (by self-drowning) of his first wife (Harriet).

Mary gave birth to four children (in and out of wedlock), only one of whom survived to adulthood. The first was a girl (unnamed), who was born prematurely in 1815 and died 11 days after birth. Her second child, William, was born in 1816 and died of malaria three years later (in 1819). Her third child, Clara Everina was born in 1817 and died of dysentery the very next year. Her last child, Percy Florence, was born in 1819, lived to adulthood, and died in 1889.

Mary Shelley suffered her greatest personal loss in 1822, when her husband (Percy) drowned in a boating accident on July 8. In a cruel twist of fate, Percy had just saved her life about a month earlier when he kept her from bleeding to death as a result of a miscarriage during her fifth pregnancy. After Percy's death

Mary never remarried. Instead, she spent her time raising her son (Percy Florence), tending to her father (until his death in 1836), and writing. A revised edition of Frankenstein was published in 1831. In this version of the classic horror story, Mary placed more emphasis on the lack of personal choice in human lives and on the power of fate. Her view of nature is more of mechanistic force that can create, preserve, and destroy. Mary Shelley's last two novels: *Lodore* (1835) and *Falkner* (1837) are viewed as being somewhat autobiographical. She died in London on February 1, 1851, succumbing to a paralyzing brain tumor. Mary Shelley's life was indeed intense, tempestuous, and conducted on the edge of the social envelop. Clearly reflecting the spontaneity of her life, *Frankenstein* emerged under rather unusual circumstances.

In May 1816, Mary Wollstonecraft Godwin (then 19 years old) and her lover, the poet Percy Bysshe Shelley visited the poet Lord Byron at his villa alongside Lake Geneva in Switzerland. Unusually cold and stormy weather that summer kept them indoors for most of the visit. One day, as a source of entertainment, the literary group decided to read a book of German ghost stories.

A few evenings later, Byron challenged each of his guests to each write a ghost story. Inspired by the legend of Prometheus, Mary Shelley was the unquestionable winner of the informal contest. In Greek mythology, Prometheus, whose name means forethought, was a very wise Titan. After creating man, he took pity on the human race, because it was so helpless compared to the other animals, which were endowed with all manner of physical gifts. So, Prometheus gave the human race the gift of fire—for which act of kindness Prometheus was severely punished by Zeus. Prometheus was bound to a rocky peak and each day an eagle would tear out his liver, which would regenerate itself overnight, since he was an immortal lesser god in Greek mythology. Prometheus remained there in torment until eventually freed by the hero Hercules.

Her story, *Frankenstein, or The Modern Prometheus*, was published in 1818 and went on to influence literature and popular culture up to this very day. Many historians view this story as much more than a Gothic novel. Mary Shelley's classic story is often treated as the first science fiction novel. This honor is not without merit because the story has stimulated a complete genre of horror stories and motion pictures.

Mary Shelley's basic theme of man disastrously tampering with nature has repeated itself for decades. In her story, Victor Frankenstein attempts to create artificial life through alchemy and the combination of body parts from corpses. Victor intends the creature to be beautiful, but when the creature awakens, Victor is horrified with the results and flees the room. The creature also flees. Mary Shelley never gave the monster a name, but rather referred to it alternately as "the creature," "the monster," or "Frankenstein's monster." In 1930, Hollywood's Universal Studios produced a motion picture based on Shelley's novel. In the movie, the actor Boris Karloff played Victor's monster. Movie audience soon began speaking of the monster as Frankenstein and the direct association of the name Frankenstein and the monster has remained ever since. The film became an instant classic of a new genre—the horror movie. The movie, and subsequent films, made Mary Shelley's "creature" a cultural icon.

In her novel Mary Shelley is silent on just how Victor Frankenstein breathes life into his creation. Saying only that success crowned "days and nights of incredible labor and fatigue," Shelley's book *Frankenstein* offers no monster-making recipes. But her famous story did not arise from the cosmic void. The scientists and physicians of her time were intrigued by the elusive boundary between life and death. A great deal of experimental work was going on attempting to resuscitate drowning victims, including the use of electricity to restore life to the recently dead.

During the 1790s, the Italian physician and scientist, Luigi Galvani was actively involved in searching for something he called animal electricity, a life-giving force presumed capable of animating inanimate matter. Galvani was encouraged in his activities by experiments with frog legs that appeared to twitch when jolted by a spark of electricity from an electrostatic machine. Although his contemporary and intellectual adversary, Alessandro Volta, proved animal electricity did not exist, Galvani's research did anticipate the discovery of nerve impulses, which travel throughout the human body, and the existence of tiny electric currents in the brain, which are noninvasively measured in research. However, at the start of the nineteenth century, Volta demonstrated that the flow of electricity was separate from biological activity (life), and to prove his point, he invented the voltaic pile—the ancestor of the modern battery.

Historians suggest that perhaps discussions by Lord Byron or Percy Shelley about the work of Galvani might have provided Mary Shelley some of the background for *Frankenstein*. Another experience, much more personal and painful, is also considered a stimulus for this story. In March 1815, Mary Shelley dreamed of her dead infant daughter held before a fire, rubbed vigorously, and restored to life. At the time, scientists would not have entirely dismissed such a possibility. In fact, some of the research areas of the day involved resuscitation experiments. Another area of scientific interest was discovering how life could arise in inanimate matter. The newly founded science of electricity was attracting a great deal of interest in this regard.

Shelley's story then takes several interesting twists. The monster, originally born innocent, turns evil because Victor rejects it. In revenge, the creature kills Victor's youngest brother William, and frames an innocent maid Justine for the crime. After Victor fails in his attempt to create a female companion for the creature, the monster seeks retribution by killing Victor Frankenstein's best friend (Clerval) and Victor's wife on their wedding night. An enraged Victor pursues the monster into the Arctic.

The trip and encounter with the creature prove fatal for Victor Frankenstein. As he lies dying aboard the explorer Robert Walton's ship, the scientist assesses his own conduct. Through Victor Frankenstein's dying words, Mary Shelley suggests that the scientist's misfortune did not arise from his Promethean ambition of creating life, but in the mistreatment of his creature. (Mary Shelley made some adjustments in this theme in the 1831 revision, but her basic message concerning the hazards of tampering with nature was basically retained.) The story ends when the explorer also encounters the monster, which expresses remorse for its deeds and then commits suicide by disappearing in the icy waves. The tragedy of Frankenstein and his monster is complete.

There is one significant difference between Mary Shelley's story and the 1930 movie version. Spurned by his creator, Mary Shelley's monster kills for revenge. The movie monster (played by Boris Karloff), on the other hand, kills because he has been given the brain of a criminal. This reflects the trend of "biological determinism" and eugenics that was the trend in science in the early part of the twentieth century. Biological determinism suggested that heredity, more than environment or education, causes social problems.

George Westinghouse (1846–1914)

The visionary American engineer and entrepreneur, George Westinghouse, helped create the modern electric power industry, by financially supporting Nikola Tesla's development of alternating current (AC) generators, motors, and transformers. Prior to that, Westinghouse developed the air brake, which vastly improved railroad safety.

Westinghouse was born in Center Bridge, New York, on October 6, 1846. He moved with his family to Schenectady, New York, where his father opened a shop for agricultural machinery and small steam engines. At age 15, Westinghouse started serving in the Union Army during the American Civil War (1861–1865). After the war he attended Union College, but soon departed the campus to return to his father's business in Schenectady. There, while working in his father's shop, Westinghouse developed and patented a rotary steam engine. Later that year he invented a device for placing derailed railroad freight cars back on their tracks.

At this point in American history, the railroads served as the spine and backbone of a growing nation that would eventually become the dominant industrial power of the planet. So, many of Westinghouse's inventions dealt with railroad efficiency and safety. Most famous, perhaps, is his development and patenting of the first successful compressed air brake system (in 1869). Westinghouse's device proved much more efficient and effective than manual braking. This invention gave rise to the Westinghouse Air Brake Company. Over time, air brakes became standard safety equipment on all American trains.

Westinghouse recognized that railroads contributed significantly to the industrialization and growth of the United States. To help overcome the growing problem of railroad traffic jams, in 1882 Westinghouse developed a system of signals and interlocking switches, which used a combination of electricity and compressed air. To market this invention and other similar ideas, Westinghouse founded another company, the Union Switch and Signal Company.

In 1886, he founded Westinghouse Electric in Pittsburgh, Pennsylvania. The mission of this company was to create the equipment necessary to deliver alternating current (AC) to the growing electric power market. Westinghouse's decision to back the AC power system concept of Serbo-American engineer, Nikola Tesla put him in direct conflict with Thomas Edison, whose company was investing large sums of money to generate and deliver direct current (DC) electricity in New York City. At that time, the effective range of delivery of DC electricity was only 5 kilometers (at the very best). Westinghouse, himself an excellent engineer, strongly believed in the viability of the AC system and so joined with Nikola Tesla to create the generators, transformers, and motors necessary to deliver AC electricity to a much larger number of customers over greater distances. In 1888, Tesla received U.S. patents for his three-phase (polyphase) system of AC generators, transformers, and motors. That same year, Tesla sold his patents for the AC motor and dynamo (generator) to Westinghouse, who hired Tesla and funded his research related to commercializing the AC system. Their pioneering efforts would make AC the standard for commercial electric power generation and transmission.

After a number of impressive public demonstrations at the end of the nineteenth century, the AC system advocated by Tesla and Westinghouse soundly defeated Edison's DC system. For example, with Westinghouse's support, Tesla's AC equipment was used to illuminate the 1893 World Columbian Exhibition in Chicago. Even more significantly, in 1895, Westinghouse won a coveted contract to use Niagara Falls to generate electricity and to deliver the generated (AC) electricity to the city of Buffalo, New York, a city about 35 kilometers away.

Soon after these impressive demonstrations, 95 percent of public electricity switched to the AC system. The transition was so complete that by 1896, even General Electric (Edison's electric company) was forced to cross-license Westinghouse's patents. George Westinghouse died in New York City on March 12, 1914. One of the most eloquent tributes to Westinghouse was penned by Nikola Tesla, who wrote:

> George Westinghouse was, in my opinion, the only man on this globe who could take my alternating-current system under the circumstances then existing and win the battle against prejudice and money power. He was one of the world's true noblemen, of whom Americans may well be proud and to whom humanity owes an immense debt of gratitude.

Westinghouse was a true hero in an age of heroes. In an exciting era exploding with scientific discovery and creativity, he proved to be one of the world's leading inventor-engineers. His efforts helped establish the world of AC electricity, which is now so much a part of modern living. In 1957, he was inducted into the Hall of Fame for Great Americans and, then, in 1989, into the National Inventors Hall of Fame.

Thomas Alva Edison (1847–1931)

Nicknamed the "Wizard of Menlo Park," Thomas Alva Edison was the greatest inventor of the modern era. His 1,093 U.S. patents and numerous inventions profoundly influenced the lives of nearly everyone in the world. Most notable of Edison's inventions were the phonograph (1877), the durable incandescent electric light (1878), and the motion picture camera and projector (1889). Early in his career, he made numerous improvements in the telegraph and the telephone, greatly enriching the use of both of these information technology systems.

Edison was born in Milan, Ohio, on February 11, 1847. His family moved to Port Huron, Michigan, in 1854. As a young boy, he was inattentive in formal schooling, so his mother (a former schoolteacher) removed him from elementary school and provided an enriched home-schooling experience. Edison spent a great deal of his free time reading technical and scientific books. At age 13, he worked as a newsboy and three years later found employment as a telegrapher. The telegraph was causing an information revolution throughout the United States and Edison soon traveled around the country working in this industry and performing scientific experiments in his spare time.

In 1868, Edison arrived in Boston and decided to change his profession from telegrapher to inventor. During this period, he patented his first invention, an electric vote recorder for use by members of Congress. Although the device was a technical success, it was an economic failure because—as Edison soon learned—members of Congress frequently wanted to delay and stall the voting process, rather than to speed it up. At this point, he vowed never to invent something that people did not want. His future career as an inventor would contain many spectacular successes, as well as some really dismal failures. But Edison always learned from his mistakes and remained persistent in reaching the lofty goals and often extremely tight invention schedules (typically a minor invention each month and one major invention every six months) he placed upon himself and his staff.

In 1869, Edison moved to New York City and had the opportunity to meet people influential in the telegraph industry and in the stock market. His invention of an improved stock ticker was sold to the president of a large Wall Street firm. Money from this device and the sale of Edison's quadruplex telegraph (invented in 1874) allowed him to establish his own "invention factory" in 1876 in Menlo Park, New Jersey. The industrial research and development laboratory was one of Edison's most enduring legacies and became the model of other important, privately owned and operated research and development facilities, like the Bell Laboratories.

At Menlo Park, Edison set out improving the telephone (in 1877), after it was invented by Alexander Graham Bell. Edison also invented the phonograph (1877) and a durable incandescent light bulb (1878) at this facility. On January 27, 1880, Edison filed for a patent for the electric incandescent lamp. In 1880, Edison set up an electric distribution system in the Wall Street area of New York City. Edison pursued the use of direct current (DC) and switched on the world's first electric power distribution system in the lower part of Manhattan on September 4, 1882.

During the early years of electric lighting, Edison's DC generating and distribution system served as the standard for the emerging electric power industry. However, the distribution of DC electricity had inherent inefficiencies and limitations and Edison soon found himself locked in a bitter "current war" with George Westinghouse and Nikola Tesla, who were championing an AC electricity approach to electric power distribution. Despite Edison's enormous attempt to defeat the use of AC electricity, by 1896 the tempestuous battle of AC versus DC electricity ended in favor of Westinghouse and Tesla. With 95 percent of public electricity switching to the AC system, even the General Electric Company (an Edison company) decided to cross-license Westinghouse Electric and Manufacturing Company's AC system patents.

Undaunted by his defeat over DC electric power generation, Edison turned his attention to the use of electricity in the entertainment industry. In 1877, Edison miniaturized one of his inventions (the phonograph) and then integrated this device into a "talking" doll. Though a bit too fragile for children to play with, Edison's factory in Orange, New Jersey, turned out about five hundred of these leading-edge automatons. All a child or adult had to do was to turn the mechanical crank in the doll's back, and the doll recited "Mary had a little lamb."

While the talking doll may not have been an economic success, Edison's next invention made a much more indelible mark on the entertainment industry. In 1889, Edison developed the motion picture camera at his much larger research complex in West Orange, New Jersey. He also created the first film studio, called the Black Maria, in New Jersey and then produced and distributed motion pictures through the Edison Trust—a conglomerate of nine major film studios. Edison's movie studio produced *The Great Train Robbery* (1903), the first motion picture to tell a story. However, Edison's hold on the emerging U.S. motion picture industry was loosened by antimonopoly legal actions in the early 1900s.

A creative legend in his own time, Edison was married twice and had six children (three by each wife). His first wife (Mary Stilwell) died in 1884. In the 1880s, Edison purchased property in Fort Myers, Florida, and used this property as a winter retreat. Edison was a friend of Henry Ford and they were also "snowbird" neighbors in Florida. Beyond 1911, as sickness and old age began to take their toll on his inventive genius, Edison grew more and more dissatisfied with his inventive efforts. He died in Fort Myers, Florida, on October 18, 1931.

Nikola Tesla (1856–1943)

The Croatian-born, Serbo-American electrical engineer, Nikola Tesla, was a technical genius, who helped electrify the world during the sociotechnical transformation often called the Second Industrial Revolution. Tesla's patents and research work formed the basis for modern AC power systems, including the AC motor and the polyphase (out-of-step) power distribution system.

Tesla was born of Serbian parents on July 10, 1856, in Smiljana, Gospić, in the Military Frontier of the Austro-Hungarian Empire (now in Croatia). While studying electrical engineering (about 1875) at the Austria Polytechnic in Graz, Austria, Tesla became fascinated with and began investigating the properties and applications of AC current. If the charges move around an electric circuit in the same direction at all times, engineers call the electric current a DC. Batteries connected in a circuit provide a DC current. In contrast, when the charges move first in one direction in a circuit and then the opposite way, engineers call the current an AC. The generators at modern power plants produce AC current and many modern "wall-plug power" electrical devices (including industrial robots in factories) use AC.

About 1880, without completing his degree at the University of Prague, Tesla moved to Budapest (Hungary) to accept a position within the European office of the American Telephone Company. He quickly rose to the position of chief electrician of the company and later supported the Yugoslav government in the establishment of that country's first telephone system. The following year (1882), Tesla moved to Paris to work as an engineer for the Continental Edison Company. It was about this time that he also conceived the basic ideas for his greatest invention, the electromagnetic motor—an AC electrical device that would transform the world.

The DC electric motor, as independently discovered in about 1831 by both Michael Faraday (1791–1867) and Joseph Henry (1797–1878), converts electric energy to mechanical energy by using DC to make a metallic loop (the armature or rotor) spin around a central shaft. Tesla was convinced that he could modify the DC electric motor to operate without a commutator—that portion of the armature of an electric motor (or generator) through which connections are made to external circuits. The commutator functions as an external switch, which reverses the direction of the (direct) current in the rotor every 180 degrees to keep it spinning in one direction. Tesla's technical instincts would soon lead to the electromagnetic motor, a machine that efficiently used AC power.

While working for the Continental Edison Company, Tesla was an engineer responsible for making improvements to electric equipment. Sometime in 1883, during an assignment in Strasbourg, France, he used his free time to construct the world's first polyphase (out-of-step) AC motor. Using his genius for invention, Tesla arranged the coils in this motor so that when the coils were energized by out-of-phase alternating currents, the resulting magnetic field rotated at a predetermined speed.

Tesla came to the United States in 1884 to accept a position with Thomas Edison's company in New York City. However, the initially cordial relationship between the two geniuses soon soured. Tesla did not respond well to Edison's authoritative management style, especially when it involved the great nineteenth-century controversy within the emerging electric power industry: AC or DC? Edison had committed himself thoroughly to the use of DC. The intellectually gifted, but rebellious, Tesla was the world's most talented advocate for the use of AC. In just a very short time, sparks began to fly as this controversy heated up.

The big problem facing the nascent electric power industry in the late nineteenth century was how best to transport electricity over transmission wires without incurring too great an energy (heating) loss. Tesla recognized that transporting electricity at high voltage using transformers at both the generating station (to raise the voltage) and then at the consumer end (to lower the voltage) solved the problem. But, transformers only work with AC systems. In addition to the growing professional disagreement over the choice of DC versus AC for commercial electric power systems, Tesla was also embittered with Edison because Edison apparently failed to make good on a promised bonus payment for the special work Tesla performed for Edison to improve the efficiency of Edison's DC generators.

So, after just a year, Tesla quit his job with Edison's company and set out to prove AC was best. His departure marked the start of a bitter feud between the two geniuses, a feud that raged for decades. One example of the bitterness of this dispute involved the Nobel Prize Awards Committee, which was considering coawarding the 1912 physics prize to both Edison and Tesla. Because of Tesla's adamant refusal to be associated with Edison, the committee members squashed the nomination. Instead, the prestigious 1912 Nobel Prize in physics went to the selection committee's more distant secondary choice, the Swedish engineer, Nils Gustaf Dalén, for inventing automatic gas regulators.

Responding to his disappointing relationship with Edison, Tesla formed his own company in 1886 and called it Tesla Electric Light and Manufacturing. His goal was to develop and market an AC motor. Unfortunately, this goal caused a major disagreement between Tesla and his financial backers, who soon relieved him of his duties at the fledging company which bore his name. Undeterred, Tesla worked as a common laborer for the next year to feed himself and to save enough money for the construction of the electromagnetic induction motor (for which he eventually received U.S. patent # 381,968). Tesla's revolutionary motor used a rotating magnetic field, rather than mechanical switches (that is, commutators) to spin the armature or rotor. Tesla's device opened the way for the modern three-phase AC power system (generator, transformers, and motors) and for the common electrical devices found in most factories, offices, and homes.

In 1886, the American engineer and industrialist, George Westinghouse, founded Westinghouse Electric in Pittsburgh and entered head-to-head commercial competition with Thomas Edison, who was convinced the DC electric power system he had installed in New York City could not be outdone. However, at that time, the effective range of delivery of DC electricity was only 5 kilometers (at the very best). Westinghouse believed in the viability of the AC and joined with Nikola Tesla to create the generators, transformers, and motors necessary to deliver AC electricity to a much larger number of customers over greater distances. In 1888, Tesla received U.S. patents for his three-phase (polyphase) system of AC generators, transformers, and motors. That same year, Tesla sold his patents for the AC motor and dynamo (generator) to Westinghouse, who hired Tesla and funded his research related to commercializing AC. Their pioneering efforts would make AC the standard for commercial electric power generation and transmission.

After a number of impressive public demonstrations at the end of the nineteenth century, the AC system advocated by Tesla and Westinghouse soundly defeated Edison's DC system. For example, with Westinghouse's support, Tesla's AC equipment was used to illuminate the 1893 World Columbian Exhibition in Chicago. Even more significantly, in 1895, Westinghouse won a coveted contract to use Niagara Falls to generate electricity and to deliver the generated (AC) electricity to the city of Buffalo, New York, a city about 35 kilometers away. In a certain sense, these demonstrations and the victory of the Tesla-Westinghouse AC electric power system over Edison's DC electric power system serve as the highwatermark in Tesla's professional life. He gained worldwide notoriety, some short–term wealth, and expanded his social circle to include notables, like the American author Mark Twain.

In 1899, Tesla moved to Colorado Springs, where he constructed a large laboratory so he could investigate lightning and conduct high frequency, high-voltage experiments exploring the possibilities of wireless telegraphy, telephony, and even the wireless transmission of electric power. His laboratory notes suggest Tesla thought he had recorded extraterrestrial radio signals, possibly from Mars. Tesla left Colorado Springs in early January 1900 and began planning his next project, called the Wardenclyffe Tower facility in Shoreham, Long Island, New York.

In the early summer of 1902, Tesla moved his laboratory operations from Houston Street in New York City to the Wardenclyffee facility. The site's huge 57-meter tall antenna was to serve as the first station in Tesla's envisioned worldwide wireless telecommunications system. Although the structure for the radio tower was completed in 1904, Tesla never completed his planned transceiver, because his financial supporters pulled out. Faced with rising debts, Tesla abandoned the project. The tower was demolished and sold for scrap in 1917. The entire experience left Tesla in permanent financial distress and a state of deep depression from which he would never recover.

Soon after the German physicist, Heinrich Rudolf Hertz (1857–1894) produced and detected radio waves for the first time in 1888, Tesla, Guglielmo Marconi (1874–1937), and other late nineteenth-century researchers began exploring the possibility of wireless communications. Tesla's radio wave research put him on a direct collision course with Marconi. In fact, Tesla always disputed the claim that Marconi invented radio. Up to World War I, Tesla engaged in an expensive (but unsuccessful) legal battle against Marconi. By 1916, Tesla was forced to file for bankruptcy and spent the rest of his life in poverty. Tesla died in New York City on January 7, 1943. Ironically, in the year of Tesla's death (1943), the U.S. Supreme Court ruled that Tesla's patents for the radio superseded those of Marconi.

Tesla was clearly one of the greatest engineers of all time. His inventions helped to electrify the modern world, yet his genius, with its obsessive-compulsive dark side, brought him neither wealth nor contentment. In the mid-1950s, the international scientific community named the SI unit of magnetic flux density, the tesla (T), in his honor. One tesla is equal to one weber per square meter.

Heinrich Rudolf Hertz (1857–1894)

In 1888, the German physicist, Heinrich Rudolf Hertz produced and detected radio waves for the first time. He also demonstrated that this form of electromagnetic radiation, like light, propagates at the speed of light. His discoveries form the basis of the global telecommunications industry (including communications satellites), radio astronomy, telecommunications with distant space robots, and radio control of a variety of mobile robots. The hertz (Hz) is the SI unit of frequency named in his honor. One hertz is equal to one cycle per second.

Hertz was born on February 22, 1857, in Hamburg, Germany, into a prosperous and cultured family. Following a year of military service from 1876 to1877, he entered the University of Munich to study engineering. However, after just one year he found engineering not to his liking and began to pursue a life of scientific investigation as a physicist in academia. Consequently, in 1878, he transferred to the University of Berlin and started studying physics with the famous German scientist Herman von Helmholtz (1821–1894) as his mentor. Hertz graduated magna cum laude with his Ph.D. in physics in 1880. Following graduation, he continued working at the University of Berlin as an assistant to Helmholtz for the next three years.

He left Berlin in 1883 to work as a physicist at the University of Kiel. There, following suggestions from his mentor, Hertz began investigating the validity of the electromagnetic theory recently proposed by Scottish physicist, James Clerk Maxwell (1831–1879). As a professor of physics at the Karlsruhe Polytechnic from 1885 to 1889, Hertz finally gained access to the equipment he needed to perform the famous experiments that demonstrated the existence of electromagnetic waves and verified Maxwell's equations. During this period, Hertz not only produced electromagnetic (radio frequency) waves in the laboratory, but also measured their wavelength and velocity. Of great importance to modern physics and the fields of robotics, telecommunications, and radio astronomy, Hertz showed that his newly identified radio waves propagated at the speed of light, as predicted by Maxwell's theory of electromagnetism. He also discovered that radio waves were simply another form of electromagnetic radiation, similar to visible light and infrared radiation, save for their longer wavelengths and shorter frequencies. Hertz's experiments verified Maxwell's electromagnetic theory and set the stage for others like Guglielmo Marconi (1874–1937) to use the newly discovered "radio waves" to transform the world of communications in the twentieth century.

In 1887 while experimenting with ultraviolet radiation, Hertz observed that incident ultraviolet radiation was releasing electrons from the surface of a metal. Unfortunately, he did not recognize the significance of this phenomenon nor did he pursue further investigation of the photoelectric effect. In 1905, Albert Einstein (1879–1955) wrote a famous paper describing this effect, linking it to Max Karl Planck's (1858–1947) idea of photons as quantum packets of electromagnetic energy. Einstein earned the 1921 Nobel Prize in physics for his work on the photoelectric effect.

Hertz performed his most famous experiment in 1888 with an electric circuit in which he oscillated the flow of current between two metal balls separated by an air gap. He observed that each time the electric potential reached a peak in one direction or the other, a spark would jump across the gap. Hertz applied Maxwell's electromagnetic theory to the situation and determined that the oscillating spark should generate a very long electromagnetic wave that traveled at the speed of light. He also used a simple loop of wire, with a small air gap at one end, to detect the presence of electromagnetic waves produced by his oscillating spark circuit. With this pioneering experiment, Hertz produced and detected *Hertzian waves*—later called *radiotelegraphy waves* by Marconi and then simply radio waves. By establishing that Hertzian waves were electromagnetic in nature, the young German physicist extended human knowledge about the electromagnetic spectrum, validated Maxwell's electromagnetic theory, and identified the fundamental principles for wireless communications.

In 1889, Hertz accepted a professorship at the University of Bonn. There, he used cathode ray tubes to investigate the physics of electric discharges in rarified gases, again just missing another important discovery—the discovery of X-rays, which was accomplished by the German physicist Wilhelm Conrad Roentgen (1845–1923) at Würzburg in 1895.

Hertz was an excellent physicist whose pioneering research with electromagnetic waves gave physics a solid foundation upon which others could build. His major publications included *Electric Waves* (1890) and *Principles of Mechanics* (1894). He suffered from lingering ill health due to blood poisoning and died as a young man (in his late thirties) on January 1, 1894, in Bonn, Germany. The international scientific community named the basic unit of frequency the hertz (symbol Hz) in his honor.

Herman Hollerith (1860–1929)

The American engineer and inventor, Herman Hollerith patented a punch card tabulating machine in 1889 that transformed the handling of large quantities of statistical data and became the basis for the modern data tabulating and processing industry. Used with great success during the 1890 United States Census, Hollerith's automated system for storing data on punched cards also became the initial choice for storing the data and programs in the early high-speed digital computers that emerged in the mid-1940s.

Hollerith was born on February 29, 1860, in Buffalo, New York. His parents had immigrated to the United States from Germany in 1848. Although a bright child, he was not comfortable with formal schooling and so was eventually taken out of elementary school and tutored privately at home. In 1875, he entered the City College of New York on scholarship and went on to complete his undergraduate education at Columbia University in 1879, receiving an engineering degree from the university's School of Mines.

One of Hollerith's professors received an appointment as Chief Special Agent to the U.S. Census Bureau and invited the young engineer to become his assistant as a statistician. This fortuitous opportunity exposed Hollerith to the problem of data collection that was encountered during the 1880 U.S. Census. As a result of this job, Hollerith started thinking about automated ways of performing the tedious task more efficiently.

In 1882, Hollerith joined the faculty of the Massachusetts Institute of Technology (MIT) as an instructor in mechanical engineering. Hollerith did not enjoy teaching very much, but his time at MIT allowed him to further explore the concept of automated data collection. For one thing, he was exposed to the automated Jacquard loom, constructed in the early nineteenth century by the French textile manufacturer, Joseph Marie Jacquard. The automated loom was programmed through the use of information on a series of punched cards. In this period, Hollerith also received creative insight when he observed how railroad conductors would form an information profile about each passenger by punching each individual ticket in a certain way. Hollerith decided that he could build an automated electromechanical system that used punched cards to record and tabulate statistical data.

Hollerith began experimenting with different ways to tabulate and process data at MIT in the early 1880s. First, he tried paper tapes, with pins that would go through the punched holes in the tape to complete an electrical contact. He was definitely on the right track, but the paper tape proved troublesome, since the tape would have to continually stop in its motion to allow the pins to pass through and the data to be read. So, he settled on stiff paper cards unto which the data could be punched. Another critical idea he came up with was the fact the punched cards could have a variety of data stored on them as numeric entries in an orderly fashion in specific columns. He would then instruct his automated tabulating machine to examine each card for these data. In the case of census data, for example, certain columns of the punched card could relate to the geographic region or census office, while other columns could efficiently contain information about the person's profession, marital status, number of children, and so on.

Hollerith left MIT in 1884 and accepted a position at the U.S. Patent Office. This new position allowed him to see how the patent process worked. Later that year, he applied for a patent (granted on January 8, 1889) for his method of compiling statistics by combinations of holes punched in cards that were then read by an electromechanical tabulating device. This patent was the first of 30 patents on data-processing devices that Hollerith obtained over the next two decades or so. His efforts created the modern information processing industry and set the stage for the use of similar punch card systems to enter data and instructions into the early digital computer systems of the mid-1940s.

The admiral's test for Hollerith came in 1890, when he constructed tabulating machines under contract for the U.S. Census Bureau as part of the official census. Hollerith's machines tabulated the 1890 Census data far more efficiently and in much less time than had occurred during the 1880 Census, when data were essentially processed by hand. Rapid and accurate processing of the 1890 Census data was especially important to the government in 1890, because the United States had developed into one of the world's leading industrial powers, with a population of over 62 million people.

In 1890, Hollerith renewed his academic ties with Columbia University and submitted a dissertation entitled "The Electric Tabulating System." His doctoral committee approved his innovative work on automated data processing during the 1890 U.S. Census and he received his Ph.D. that year.

Once he successfully demonstrated the value of automated data tabulation during the 1890 Census, Hollerith proceeded to harvest wealth from his ideas through a commercial enterprise. In 1896, he founded the Tabulating Machine Company and continued to make improvements in the electromechanical mechanisms by which cards were fed, punched, and read in an automated manner. His machines became famous around the world. Canada, Norway, and Austria used Hollerith's machines during census activities in 1891, and the United Kingdom in that country's 1911 Census.

The U.S. Census Bureau also used Hollerith's machines during the 1900 Census. But, this time Hollerith's company had charged too much for the use of its equipment, so the government decided to create its own equipment, working around Hollerith's patents and essentially going into competition with him. Other commercial competitors also appeared and pressured his company's once dominant market position.

Faced with failing health and vigorous economic competition, in 1911 Hollerith sold the Tabulating Machine Company. Hollerith's former company then merged with the International Time Recording Company and the Computing Scale Company of America. Under the presidency of Thomas J. Watson this new company turned its declining fortunes around. In 1924, the company was renamed, International Business Machines (IBM).

Anchored in an economically comfortable life by his stock and proceeds from the sale of the Tabulating Machine Company, Hollerith continued to work as a consultant to the data-tabulating industry. On November 17, 1929, he died of a heart attack in Washington, DC.

Karel Čapek (1890–1938)

The Czech writer Karel Čapek gave the world the term robot when he wrote the play *Rossum's Universal Robots* (*R.U.R.*) in 1920. Robata is the Czech word for forced labor or servitude. Some historians suggest that it was actually his brother, Josef Čapek (a painter and writer) who suggested use of the word robot to mean a serving machine in the play. However, since the two brothers often collaborated on literary projects and the word first appeared in Karel Čapek's play *R.U.R.*, he generally gets the credit for coining the word robot. The play premiered in Prague in 1921 and was then translated into English and first appeared on the English stage in 1923.

The play is a satire on the mechanization of civilization. Although some of the impact of Karel Čapek's play was lost in translation from Czech to English, the concept of the robot as derived from *robata*, a Czech word meaning compulsory labor or servitude was not. In fact, following the appearance of Čapek's play, the word robot began replacing such older words as android or automaton.

Karel Čapek was one of the most important Czech writers of the twentieth century. He wrote on a variety of subjects. He focused a portion of his literary energies on nonspace travel-related science fiction—a literary genre that looked at the impact of technology on human society and civilization. In the first few decades of the twentieth century, the works of George Orwell and Aldous Huxley also popularized this type of futuristic fiction. Some of Čapek's other works include *The Absolute at Large (1922)*, *Krakatit* (1924), and *War With the Newts* (1936).

A fiercely loyal and patriotic Czech, Karel Čapek was a close associate of the first president of the initial Republic of Czechoslovakia (1918–1938), Thomas G. Masaryk. In the 1930s, Karel Čapek shifted the focus of his writings to oppose the rising dictatorships in Europe, especially the great threat posed by Nazi Germany. His anti-Nazi works included *The White Disease* and *The Mother*. When the United Kingdom, France, and other nations did not oppose the Nazi Germany invasion of Czechoslovakia in 1938, Karel Čapek refused to leave his occupied native land and went on a hunger strike instead. His political opposition to the Nazi regime earned him the "number two public enemy" ranking from the German secret police (the Gestapo). Čapek soon contracted double pneumonia and died on December 25, 1938, in Prague. His brother, Josef, died several years later in the Bergen-Belsen concentration camp.

Because of its significance in the field of robotics, a very brief synopsis of Čapek's play *R.U.R.* is provided here. In essence, Čapek's famous play is a contemporary version of the Golem legend, which appeared in the Middle Ages and remained quite popular in central Europe. The opening dialog tells the tale of an eccentric old scientist named Rossum who uses biological and electrical technology to create an artificial substitute for flesh and bones. (Today, his fictional effort would be called some form of genetic engineering.) Rossum then uses this new material in the pursuit of his lifelong vision of creating artificial life. However, Rossum's experiments fail and the old man goes insane. His son, a practical engineer and industrialist decides to salvage something from the elder Rossum's efforts and comes up with mass-produced, human-like workers, that is, robots, who remember everything, but think of nothing new. (Neither the Old or Young Rossum actually appear in Čapek's play.)

The story begins when an idealistic young female, named Helena Glory, arrives at the remote island factory, which manufactures Rossum's Universal Robots. She is president of the Humanity League and is on a mission to liberate the robots from their misery. Upon arrival at the island facility, she meets the general manager of R.U.R.—a character named Harry Domin (sometimes translated as Domain). Domin tells her all the about the Rossums (old and young) and how they tinkered with artificial life. The young Rossum's end product was an inexpensive, human-like robot capable of doing work. But the young Rossum's efforts have exceeded beyond his wildest expectations. His robots have been sold all over the world. Soon, nations begin using Rossum's robots to form ruthless and efficient armies that kill all living things in their paths.

The humans at the island factory faced several key questions. First, could the next generation of robots be modified with some type of conscience or soul? Or, second, should the factory be shut down and robot production stopped? That way these dangerous, soulless robot armies would soon die out.

Then, somewhat out of the blue in the play, Domin asks Helena to marry him. She agrees, possibly so she can continue her work to help free the robots from their life of work. Along these lines, she asks a scientist on the island to modify some of the robots so that their consciousness might emerge more fully. This proves to be a fatal mistake. Human fertility drops around the world as the race toward mechanization makes people superfluous.

Things come to a climax when one of the modified robots rises up and issues a manifesto for the robots around the world to rise up and kill all humans. The last few humans are holed up in the original robot factory. Domin holds an important bargaining chip, namely Rossum's original formula for producing robots. But his wife Helena decides to burn the formula, thinking if she stops the production of robots the ongoing political chaos, slaughter of humans, and social collapse will cease. The robots swarm into the factory and kill all remaining humans, save one—a character named Alquist, the only human who actually performed manual work in the play.

Without Rossum's old formula, the robot-producing machines end up turning out chunks of bloody meat. As the play nears the end, the soulless generation of rampaging robots begins to die out, having exterminated the human race. At this rather dismal point, Čapek's play takes a positive turn. Two specially modified domestic robots (named Primus and Helena) appear on stage. These robots recognize the chaos around them, are concerned for each other, and fall deeply in love. With a nice touch from the *Book of Genesis*, the last human (Alquist) renames these two robots, Adam and Eve, and sends them out into the chaotic world, with instructions to avoid the mistakes that destroyed their predecessors.

Norbert Wiener (1894–1964)

In 1948, the American mathematician Norbert Wiener formally introduced the science of cybernetics, when he published the book *Cybernetics or Control and Communication in the Animal and Machine*. As a scientific discipline cybernetics investigates communication and control processes in living systems, as well as in machines built by human beings. Wiener coined the word *cybernetics* from the ancient Greek word (kubernētēs meaning steersman, pilot, or governor). Within a mathematical framework, cybernetics deals with how regulatory feedback signals are communicated and controlled in electronic, mechanical, and biological systems.

Feedback is the process by which the output of a system is used to control its performance. In negative feedback, a return signal associated with the output of the system is used to reduce the input. Similarly, in positive feedback, a return signal associated with the output is used to increase the input. A simple example of a negative feedback device is an electromechanical governor (intended to keep a vehicle moving at some optimum speed), which reduces the fuel supply to an internal combustion engine as the vehicle's speed increases. For an industrial robot, feedback often is used to locate and precisely control certain moving parts. The design of the robot usually includes a specific subsystem that indicates the current position of the moving part (or parts). The signal representing the current position is compared to the desired or target position (for a particular operation or activity) and adjustments are made until the difference between the target position and the actual position of the moving part are zero—or at least within some established level of tolerance. This important area of control system theory was greatly organized and intellectually stimulated by Wiener's work at MIT in the 1940s.

Wiener was born in Columbia, Missouri, on November 26, 1894. From the start, his father, a professor of Slavic languages at Harvard, forced young Norbert to excel in his studies, especially in mathematics. Primarily home schooled, Wiener epitomized the high-pressure life of a child prodigy. In the fall of 1906, when Wiener was only 11 years old, his father enrolled him in Tufts College. In 1909 (just 14 years old), he received a bachelor's degree in mathematics. After somewhat disappointing attempts at graduate school study in zoology (at Harvard) and philosophy (at Cornell), he returned to Harvard and received his doctoral degree in mathematics in 1912. The newly minted Ph.D. was just 18 years old.

In the years following his graduation from Harvard, Wiener traveled extensively in Europe, where he interacted with some of the intellectual giants of the day, including the Welsh philosopher and mathematician Bertrand Arthur Russell (1872–1970). At the onset of World War I, Wiener returned to the United States and did work on ballistics for the U.S. Army at the Aberdeen Proving Ground in Maryland. Following this job, he received a position as an instructor of mathematics at the Massachusetts Institute of Technology (MIT). He remained an instructor in mathematics at MIT from 1919 to 1960. By student accounts, Wiener was noted for his absentmindedness, humor, and very poor lecture style.

In 1926, he met and married Margaret Engemann, a German immigrant, who bore him two daughters. During World War II, Wiener worked on gunnery control projects and this work encouraged him to formulate the theory of cybernetics. Although he personally never used computers extensively, his pioneering mathematical work would eventually touch several emerging computer science fields, including artificial intelligence, advanced automation, control of machines by computers, and computer-supported robotics.

After World War II, he began to speak out against the militarization of science. To emphasize his position, Wiener refused to work on military projects and would no longer accept government funding.

The synthesis of Wiener's work involving communication theory and control theory appeared in 1948 in the form of his classic book *Cybernetics or Control and Communications in the Animal and the Machine*. During the post-World War II period he also expanded his book writing efforts.

Much of his later works included heavy doses of philosophy, which he freely shared with the world as a so-called "child prodigy."

For example, in his 1950 work entitled *The Human Use of Human Beings*, Wiener warned against the possible misuse of computers to control people. Three other books were *Ex-Prodigy* (1953), *I am a Mathematician* (1956), and *God and Golem, Inc.: A Comment on Certain Points Where Cybernetics Impinges on Religion* (1964). In the last title, Wiener provides his view on the place of machines in society and he also presents a variety of other machine-related issues, focusing on the potential consequences to society and the proper role of technology. In 1964, the American president, Lyndon Baines Johnson (1908–1973), presented Wiener the National Medal of Technology. Later that year, the mathematician and philosopher died in Stockholm, Sweden, on March 18.

John von Neumann (1903–1957)

The Hungarian-American mathematician, John von Neumann, played a critical role in the development of the American atomic and hydrogen bombs, founded game theory, made important contributions to the development of high-speed digital computers, and explored the fascinating concept of self-replicating machines, which he called universal constructors.

Von Neumann was born on December 28, 1903, in Budapest, Hungary. His family was prosperous as a result of banking interests, and provided him an intellectually cultured environment in which to develop. As a youth, he soon demonstrated strong talents in mathematics. In the early 1920s, von Neumann studied at various universities in Germany and Switzerland. When he was just 23, he earned a Ph.D. in mathematics from the University of Budapest. Following graduation (in 1926) and up until 1930, von Neumann worked as a private lecturer in mathematics in Germany.

In 1930, Princeton University invited him to teach mathematics and physics. Von Neumann took advantage of this opportunity to immigrate to the United States, where he became a naturalized American citizen (in 1937) and spent the remainder of his life. In 1933, the newly founded Institute for Advanced Study (IAS) at Princeton offered von Neumann a prestigious position as professor of mathematics. (As a frame of reference to the significance of this appointment, Albert Einstein was one of the other initial fellows of the IAS.) Von Neumann retained this IAS appointment for the remainder of his life. Married twice, von Neumann had a zest for thinking, mathematics, and elaborate parties and social interactions.

Von Neumann founded game theory. In 1928, he began exploring the minimax theorem as found in certain zero sum games in game theory. He developed this interesting area of applied mathematics and made many contributions by applying game theory to economic problems, strategic planning, and decision making in the area of national defense. His efforts in game theory culminated with the publication of the classic book *The Theory of Games and Economic Behavior*, which he coauthored in 1944 with Oskar Morgenstern.

In 1932, von Neumann published *Mathematical Foundations of Quantum Mechanics*—an important work, in which he applied operator theory to quantum mechanics. Von Neumann used elegant mathematics to demonstrate that the matrix mechanics of Werner Heisenberg and the wave mechanics of Erwin Schrödinger were equivalent forms of quantum mechanics.

Profoundly antifascist (and later anticommunist), von Neumann made many important contributions to the American war effort during World War II. During the Manhattan Project, he played a major role in the development of the plutonium implosion weapon and participated in the atomic bomb target selection committee. After World War II, he became one of the major scientific advocates for the development of an American hydrogen bomb. He chaired a special committee, called the Von Neumann Committee for Missiles, for President Dwight Eisenhower and vigorously advised the president to pursue development of intercontinental ballistic missiles and submarine-launched ballistic missiles, armed with nuclear warheads. A frequent consultant to a variety of top level U.S. government agencies, including the Department of Defense and the U.S. Atomic Energy Commission (USAEC), von Neumann helped develop the strategic doctrine of mutually assured destruction (MAD)—the defense strategy based on strategic nuclear weapons equilibrium that governed much of the cold war era.

In the 1940s, von Neumann was one of the first scientists to recognize the great value of the newly emerging electromechanical computers. He championed their ability to perform complex calculations in applied mathematics and the simulation of complicated physical phenomena, like the hydrodynamics and energy flow patterns encountered in nuclear detonations. While working on the ENIAC (Electronic Numerical Integrator and Calculator) project for the U.S. government in 1945, von Neumann wrote a summary report (entitled "First Draft of A Report on EDVAC") in which he proposed the concept of an internally stored program, where the step-by-step directions for computations (called instructions) are stored within the computer and so computations can progress without the need for external (human) guidance.

The following year (1946), Von Neumann wrote a more comprehensive report "Preliminary Discussion of the Logical Design of an Electronic Computing Instrument." In this more detailed report, he introduced the basic architecture and design principles of the modern computer. Von Neumann organized the electronic (digital) computer system into four main components: the central arithmetical (CA) unit, the control unit (CU), the memory (M), and various input/output (I/O) devices. The CU was responsible for controlling the proper sequencing of operations and making the individual components of the computer system operate smoothly together to conduct the specific task programmed into the computer. The memory contained both stored numerical data (including physical constants, initial conditions, boundary conditions, etc.) and also a set of numerically coded instructions.

His approach became widely known as the von Neumann architecture and has served as a model for all high-speed computing machines that use a single storage structure to hold the data required or generated in a computation as well as the set of instructions on how to perform the computations. Today, such machines are more commonly referred to as stored-program computers. Von Neumann's contributions to computer science and engineering endure, and his basic architecture is readily apparent in many of today's most modern machines.

In 1952, von Neumann oversaw construction of the MANIAC (Mathematical Analyzer, Numerical Integrator, and Computer) at the Institute for Advanced Study in Princeton. This pioneering digital computer embodied von Neumann's landmark idea of storing instructions (as distinct from data) in the computer's memory. With stored instructions the electronic computer could perform its computations without the need for external (human) guidance. Science historians suggest that von Neumann used this and other emerging high-speed computers (at the Los Alamos National Laboratory) to perform the hydrodynamic simulations needed to perfect the American hydrogen bomb.

Toward the end of his life, von Neumann became interested in the concept of self-replicating automata. A self-replicating system (SRS) is an advanced robotic device and a single SRS unit is a machine system that contains all the elements required to maintain itself, to manufacture desired products, and even (as the name implies) to reproduce itself. Von Neumann was the first person to seriously consider the problem of self-replicating machine systems. During and following World War II, he became interested in the study of automatic replication as part of his wide-ranging interests in complicated machines. From von Neumann's initial work and the more recent work of other investigators, five general classes of SRS behavior have been defined: production, replication, growth, repair, and evolution. Von Neumann's thoughts on this interesting topic were posthumously published in the book *Theory of Self-Replicating Automata*, which was edited by Arthur W. Burks and made available in 1966 by the University of Illinois Press. In advanced robotics, SRSs are sometimes referred to as von Neumann machines. This designation is sometimes improperly confused with high-speed computing machines that use von Neumann architecture.

Recognized as a mathematical genius during his lifetime, von Neumann received many awards and honors from the U.S. government. These included the Distinguished Civilian Service Award (1947), the Presidential Medal of Freedom (1956), and the prestigious Enrico Fermi Award (1956) from the USAEC. Von Neumann died of cancer in Washington, DC, on February 8, 1957.

George C. Devol, Jr. (b. 1912–)

Called the "grandfather of industrial robotics," George C. Devol, Jr. is the self-taught engineer and inventor who designed the first programmable industrial robot in 1954 and then received a patent for this pioneering device in 1961.

Devol was born in Louisville, Kentucky, on February 20, 1912. He was a self-made engineer and technically skilled businessperson. In 1932, he started the United Cinephone Corporation—a company that manufactured amplifiers and phonograph arms. Following this activity, he worked at the Sperry Gyroscope Company for several years.

Devol was a technical participant in the 1939 New York World's Fair. He developed and installed automated guest counters to tally the number of visitors arriving at the fair each day. Even though the storm clouds of World War II were looming in the background over Europe, the 1939 World's Fair proved to be an extraordinary technical showcase of tomorrow. For example, the Radio Corporation of America (RCA) (now part of the General Electric Company) used its participation in the fair to launch an interesting new technology—television.

During World War II, Devol founded and operated General Electronics Industries—a new company that had diversified interests in industrial electronics and radar test equipment. One area of particular importance to the American war effort was his company's work in electronic countermeasures and radar jamming devices.

Following the war, Devol founded yet another company, named Devol Research Associates. In 1946, he received a patent for a general-purpose playback device, which used a magnetic process recorder for controlling machines. Just about the same time that the digital computer arrived on the scene, Devol's company was developing various control systems, ultrahigh-speed printers, sensors, and a large-scale random access memory system. The stage was set for the emergence of the industrial robot.

In the late 1940s, Devol keenly recognized that about half of all factory workers spend their time in putting or taking tasks. That is, the worker picks up some part of subassembly and brings it over to an area where other workers are bringing together different parts or subassemblies in an effort to manufacture a final product. Why not invent some kind of smart machine with an adaptive mechanical arm that can do the picking and placing without getting bored, tired, or making mistakes? That is exactly what Devol did.

In 1954, Devol designed and patented the first programmable industrial robot, coining the term *universal automation* in the process. The U.S. government granted him a patent entitled "Program Controlled Article Transfer," in 1961. Devol's wife suggested that he shorten the term he had minted to simply unimation. He did, and the rest is industrial robot history.

As a result of a chance meeting with the entrepreneur and engineer, Joseph F. Engelberger, Devol found himself with a business partner and enthusiastic proponent of the programmable method of moving objects between different parts of a factory. Later that year, the two men formed the world's first industrial robot company, Unimation, Inc. The union of Devol's broad technical skills with Engelberger's enthusiastic management style allowed them to successfully combine industrial manipulator technology and the emerging computer-based control technologies of the early 1960s.

The industrial robot then followed a somewhat convoluted path before gaining worldwide acceptance and becoming the mainstay of many manufacturing operations, including and especially the automobile industry. In 1960, Unimation, Inc. was purchased by Consolidated Diesel Electric Corporation (Condec Corporation). At this point, Engelberger began to aggressively translate Devol's technical ideas into marketable hardware. For his part, Devol chose not to become an employee of the new company. Rather, he preferred to license his patents under a long-term contract.

As the demand for cars grew, automobile manufacturers looked for new ways to increase the efficiency of the assembly line through *telecherics*. This new field focused on robots that mimicked the operator's movements from a distance. In 1961, General Motors installed an applied telecherics system on the assembly line at a plant in Trenton, New Jersey. The one-armed Unimate robot unloaded die casts, cooled components, and delivered them to a trim press. This was precisely the type of dull, dirty, and dangerous pick-and-place factory task for which Devol had invented his programmable method of transferring articles in a factory in 1954.

Alan Mathison Turing (1912–1954)

The British mathematician and logician, Alan Mathison Turing conceived the idea of the modern digital computer in the mid-1930s, when he wrote a seminal paper on computability, using the hypothetical construct of a theoretical computing device, now called the Turing machine. During World War II, he used probability theory and his unique insights into the emerging field of electronic computing to assist the British government's secret code-breaking facility at Bletchley Park. By breaking the Enigma naval ciphers of Nazi Germany, Turing's efforts as a cryptographer saved countless lives and significantly shortened the war in Europe. In 1950, Turing pioneered the area of artificial intelligence when he proposed a simple test (now called the Turing test) to determine whether a computing machine is conscious and can think like a human being.

Turing was born on June 23, 1912, in London, England. His father was a member of the British civil service, posted in India. When Turing was about a year old, his mother rejoined her husband at his post in India. Turing's parents left Alan and his brother with relatives, so their two children could grow up in Great Britain. Although just an average student in Sherborne boarding school (which emphasized a classical education), Turing soon showed early signs of his amazing abilities in mathematics. He attended King's College, Cambridge, from 1931 to 1934 and was elected a Fellow at the college in 1935.

In 1936, Turing wrote what is arguably his most important technical paper entitled "On Computable Numbers, with an Application to the Entscheidungsproblem." Turing explored basic concepts in computability in this landmark paper and anticipated the development of the digital computer about a decade later. He was exploring the Austrian logician Kurt Gödel's (1906–1978) theorem of 1931, which showed that in any mathematical system there is some assertion which can neither be proved nor disproved. Turing did not use a traditional mathematical approach in his efforts. Rather, he approached the problem by postulating the existence of an ideal computing machine (now called the Turing machine), which had infinite memory capacity and performed its functions in discrete steps. At any given moment, Turing's theoretical computing device assumed one of a finite list of internal states. The Turing machine can scan an infinite tape, which is divided into squares—each square is either blank or has one of the finite number of symbols printed on it. Turing's theoretical computing machine can change the condition of a scanned square (by printing a symbol, erasing an existing symbol, or both), move the position on the scanned square one step to the left or right, or change to another internal state. Computer scientists often use the Turing machine as a convenient reference when they discuss the theory of computability. Turing's pioneering work anticipated the development of digital computers in the mid-1940s and provided a theoretical basis for their operation.

In 1936, Turing went to Princeton University as a visiting graduate student to continue his work on the theory of computability. There, in 1938, he completed his Ph.D. under the American mathematician Alonzo Church (1903–1995). Turing's dissertation involved the idea of hypercomputation in which universal Turing machines equipped with so-called oracles (black box mechanisms capable of carrying out uncomputable tasks) were able to study problems that could not be solved with algorithms. Together, Turing and Church, each in their own separate ways, established the basis of computability theory. A universal Turing machine, or simply a universal machine, is a Turing machine capable of simulating any other Turing machine. Turing's new kind of computing machine, as discussed in his dissertation, is sometimes called an O-machine because it is the result of augmenting a universal Turing machine with a black box (or oracle) that is a device capable of performing uncomputable tasks. However, Turing did not give any indication in his dissertation as to how this so-called oracle device might actually work. In the exotic mathematical realm of hypercomputation, tasks like distinguishing between arithmetic theorems and non-theorems are no longer uncomputable. The key to hypercomputation lies in the ability to successfully implement Turing's suggested oracle mechanism. If such an oracle device is ever developed, the impact on the field of modern computer science would be enormous.

Just before the start of World War II in Europe, Turing returned to Great Britain, where he became a major participant in the British government's secret code-breaking facility at Bletchley Park. At the Government Code and Cypher School, hidden away in a Victorian mansion called Bletchley Park in Buckinghamshire, Turing played a critical role in applying probability theory and in designing primitive, computer-like machines that were able to decipher the Nazi Enigma codes used by the German Navy to communicate with U-boats (submarines) in the North Atlantic. Turing's work at this secret facility, which greatly assisted in the overall British war effort, was not officially acknowledged by the government for over 30 years following his death.

In 1946, Turing went to work at the British National Physical Laboratory, where he participated in the development of a digital computer, called the Automatic Computing Machine (ACE). The design of this early British computing machine included instructions stored in the machine's memory. Turning contributed another important concept in this project, namely that the machine's programs could modify themselves by treating stored instructions like other data stored in memory. However, unable to work in a stifling postwar bureaucratic environment, Turing left this project in 1948. The ACE device was eventually completed in 1950.

In 1949, Turing became the deputy director of the University of Manchester's computing laboratory. His duties included developing software for the Manchester Mark I machine. During this period, he did pioneering work in the field of artificial intelligence. His 1950 paper, "Computing machinery and intelligence," raised the interesting question of machine intelligence and consciousness. He introduced a test, now called the Turing test, in an attempt to create a standard for determining when a machine is conscious or like the human mind in its thinking behavior. Today, computer scientists everywhere recognize the Turing test is a simple, yet clever, procedure, which examines whether a computing machine is capable of thinking like a human being.

In its simplest form, the Turing test involves a human being (called the interrogator) sitting at a teletype machine, connected to but isolated from two other correspondents. One of these correspondents is a human being, while the other is a "thinking" computer. By asking questions and examining the responses, the interrogator tries to determine, which one of the correspondents is the computer and which one is the human being. The advanced computer is programmed to give delayed answers and even deceptive answers, mimicking how the human mind would respond to questions. If it is impossible for the human interrogator to determine which correspondent is a machine and which is a human being, then the computer has passed the Turing test and is considered capable of humanlike thought. Many computer scientists regard Turing's 1950 paper as the start of the field of artificial intelligence. In 1951, Turing's contributions to mathematics and computer science were honored by his election as a Fellow to the British Royal Society,

From 1952 until his untimely death in 1954, Turing turned his attention to the area of mathematical biology, specifically morphogenesis. He was fascinated by the existence of Fibonacci numbers in plant structures and published the paper "The Chemical Basis of Morphogenesis" in 1952.

At the height of his intellectual powers and contributions to science, Turing's life came to a tragic end, as a result of a rather ugly legal episode, which was aggravated by the fact that he was admittedly gay in a society that (at the time) heavily punished such activities. In 1952, Turing stumbled into a set of circumstances that led to his arrest for homosexual activity. Found guilty, Turing was given the choice of medical "treatment" or imprisonment at the conclusion of his trial (which took place in March 1952). He accepted probation and the medical therapy option ("injections of hormones to restrain his libido"), so he could return to his academic activities. No one from the British government came forward at his trial to acknowledge his great national service in breaking the German codes during World War II. Rather, because of his former access to sensitive government secrets, the now "convicted and openly gay" Turing became the object of intense surveillance by government security officers.

The brilliant mathematician, hounded by senseless and extreme social pressure, died of potassium cyanide poisoning on June 7, 1954, in Wilmslow, Cheshire, England. A half-eaten apple, laced with the poison, suggested Turing died at his own hand. However, Turing's mother bitterly disagreed with the conclusions of the coroner's inquest and always maintained that her son's death was the result of an accidental poisoning, since he was conducting electrolysis experiments at the time.

Isaac Asimov (1920–1992)

Isaac Asimov was a premier science fiction and science fact writer, who popularized the word robotics and formulated his now-famous three laws of robotics. Asimov introduced these rules of robot behavior in the science fiction story "Runaround," which appeared in the March 1942 issue of *Astounding* magazine.

These laws have become part of the cult and culture of modern robotics. They are: (1st Law) "A robot may not injure a human being, or, through inaction, allow a human being to come to harm." (2nd Law) "A robot must obey the orders given it by human beings except where such orders would conflict with the first law." (3rd Law) "A robot must protect its own existence as long as such protection does not conflict with the first or second law."

Asimov was born on January 2, 1920, in the town Petrovichi, about 400 kilometers to the southwest of Moscow in the former Soviet Union. His family immigrated to the United States in 1923 and settled in the New York City area, where Asimov grew up. He completed his Bachelor of Science degree in chemistry in 1939 at Columbia University and then attempted (without success) to get into medical school. Rejected for medical school, he decided to pursue graduate work in chemistry. During World War II, Asimov interrupted his graduate studies to serve as a chemist in the United States Navy Yard in Philadelphia and later as a member of the armed forces. Following the war, Asimov completed his Ph.D. in chemistry at Columbia University in 1948 and in the following year became a member of faculty of Boston University School of Medicine, eventually rising to the academic rank of associate professor of biochemistry (in 1955). However, Asimov enjoyed writing science fiction and science fact much more than the duties of the academic life. So, in 1958 he left the university and pursued his writing interests on a full-time, professional basis.

Asimov sold his first science fiction story in 1939. Entitled "Marooned off Vesta," the story appeared in the March 1939 issue of *Amazing Stories*. This story launched an equally amazing literary career in which Asimov authored or edited over 450 volumes. His published works populate every major library category (as identified by the Dewey Decimal System), save for philosophy. Science fiction aficionados regard his 1941 story, "Nightfall," as his very best. He started his famous *Foundation* series (about the rise and fall of a galactic empire) in 1942. The book *I, Robot* (published in 1950) contains a collection of his robot-related short stories (which were also begun in about 1942) and helped promulgate Asimov's famous three laws of robotics.

Between 1957 and the early 1980s, Asimov focused his writing efforts primarily on a variety of science fact and history books. In a later science fiction book, entitled *Robots and Empire* (1985), he introduced the so-called zeroth law of robotics, which states that "a robot may not injure humanity, nor through inaction, allow humanity to come to harm." With this additional ethical guideline for robots, Asimov introduced several interesting "fictional" paradoxes with respect to his previously introduced three laws of robotics and the overall interaction between intelligent robots and the human race (taken as a whole). If robots proved to be stifling to the development of the human race, then under the conditions of the zeroth law, once robots recognized that their existence was harming the human race, they would be obliged to phase themselves out in order to "save humanity." Similarly, if one or several human beings acted in a way that endangered the survival of the entire human race, then a robot (responding to the zeroth law) would be obliged to neutralize or destroy the offending humans in order to save the human race. Since the action of harming or killing a particular human is in clear conflict with the basic three laws of robotics, the zeroth law is sometimes viewed as an overriding rule of machine behavior—though this rule does not enjoy the same widespread notoriety or recognition as Asimov's original three laws of robotics.

Later in life, Asimov returned to his famous Foundation series—the original trilogy of which appeared between 1951 and 1953. Asimov used *Foundation and Earth* (1986) to insert robots into the fictional Foundation universe he had previously crafted. Then, in his last two novels *Prelude to Foundation* (1988) and *Forward the Foundation* (1993), Asimov decided to portray robots as secret or covert operatives, which act for the overall benefit of the human race. Asimov died on April 6, 1992.

Joseph F. Engelberger (b. 1925)

The American entrepreneur and robot advocate, Joseph F. Engelberger, is often regarded as the "father of the industrial robot." Working with the self-educated engineer and inventor George Devol Jr., Engelberger formed Unimation, Inc., the first industrial robot company. Because of his untiring efforts to promote the use of industrial robots both in the United States and in Japan, these systems have now become an accepted and normal part of the manufacturing industry landscape around the world in the twenty-first century.

Engelberger was born in Brooklyn, New York, on July 26, 1925. Growing up in Connecticut during the Great Depression, he returned to New York City for his college education. Toward the end of World War II, he attended Columbia University, by participating in that institution's special accelerated college and officer training program. After receiving a Bachelor of Science degree in physics from Columbia University in 1946, he served briefly in the U.S. Navy. He then returned to Columbia to complete a Master of Science degree in physics in 1949. He was especially intrigued by the university's new course on servo theory. After graduate school, Engelberger joined the industrial firm of Manning, Maxwell, and Moore in Stanford, Connecticut. He remained with that firm until 1956, when a chance meeting with George C. Devol, Jr. led Engelberger to found Unimation, Inc., the world's first industrial robot company.

Two years earlier, Engelberger's business partner, Devol, had invented and patented a programmable method for transferring articles between different parts of a factory. Thanks to the technical skills of Devol and the marketing efforts and management skills of Engelberger, the industrial robot eventually became a reality.

In 1960, the Consolidated Diesel Electric Corporation (Condec Corporation) purchased Unimation, Inc. and began commercial development of the Unimate line of industrial robots. Although a test model had been placed in a General Motors' plant in 1959, the first commercial version of a Unimate industrial robot was shipped from Connecticut and installed in a General Motors plant in Trenton, New Jersey, in 1961. This pioneering industrial robot lifted up hot pieces of metal from a die-casting machine and stacked the pieces up for processing by a trim press. In 1975, Unimation, Inc. demonstrated a profit. Then, in 1978, Unimation introduced the versatile PUMA (Programmable Universal Machine for Assembly) industrial robot.

Over the next two decades, Engelberger enthusiastically pursued every reasonable opportunity to advocate the use of industrial robots. His advocacy message frequently focused on the use of industrial robots to replace workers from performing factory tasks that were either dull (that is, monotonously repetitive), dirty, or dangerous. As chief executive officer, Engelberger rarely missed an opportunity to promote his company's Unimate robots. In 1966, for example, the entertainment icon, Johnny Carson, invited Engelberger to appear on the famous late night television show (*The Tonight Show*). Engelberger accepted and proved to be an excellent spokesperson for industrial robots. As millions of Americans viewed this episode of the television show, Engelberger's Unimate robot poured a beer, led the band, and played golf—sinking a putt.

Unfortunately, the U.S. automobile industry did not immediately embrace the arrival of the industrial robot with the same foresight and enthusiasm that Engelberger experienced in Japan. Recognizing how industrial robots would replace workers in dull, dangerous, or dirty jobs, Japanese manufacturers heeded Engelberger's suggestions and invested heavily in these systems. Soon, the postwar Japanese manufacturing industry achieved a worldwide reputation for efficiency and quality. Engelberger's influence on Japan's industrial ascendancy was publicly acknowledged in 1997 when he was awarded the Japan Prize—the highest Japanese technology honor for the establishment of the robot industry in that country.

Engelberger served as the chief executive officer (CEO) at Unimation, Inc. from its founding in 1958, through several corporate purchases, until 1982 when Westinghouse acquired the robot company. Shortly after that event, Engelberger, ever the entrepreneur, founded Transitions Research Corporation, which later became HelpMate Robotics, Inc. The Helpmate is a well-known robotic hospital courier installed in over 80 U.S. hospitals, as well as medical facilities in Europe and Japan. (In 1997 Cardinal Health acquired HelpMate.) Recently, Engelberger has focused on developing a two-armed robot system for use in elder care. This articulate, mobile, and sensate robot could function as a servant-companion for elderly people who are cognitive, but mobility-impaired.

He authored two important books on the role of robots in industry, *Robotics in Practice* (1980) and *Robots in Service* (1989). The U.S. National Academy of Engineering (NAE) elected Engelberger as a member in 1984. The NAE is a member of the U.S. National Academies, which consist of the NAE, the National Academy of Sciences (NAS), the Institute of Medicine (IOM), and the National Research Council (NRC).

4

How Robot Technology Works

Robots come in a wide variety of shapes and sizes. Robots are basically smart machines and the great majority of the robotic systems currently in use have little resemblance to the fictional androids and superrobots that appear in motion pictures. This chapter starts with some of the basic principles that pertain to most robotics and then introduces the characteristics and features of interesting classes of modern robots, such as industrial robots, space robots, military robots, and medical robots. A generous collection of illustrations complements each section of the chapter and provides a better insight into the diversity and capabilities of modern robots. Remember that the systems discussed and shown in this chapter are real systems that have been built and operated to do specific jobs—jobs that are often too dangerous, too onerous, or too repetitive and boring for human beings. In some cases such as deep space exploration, the robot systems are the only way of accomplishing the mission. In other cases, like explosive ordnance disposal (EOD) operations, the robots help keep danger to human beings at a safe, standoff distance. While the robots of science fiction may still be decades away, there exists, nonetheless, an amazing and rapidly growing population of real-world robots that serve the human race in a variety of interesting and important ways.

THE FUNDAMENTAL PRINCIPLES OF ROBOTICS

Robotics is the science and technology of designing, building, and programming robots. Robotic devices, or robots, as they are usually called, are primarily smart machines with manipulators that can be programmed to do a variety of manual or human labor tasks automatically and with sensors that explore the surrounding environment, including the landscape of interesting alien worlds in outer space or the strange creatures that live on the ocean floor in inner space.

Engineers define a machine as a device with fixed and/or moving parts that modifies mechanical energy in order to do work and changes the magnitude or direction (or both) of an applied force. Some machines permit small forces to overcome heavy loads or to result in larger forces. Mechanical engineers generally design a machine to transmit and modify certain forces. The three basic machines are the lever, the inclined plane, and the wheel and axle. These simple devices lie at the heart of many machines, including fixed (stationary) and mobile robots. Other basic machines include the hammer, the pulley, the lever, and the screw. Whether a machine is a simple tool or complex and complicated mechanism, its main purpose remains the same, namely, to transform input forces into output forces.

Advanced machines often include one or more of the following components: bearings, cams, clutches, gears, shafts, and springs. Engineers define the mechanical efficiency of a machine (η) as the output work divided by the input work. An ideal (hypothetical) machine has a mechanical efficiency of 100 percent, or $\eta = 1$. Factory or industrial robots are examples of computer-driven machines that exhibit varying degrees of complexity, as they perform complicated work functions, involving the application or transmission of forces.

A robot, therefore, is simply a machine that does mechanical, routine tasks on human command. The expression robot is attributed to Czech writer Karel Čapek, who wrote the play *R.U.R. (Rossum's Universal Robots.)* This play first appeared in English in 1923 and is regarded as a dark satire on the mechanization of civilization. The word robot is derived from *robata*, a Czech word meaning compulsory labor or servitude.

Here on Earth, a typical robot normally consists of one or more manipulators (arms), end effectors (hands), a controller, a power supply, and possibly an array of sensors to provide information about the environment in which the robot must operate. Because most modern robots are used in industrial applications, their classification is traditionally based on these industrial functions. So, terrestrial robots frequently are divided into the following classes: nonservo (that is, pick-and-place), servo, programmable, computerized, sensory, and assembly robots.

The nonservo robot is the simplest type. It picks up an object and places it at another location. The robot's freedom of movement usually is limited to two or three directions.

The servo robot represents several categories of industrial robots. This type of robot has servomechanisms for the manipulator and end effector, enabling the device to change direction in midair (or mid stroke) without having to trip or trigger a mechanical limit switch. Five to seven directions of motion are common, depending on the number of joints in the manipulator.

The programmable robot is essentially a servo robot that is driven by a programmable controller. This controller memorizes (stores) a sequence of movements and then repeats these movements and actions continuously. Often, engineers program this type of robot by "walking" the manipulator and end effector through the desired movement.

The computerized robot is simply a servo robot run by computer. This kind of robot is programmed by instructions fed into the controller electronically. These smart robots may even have the ability to improve upon their basic work instructions.

The sensory robot is a computerized robot with one or more artificial senses to observe and record its environment and to feed information back to the controller. The artificial senses most frequently employed are sight (robot or computer vision) and touch. Finally, the assembly robot is a computerized robot, generally with sensors, that is designed for assembly line and manufacturing tasks.

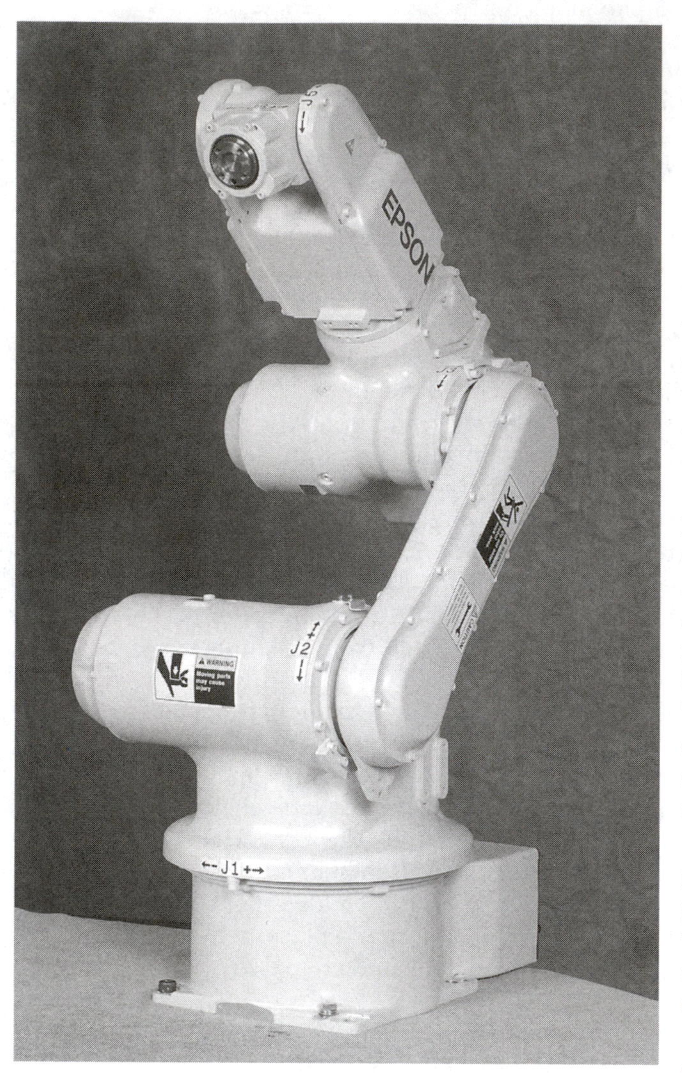

Figure 4-1 This is the EPSON Pro Six Model PS5DSC robot—an example of a computer-controlled, modern industrial robot with electric motors and six degrees of freedom. (Credit: Photograph courtesy of EPSON Robots.)

In industry, robots are designed mainly for manipulation purposes. The actions that can be produced by the end effector or hand include: (1) motion (from point to point, along a desired trajectory or along a contoured surface); (2) a change in orientation; and (3) rotation.

Nonservo robots are capable of point-to-point motions. For each desired motion, the manipulator moves at full speed until the limits of its travel are reached. As a result, nonservo robots often are called limit-sequence, bang-bang, or pick-and-place robots. When nonservo robots reach the end of a particular motion, a mechanical stop or limit switch is tripped, stopping the particular movement.

Servo robots are also capable of point-to-point motions; but their manipulators move with controlled variable velocities and trajectories. Servo robot motions are controlled without the use of stop or limit switches.

Many different types of manipulator arms have been developed to accomplish robot motions. The four major types of manipulator arms are the rectangular, cylindrical, spherical, and anthropomorphic (articulated or jointed arm). Each of these manipulator arm designs features two or more degrees of freedom (DOF)—a term that refers to the direction a robot's manipulator arm is able to move. For example, simple straight line or linear movement represents one DOF. If the manipulator arm is to follow a two-dimensional curved path, it needs two degrees of freedom: up and down and right and left. Of course, more complicated motions will require many DOF. To locate an end effector at any point and to orient this effector in a particular work volume requires six DOF. If the manipulator arm needs to avoid obstacles or other equipment, even more degrees of freedom are required. For each DOF, one linear or rotary joint is needed. Robot designers sometimes combine two or more of these four basic manipulator arm configurations to increase the versatility of a particular robot's manipulator.

Actuators are used to move a robot's manipulator joints. Three basic types of actuators currently are used in contemporary robots: pneumatic, hydraulic, and electrical. Pneumatic actuators employ a pressurized gas to move the manipulator joint. When a pump propels the gas through a tube to a particular joint, the process triggers or actuates movement. Pneumatic actuators are inexpensive and simple, but their movement is not precise. Therefore, this kind of actuator is found most often in nonservo, or pick-and-place robots. Hydraulic actuators are quite common and capable of producing a large amount of power. The main disadvantages of hydraulic actuators are their accompanying apparatus (fluid pumps and storage tanks) and problems with fluid leaks. Electric motor-driven actuators provide smoother movements, can be controlled very accurately, and are very reliable. However, these actuators cannot deliver as much power as hydraulic actuators of comparable mass. Nevertheless, for modest power actuator functions, electrical actuators often are preferred.

Many industrial robots are fixed in place or move along rails and guide ways. Some terrestrial robots are built into wheeled carts, while others use their end effectors to grasp handholds and pull themselves along. Advanced robots use articulated manipulators as legs to achieve a walking motion.

A robot's end effector (hand or gripping device) generally is attached to the end of the manipulator arm. Typical functions of this end effector include grasping, pushing and pulling, twisting, using tools, performing insertions, and various types of assembly activities. End effectors can be mechanical, vacuum, or magnetically operated, can use a snare device or have some other unusual design feature. The shapes of the objects that the robot must grasp determine the final design of the end effector. Usually most end effectors are some type of gripping or clamping device.

Robots can be controlled in a wide variety of ways, from simple limit switches tripped by the manipulator arm to sophisticated computerized remote sensing systems that provide machine vision, touch, and hearing. In the case of a computer-controlled robot, the motions of its manipulator and end effector are programmed, that is, the robot memorizes what it is supposed to do. Sensor devices on the manipulator help to establish the proximity of the end effector to the object to be manipulated and feed information back to the computer controller concerning any modifications needed in the manipulator's trajectory.

Most industrial robots use "wall power," that is, they derive their primary power from the available electric grid that services the factory, laboratory, or facility. Providing electric power to mobile robots is a bit more challenging for engineers. On Earth, mobile robots that are remotely operated by cable links can be provided electric power (via tethered lines) from a portable generator in the field or from a source of commercial (wall-power) in laboratory, industrial, or urban operational environments. Untethered, remotely operated mobile robots (those that use radio signals for control) must contain their own source of primary power, usually batteries or fuel cells. In some cases—such as remotely

Figure 4-2 A close-up view of the gripper (end effector) of an explosive ordnance disposal (EOD) robot opening a door latch during a bomb disposal test at Sandia National Laboratories in Albuquerque, New Mexico (2001). (Credit: Photograph courtesy of the U.S. department of Energy/SNLA.)

operated, heavy-duty mobile robots—the robot vehicle is propelled by a diesel or gasoline-powered engine, which also generates onboard electric power. An automobile operates in much the same way, with the gasoline or diesel engine providing primary motive power, as well as the shaft work necessary to operate an electric generator. A heavy-duty battery is used for motor startup and auxiliary (emergency) electric power.

Submersible mobile robots either obtain their electric power via cables from the surface ship, as is the case for most remotely operated vehicles (ROVs), or else operate on self-contained sources of electric power, such as batteries or fuel cells, as is the case for autonomous underwater vehicles (AUVs). Space robots of all types must provide their own electric power. For these robot systems, electric power is generated in a variety of ways, including the use of long-lived batteries, solar cells, or radioisotope thermoelectric generators—depending on the needs and location of the specific mission.

INDUSTRIAL ROBOTS

Exactly what is an industrial robot? This question is one of the most frequently asked questions in robotics and one of the most elusive to answer precisely. The reason is because industrial robots now come in a wide variety of sizes, shapes, number of axes, degrees of freedom, and design configurations. These factors influence the dimensions of the robot's working envelope or volume of space within which the robot can move and perform its designated tasks. Over the years since they were first introduced in the 1960s, engineers have characterized industrial robots by their motions, configurations, and/or applications. Such actions resulted in an interesting, but rather confusing, collection of names—such as pick-and-place robot, spray-painting robot, Cartesian robot, laboratory robot, dispensing robot, and so forth. Some of the names are self-evident, others like the selective compliance assembly robot arm (SCARA), need a little explaining.

To referee the semantic confusion, the International Organization of Standardization (ISO) came up with a formal, or "official," definition of the industrial robot. As specified in ISO 8373:1994 (*Manipulating industrial robots—Vocabulary*), the industrial robot is "an automatically controlled, reprogrammable, multipurpose manipulator in three or more axes." This precise definition is helpful to engineers and other professionals who build, sell, or use industrial robots, but the words may leave most readers still somewhat uncertain. The remainder of this chapter will try to remove any uncertainty by expanding on the concept of a computer-controlled, programmable smart machine that is capable of moving or manipulating things in a precise, repetitive, and dependable manner.

Industrial engineers sometimes use the terms hard (or stiff) automation and flexible automation to describe manufacturing processes performed by modern, computer-controlled machines. In *hard automation* the machine (often computer-controlled) has been specifically designed to mass-produce a certain product at a certain rate and with a certain level of quality control and reliability. The modern machine that carefully squirts just the right amount of soft drink into each bottle on a moving belt, carefully labels and seals each bottle, and then places the filled units in proper quantities in cases is an example of hard automation. There may also be an automated final inspection step to make sure that each bottle is properly filled and sealed and does not contain any foreign objects (broken equipment) or furry little critters (wandering rodents). Such machines, while marvels of automation and mass production, are usually not regarded as industrial robots.

In *flexible automation*, one or several industrial robots assist in the customized production, assembly, packaging, and inspection of products. The human operator can specify (usually by means of the control station computer) the specific characteristics of each unit or batch of units. Once instructed, the machines (including any programmable industrial robots) do the rest. Flexible automation is the hallmark of computer-integrated manufacturing (CIM). The Toymaker 3000 exhibit at the Museum of Science and Industry in Chicago is an accessible and entertaining example of how industrial robots can be integrated with other machines to create one giant, flexible manufacturing facility that can build, inspect, package, and deliver a customer-customized product. (This CIM exhibit is discussed in Chapters 5 and 10.)

The ISO has published several standards pertaining to the characterization, components, applications, and safety of manipulating industrial robots. The ISO standards and documents have proven very helpful in the rapid growth of the global population of modern industrial robots. ISO is a network of the national standards institutes of 156 countries, formed on the basis of one member per country. For example, the National Institute of Standards and Technology (NIST) is the nonregulatory, federal agency within the U.S. Commerce Department's Technology Administration that is responsible for promoting national standards. The Central Secretariat of the ISO is located in Geneva, Switzerland, and coordinates the international system. ISO is a nongovernment organization (NGO). The organization's official name is "International Organization for Standardization." Since this name would translate as different acronyms in different languages, the ISO leadership decided to avoid any possible confusion. Inspired by the Greek word isos, meaning equal, the Central Secretariat selected the acronym ISO as the official short form for the organization.

Types and Classifications of Industrial Robots

The modern industrial robot was introduced to the world in the early 1960s by the American entrepreneurs George C. Devol, Jr. and Joseph F. Engelberger. One historic milestone occurred in 1961, when the Consolidated Diesel Electric Corporation (Condec Corporation) shipped the first commercial version of a Unimate industrial robot from Connecticut and installed the device in a General Motor's plant in New Jersey. Despite aggressive marketing efforts the application of robots in industry (American and foreign) did not become widespread until the 1970s. In 1978, Unimation introduced the versatile industrial robot called PUMA, for Programmable Universal Machine for Assembly.

With the introduction of its versatile PUMA robot, Unimation Incorporated became one of the world's leading manufacturers of industrial robots and was acquired by Westinghouse Electric Corporation in February 1983. The Stäubli Group then purchased Unimation from Westinghouse in 1988 and proceeded to expand the company's line of industrial robots. Today, Stäubli Robotics offers an extended product range of industrial robots, ranging from a variety of SCARA robots to heavy payload six-axis robots, many of which can be operated using a personal computer (PC)-based control platform.

While American industry (especially the automobile manufacturers) reluctantly acknowledged the arrival of the industrial robot, Japanese engineers and industrial leaders embraced the new technology and went on to redefine the concept of modern manufacturing. As more and more industrial robots populated Japanese plants, the postwar Japanese manufacturing industry achieved a worldwide reputation for efficiency and quality. The robot's influence on Japan's industrial ascendancy in the 1980s was publicly acknowledged in 1997, when the American robot engineer and entrepreneur, Joseph Engelberger was awarded the Japan Prize (the highest Japanese technology honor)—for helping establish the robot industry in that country.

Many industrial robots used today in the United States support light manufacturing, heavy manufacturing, casting/foundry operations, metal machining, and automobile manufacturing. Sometimes referred to as "steel collar" workers, industrial robots now perform a wide variety of tasks including loading and unloading, paint spraying, arc welding, cutting, grinding, forging, palletizing products, and inspecting products.

Robots have been used very successfully for tasks that involve boring or repetitive work—tasks that quickly sap a human worker's motivation or attention. Drill press operations, visual inspection of printed circuit boards, the

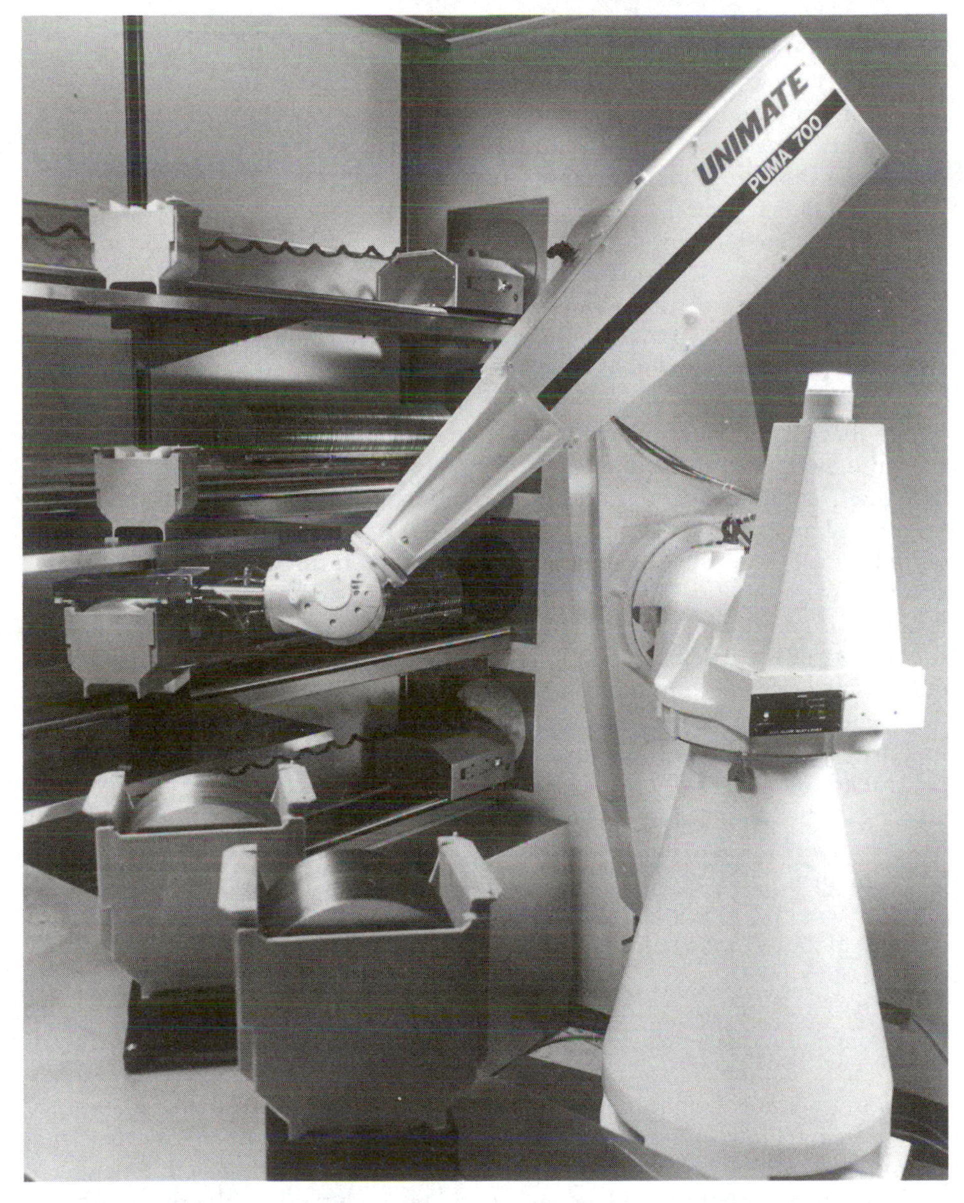

Figure 4-3 This photograph shows a Unimate PUMA 700 robot. The versatile pick-and-place industrial robot was designed (in the 1980s) to perform a variety of tasks in assembly, material handling and transfer, joining, inspection, and palletizing. (Credit: Photograph courtesy of Unimation Incorporated, a Westinghouse Company.)

continuous loading or unloading of bins are examples of jobs that promote inattention, excessive absenteeism, or poor quality results when performed repetitively by human workers and yet are ideally suited for robots. Similarly, die casting operations are hazardous tasks more safely performed by robots. Die casting involves close proximity to molten plastic and metal. Human workers must wear thick thermal protective gloves, sometimes get splashed by spurts of liquid metal as it pours, and must breathe air loaded with noxious fumes from the die lubricant. The properly designed "steel collar" worker (robot) is generally unaffected by such workplace hazards and goes about performing the die casting tasks with a constant level of efficiency.

As previously mentioned industrial robots are usually classified as either servo or nonservo controlled. Servo robots are controlled through the use of sensors, which continually monitor the robot's axes for positional and velocity feedback information. The robot system compares this feedback information with the data programmed and stored in its memory. Conditions that fall outside the desired levels programmed into the machine are corrected (within the tolerance limits of the system). For example, if the robot's arm attempts to extend out a bit too far (because of the mass and/or momentum of a gripped load), the servo robot automatically adjusts the arm's position to the proper (programmed) location. Depending on the magnitude of the necessary adjustment, the servo robot's use of feedback can result in just one correction or a series of small, iterative corrections—automatically and carefully performed until the location of the robot, and other parameters are within the desired value. (This process is limited only by the inherent tolerances of the machine.) The servo robot (or servo-controlled robot) actually represents several categories of industrial robots, including the point-to-point robot and the continuous path robot. This type of robot has servomechanisms for the manipulator and end effector to enable it to change direction in midair (or mid stroke) without having to trip or trigger a mechanical limit switch. Five to seven directions of motion are common, depending on the number of joints in the manipulator. The servo robot is the most common industrial robot in use today.

In 1948, the American mathematician Norbert Wiener formally introduced the science of cybernetics, when he published the book *Cybernetics or Control and Communication in the Animal and Machine*. As a scientific discipline, cybernetics investigates communication and control processes in living systems as well as in machines built by human beings. Wiener coined the word *cybernetics* from the ancient Greek word (kubernẽtẽs meaning steersman, pilot, or governor). Within a mathematical framework, cybernetics deals with how regulatory feedback signals are communicated and controlled in electronic, mechanical, and biological systems. Feedback is the process by which the output of a system is used to control its performance. In negative feedback, a return signal associated with the output of the system is used to reduce the input. Similarly, in positive feedback, a return signal associated with the output is used to increase the input. A simple example of a negative feedback device is an electromechanical governor (intended to keep a vehicle moving at some optimum speed), which reduces the fuel supply to an internal combustion engine as the vehicle's speed increases. For an industrial robot, feedback often is used to locate and precisely control certain moving parts. The design of the robot usually includes a specific

Figure 4-4 The picture (taken April 15, 2004) shows a robotic water-blasting system developed by NASA to automatically remove thermal protection materials and coatings from recovered space flight hardware. During refurbishment operations, the robot system uses a stream of high-pressure water to carefully remove expended thermal protection materials without damaging the underlying structure and without resorting to the use of hazardous chemical strippers. The use of this automated robotic system improves worker safety by reducing human exposure to high-pressure water and high-velocity debris fragments. (Credit: Photograph courtesy of NASA/MSFC.)

subsystem that indicates the current position of the moving part (or parts). The signal representing the current position is compared to the desired or target position (for a particular operation or activity) and adjustments are made until the difference between the target position and the actual position of the moving part are zero—or at least within some established level of tolerance). This important area of control system theory was greatly organized and intellectually stimulated by Wiener's work at MIT in the 1940s.

The nonservo robot is the simplest type of industrial robot. This kind of robot simply picks up an object and places it at another location. The robot's freedom of movement usually is limited to two or three directions. Nonservo robots do not use feedback from position sensors. Instead, this type of robot uses a system of mechanical stops and/or limit switches to control the motion of its arm(s) and the position of its axes.

A robot's degrees of freedom are the directions of motion inherent in the design of the robot's mechanical system. Industrial robots can have up to six DOF or types of movement. Engineers define these motions as: *vertical tranverse*—up and down motion of the robot arm caused by pivoting the entire arm about a horizontal axis or moving the arm along a vertical slide; *radial traverse*—retraction and extension of the arm (in and out movements); *rotational traverse*—rotation about the vertical axis (right or left swivel of the robot arm); *wrist pitch*—up and down movement of the wrist; *wrist yaw*—right and left swivel of the wrist; and *wrist roll*—rotation of the wrist.

Each actuator usually causes linear or rotary motion with respect to an axis. The number of axes is normally the same as the number of degrees of freedom of the robot. However, the motions of actuators in the end effector (such as closing a robot's grippers) do not constitute additional degrees of freedom. Depending on the robot's geometry, motion of one or more axes may be needed if the robot is to move to a new location in space.

Engineers usually employ three axes to move the robot's wrist-end effector interface to a new position in space. Additional axes accommodate rotation at that point to permit flexibility in orientation. The robot may have one, two, or three rotating axes at the wrist, depending on how sophisticated engineers want the robot to be. Articulated arm and gantry robots usually have six DOF available. The SCARA configuration typically provides four DOF. Engineers give a robot seven or more DOF (axes) for some special applications. One example is an articulated arm-welding robot that must work on the far side of an automobile body. Engineers can also add a seventh DOF to a robot system by placing a six-degree-of-freedom robot on rails, thereby giving the robot an extended longitudinal range.

Sometimes engineers prefer to use more specific terms (rather than servo or nonservo) to describe a particular industrial robot. Some of these more specialized terms are provided here, while others appear in the glossary (Chapter 8). A Cartesian robot is a robot that has its tooling mounted to an arm, which travels with Cartesian coordinate motion—that is, along the *x*-, *y*-, and *z*-axes. Unlike other types of industrial robots, the Cartesian coordinate robot does not revolve around a stationary rotary axis. This type of robot tends to have greater accuracy and repeatability than other types of industrial robots, especially for heavy loads. Engineers sometimes call this type of robot a rectangular coordinate robot or Cartesian coordinate robot. A continuous path robot is one of two basic types of servo-controlled robots. To teach a continuous path robot its path, a human being must physically move the robot's manipulator arm through whatever series of motions it is expected to perform. These learned or rehearsed motions are then stored in the robot's computer for future recall. A cylindrical coordinate robot is a robot that has a horizontal shaft, which goes in and out. The robot also rides up and down on a vertical shaft, which (shaft) also rotates about the base.

A hydraulic robot is an industrial robot that uses hydraulic power to move its arm, wrist, and end effector. The hydraulic power supply is often located some distance away from the robot's work site and generally consists of a motor-driven pump, reservoir for the hydraulic fluid, a filter, heat exchanger, and pipes to deliver the pressurized hydraulic fluid to the robot. High-pressure fluid leaks are a major problem with hydraulic robots. The pick-and-place robot is one of the two basic types of industrial robot (the other being the servo robot). This type of robot has direction control stops or valves, which are either fully opened or closed, thereby limiting positioning capability and program capacity. Engineers also call this type of robot a bang-bang robot, an end-point robot, a limited-sequence robot, or a non-servo-controlled robot. A pneumatic robot is an industrial robot that is pneumatically actuated. The power for pneumatic actuation is usually provided by a remote compressor, which may provide pressurized working fluid (e.g., compressed air) to other equipment at the industrial facility. Sometimes engineers refer to this type of robot as an air-logic robot.

A point-to-point robot is an industrial robot, representing one of the two basic types of servo robots. The expression "point-to-point" refers to the fact that this type of robot must be taught to perform its assigned task one step (or point) at a time. The human robot technician positions the robot's arm (especially the end effector or hand) at a particular point in space and then instructs (programs) the robot to store that particular position in its computer memory. The technician repeats this procedure on a point-by-point basis, until the robot has stored in its memory the complete sequence of motions and actions it is expected to perform. A programmable robot is a type of industrial robot that is essentially a servo robot, which is driven by a programmable controller. The controller memorizes (stores) a sequence of movements and then repeats these movements and actions continuously. Often, engineers program this type of robot by "walking" the manipulator and end effector through the desired movement.

The selective compliant assembly (or articulated) robot arm (SCARA) is a four-axis industrial robot, which can move to any x-y-z position (coordinate) within its working envelope. The SCARA also provides a fourth axis of motion, namely the rotation of its wrist. Since the SCARA's jointed two-link arm layout is similar to a human being's arm, engineers often refer to the robot's first joint as the "shoulder" and to the second joint as the "elbow." The robot's shoulder and elbow joints accommodate movement along the x- and y-axes. The third joint is regarded as the translation joint (z-axis motion), while the fourth joint is considered the wrist (providing rotation around the z-axis). Engineers also use the term "articulated" to describe how this robot's mechanical arm can extend into confined areas and then retract (fold up) out of the way. SCARA robots are often used in the electronics industry to assemble and handle circuit boards.

A sensory robot is a computerized robot with one or more artificial senses to observe and record its environment and to feed information back to the controller. The artificial senses most frequently employed are sight (robot or computer vision) and touch (tactile sensors). The sensory robot can be an advanced industrial robot or a sophisticated mobile robot that supports such missions, as national defense activities, search and rescue operations, environmental monitoring and cleanup activities, and civilian law enforcement. Mobile robots are discussed in subsequent sections of this chapter.

A spherical coordinate robot is an industrial robot that has a configuration similar to a tank turret. The robot's arm can move (or slide) in and out. The arm can also be raised and lowered in an arc (much like a tank's cannon can be raised or lowered in adjusting the firing elevation). Finally, the robot's arm can rotate about the base (much like a tank's turret rotates about the weapon system's treaded chassis).

Paths Generated by Industrial Robots

Engineers and technicians can program industrial robots to move over paths generated with different types of control. This section discusses three major types of path generation. In the following discussion, movement of the robot refers to movement of the end effector or wrist to which tooling is attached.

With *point-to-point programming*, the robot moves from one discrete point to another within its working envelope. However, motion between the points is generally not performed in a straight line, and the orientation of any held object may vary as the joint actuators operate independently to arrive at their new positions. Because it is difficult to predict the robot's exact path, robot movement between points has the potential of resulting in a safety hazard for human workers or peripheral equipment.

With *controlled-path programming*, the robot travels from point to point by moving along a predictable, computer-generated path. This computer-generated path may be a straight line with respect to end-effector orientation or it may involve a curved path through successive points—possibly accompanied by gradual changes in end-effector orientation. The coordinate transformations that control these precise movements are calculated by the robot's control system computer. The movement of the robot is more precise than what occurs with point-to-point programming and less likely to represent a hazard to human workers or peripheral equipment.

With *continuous-path programming*, the robot's path is controlled by storing a large number (or close succession) of spatial points in the robot's memory, during the system's teach sequence. During the teach sequence, while the robot is being moved by the human trainer, the corresponding spatial coordinate points of each axis are continually monitored and placed into the control system's computer memory. Then, when the robot is placed in the automatic mode, engineers or technicians can replay the program from memory and the robot duplicates the original path.

Basic Components of an Industrial Robot

An industrial robot typically consists of four major subsystems: the mechanical unit, the drive, the control system, and appropriate tooling. When engineers speak of the robot's mechanical unit, they are referring to the robot's manipulator arm and its base. Tooling such as end effectors, tool changers, and grippers are attached to the wrist-tooling interface. The mechanical unit generally consists of a fabricated structural frame, which has provisions for supporting mechanical linkage and joints, guides, actuators, control valves, limiting devices, and sensors. Application requirements determine the robot's physical dimensions, loading capability, and design.

The vast majority of new industrial robots use electric drives. Engineers have used pneumatic drives for high speed, nonservo robots. Pneumatic drives are frequently used for power tooling, such as grippers. Robot engineers have used hydraulic drives for systems required to do heavy lifting, typically where accuracy is not an important performance parameter. Electric drive systems can provide both lift and precision, depending on the engineer's choice of motor and servo system selection. Depending on the robot's design and intended applications, electric drives may involve alternating current (a.c.) or direct current (d.c.) powered motors.

The majority of industrial robots incorporate computer or microprocessor-based controllers. These controllers perform computational functions and interface with and control sensors, grippers, tooling, and other peripheral equipment. The control system also performs the necessary sequencing and memory functions associated with communication and interfacing for on-line sensing, branching, and the integration of other equipment.

Engineers or technicians may accomplish controller programming on-line or from remote, off-line control stations. Programs may be loaded on cassettes, disks, internal drive, or in memory; and may be loaded or downloaded by cassettes, disks, or telephone modem. Robot controllers may also have a self-diagnostic capability—a design feature that can reduce the downtime of the robot system. Some robot controllers have sufficient computational ability, memory capacity, and input/output capability to serve as system controllers for other equipment and processes. Furthermore, the robot controller may be placed in a control hierarchy such that it receives instructions, reports positions, and/or gives directions. In most cases, industrial robot manufacturers use proprietary languages for programming robot systems and their controllers.

The industrial robot manipulates its assigned tooling to perform the functions and applications for which the system was designed and constructed. Depending on the application, the robot might have just one primary functional capability, such as spray painting or making spot welds. Engineers can integrate these capabilities into the robot's mechanical system, or else they can add the capability through tooling attached at the robot's wrist-end effector interface. Other industrial robots may use multiple tools, which can be changed manually (by human technicians as part of the setup for a new operation) or else automatically, while the robot is performing its work cycle.

Safety engineers caution that tooling and objects carried by a robot's gripper can significantly increase the working envelope in which human workers or peripheral equipment can be struck. On many occasions, the tooling being manipulated by the robot or the objects being carried in the robot's gripper can represent more of a potential hazard to workers or peripheral equipment than the motion of the bare robot system.

Teaching Industrial Robots

Industrial robots perform tasks for a particular application by following a programmed sequence of directions from the control system. The robot's program establishes a physical relationship between the robot and other equipment. This program consists of a sequence of positions for the axes of movement and any end-effector operation, path information, velocities, timing, sensor data reading, external data source reading, and commands or output to externally connected systems. Robot technicians can teach the robot its program by manually commanding the robot to learn a series of positions and operations (such as closing its gripper). Collectively the series of positions and operations make up the robot's work cycle. The robot then converts these positions and operations into its programming language.

As an alternative, an engineer or technician can program a robot directly by inputting its programming language at an appropriate control terminal. The terminal can either be the robot's own controller or else another, separate computer system. When a human worker performs robot programming, verification and some modifications are often necessary. This procedure is called program touchup. It is normally performed in the teach mode of operation, with the human teacher manually leading the robot through the preprogrammed steps.

Robot technicians and engineers use three basic teaching or programming techniques for industrial robots. These three teaching techniques are called lead-through programming; walk-through programming; and off-line programming.

In the lead-through programming technique, the robot technician usually employs a teach pendant. This approach allows the teacher to direct the robot through a series of positions and to enter associate commands and other important information, such as velocities. In this technique, the robot technician teaches the robot the proper physical positions. The robot's controller then generates the programming commands needed to move between positions, when the operational program is played. When using this robot-teaching technique, the robot technician (teacher) may need to enter the robot's working envelope. This circumstance introduces a high potential for accidents, because safeguarding devices (such as light curtains and interlocks) will probably have been deactivated to allow the technician's entry into the robot's working envelope. As a safety measure, the technician should use only the teach pendant to program the robot.

While a human teacher is in the working envelope with a teach pendant, the simultaneous use of another programming console or the robot's own controller by a different technician is to be avoided. Conflicting commands and activities could create a very hazardous situation for the worker who is in close proximity to the robot. Imagine being the mechanic, who is working under the hood of an automobile that has its engine running and another mechanic sitting in the driver's seat with his foot dangerously close to the automobile's accelerator. A similar dangerous situation occurs, when two robot technicians simultaneously attempt to program (teach) a robot using different controllers and locations at the same time. The teach pendant should have the capability of overriding all other inputs and commands to a robot.

Another approach to teaching an industrial robot is called the walk-through programming technique. The teacher (robot technician) moves (or "walks") the robot through the desired positions within the robot's working envelope. As this is manually taking place, the robot's controller may scan and store spatial coordinate values on a fixed time-interval basis. Next, these coordinate values and other functional information are replayed in the automatic mode, but most likely at a different (higher) speed than the teacher (robot technician) used in the more cautious and deliberate walk-through procedure.

The walk-through programming technique employs triggers on manual handles that move the robot. When the technician depresses the trigger, the controller "remembers" the spatial coordinates of the position. Later, when the program is played, the robot's controller will generate the appropriate movement between these points. The walk-through teaching technique requires that the human worker (teacher) be within the robot's working envelope with the robot's controller energized—maybe not fully, but at least with respect to its position sensors. Since safeguarding devices and interlocks may also be deactivated during teaching activities, any human being in the robot's working envelope must exercise heightened vigilance to avoid experiencing serious injury or a fatal accident. With all the safeguards deactivated to allow him access to the robot's working envelope, the technician must ensure that only the position sensors are energized in the robot's controller and nothing else that would allow the robot to undergo sudden, potentially dangerous movements.

In the off-line programming technique, the teacher (robot technician) uses a remote programming computer. The teacher establishes the required sequence of functional and positional steps. Then, the program is transferred to the robot's controller by disk, cassette, or network link. Generally, positional references are established on the robot to transform (or calibrate) the coordinates used in the remote programming for the actual setup.

SPACE ROBOTS

This section discusses space robots and provides a special insight as to how some of the most important mobile robots ever built function. A robot spacecraft is an unmanned platform that aerospace engineers have designed to be placed into an orbit about Earth or on an interplanetary trajectory to another celestial body or into deep space. Aerospace engineers custom-design scientific spacecraft to meet the widely varying needs of the scientific community. The result is an assortment of interesting platforms that come in all sizes and shapes.

The space robot is essentially a combination of hardware that forms a mission-oriented spacecraft. The collection of hardware that makes up a robot spacecraft includes structure, thermal control, wiring, and subsystem functions, such as attitude control, command, data handling, and power. NASA engineers often refer to a robot spacecraft as a flight system to distinguish it from equipment that remains on Earth as part of the ground system for a particular project or mission.

Figure 4-5 This is a picture of NASA's Ranger spacecraft—a family of space robots that were sent to the Moon in the early to mid-1960s to pave the way for the lunar landings by the Apollo astronauts at the end of the decade. Engineers designed these attitude-controlled robot spacecraft to photograph the Moon's surface at close range before crashing (impact). (Credit: Photograph courtesy of NASA/JPL.)

The robot spacecraft itself might contain 10 or more subsystems, including an attitude control subsystem, which in turn contains numerous assemblies, such as reaction wheel assemblies or inertial reference assemblies. In certain instances, like those involving the telecommunications system, there are transmitter and receiver subsystems on both the spacecraft (as part of the flight

Figure 4-6 The top panel in this picture shows the robot eyes (panoramic camera) used by the NASA's twin Mars Exploration Rovers (MERs), called *Spirit* and *Opportunity*. The panel on the lower left shows *Spirit* in the clean room at NASA's Jet Propulsion Laboratory before its flight to Mars. The smaller robot rover is the flight spare from NASA's very successful Mars Pathfinder mission (1997–1998). The panel on the lower right highlights the multicolored filter wheel that allows the robot's camera to see a rainbow of colors in addition to infrared bands of light. By seeing Mars in all its colors, scientists can gain insight into the different minerals that constitute its rocks and soil. (Credit: Composite photograph courtesy of NASA/JPL.)

system) and equipment back on Earth (as part of the ground system). So, when engineers use the terms system and subsystem for the same piece of equipment, the use of nomenclature can appear a bit confusing. There are even times when systems are contained in subsystems, as, for example, an *imaging subsystem* that contains a *lens system*. It is probably best to remember that the hierarchy of aerospace hardware is: system, subsystem, assembly, and component (or part) in that descending order.

Because of the complexity of a robot spacecraft and the interdependent nature of many of its systems and subsystems, engineers and scientists sometimes appear very arbitrary in their use of this hardware classification scheme. Fortunately, any apparent confusion in nomenclature in no way detracts from the quality of the hardware that makes the robot spacecraft function and perform important feats of automated exploration and scientific data collection. And that, after all, is the main reason why the scientists and engineers have constructed these fascinating exploring machines in the first place.

Individual robot spacecraft can be very different from one another in their design and level of complexity, including the type and number of subsystems and component parts and assemblies found in each individual subsystem. Not all the different types of robot spacecraft discussed in this section need the same subsystems. For example, a robot probe, which descends into a planetary atmosphere on a one-way scientific mission, will generally not require a propulsion subsystem or an attitude control subsystem. But the probe has a collection of scientific instruments, needs electric power, requires a structure, needs an effective thermal control system, uses an onboard computer, and transmits the data it collects. The discussion focuses on the basic subsystems that satisfy mission requirements of modern, complex flyby or orbiter-class robot spacecraft. However, the treatment is sufficiently broad so as to embrace the often less complex (from a spacecraft engineering perspective) types of space robots, such as landers and rovers.

Different space robots possess different levels of machine (or artificial) intelligence. A robot's level of machine intelligence determines the degree of autonomous operation possible and the amount of human supervision required. For deep space missions, direct human supervision is usually impractical or impossible, so a space robot engaged in this type of mission needs an appreciable level of AI. Specifically, at a great distance from Earth, the robot spacecraft must have the autonomy and AI capability necessary to monitor and control itself. When a space robot is light-minutes away from Earth, human members of the mission cannot respond to problems or anomalies in time. All of a robot spacecraft's subsystems must contain and run fault protection algorithms, which can quickly detect and respond to a problem without direct human assistance. When a fault protection algorithm detects a problem, it can respond by putting the subsystem in difficulty in a safe condition.

Safing is the process by which a robot spacecraft automatically shuts down or reconfigures components to prevent damage either from within or due to changes in the external environment. Many terrestrial machines and home appliances have similar safing features that have been engineered into the device. The thermal limit switch on the electric motor of a vacuum cleaner is an example. When the motor works too hard and starts getting a bit too hot, the thermal limit switch shuts down the device before any permanent damage can occur. When the motor cools to a safe level, the thermal limit switch resets and a person can again use the vacuum cleaner. Well-designed robot spacecraft have many such safing features engineered into their complex subsystems.

Scientific Space Robots

Robot spacecraft come in all shapes and sizes. Each space robot is usually custom-designed and carefully engineered to meet the specific needs and environmental challenges of a particular space exploration mission. For example, lander spacecraft are designed and constructed to acquire scientific data and to function in a hostile planetary surface environment. Since the complexity of space robots varies greatly, engineers and space scientists find it convenient to

Figure 4-7 Technicians from the Jet Propulsion Laboratory (JPL) clean and prepare the upper equipment module for mating with the propulsion module of the Cassini orbiter spacecraft at the Kennedy Space Center (1997). NASA successfully launched the large *Cassini/Huygens* spacecraft configuration on October 15, 1997, using a powerful Titan IV-Centaur rocket vehicle configuration. The sophisticated robot spacecraft arrived at Saturn in July 2004, after a long journey through interplanetary space, including gravity-assist flybys of Venus (April 1998 and June 1999), Earth (August 1999), and Jupiter (December 2000). (Credit: Photograph courtesy of NASA/JPL.)

categorize robot spacecraft according to the missions they are intended to fly. This section discusses the broad general classes of robot spacecraft.

Most interplanetary missions are flown to collect scientific data. However, some space robot missions, like NASA's *Deep Space-One* (DS-1), have as their primary objective the demonstration of new space technologies. On technology demonstration missions, the collection of scientific data remains an important, though secondary objective. When the collection of scientific data is the primary mission of a robot spacecraft, then all the subsystems and components that engineers place onboard the spacecraft are there in support of that single purpose. Simply stated, the scientific space robot is carefully designed and constructed so as to gather the most scientific data at the target interplanetary location or celestial object.

The robot spacecraft exists to deliver its scientific instruments to a particular interplanetary destination; to allow these instruments to make their measurements, perform their observations, and/or conduct their experiments under the most favorable achievable conditions; and then to return data from the instruments back to scientists on Earth. In the interesting case of a sample return mission, the robot spacecraft must collect and then return material samples from an alien world. Once the space robot delivers its extraterrestrial cargo to Earth, scientists perform detailed investigations on the alien materials in a special, biologically isolated (quarantine) facility.

There are many different types of scientific instruments that a robot spacecraft can carry. For convenience, scientists and engineers usually divide these instruments into two general classes: direct-sensing instruments and remote-sensing instruments. A direct-sensing instrument interacts with the phenomenon (of interest) in the immediate vicinity of the instrument. Examples include a radiation detection instrument and a magnetometer. In contrast, a remote-sensing instrument examines an object or phenomenon at a distance without being in direct contact with that object. The passage of electromagnetic radiation from the object to instrument supports information transfer and data collection. Remote-sensing instruments usually form some type of image of the object being studied or else collect characteristic data from the object, such as its temperature, luminous intensity, or energy level at a particular wavelength.

Scientists also find it helpful to classify scientific instruments as either passive or active. A passive instrument detects radiation, particles, or other information naturally emitted by the object or phenomenon under study. A magnetometer is a passive, direct-sensing scientific instrument carried by many robot spacecraft to detect and measure the interplanetary magnetic fields in the vicinity of the spacecraft. Imaging instruments are examples of passive remote-sensing instruments, which collect the electromagnetic radiation emitted by or reflected from a planetary body. Sunlight serves as the natural source of illumination for the observed reflected radiation from a planetary body. An active instrument supplies its own source of electromagnetic radiation or particle radiation to stimulate a characteristic response from the target being illuminated or irradiated. A synthetic aperture radar, as carried by NASA's *Magellan* orbiter spacecraft, and the alpha proton X-ray spectrometer (APXS) used by NASA's *Mars Pathfinder* rover are examples of active scientific instruments.

Figure 4-8 NASA's *Magellan* spacecraft with its attached inertial upper stage (IUS) rocket in the payload bay of the space shuttle *Atlantis* prior to launch in April 1989. On May 4, the *Atlantis* delivered and deployed *Magellan* into low Earth orbit. The IUS rocket then sent the robot orbiter spacecraft on an interplanetary trajectory to Venus. From 1990 to 1994, *Magellan* used its sophisticated imaging radar system to make the most detailed maps of the cloud-enshrouded planet ever collected. (Credit: Photograph courtesy of NASA/KSC.)

General Classes of Scientific Spacecraft

Scientific space robots include: flyby spacecraft, orbiter spacecraft, atmospheric probe spacecraft, atmospheric balloon packages, lander spacecraft, surface penetrator spacecraft, surface rover spacecraft, and observatory spacecraft.

There are three basic possibilities for a robot spacecraft's trajectory when it encounters a planet. The first possible trajectory involves a direct hit or hard landing. This is an impact trajectory. A hard landing involves a relatively high-velocity impact landing of the robot spacecraft on the surface of a planet or moon that usually destroys all equipment, except perhaps for a very rugged instrument package or payload container. The hard landing could be intentional, as occurred during NASA's Ranger spacecraft missions, which were designed to crash into the lunar surface; or unintentional, as when a retrorocket system fails to fire or a parachute system fails to deploy, and the robot lander strikes the planetary surface at an unexpected and unplanned high speed.

Aerospace engineers design lander spacecraft to follow an impact trajectory to a planet's surface. They also want the robot to survive by touching down on the surface at a very low speed. Sometimes, a lander spacecraft is sent on a direct impact trajectory; other times the robot is carried through interplanetary space by a mother spacecraft and then released on an impact trajectory after the mother spacecraft has achieved orbit around the target planet. Following separation from the orbiting mother spacecraft, the lander travels on a carefully designed impact trajectory to the target planet's surface. NASA's Surveyor spacecraft to the Moon are an example of the former soft-landing mission approach, while the *Viking 1* and *2* lander missions to Mars are an example of the latter design approach.

The *Viking 1* and *2* lander spacecraft placed on the Martian surface in 1976 represent one of the great early triumphs of robotic space exploration. After separation from the Viking orbiter spacecraft, the lander (protected by an aeroshell) descended into the thin Martian atmosphere at speeds of approximately 16,000 kilometers per hour. As it descended, the lander was slowed down by aerodynamic drag until its aeroshell was discarded. Each robot lander spacecraft then slowed down further by releasing a parachute. Finally, the robot achieved a gentle landing by automatically firing retrorockets. Of special significance is the fact that both Viking landers successfully accomplished the entire soft-landing sequence automatically without any direct human intervention or guidance.

In another lander/probe mission scenario, the mother spacecraft releases the lander or robot probe, while the cojoined spacecraft pair is still some distance from the target planetary object. Following release and separation, the robot probe follows a ballistic impact trajectory into the atmosphere and unto the surface of the target body. This scenario occurred when the *Cassini* mother spacecraft released the hitchhiking *Huygens* probe on December 25, 2004, as *Cassini* orbited around Saturn. Following separation, the *Huygens* probe traveled for about 20 days along a carefully planned ballistic trajectory to Saturn's moon Titan. When it arrived at Titan on January 14, 2005, the *Huygens* probe entered the moon's upper atmosphere, performed a superb data-collecting descent, and successfully landed on the moon's surface.

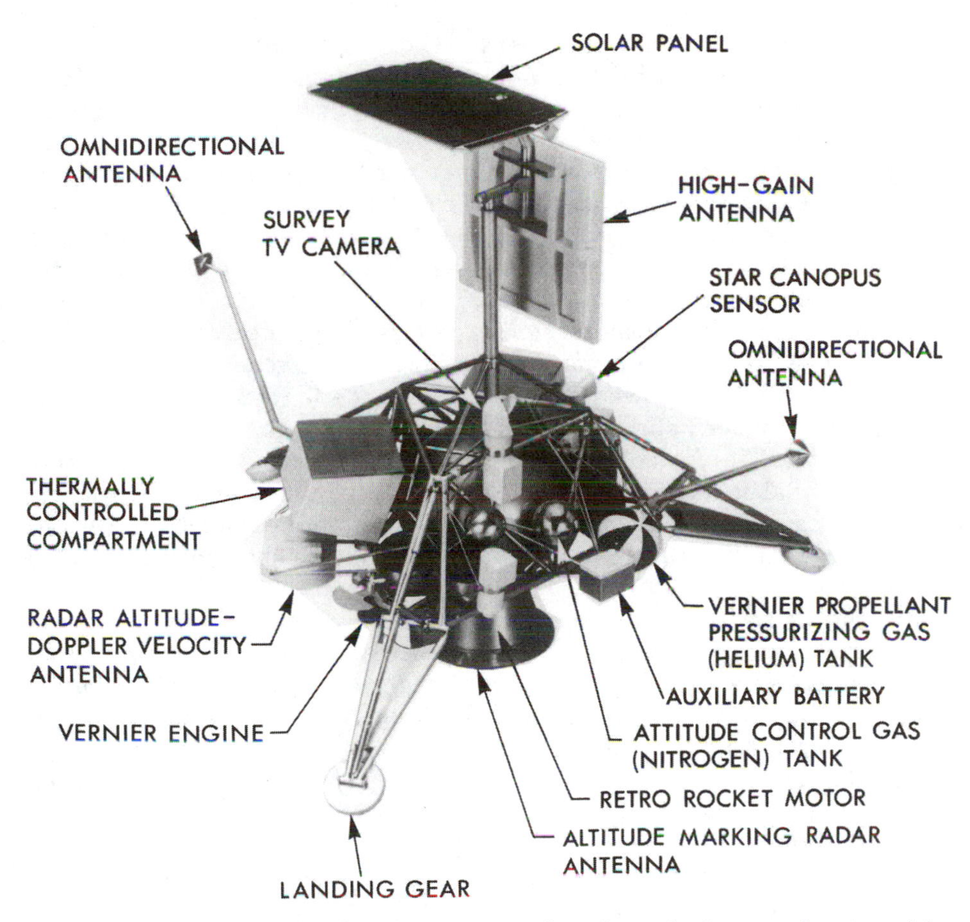

Figure 4-9 NASA's family of *Surveyor* spacecraft performed robotic exploration of the Moon's surface from 1966 to 1968 in preparation for the lunar landings missions by the Apollo astronauts (1969–1972). (Credit: Photograph courtesy of NASA.)

The second type of trajectory is an orbital-capture trajectory. The spacecraft is simply captured by the gravitational field of the planet and enters orbit around it. Depending on its precise speed and altitude (and other parameters), the robot spacecraft can enter this captured orbit from either the trailing edge or the leading edge of the planet. In the third type of trajectory, called a flyby trajectory, the spacecraft remains far enough away from the planet to avoid capture but passes close enough to be strongly affected by its gravity. In this case, the speed of the spacecraft will be increased if it approaches from the trailing side of the planet and diminished if it approaches from the leading side. In addition to changes in speed, the direction of the spacecraft's motion also changes.

The increase in speed of the flyby spacecraft actually comes from a decrease in speed of the planet itself. In effect, the spacecraft is being "pulled along" by the planet. Of course, this is a greatly simplified discussion of complex encounter phenomena. A full account of spacecraft trajectories must consider the speed and actual trajectory of the spacecraft and planet, how close the spacecraft will come to the planet, and the size (mass) and orbital speed of the planet in order to make even a simple calculation. Aerospace engineers make good use of this natural planetary tug on a flyby spacecraft and they call this important orbital mechanics technique, a gravity-assist maneuver.

Flyby spacecraft follow a continuous trajectory and are not captured into a planetary orbit. These spacecraft have the capability of using their onboard instruments to observe passing celestial targets (for example, a planet, a moon, an asteroid), even compensating for the target's apparent motion in an optical instrument's field of view. They must be able to transmit data at high rates back to Earth and also must be capable of storing data onboard for those periods when their antennas are not pointing toward Earth. Flyby spacecraft must also be capable of surviving in a powered-down, cruise mode for many years of travel through interplanetary space and then of bringing all their sensing systems to focus rapidly on the target object, during an encounter period that may last only for a few crucial hours or minutes. NASA's *Pioneer 10* and *11* and the *Voyager 1* and *2* are examples of highly successful flyby scientific spacecraft. NASA uses the flyby spacecraft during the initial, or reconnaissance phase, of solar system exploration.

An orbiter spacecraft is designed to travel to a distant planet and then orbit around that planet. This type of scientific spacecraft must possess a substantial propulsive capability to decelerate at just the right moment in order to achieve a proper orbit insertion. Aerospace engineers design an orbiter spacecraft recognizing the fact that solar occultations will occur frequently as it orbits the target planet. During these periods of occultation, the spacecraft is shadowed by the planet, cutting off solar array production of electric power and introducing extreme variations of the spacecraft's thermal environment. Generally, a rechargeable battery system augments solar electric power. Active thermal control techniques (e.g., the use of tiny electric-powered heaters) are used to complement traditional passive thermal control design features. The periodic solar occultations also interrupt uplink and downlink communications with Earth, making onboard data storage a necessity. NASA uses orbiter spacecraft as part of the second, in-depth study phase of solar system exploration. The *Lunar Orbiter*, *Magellan*, *Galileo*, and *Cassini* spacecraft are examples of successful scientific orbiters.

Some scientific exploration missions involve the use of one or more smaller, instrumented spacecraft, called atmospheric probe spacecraft. These probes separate from the main spacecraft prior to closest approach to a planet in order to study the planet's gaseous atmosphere as they descend through it. Usually an atmospheric probe spacecraft is deployed from its mother spacecraft (that is, the main or carrier spacecraft) by the release of springs or other devices that simply separate it from the mother spacecraft without making a significant modification of the probe's trajectory. Following probe release, the mother spacecraft usually executes a trajectory correction maneuver to prevent its own atmospheric entry and to help the main spacecraft continue on with its flyby or orbiter mission activities. NASA's *Pioneer Venus* (four probes), *Galileo* (one probe), and *Cassini* (*Huygens* probe) missions involved the deployment of a probe or probes into the target planetary body's atmosphere (that is, Venus, Jupiter, and Saturn's moon Titan, respectively).

An aeroshell protects the atmospheric probe spacecraft from the intense heat caused by atmospheric friction during entry. At some point in the descent trajectory, the aeroshell is jettisoned and a parachute then is used to slow the probe's descent sufficiently so it can perform its scientific observations. Data usually are telemetered from the atmospheric probe to the mother spacecraft, which then either relays the data back to Earth in real time or else records the data for later transmission to Earth.

An atmospheric balloon package is designed for suspension from a buoyant gas-filled bag that can float and travel under the influence of the winds in a planetary atmosphere. Tracking of the balloon package's progress across the face of the target planet yields data about the general circulation patterns of the planet's atmosphere. A balloon package needs a power supply and a telecommunications system (to relay data and support tracking). Scientists can also equip the balloon package with a variety of scientific instruments to measure the planetary atmosphere's composition, temperature, pressure, and density.

During their flyby of Venus in June 1985, the Russian *Vega 1* and *2* robot spacecraft deployed constant-pressure instrumented balloon aerostats. Each 3.4-meter diameter balloon had a 5-kilogram science payload suspended beneath it by a 12-meter-long cable. The aerostats floated at an altitude of approximately 50 kilometers, the most active layer of Venus's three-tiered cloud system. Data (such as temperature, pressure, and wind velocity) from each balloon's science instruments were transmitted directly to Earth for the 47-hour lifetime of the aerostat mission. After two days of operation and floating almost 9,000 kilometers through the atmosphere of Venus, each balloon entered the sunlit dayside of the planet, experienced overexpansion due to solar heating, and burst.

Lander spacecraft are designed to reach the surface of a planet and survive at least long enough to transmit back to Earth useful scientific data, such as imagery of the landing site, measurement of the local environmental conditions, and an initial examination of soil composition. For example, the Russian Venera lander spacecraft have made brief scientific investigations of the inferno-like Venusian surface. In contrast, NASA's Surveyor lander craft extensively explored the lunar surface at several landing sites in preparation for the human Apollo Project landing missions, while NASA's *Viking 1* and *2* lander craft investigated the surface conditions of Mars at two separate sites for many months.

A surface penetrator spacecraft is designed to enter the solid body of a planet, an asteroid, or a comet. It must survive a high-velocity impact and then transmit subsurface information back to an orbiting mother spacecraft.

NASA launched the *Mars Polar Lander* (MPL) spacecraft in early January 1999. MPL was an ambitious mission to land a robot spacecraft on the frigid surface of Mars near the edge of the planet's southern polar cap. Two small penetrator probes (called *Deep Space 2*) piggybacked with the lander spacecraft on the trip to Mars. After an uneventful interplanetary journey, all contact with the MPL and the *Deep Space 2* penetrator experiments was lost as the spacecraft arrived at the planet on December 3, 1999. The missing lander was equipped with cameras, a robotic arm, and instruments to measure the composition of the Martian soil. The two tiny penetrators were to be released as the lander spacecraft approached

Mars and then follow independent ballistic trajectories, impacting on the surface and plunging below it in search of water ice.

The exact fate of the lander and its two tiny microprobes remains a mystery. Some NASA engineers believe that the MPL might have tumbled down into a steep canyon, while others speculate the MPL may have experienced too rough a landing and become disassembled. A third hypothesis suggests the MPL may have suffered a fatal failure during its descent through the Martian atmosphere. No firm conclusions could be drawn because the NASA mission controllers were completely unsuccessful in communicating with the missing lander or either of its hitchhiking planetary penetrators.

Finally, a surface rover spacecraft is carried to the surface of a planet, soft-landed, and then deployed. The rover can either be semiautonomous or fully controlled (through teleoperation) by scientists on Earth. Once deployed on the surface, the electrically powered rover can wander a certain distance away from the landing site and take images and perform soil analyses. Data then are telemetered back to Earth by one of several techniques; via the lander spacecraft, via an orbiting mother spacecraft, or (depending on size of rover) directly from the rover vehicle. The Soviet Union deployed two highly successful robot surface rovers (called *Lunokhod 1* and *2*) on the Moon in the 1970s. In December 1996, NASA launched the Mars Pathfinder mission to the Red Planet. From its innovative airbag-protected bounce and roll landing on July 4, 1997, until the final data transmission on September 27, the robot lander/rover team returned numerous close-up images of Mars and chemical analyses of various rocks and soil found in the vicinity of the landing site. The *Spirit* and *Opportunity* (2003) Mars Exploration Rovers are the first of many robot rovers that will scamper across the Red Planet this century. *Spirit* landed successfully on Mars on January 3, 2004, and *Opportunity* successfully on January 24, 2004. Both rovers exceeded their mission design lifetime manifold and, as of July 1, 2006, still functioned on the surface of the Red Planet.

An observatory spacecraft is a space robot that does not travel to a destination to explore. Instead, this type of robot spacecraft travels in an orbit around Earth or around the Sun, from where the observatory can view distant celestial targets unhindered by the blurring and obscuring effects of Earth's atmosphere. NASA's *Hubble Space Telescope* (HST) and *Spitzer Space Telescope* (SST) are examples.

The SST is the final mission in NASA's Great Observatories Program—a family of four orbiting observatories each studying the universe in a different portion of the electromagnetic spectrum. The SST—previously called the *Space Infrared Telescope Facility* (SIRTF)—consists of a 0.85-meter diameter telescope and three cryogenically cooled science instruments. NASA renamed this space-based infrared telescope to honor the American astronomer Lyman Spitzer, Jr. (1914–1997) The SST represents the most powerful and sensitive infrared telescope ever launched. The orbiting facility obtains images and spectra of celestial objects at infrared radiation wavelengths between 3 and 180 micrometers (μm)—an important spectral region of observation mostly unavailable to ground-based telescopes because of the blocking influence of Earth's atmosphere. Following a successful launch (August 25, 2003) from Cape Canaveral, SST traveled to an Earth-trailing heliocentric orbit that allowed the telescope to cool rapidly with

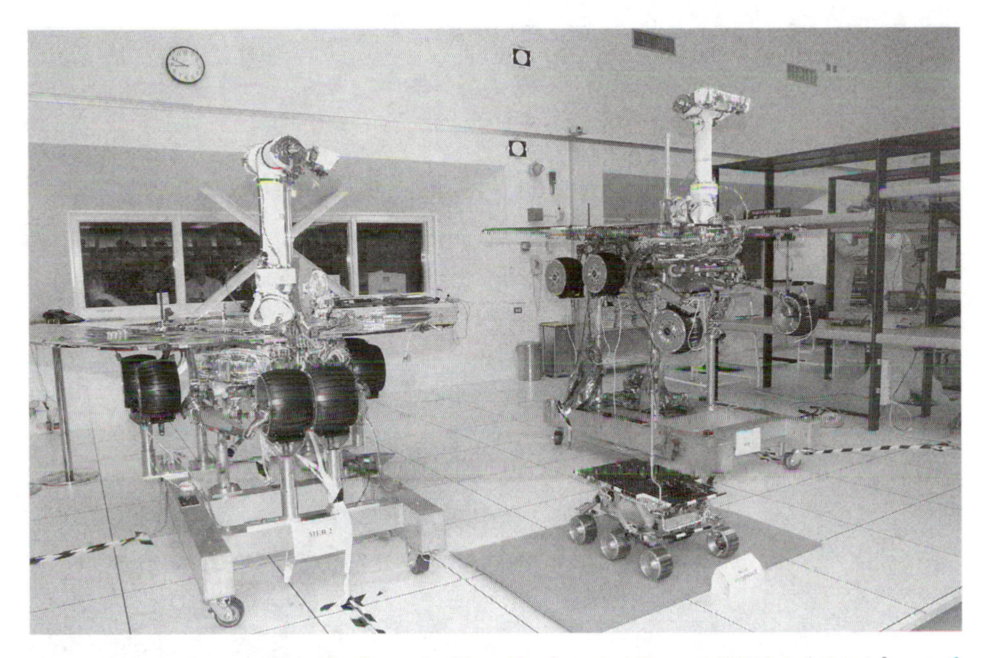

Figure 4-10 Prior to launch, the twin Mars Exploration Rovers (MERs), *Spirit* (elevated on right) and *Opportunity* (elevated on left) pose with the flight spare of their groundbreaking robot predecessor, the *Mars Pathfinder* minirover (on floor) at NASA's Jet Propulsion Laboratory in February 2003. (Credit: Photograph courtesy of NASA/JPL.)

a minimum expenditure of onboard cryogen (cryogenic coolant). With a projected mission lifetime of at least 2.5 years, SST has taken its place alongside NASA's other great orbiting astronomical observatories and is now collecting high-resolution infrared data that help scientists better understand how galaxies, stars, and planets form and develop. Other robot observatory missions in this program included the *Hubble Space Telescope* (HST), the *Compton Gamma Ray Observatory* (CGRO), and the *Chandra X-ray Observatory* (CXO).

Functional Subsystems for Robot Spacecraft

A robot spacecraft's functional subsystems support the mission-oriented science payload and allow the space robot to operate, collect data, and communicate back with Earth. Aerospace engineers attach all the other spacecraft components on the structural subsystem. Aluminum is the most common spacecraft structural material. The engineer can select from a wide variety of aluminum alloys, providing the spacecraft designer a broad range of physical characteristics, such as strength and machinability. A space robot's structure may also contain magnesium, titanium, beryllium, steel, fiberglass, or low-mass and high-strength carbon composite materials.

How much power does a robot spacecraft need? Engineers have learned from experience that a complex robot spacecraft needs between 300 and 3,000 watts (electric) to properly conduct its mission. Small short-lived robot spacecraft, such

Figure 4-11 This interesting picture shows one famous space robot interacting with another famous space robot. The space shuttle *Discovery's* remote manipulator system (RMS) is shown carefully lifting the *Hubble Space Telescope* (HST) out of the shuttle's cargo bay. This event took place in February 1997 during the STS-82 mission, also known as the second HST servicing mission (HST SM-02). (Credit: Photograph courtesy of NASA/JSC.)

as an atmospheric probe and minirover might need only 25 to 100 watts (electric), which can often be supplied by long-lived batteries. However, the less power available, the less performance and flexibility the engineers can give the space robot.

The power subsystem must satisfy all the electric power needs of the robot spacecraft. Engineers commonly use a solar-photovoltaic (solar-cell) system in combination with rechargeable batteries to provide a continuous supply of electricity. The spacecraft must also have a well-designed, built-in electric utility grid, which conditions and distributes power to all onboard consumers.

Solar photovoltaic conversion is the direct conversion of sunlight (solar energy) into electrical energy by means of the photovoltaic effect. A single photovoltaic (PV) converter cell is called a solar cell, while a combination of cells, designed to increase the electric power output, is called a solar array or a solar panel.

Since 1958, solar cells have been used to provide electric power for a wide variety of spacecraft. The typical spacecraft solar cell is made of a combination of n-type (*negative*) and p-type (*positive*) semiconductor materials (generally silicon). When this combination of materials is exposed to sunlight, some of the incident electromagnetic radiation removes bound electrons from the semiconductor material atoms, thereby producing free electrons. A hole (positive charge) is left at each location from which a bound electron has been removed. Consequently, an equal number of free electrons and holes are formed. An electrical barrier at the p-n junction causes the newly created free electrons near the barrier to migrate deeper into the n-type material and the matching holes to migrate further into the p-type material.

If electrical contacts are made with the n- and p-type materials and these contacts connected through an external load (conductor), the free electrons will flow from the n-type material to the p-type material. Upon reaching the p-type material, the free electrons will enter existing holes and once again become bound electrons. The flow of free electrons through the external conductor represents an electric current that will continue as long as more free electrons and holes are being created by exposure of the solar cell to sunlight. This is the general principle of solar photovoltaic conversion.

Solar arrays work very well on Earth-orbiting spacecraft and on spacecraft that operate in the inner solar system (within the orbit of Mars and outside the orbit of Mercury). Solar cells do not work well on spacecraft that must fly very close to the Sun, because of the severe thermal environment encountered. Also, the ionizing radiation environment that a spacecraft experiences in interplanetary space (for example, a large solar flare) or while orbiting in a planet's trapped radiation belt can damage the solar cells and significantly reduce their useful lifetime.

Some robot spacecraft must operate for years in deep space or in very hostile planetary environments, where a solar photovoltaic power subsystem becomes impractical if not altogether infeasible. Under these mission circumstances, the engineer selects a long-lived nuclear power supply called a radioisotope thermoelectric generator (RTG). The RTG converts the decay heat from a radioisotope directly into electricity by means of the thermoelectric effect. The United States uses the radioisotope plutonium-238 as the nuclear fuel in its RTGs.

A spacecraft's attitude control subsystem includes the onboard system of computers, low-thrust rockets (thrusters), and mechanical devices (such as a momentum wheel) used to keep the spacecraft stabilized during flight and to precisely point its instruments in some desired direction. Stabilization is achieved by spinning the spacecraft or by using a three-axis active approach that maintains the spacecraft in a fixed reference attitude by firing a selected combination of thrusters when necessary.

Stabilization by spinning the spacecraft, as was done on the *Pioneer 10* and *11* spacecraft during their missions to the outer solar system employs the gyroscopic action of the rotating spacecraft mass as the stabilizing mechanism. Propulsion system thrusters are fired to make any desired changes in the spacecraft's spin-stabilized attitude.

Spacecraft designed for active three-axis stabilization, as were the *Voyager 1* and *2* spacecraft which explored the outer solar system and beyond, rely on small propulsion system thrusters gently nudging the spacecraft back and forth within a deadband of allowed attitude error. Another method of achieving active three-axis stabilization is to use electrically powered reaction wheels, which are also called momentum wheels. These massive wheels are mounted in three orthogonal axes onboard the spacecraft. To rotate the spacecraft in one direction, the proper wheel is spun in the opposite direction. To rotate the vehicle back, the wheel is slowed down. Excessive momentum, which builds up in the system due to internal friction and external forces, occasionally must be removed from the system; this usually is accomplished with propulsive maneuvers.

Either general approach to spacecraft stabilization has basic advantages and disadvantages. Spin-stabilized vehicles provide a continuous "sweeping motion," that is generally desirable for fields and particles science instruments. However, such spacecraft may then require complicated systems to despin antennas or optical instruments that must be pointed at targets in space. Three-axis controlled spacecraft can point antennas and optical instruments precisely (without the necessity for despinning), but these robot craft may then have to perform rotation maneuvers to use their fields and particles instruments properly.

Some robot spacecraft have an articulation control subsystem, which is closely associated with a spacecraft's attitude control subsystem. The arti-culation control subsystem controls the movement of jointed or folded components and assemblies. Examples include a packaged solar array that is unfolded following launch; a robot arm that extends from a lander and scoops up soil; and an electro-optical imaging system on a steerable platform, which can track a planetary target during a flyby encounter.

Scientists and engineers have used two families of detectors to perform electro-optical imaging from scientific spacecraft. These are vidicons and the newer charge coupled devices (CCDs). Although the detector technology differs, in each case an image of the target celestial object is focused by a telescope onto the detector, where it is converted to digital data. Color imaging requires three exposures of the same target, through three different color filters selected from a filter wheel. Ground processing combines data from the three black-and-white images, reconstructing the original color by using three values for each picture element (pixel).

A vidicon is a vacuum tube resembling a small cathode ray tube (CRT). An electron beam is swept across a phosphor coating on the glass where the image is focused, and its electrical potential varies slightly in proportion to the levels of light it encounters. This varying potential becomes the basis of the video signal produced. *Viking*, *Voyager*, and many earlier NASA spacecraft used vidicon-based electro-optical imaging systems to send back spectacular images of Mars (*Viking1* and *2* orbiter spacecraft) and the outer planets: Jupiter, Saturn, Uranus, and Neptune (*Voyager 1* and *2* flyby spacecraft).

The newer CCD imaging system is typically a large-scale integrated circuit that has a two-dimensional array of hundreds of thousands of charge-isolated wells, each representing a pixel. Light falling on a well is absorbed by a photoconductive substrate (for example, silicon) and releases a quantity of electrons proportional to the intensity of the incident light. The CCD then detects and stores accumulated electrical charges, which represent the light level on each well. These charges subsequently are read out for conversion to digital data. CCDs are much more sensitive to light over a wider portion of the electromagnetic spectrum than vidicon tubes; they are also less massive and require less energy to operate. In addition, they interface more easily with digital circuitry, simplifying (to some extent) onboard data processing and transmission back to Earth. The *Galileo* spacecraft's solid state imaging (SSI) instrument contained a CCD with an 800 × 800 pixel array.

Not all CCD imagers have two-dimensional arrays. The imaging instrument on NASA's *Mars Global Surveyor* orbiter spacecraft has a detector, called the Mars Orbiter Camera (MOC), consisting of a single line of CCD sensors. As the spacecraft moves in orbit around Mars, the single line of CCD sensors creates a two-dimensional image of the Martian surface in a push-broom effect due to the spacecraft's motion.

The attitude control subsystem works closely with a robot spacecraft's propulsion subsystem and makes sure that the space robot points in the right direction before a major rocket engine burn or a sequence of tiny thruster firings occurs. Minor attitude adjustments usually take place automatically, as a smart space robot essentially drives itself through interplanetary space. Some major rocket engine burns take place under the supervision of mission controllers on Earth, who uplink precise firing instructions to the spacecraft's computer/clock through the telecommunications subsystem. Other major propulsion system firings, like an orbit injection burn, involve a totally automated sequence of events.

The process of planetary orbit insertion places the robot spacecraft at precisely the correct location at the correct time to enter into an orbit about the target planet. Orbit insertion requires not only the precise position and timing of a flyby mission, but also a controlled deceleration. As the spacecraft's trajectory is bent by the planet's gravity, the command sequence within the onboard computer/clock subsystem fires the spacecraft's retroengine(s) at the proper moment and for the proper duration. Once this retroburn (or retrofiring) has been completed successfully, the spacecraft is captured into orbit by its target planet. If the retroburn fails (or is improperly sequenced), the spacecraft will continue to fly past the planet. It is quite common for this retroburn to occur on the farside of a planet as viewed from Earth—requiring this portion of the orbit insertion sequence to occur essentially automatically (based entirely on onboard commands and machine intelligence) and without any interaction with the flight controllers on Earth.

The thermal control subsystem maintains the temperature of a robot spacecraft and keeps it from getting too hot or too cool. Thermal control is a complex problem because of the severe temperature extremes a space robot experiences during a typical scientific mission. In the vacuum environment of outer space, radiation heat transfer is the only natural mechanism for exchanging thermal energy (heat) into or out of a spacecraft. Under some special circumstances, a gaseous or liquid working fluid might be dumped from the spacecraft to provide a temporary solution to a transient heat load—but this is an extreme exception rather than the generally accepted design approach to thermal control. The overall thermal energy balance for a spacecraft near a planetary body is determined by several factors: thermal energy sources within the spacecraft; direct solar radiation (the Sun has a characteristic blackbody temperature of about 5,770 K); direct thermal (infrared) radiation from the planet (e.g., Earth has an average surface temperature of about 288 K); indirect (reflected) solar radiation from the planetary body; and thermal radiation emitted from the surface of the spacecraft to the low-temperature sink of outer space (deep space has a temperature of about 3 K).

Under these conditions, thermally isolated portions of a spacecraft in orbit around Earth could encounter temperature variations from about 200 K, during Earth-shadowed or darkness periods, to 350 K while operating in direct sunlight. Spacecraft materials and components can experience thermal fatigue due to repeated temperature cycling during such extremes. Consequently, engineers use great care in providing the proper thermal control for a spacecraft. As previously mentioned, radiation heat transport is the principal mechanism for heat flow into and out of the spacecraft, while conduction heat transfer generally controls the flow of heat within the spacecraft.

There are two major approaches to spacecraft thermal control, passive and active. Passive thermal control techniques include the use of special paints and coatings, insulation blankets, radiating fins, sun shields, heat pipes, as well as the careful selection of the spacecraft's overall geometry (that is, both the external and internal placement of temperature-sensitive components). Active thermal control techniques include the use of heaters (including small radioisotope sources) and coolers, louvers and shutters, or the closed-loop pumping of cryogenic materials.

An open-loop flow (or overboard dump) of a rapidly heated working fluid might be used to satisfy a one-time or occasional special mission requirement to remove a large amount of thermal energy in a short period of time. Similarly, a sacrificial ablative surface could be used to handle a singular, large transitory external heat load. But these transitory (essentially one-shot) thermal control approaches are the exception rather than the engineering norm.

For interplanetary spacecraft, engineers often use passive thermal control techniques such as surface coatings, paint, and insulation blankets to provide an acceptable thermal environment throughout the mission. Components painted black will radiate more efficiently. Surfaces covered with white paint or white thermal blankets will reflect sunlight effectively and protect the spacecraft from excessive solar heating. Engineers also use gold (that is, gold-foil surfaces) and quartz mirror tiles on the surfaces of special components.

Active heating can be used to keep components within tolerable temperature limits. Resistive electric heaters controlled either autonomously or on command from Earth can be applied to special components to keep them above a certain minimum allowable temperature during the mission. Similarly, radioisotope heat sources (generally containing a small quantity of plutonium-238) can be installed where necessary to provide at-risk components with a small, essentially permanent supply of thermal energy. The small radioisotope heat sources are especially useful for specific components on lander and rover robots that must stay within certain temperature limits in order to survive the frigid nighttime conditions experienced on the surface of the Moon or Mars.

Spacecraft Clock and the Data Handling Subsystem

A modern clock is generally an electronic circuit, often involving a fairly sophisticated integrated circuit, which produces high-frequency timing signals. One common application for high-precision electronic clocks is synchronization of the operations performed by a computer or microprocessor-based system. Typical clock rates in microprocessor circuits are in the one megahertz range, with 1 megahertz (1 MHz) corresponding to 1 million cycles per second.

Aerospace engineers usually make a robot spacecraft's clock an integral part of the command and data handling subsystem. The spacecraft clock is very important because it meters the passing time during the life of the space robot mission and regulates nearly all activity within the spacecraft. The clock may be very simple (for example, incrementing every second and bumping its value up by one), or it may be much more complex (with several main and subordinate fields of increasing temporal resolution down to milliseconds, microseconds, or less). In aerospace operations, many types of commands that are uplinked to the spacecraft are set to begin execution at specific spacecraft clock counts. In downlinked telemetry, spacecraft clock counts (which indicate the time a telemetry frame was created) are included with engineering and science data to facilitate processing, distribution, and analysis.

The data handling subsystem is the onboard computer responsible for the overall management of a robot spacecraft's activity. Aerospace engineers often refer to this type of multifunctional spacecraft computer, as the command and data handling subsystem. The important subsystem is usually the same computer that maintains timing, interprets commands from Earth, collects, processes, and formats the telemetry data that are to be returned to Earth, and manages high-level fault protection and fail-safe routines. Under fail-safe design philosophy, engineers try to design aerospace system hardware that avoids compounding failures. Should a component fail, the subsystem moves into a predetermined "safe" position, before the failure can cause further damage. Fail-safe design allows a robot spacecraft to sustain a failure and still retain the capability to accomplish most, if not all, its planned mission.

Fault tolerance is the capability of a robot spacecraft (or one of its major subsystems) to function despite experiencing one or more component failures or software glitches. Engineers use redundant circuits or functions, as well as components that can readily be reconfigured, to construct spacecraft that are very fault tolerant. For the robot flight system to enjoy an effective level of fault tolerance, the spacecraft's main computer must be robust and contain a great deal of internal redundancy. The computer must also possess a high level of machine intelligence, so it can monitor the health and status of all spacecraft subsystems, quickly detect imminent failures, and then promptly take effective action to curtail the problem—without direct human guidance or supervision. For example, the spacecraft's computer could issue commands to the affected subsystem, activating standby hardware or making software changes—either or both of which steps constitute a viable workaround repair or safe isolation of the fault. Prompt isolation of the troublesome equipment, or misbehaving software, prevents the original fault from rippling through the spacecraft.

Navigation—Helping a Robot Find Its Way in Space

There are two main aspects to the navigation of a robot in space. First is orbit determination—the task involving knowledge and prediction of the spacecraft's position and velocity. The second aspect of spacecraft navigation is flight path control—the task involving the firing a spacecraft's onboard propulsion systems (such as a retrorocket motor or tiny attitude control rockets) to alter the spacecraft's velocity.

Navigating a robot spacecraft in deep space is a challenging task. For example, no single measurement directly provides mission controllers information about the lateral motion of a spacecraft as it travels on a mission deep in the solar system. Aerospace engineers define lateral motion as any motion except motion directly toward or away from Earth (which motion is called radial motion). Spacecraft flight controllers use measurements of the Doppler shift of telemetry (particularly a coherent downlink carrier) to obtain the radial component of a spacecraft's velocity relative to Earth. Spacecraft controllers add a uniquely coded ranging pulse to an uplink communication with a spacecraft and record the transmission time. When the spacecraft receives this special ranging pulse, it returns a similarly coded pulse on its downlink transmission. Engineers know how long it takes the spacecraft's onboard electronics to "turn" the ranging pulse around. For example, the *Cassini* spacecraft takes 420 nanoseconds (ns) ± 9 ns to turn the ranging pulse around. There are other known and measured (calibrated) delays in the overall transmission process, so when the return pulse is received back on Earth—at NASA's Deep Space Network (DSN), for example—spacecraft controllers can then calculate how far (radial distance) the spacecraft is away from Earth. Mission controllers also use angular quantities to express a spacecraft's position in the sky.

Robot spacecraft that carry electro-optical imaging instruments can use these instruments to perform optical navigation. They can observe the destination (target) planet or moon against a known background star field. Mission controllers will often carefully plan and uplink appropriate optical navigation images as part of a planetary encounter command sequence uplink. When the spacecraft collects optical navigation images, it immediately downlinks (transmits) these images to the human navigation team at mission control. The mission controllers then rapidly process the optical imagery and use these data to obtain precise information about the spacecraft's trajectory as it approaches its celestial target.

Once a spacecraft's solar or planetary orbital parameters are known, these data are compared to the planned mission data. If there are discrepancies, mission controllers plan for and then have the spacecraft execute an appropriate trajectory correction maneuver (TCM). Similarly, small changes in a spacecraft's orbit around a planet may become necessary to support the scientific mission. In that case, the mission controllers plan for and instruct the spacecraft to execute an appropriate orbit trim maneuver (OTM). This generally involves having the spacecraft fire some of its low-thrust, attitude-control rockets. Trajectory correction and orbit trim maneuvers use up a spacecraft's onboard propellant supply, which is often a very carefully managed, mission-limiting consumable.

Telecommunications—Helping Space Robots Phone Home

Aerospace engineers use the word telecommunications to describe the flow of data and information (usually by radio signals) between a spacecraft and an Earth-based communications system. A robot spacecraft generally has only a limited amount of power available to transmit a signal that sometimes must travel across millions or even billions of miles (kilometers) of space before reaching Earth. A deep space exploration spacecraft often has a transmitter that has no more than 20 watts of radiating power.

One part of aerospace engineering's solution to this problem is to concentrate all available radiating power for signal generation into a narrow radio beam and then to send this narrow beam in just one direction, instead of broadcasting the radio signal in all directions. This is often accomplished by using a parabolic dish antenna on the order of 1–5 meters in diameter.

When these concentrated radio signals reach Earth, however, they have very small power levels. The other portion of the solution to the telecommunications problem is to use special, large-diameter radio receivers on Earth, such as found in NASA's Deep Space Network, which is discussed shortly. These sophisticated radio antennas are capable of detecting the very-low-power signals from distant spacecraft.

In telecommunications, the radio signal transmitted to a spacecraft is called the uplink. The transmission from the spacecraft to Earth is called the downlink. Uplink or downlink communications may consist of a pure radio-frequency (RF) tone (called a carrier), or these carriers may be modified to carry information in each direction. Engineers sometimes refer to commands transmitted to a spacecraft as an upload. Communications with a spacecraft involving only a downlink are called one-way communications (OWC). When the spacecraft is receiving an uplink signal at the same time that a downlink signal is being received on Earth, the telecommunications mode is often referred to as two-way communications (TWC).

Engineers usually modulate spacecraft carrier signals by shifting each waveform's phase slightly at a given rate. One scheme is to modulate the carrier with a frequency, for example, near one megahertz (MHz). This one MHz modulation is then called a subcarrier. The subcarrier is modulated to carry individual phase shifts that are designated to represent binary ones (1s) and zeros (0s)—the spacecraft's telemetry data. The amount of phase shift used in modulating data onto the subcarrier is referred to as the modulation index and is measured in degrees. This same type of communications scheme is also on the uplink. Binary digital data modulated onto the uplink are called command data. They are received by the spacecraft and either acted upon immediately or stored for future use or execution. Data modulated onto the downlink are called telemetry and include science data from the spacecraft's instruments and spacecraft state-of-health data from sensors within the various functional subsystems (such as, power, propulsion, thermal control, and so forth).

Demodulation is the process of detecting the subcarrier and processing it separately from the carrier, detecting the individual binary phase shifts, and registering them as digital data for further processing. The device used for this is called a modem, which is short for modulator/demodulator. These same

Figure 4-12 A view of the 70-meter-diameter antenna of the Canberra Deep Space Communications Complex, located outside Canberra, Australia. This facility is one of three complexes that comprise NASA's Deep Space Network (DSN). The other complexes are located in Goldstone, California, and Madrid, Spain. National flags representing the three DSN sites appear in the foreground of this image. (Credit: Courtesy of NASA/JPL.)

processes of modulation and demodulation are often used with Earth-based computer systems and facsimile (fax) machines to transmit data back and forth over a telephone line. Before the era of high-speed cable connections, when a person used his or her personal computer to chat over the Internet, their dial-up modem would employ a familiar audio frequency carrier that the telephone system could handle.

The dish-shaped, high-gain antenna (HGA) is the type of antenna frequently used by robot spacecraft for communications with Earth. The amount of gain achieved by an antenna refers to the amount of incoming radio signal power it can collect and focus into the spacecraft's receiver(s). In the frequency ranges used by spacecraft, the HGA incorporates a large parabolic reflector. Such an antenna may be fixed to the spacecraft bus or may be steerable. The larger the collecting area of the HGA, the higher the gain, and the higher the data rate it will support. However, the higher the gain, the more highly directional the antenna becomes. Therefore, when a spacecraft uses an HGA, the antenna must be pointed within a fraction of a degree of Earth for communications to occur. Once this accurate antenna pointing is achieved, communications can take place at a high rate, using a highly focused radio signal.

The low-gain antenna (LGA) provides wide-angle coverage at the expense of gain. Coverage is nearly omni-directional, except for areas that may be shadowed by the spacecraft structure. The LGA is designed for relatively low data rates. It is useful as long as the spacecraft is relatively close to Earth (for example, within a few astronomical units). Sometimes a spacecraft is given two LGAs to provide full omni-directional coverage, since the second LGA will avoid the spacecraft structure blind spots experienced by the first LGA. Engineers often mount the LGA on top of the HGA's subreflector.

The medium-gain antenna (MGA) represents a design compromise in spacecraft engineering. Specifically, the MGA provides more gain than the LGA and has wider-angle antenna-pointing accuracy requirements (typically 20 to 30 degrees) than the HGA.

The majority of NASA's scientific investigations of the solar system have been accomplished through the use of robot spacecraft. The Deep Space Network (DSN) provides the two-way communications link that guides and controls these spacecraft and brings back the spectacular planetary images and other important scientific data they collect.

The DSN consists of telecommunications complexes strategically placed on three continents—providing almost continuous contact with scientific spacecraft traveling in deep space as Earth rotates on its axis. The DSN is the largest and most sensitive scientific telecommunications system in the world. It also performs radio and radar astronomy observations in support of NASA's mission to explore the solar system and the universe. The Jet Propulsion Laboratory (JPL) in Pasadena, California, manages and operates the DSN for NASA.

The JPL established the predecessor to the DSN. Under a contract with the United States Army in January 1958, the laboratory deployed portable radio tracking stations in Nigeria (Africa), Singapore (Southeast Asia), and California to receive signals from and plot the orbit of *Explorer 1*—the first American satellite to successfully orbit Earth. Later that year (on December 3, 1958), as part of the emergence the new federal civilian space agency, JPL was transferred from U.S. Army jurisdiction to that of NASA. At the very onset of the nation's civilian space program, NASA assigned JPL responsibility for the design and execution of robotic lunar and planetary exploration programs. Shortly afterward, NASA embraced the concept of the DSN as a separately managed and operated telecommunications facility that would accommodate all deep space missions. This management decision avoided the need for each space flight project to acquire and operate its own specialized telecommunications network.

Today, the DSN features three deep space communications complexes placed approximately 120 degrees apart around the world: at Goldstone in California's Mojave Desert; near Madrid, Spain; and near Canberra, Australia. This global configuration ensures that, as Earth rotates, an antenna is always within sight of a given spacecraft, day and night. Each complex contains up to 10 deep space communication stations equipped with large parabolic reflector antennas.

Every deep space communications complex within the DSN has a 70-meter diameter antenna. These antennas, the largest and most sensitive in the DSN, are capable of tracking robot spacecraft that are more than 16 billion kilometers away from Earth. The 3,850 square meter surface of the 70-meter diameter reflector must remain accurate within a fraction of the signal wavelength, meaning that the dimensional precision across the surface is maintained to within one centimeter. The dish and its mount have a mass of nearly 7.2 million kilograms.

There is also a 34-meter diameter, high-efficiency antenna at each complex, which incorporates advances in radio frequency antenna design and mechanics. The reflector surface of the 34-meter diameter antenna is precision shaped for maximum signal-gathering capability.

The most recent additions to the DSN are several 34-meter beam waveguide antennas. On earlier DSN antennas, sensitive electronics were centrally mounted on the hard-to-reach reflector structure, making upgrades and repairs difficult. On beam waveguide antennas, the sensitive electronics are now located in a below ground pedestal room. Telecommunications engineers bring an incident radio signal from the reflector to this room through a series of precision-machined radio frequency reflective mirrors. Not only does this architecture provide the advantage of easier access for maintenance and electronic equipment enhancements, but the new configuration also accommodates better thermal control of critical electronic components. Furthermore, engineers can place more electronics in the antenna to support operation at multiple frequencies. Three of these new 34-meter beam waveguide antennas have been constructed at the Goldstone, California complex, along with one each at the Canberra and Madrid complexes.

There is also one 26-meter diameter antenna at each complex for tracking Earth-orbiting satellites, which travel primarily in orbits 160–1,000 kilometers above Earth. The two-axis astronomical mount allows these antennas to point low on the horizon to acquire (pick up) fast-moving satellites as soon as they come into view. The agile 26-meter diameter antennas can track (slew) at up to three degrees per second.

Finally, each complex also has one 11-meter diameter antenna to support a series of international Earth-orbiting missions under the Space Very Long Baseline Interferometry project.

All of the antennas in the DSN communicate directly with the Deep Space Operations Center (DSOC) at JPL in Pasadena, California. The DSOC staff directs and monitors operations, transmits commands, and oversees the quality of spacecraft telemetry and navigation data delivered to network users. In addition to the DSN complexes and the operations center, a ground communications facility provides communications that link the three complexes to the operations center at JPL, to space flight control centers in the United States and overseas, and to scientists around the world. Voice and data communications traffic between various locations is sent via landlines, submarine cable, microwave links, and communications satellites.

The DSN's radio link to scientific robot spacecraft is basically the same as other point-to-point microwave communications systems, except for the very long distances involved and the very low radio frequency signal strength received from the robot spacecraft. The total signal power arriving at a network antenna from a typical robot spacecraft encounter among the outer planets can be 20 billion times weaker than the power level in a modern digital wristwatch battery.

The extreme weakness of these radio frequency signals results from restrictions placed on the size, mass, and power supply of a particular spacecraft by the payload volume and mass-lifting limitations of its launch vehicle. Consequently, the design of the radio link is the result of engineering tradeoffs between spacecraft transmitter power and antenna diameter, and the signal sensitivity that engineers can build into the ground receiving system.

Typically, a spacecraft signal is limited to 20 watts, or about the same amount of power required to light the bulb in a refrigerator. When the spacecraft's transmitted radio signal arrives at Earth—from, for example, the neighborhood of Saturn—it has spread over an area with a diameter equal to about 1,000 Earth diameters. (Earth has an equatorial diameter of 12,756 kilometers.) As a result, the ground antenna is able to receive only a very small part of the signal power, which is also degraded by background radio noise, or static.

Radio noise is radiated naturally from nearly all objects in the universe, including Earth and the Sun. Noise is also inherently generated in all electronic systems, including the DSN's own detectors. Since noise will always be amplified along with the signal, the ability of the ground receiving system to separate noise from the signal is critical. The DSN uses state-of-the-art, low-noise receivers and telemetry coding techniques to create unequalled sensitivity and efficiency.

Telemetry is basically the process of making measurements at one point and transmitting the data to a distant location for evaluation and use. A robot spacecraft sends telemetry to Earth, by modulating data onto its communications downlink. Telemetry includes state-of-health data about the spacecraft's subsystems and science data from its instruments. A typical scientific spacecraft transmits its data in binary code, using only the symbols 1 and 0. The spacecraft's data handling subsystem (telemetry system) organizes and encodes these data for efficient transmission to ground stations back on Earth. The ground stations have radio antennas and specialized electronic equipment to detect the individual bits, decode the data stream, and format the information for subsequent transmission to the data user (usually a team of scientists).

Data transmission from a robot spacecraft can be disturbed by noise from various sources that interferes with the decoding process. If there is a high signal-to-noise ratio, the number of decoding errors will be low. But if the signal-to-noise ratio is low, then an excessive number of bit errors can occur. When a particular transmission encounters a large number of bit errors, mission controllers will often command the spacecraft's telemetry system to reduce the data transmission rate (measured in bits per second) in order to give the decoder (at the ground station) more time to determine the value of each bit.

To help solve the noise problem, a spacecraft's telemetry system might feed additional or redundant data into the data stream, which additional data are then used to detect and correct bit errors after transmission. The information theory equations used by telemetry analysts in data evaluation are sufficiently detailed to allow the detection and correction of individual and multiple bit errors. After correction, the redundant digits are eliminated from the data, leaving a valuable sequence of information for delivery to the data user.

Error-detecting and encoding techniques can increase the data rate many times over transmissions that are not coded for error detection. DSN coding techniques have the capability of reducing transmission errors in spacecraft science information to less than one in a million.

Telemetry is a two-way process, having a downlink as well as an uplink. Robot spacecraft use the downlink to send scientific data back to Earth, while mission controllers on Earth use the uplink to send commands, computer software, and other crucial data to the spacecraft. The uplink portion of the telecommunications process allows human beings to guide spacecraft on their planned missions, as well as to enhance mission objectives through such important activities as upgrading a spacecraft's onboard software while the robot explorer is traveling through interplanetary space. When large distances are involved, human supervision and guidance is limited to non-real-time interactions with the robot spacecraft. That is why deep space robots must possess high levels of machine intelligence and autonomy.

Data collected by the DSN are also very important in precisely determining a spacecraft's location and trajectory. Teams of human beings (called the mission navigators) use these tracking data to plan all the maneuvers necessary to ensure that a particular scientific spacecraft is properly configured and at the right place (in space) to collect its important scientific data. Tracking data produced by the DSN let mission controllers know the location of a robot spacecraft that is billions of kilometers away from Earth, to an accuracy of just a few meters.

MILITARY ROBOTS

Military robots fall into several major categories: guided missiles, military spacecraft, unmanned aerial vehicles (UAVs), unmanned ground vehicle (UGVs), and underwater submersibles, both remotely operated vehicles (ROVs) and autonomous underwater vehicles (AUVs). This section provides a brief discussion about guided missiles and unmanned military spacecraft and focuses on UAVs and UGVs. Because of their dual military and civilian uses, underwater submersibles are treated in a subsequent section of this book.

Guided Missiles

The guided missile (GM) is an unmanned, self-propelled robot weapon that moves above the surface of Earth. Depending on the level of sophistication of its onboard computer and navigation equipment, this type of military robot can control (to some degree) its trajectory or course while in flight. An air-to-air guided missile (AAGM) is an air-launched vehicle for use against aerial targets. An air-to-surface guided missile (ASGM) is an air-launched missile for use against surface targets. A surface-to-air guided missile (SAGM) is a surface-launched guided missile for use against targets in the air. Finally, a surface-to-surface guided missile (SSGM) is a surface-launched missile for use against surface targets. SSGM are either cruise missiles or ballistics missiles.

A cruise missile is a guided missile (flying military robot) that travels within the atmosphere at aircraft speeds. The cruise missile usually flies at low altitude with a trajectory that is either preprogrammed or capable of changing as a result of environmental data or updated targeting information provided to the system while it is in flight. Because of its sophisticated onboard computer, this robot weapon is capable of achieving high accuracy in striking a distant target. It is maneuverable during flight, is constantly propelled, and, therefore, does not follow a ballistic trajectory. Cruise missiles may be armed with nuclear weapons or with conventional warheads (that is, high explosives). Cruise missiles are essentially technical first cousins to unmanned aerial vehicles (UAVs) like the Predator. The major difference is that the cruise missile is designed to fly on a one-way mission to destroy a target, while the UAV is designed to return to a home base, land, and then fly other aerial surveillance and intelligence collection operations.

The Tomahawk is a long-range, subsonic cruise missile used by the U.S. Navy for land attack and for antisurface warfare. Tomahawk is an all-weather submarine or ship-launched antiship or land-attack cruise missile. After launch, a solid-propellant rocket engine propels the missile until a small turbofan engine takes over for the cruise portion of the flight. This cruise missile is a highly survivable weapon. Radar detection is difficult because of its small cross section and low-altitude flight. Similarly, infrared detection is also difficult because the turbofan emits little heat.

The antiship variant of Tomahawk uses a combined active radar seeker and passive system to seek out, engage, and destroy a hostile ship at long range. Its modified Harpoon cruise missile guidance system permits the Tomahawk to be launched and fly at low altitudes in the general direction of an enemy warship to avoid radar detection. Then, at a programmed distance, the missile begins an active radar search to seek out, acquire, and hit the target ship.

The land-attack version has inertial and terrain contour matching (TERCOM) guidance. The TERCOM guidance system uses a stored map reference to compare with the actual terrain to help the missile determine its position. If necessary, a course correction is made to place the flying robot weapon on course to the target.

The basic Tomahawk is 5.56 meters long and has a mass of 1,192 kilograms, not including the booster. It has a diameter of 51.81 centimeters and a wingspan (when deployed) of 2.67 meters. This missile is subsonic and cruises at about 880 kilometers per hour. It can carry a conventional or nuclear warhead. In the land-attack (conventional warhead) configuration, it has a range of 1,100 kilometers; while in the land-attack (nuclear warhead) configuration, it has a range of 2,480 kilometers. In the antiship role, the Tomahawk CM has a range of over 460 kilometers. This guided missile was first deployed in 1983.

Undoubtedly, the most significant and powerful robot weapon developed in the twentieth century is the long-range ballistic missile armed with one or more nuclear warheads. This essentially unstoppable—program, shoot, and forget—robot weapon system has transformed military strategy and influences world politics to the present day. The ballistic missile is propelled by rocket engines and guided only during the initial (thrust producing) phase of its flight. In the non-powered and nonguided phase of its flight, it assumes a ballistic trajectory similar to that of an artillery shell. After thrust termination, reentry vehicles (RVs)

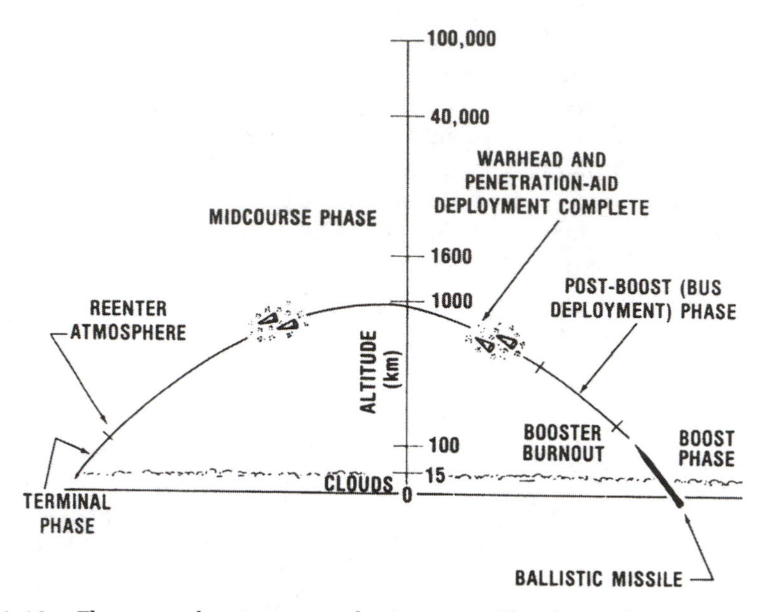

Figure 4-13 The general trajectory and mission profile of a modern intercontinental ballistic missile (ICBM). (Credit: Image courtesy of the U.S. Department of Defense.)

can be released and these RVs also follow free-falling (ballistic) trajectories toward their targets.

Aerospace analysts within the U.S. Department of Defense (DOD) often classify ballistic missiles by their maximum operational ranges, using the following scale: short-range ballistic missiles (SRBMs) are those that have a maximum operational range of about 1,100 kilometers; medium-range ballistic missiles (MRBMs) have an operational range between 1,100 and 2,750 kilometers; intermediate-range ballistic missiles (IRBMs) have an operational range between 2,750 and 5,500 kilometers; and intercontinental ballistic missiles (ICBMs) have operational ranges in excess of 5,500 kilometers. While somewhat arbitrary for an aerospace engineering perspective, this widely recognized classification scheme has proven quite useful in arms-control negotiations, ballistic missile treaty discussions, and international initiatives focused on limiting regional arms races and preventing the emergence of far-reaching ballistic missile threats from rogue nations.

The United States Air Force's Minuteman (also called the LGM-30) is a three-stage solid-propellant ICBM that is guided to its target by an all-inertial guidance and control system. These strategic missiles are equipped with nuclear warheads and designed for deployment in hardened and dispersed underground silos. The LGM-30 (Minuteman) ICBM is an element of the U.S. strategic deterrent force. The "L" in LGM stands for silo-configuration; "G" means surface attack; and "M" means guided missile.

The Minuteman robot weapon system was conceived in the late 1950s and deployed in the mid-1960s. Minuteman was a revolutionary concept and an extraordinary technical achievement. Both the missile and basing components incorporated significant advances beyond the relatively slow-reacting, liquid-fueled, remotely controlled ICBMs of the previous generation of missiles (such as the Atlas and the Titan). From the beginning, Minuteman missiles have provided a quick-reacting, inertially guided, highly survivable component of America's nuclear Triad. Minuteman's maintenance concept capitalizes on high reliability and a "remove-and-replace" approach to achieve a near 100 percent alert rate.

Through state-of-the-art improvements, the Minuteman system has evolved over three decades to meet new challenges and assume new missions. Modernization programs have resulted in new versions of the missile, expanded targeting options, and significantly improved accuracy. For example, when the Minuteman I became operational in October 1962, it had a single-target capability. The Minuteman II became operational in October 1965. While looking similar to the Minuteman I, the Minuteman II had greater range and targeting capability. Finally, the Minuteman III became operational in June 1970. This missile, with its improved third stage and the postboost vehicle, can deliver multiple independently targetable reentry vehicles and their penetration aids onto multiple targets. Over 500 Minuteman III's are currently deployed at bases in the United States.

Military Satellites

The military satellite is fundamentally a robot spacecraft in orbit around Earth that is used for military or defense purposes such as missile surveillance, navigation, intelligence gathering, battle damage assessment, and treaty monitoring. Since their arrival in the early 1960s, this collection of space robots have become an integral component of national defense strategies, supported military operations, and expanded information gathering and dissemination on a global basis. The expanded collection and flow of information essential to national security by space-based military systems now represents an essential factor in the preservation of a stable global civilization.

Space-based reconnaissance involves the acquisition of detailed information of a specific type that supports either strategic or tactical intelligence needs. Surveillance involves the use of sensors on satellites to support some type of continuous monitoring activity. Since the specific meaning of the two terms is sometimes difficult to separate, the following analogy is provided. A reconnaissance satellite is much like a military scout, traveling through hostile territory in an effort to gather certain important pieces of information. Where is the enemy? How numerous are the hostile forces? What type of weapons do they have? And so forth.

A surveillance satellite is similar to a guard or sentinel, who keeps watch from a tall tower and peers out across landscape to the distant horizon for signs of hostile activities. At the first sign a trouble, the sentinel sounds the alarm, thereby giving the friendly forces time to take appropriate defensive actions. Modern military satellites perform surveillance in three general categories: early warning (especially against ballistic missile attack), nuclear detonation detection (especially in support of nuclear test ban treaties), and weather monitoring (especially at the tactical or regional level). There are also military communication satellites and navigation satellites, whose functions assist peacekeeping and combat operations. For example, by sensing the hot plumes of ballistic missiles rising through the atmosphere, an early warning (surveillance) satellite can add crucial minutes to the strategic warning process and confirm the information being gathered by land-based radar early warning systems.

The U.S. Air Force launched the first Defense Support Program (DSP) missile surveillance satellite on November 6, 1970, using a Titan IIIC rocket, which lifted off from Launch Complex 40 at Cape Canaveral AFS, Florida. The satellite traveled to its operational (geostationary) orbit at an altitude of approximately 35,780 kilometers over the equator and began its vital mission to provide early warning of hostile intercontinental ballistic missile launches. Placed in geosynchronous orbit, a constellation of these surveillance satellites can detect missile launches, space launches, and nuclear detonations occurring around the world.

The primary (infrared) sensor of each DSP satellite supports near–real-time detection and reporting of missile launches against the United States and/or allied forces, interests, and assets worldwide. DSP satellites use an infrared sensor to detect heat from missile and booster plumes against Earth's background thermal signal. Other sensors on each satellite support the near-real-time detection and reporting of endoatmospheric (0–50 kilometers), exoatmospheric (50–300 kilometers), and deep space (>300 kilometers) nuclear detonations, worldwide.

Similarly, *Vela* spacecraft were part of a family of research and development satellites launched by the United States in the 1960s and early 1970s to detect nuclear detonations in the atmosphere down to Earth's surface, or in outer space at distances of more than 160 million kilometers. These spacecraft were jointly developed by the U.S. Department of Defense and the U.S. Atomic Energy Commission (now the Department of Energy) and were placed in pairs, 180 degrees apart in very high altitude (about 115,000 kilometers) orbits around Earth. The first pair of Vela spacecraft called *Vela 1A* and *Vela 1B* were launched successfully on October 17, 1963. The last pair of these highly successful, 26-sided (polyhedron-shaped) spacecraft called *Vela 6A* and *Vela 6B* were launched successfully on April 8, 1970.

It is interesting to note that the United States, the former Soviet Union, and the United Kingdom signed the Limited Nuclear Test Ban Treaty in October 1963. This treaty prohibits the signatories from testing nuclear weapons in Earth's atmosphere, underwater, or in outer space. In addition to supporting important U.S. government nuclear test monitoring objectives, the *Vela* satellites also supported a modest revolution in astrophysics. Between 1969 and 1972, the *Vela* satellites detected 16 very short bursts of gamma ray photons with energies of 0.2 to 1.5 million electron volts. These mysterious cosmic gamma ray bursts lasted from less than a tenth of a second to about 30 seconds. Although the *Vela* instruments were not designed primarily for astrophysical research, simultaneous observations by several spacecraft started astrophysicists on their contemporary hunt for "gamma ray bursters."

Unmanned Aerial Vehicles

Stimulated by dramatic increases in sensor and computer-processing capabilities, robotic system technologies, the need for continuous, or persistent,

Figure 4-14 As part of Operation Iraqi Freedom, a U.S. Air Force MQ-1 Predator unmanned aerial vehicle (UAV) takes off from Balad Air Base, Iraq, on June 14, 2006. The UAV is on an armed reconnaissance and interdiction mission against critical, time-sensitive terrorist targets. The Predator carries two laser-guided Hellfire anti-tank missiles. (Credit: Photograph courtesy of the U.S. Air Force.)

surveillance and intelligence operations over hostile areas, and a strong desire to minimize aircrew casualties, unmanned aerial vehicles (UAVs), formerly called remotely piloted vehicles (RPVs), have moved center stage in military planning. Small UAVs, many hand-launched, are providing important tactical intelligence and surveillance capabilities to American soldiers during antiterrorism combat operations in Iraq and Afghanistan. Larger UAVs like the Predator and Global Hawk have also earned high marks as a result of numerous successful long-duration intelligence, surveillance, and reconnaissance (ISR) missions over Iraq and Afghanistan. Predators armed with two laser-guided Hellfire antitank missiles have also taken the battle to an elusive enemy and demonstrated the efficacy of a very important new role for robot systems, that of the unmanned combat aerial vehicle (UCAV). As discussed below the MQ-1 Predator's primary mission is interdiction and conducting armed reconnaissance against critical, perishable targets.

MQ-1 Predator Unmanned Aerial Vehicle

The MQ-1 Predator is a medium-altitude, long-endurance, unmanned aerial vehicle (UAV), which is sometimes referred to as a remotely piloted aircraft (RPV). The MQ-1's primary mission is interdiction and conducting armed reconnaissance against critical, perishable targets. In the war on terrorism, the term "perishable" refers to a rapidly forming or dispersing elusive concentration of enemy fighters (terrorists), who can quickly disappear into the surroundings when challenged by conventional military forces. A relatively silent, high-altitude UAV can detect such a concentration as it begins to form and then American pilot controlling the UAV from the ground control station can make a rapid decision (based on the existing rules of engagement) whether to attack that target or not. Unlike conventional combat operations, in counterterrorism operations, if there is too much delay in striking a mercurial target of opportunity, the elusive target quickly vanishes into the background. When the MQ-1 is not actively pursuing its primary mission, it acts as the Joint Forces Air Component Commander-owned theater asset for reconnaissance, surveillance, and target acquisition in support of the Joint Forces commander.

The MQ-1 Predator is a system, not just a flying robot. A fully operational system consists of four aircraft (with sensors), a ground control station, a Predator Primary Satellite Link (PPSL), and approximately 55 personnel, who are deployed for continuous (24-hour per day) operations. The basic crew for the Predator is one pilot and two sensor operators. They fly the aircraft from inside the ground control station via a C-Band line-of-sight data link or a Ku-Band satellite data link for beyond line-of-sight flight. The aircraft is equipped with a color nose camera (generally used by the pilot for flight control), a day variable-aperture TV camera, a variable-aperture infrared camera (for low light/night), and a synthetic aperture radar (SAR) for looking through smoke, clouds or haze. The cameras produce full motion video while the SAR produces still frame radar images.

The MQ-1 Predator carries the multispectral targeting system (MTS) with inherent AGM-114 Hellfire missile targeting capability and integrates electro-optical, infrared, laser designator, and laser illuminator into a single sensor package. The aircraft can employ two laser-guided Hellfire antitank missiles with the MTS ball.

The Predator system is composed of four major components, which can be deployed for worldwide operations. The Predator aircraft can be disassembled and loaded into a shipping container, often humorously referred to by Air Force personnel as the "coffin." The ground control system (GCS) is transportable in a C-130 (or larger) transport aircraft. The Predator can operate on a 1,524 meters by 23 meters, hard surface runway with clear line-of-sight. The ground data terminal antenna provides line-of-sight communications for takeoff and landing. The PPSL provides over-the-horizon communications for the aircraft. An alternate method of employment, called Remote Split Operations, uses a smaller version of the GCS called the Launch and Recovery GCS. The Launch and Recovery GCS conducts takeoff and landing operations at the forward deployed location, while the GCS, based in the continental United States (CONUS), conducts the mission via extended communications links. The unmanned aircraft includes an upgraded turbocharged engine and glycol-weeping "wet wings" for ice mitigation. The latest upgrade (sometimes referred to as the Predator B) includes fuel injection, longer wings, dual alternators and other improvements.

What does the designation MQ-1 mean? The "M" is the Department of Defense designation for multirole and "Q" means unmanned aircraft system. The "1" refers to the aircraft being the first of a series of purpose-built remotely piloted aircraft systems. The Predator system was designed in response to a Department of Defense requirement to provide persistent ISR information to the warfighter. In April 1996, the secretary of defense selected the U.S. Air Force as the operating service for the RQ-1 Predator system. A change in designation from "RQ-1" to "MQ-1" occurred in 2002 with the addition of the armed reconnaissance role, an action that made the Predator a robot weapon, sometimes called an unmanned combat aerial vehicle (UCAV). The operational Predator squadrons are the 11th, 15th, and 17th Reconnaissance Squadrons, located at Indian Springs Air Force Auxiliary Field, Nevada.

The Predator UAV has the following basic characteristics. The robot's primary function is that of armed reconnaissance, airborne surveillance, and target acquisition. The UAV has a length of 8.22 meters, a height of 2.1 meters, an empty mass of 512 kilograms, a takeoff mass of 1,020 kilograms, and a wingspan of 14.8 meters. The Predator's contractor is the General Atomics Aeronautical Systems Incorporated and the UAV has a Rotax 914 four-cylinder engine, capable of producing 101 horsepower. The Predator has a cruise speed around 135 kilometers per hour, up to 217 kilometers per hour. The UAV's range is 730 kilometers and its maximum altitude (ceiling) is 7,620 meters. The Predator can carry a payload of 204 kilograms and a maximum fuel load of 379 liters. In March 2005, the U.S. Air Force declared that the MQ-1 Predator had achieved initial operational capability (IOC). This meant that the teleoperated military robot had come of age—marking a major milestone in the evolution of robotic system technology.

Global Hawk Unmanned Aerial Vehicle

The Global Hawk UAV provides U.S. Air Force and joint battlefield commanders near-real-time, high-resolution, intelligence, surveillance, and reconnaissance imagery. In 2005, the Global Hawk provided American military commanders more than 15,000 of these images to support Operation Enduring Freedom. The UAV has completed more than 50 missions and accumulated more than 1,000 combat hours.

Cruising at extremely high altitudes, Global Hawk can survey large geographic areas with pinpoint accuracy, to give military decision makers the most current information about enemy location, resources, and personnel. Once mission parameters are programmed into Global Hawk, the UAV can autonomously taxi, take off, fly, remain on station capturing imagery, return, and land. Ground-based human operators monitor UAV health and status, and can change navigation and sensor plans during flight as necessary.

Global Hawk began as an Advanced Concept Technology Demonstration in 1995 to give warfighters a rapidly developed prototype that could be used for Military Utility Assessment and early operational activities. In June 1999, Global Hawk began a series of exercises sponsored by U.S. Joint Forces Command to determine its future military utility. On April 20, 2000, Global Hawk (Air Vehicle No. 4) deployed to Eglin Air Force Base, Florida, to participate in two exercises that included its first transoceanic flight to Europe, and first mission flown in one theater of operations while under control from another.

The first exercise, called Linked Seas 00, took place in early May 2000 and involved joint command and individual service warfighters, and participants

Figure 4-15 The U.S. Air Force's high- altitude, unmanned aerial vehicle (UAV), called Global Hawk, being prepared for a surveillance, and reconnaissance mission at an air base in Southwest Asia (2005). (Credit: Photograph courtesy of the United States Air Force.)

from NATO. In an important demonstration of its capabilities, Global Hawk provided direct support to amphibious operations in a joint-force environment involving air, sea, subsurface, and land-based assets. During the second exercise, called Joint Task Force Exercise 00-02, which took place from May 14 to 26, 2000, Global Hawk provided direct support for the joint maritime mission of a Navy Carrier Battle Group and an Amphibious Ready Group/Marine Expeditionary Unit in a littoral (land-sea) environment. Global Hawk returned to Edwards AFB, California, on June 19, 2000, concluding the deployment exercise demonstration program. During these exercises, Global Hawk proved its military worth by providing critical intelligence, surveillance, and reconnaissance capabilities to the warfighting community. To demonstrate interoperability between U.S. and Australian military systems, Global Hawk flew more than 12,000 kilometers nonstop across the Pacific to Australia on April 22–23, 2001, setting new world records for UAV endurance. U.S. and Australian Defense Science Technology Organization officials evaluated UAV performance and future military potential during 11 sorties in the land-sea environment before the UAV flew home to Edwards AFB, six weeks later.

In March 2001, Global Hawk entered the Engineering, Manufacturing, and Development phase of the defense acquisition. Global Hawk is currently deployed supporting Operation Enduring Freedom.

Global Hawk, which has a wingspan of 35.3 meters and is 13.4 meters long, can range as far as 22,225 kilometers, at altitudes up to 19,810 meters, flying at speeds approaching 645 kilometers per hour for as long as 35 hours. During a typical mission, the UAV can fly 1,930 kilometers to an area of interest and remain on station for 24 hours. Its cloud-penetrating, synthetic aperture radar

(SAR)/ground moving target indicator (GMTI), electro-optical and infrared sensors can image an area the size of Illinois (137,000 square kilometers) in just 24 hours. Through satellite and ground systems, the imagery can be relayed in near real-time to battlefield commanders.

When fully fueled for flight, Global Hawk has a mass of approximately 211,610 kilograms. More than half the UAV's components are constructed of lightweight, high-strength composite materials, including its wings, wing fairings, empennage, engine cover, engine intake and three radomes. Its main fuselage is standard aluminum, semimonocoque construction. Northrop Grumman's Ryan Aeronautical Center in San Diego, California, serves as the prime contractor.

Unmanned Ground Vehicles

A variety of unmanned ground vehicles (UGVs) have been developed within the Department of Defense to extend the capabilities of fighting forces, while reducing the risk to American military personnel. Prior to Operation Enduring Freedom, a great deal of the attention was placed on the development of remotely operated ground robots that perform surveillance and collect information. Since the start of Operation Enduring Freedom and a more massive combat thrust against armed terrorist groups around the world interest in military UGVs has expanded to include the detection and destruction of hidden explosives, especially the infamous improvised explosive device (IED), which has caused so many casualties in Iraq and Afghanistan. It turned out, that modern mobile robots, originally developed to support law enforcement operations or hazardous material (HAZMAT) responses could be pressed into service as explosive ordnance disposal (EOD) robots. American soldiers now remotely operate these mobile robots to inspect, identify and dispose of IEDs and other suspicious things they encounter on a daily basis in Iraq and Afghanistan. Certain mobile robots have even proven ideal for exploring caves, urban hiding places, and the hidden areas of bunkers.

One example of the effective use of the UGV in the EOD robot role is the Talon robot, built for the Department of Defense for EOD activities by Foster-Miller, Inc. When used in bomb disposal activities the Talon robot is operated by radio frequency and equipped with four video cameras that enable troops to determine which areas enemy soldiers occupy. In addition, the Talon robot is waterproof down to a depth of about 30 meters, so the robot can search for bombs and hidden IEDs off-land in lakes, rivers, and shallow coastal regions. According to data provided by Foster-Miller, Inc., the Talon robot completed over 20,000 EOD missions in Iraq and Afghanistan (as of May 2006).

The family of Talon robots are designed to conduct a variety of UGV missions, including EOD/IED missions, reconnaissance missions, armed reconnaissance, and HAZMAT operations for first responders. The basic Talon robot is a rugged, man-portable device that has a mass of less than 45 kilograms and can easily be transported and then made ready for operation very quickly. The robots are mobile and can climb stairs and negotiate piles of rock and rubble, as frequently encountered in an urban warfare environment. Military personnel operate the robot using a control panel with a joystick. The robots run on long-lived batteries.

Figure 4-16 A Talon 3B robot unit climbs a flight of stairs looking for explosives during a training session at a Bahrain training range. Military EOD technicians operate the robot from safe locations through the use of monitors and video equipment attached to the robot. The Talon 3B robot is designed for the search and destruction of improvised explosive devices (IEDs), as well as other forms of ordnance found in Iraq. Because it is waterproof to a depth of about 30-meters, U.S. Navy personnel also use this robot while performing maritime security operations (MSO). (Credit: Photograph courtesy of the U.S. Navy.)

OTHER TYPES OF ROBOTS

Medical Robots

Robots are entering the field of medicine and promise to improve the quality of care. One interesting medical robot is called the da Vinci™ Surgical System. It is manufactured and marketed by Intuitive Surgical, Inc., with headquarters in Sunnyvale, California. The da Vinci™ Surgical System is a 2-meter-long machine assistant with three mechanical arms that allows physicians to perform delicate surgeries, using procedures, which extend beyond the limits of the human hand. The surgeon uses handgrips and foot pedals attached to the computer console to control three robotic arms that perform the surgery using a variety of surgical tools. The robotic arms, which have a "wrist" built into the end of the surgical tools, give surgeons additional manipulation ability during minimal invasive laparoscopic surgery, enabling easier, more intricate motion and better control of surgical tools. The robot system gives doctors unprecedented control over the tiny instruments they use during minimally invasive surgery, also known as keyhole surgery. Furthermore the use of this medical robot system permits a more detailed view of the surgical site than unaided human eyes allow.

The da Vinci™ Surgical System consists of three components: the surgical cart, a computerized-vision system, and a surgeon's console. The surgical cart, stationed adjacent to the operating table, has three robot arms—one for the surgeon's right hand, one for the surgeon's left hand, and a middle mechanical arm to hold the laparoscope that the surgeon uses to "see" inside the patient's body. Effectively, the medical robot becomes the mechanical hands and eyes of the surgeon. The computerized vision system transforms the images captured by the tiny camera inside the patient into a three-dimensional (3-D), real-time image that the surgeon views as he or she sits at the surgical console. Robotic controls allow the surgeon to make natural hand movements—in contrast to the "counterintuitive" instrument movements, which are characteristic of non-robotic, standard laparoscopic surgery. Counterintuitive instrument movement involves an operating condition similar to the surgeon working on the patient while looking in a mirror. The da Vinci™ Surgical System provides the surgeon with nearly all-natural movements of the human wrist, making the minimally invasive surgery more like open surgery. The system also eliminates natural hand tremor and improves dexterity thereby allowing the surgeon to perform ever-finer surgery in a more controlled manner. Selected hospitals and medical centers around the world now use the da Vinci™ Surgical System to perform minimally invasive surgery (MIS).

Rescue Robots

First responders often face daunting challenges during search and rescue tasks in dangerous environments. The concept of including robots as a part of the responders' tool cache is being accepted, since robots have the potential of taking responders out of harm's way and augmenting their capabilities. The Department of Homeland Security (DHS) Science and Technology (S&T) Directorate has initiated an effort with the National Institute of Standards and

Figure 4-17 This picture shows a team of mini-robots from Sandia National Laboratories demonstrating how the laboratory's swarm-algorithm-based computer program would help rescuers quickly locate a skier buried under an avalanche. (Credit: Photograph courtesy of U.S. department of Energy/Sandia National Laboratories.)

Technology (NIST) to develop comprehensive standards related to the development, testing, and certification of effective technologies for Urban Search and Rescue (US&R) robotics.

These US&R robotic performance standards cover sensing, mobility, navigation, planning, integration, and operator control in order to ensure that the robots can meet operational requirements under the extremely challenging conditions that rescuers are faced with, including long endurance missions. Where appropriate, the standards will also address issues of robotic component interoperability to reduce costs. The US&R robotic standards effort focuses on fostering collaboration between first responders, robot vendors, other government agencies, and technology developers to advance consensus standards for task specific robot capabilities and interoperability of components. These standards will allow DHS to provide guidance to local, state, and Federal homeland security entities regarding the purchase, deployment, and use of robots for US&R applications.

The problem of search and rescue is not limited to just urban areas. Rescuers seeking a skier or snowboarded buried under an avalanche face a major problem: how to find the unfortunate victim before suffocation, or frostbite and hypothermia prove fatal. Under such dire circumstances death can come in 30 minutes. To help improve the search and rescue efforts of emergency rescue teams, researchers at the Sandia National Laboratories in Albuquerque in 1999 developed a computer program, which provided group intelligence for a swarm of minirobots to rapidly pinpoint a source of contagion of a skier buried in whiteness. Computationally, finding a snow-buried skier is remarkably similar to locating the point source of a chemical or biological weapons attack. The search algorithm enables a swarm of cockroach-sized robots to "talk" to each other through radio transmitters and home in on a target more quickly than solitary searchers using more conventional means. The group search technique, called swarming, relies upon neither a central intelligence telling the searchers what to do nor the intuition of individuals. Rather, each robot continually informs others of its position and of the strength of the signal received at that position from the sought-for source. The steady streams of information from multiple sources allow each member of the swarm to continually refine the direction of its search.

Skiers in avalanche country routinely carry radio beacons as standard operating procedure, though search techniques to locate the beacon usually are not particularly advanced. A standard approach used by rescuers is to exhaustively search every inch of ground—a time-consuming procedure when the victim is in a time-critical life or death situation. Based on rescue experience, a victim's chance of survival decreases markedly after 30 minutes under the snow. Another approach requires the human searcher to make a right-angle turn when the signal strength decreases. Such searches are difficult because buried obstacles mask the strength of the radio signals, and the transmitter's physical orientation is unknown. Because finding the location of a radio frequency transmitter and finding the center of a region from which some form of lethality is emanating are essentially the same search and detection activity (from a computational perspective), the researchers at Sandia National Laboratories believe their approach can solve a whole class of similar problems with the same robotic swarm search algorithms.

5

Impact

The robot is a machine that is impacting the trajectory of human civilization in a variety of interesting and important ways. This chapter describes how the robot is already transforming industry, redefining warfare, enabling the exploration of inner and outer space, supporting important environmental cleanup initiatives, and pushing the frontiers of medicine.

The robot's impact on some areas of human activity, like manufacturing, is relatively easy to understand. Each day, just about anywhere in the world, people can see television commercials, which portray modern factory robots hard at work spray painting or spot welding automobiles on highly automated assembly lines. The existence of the factory robot and its ability to replace human beings in the performance of hazardous, repetitive, or dirty jobs should come as no great technical shock. The real impact of the industrial robot, however, involves deeper social dimensions—typically in the form of displaced blue-collar workers and in the further spread of industrialization around the planet.

The impact of robotic technology on other areas, like national defense and modern warfare, requires a little more reflection. Most people do not look upon the nuclear weapon-armed intercontinental ballistic missile (ICBM) as a robot—yet it is the most powerful robot weapon ever devised by the human race. Military leaders regard the guided missile as a scripted autonomous system—that is a weapon system that has no further human interaction once it is deployed. As discussed here, such "point, fire, and forget" weapons include the entire family of smart military robots, ranging from cruise missiles and tactical ballistic missiles to very long-range ICBMs. These robot weapons, especially the ICBM, have completely transformed strategic warfare.

Smart unmanned aerial vehicles (UAVs), autonomous underwater vehicles (AUVs), and unmanned ground vehicles (UGVs) are also transforming tactical warfare and transforming the contemporary battlespace. The word *battlespace* has been used deliberately in place of the more traditional expression *battlefield* because military space robots have completely transformed warfare by creating a three-dimensional information-based combat environment in which space systems play an integral role in leveraging the effectiveness of land, sea, and air forces. Robot sentinels in space are often used to assist military personnel as they employ UAVs, UGVs, and AUVs in modern conflicts.

The influence of robots is present in many other aspects of modern life. Robotic systems, specifically mechanical and electromechanical slave/master systems, enabled the development of nuclear technologies for both military and civilian applications in the 1940s and 1950s. Today, nuclear workers are assisted by both very sophisticated manipulator systems and advanced remotely operated mobile robots, as they perform hazardous nuclear cleanup operations. Many of these cleanup operations involve remediation of the environmental legacy of the cold war's nuclear arms race. Robots are also helping human beings operate and maintain civilian reactors. Hazmat (hazardous materials) robots assist first responders in the assessment and containment of toxic material spills and incidents. Law enforcement agencies regularly use mobile robots to reduce the risk to human beings during bomb threats standoff situations, and hostage negotiations. Underwater robots now assist in the exploration of the depths of the world's oceans for science, mineral prospecting (including oil and natural gas), and archaeology. Similarly, robot spacecraft and observatories have opened the solar system and the universe beyond to detailed study by scientists. Interest in sports robots and hobby robots (especially competitive, gladiator-like robots, called battlebots) has triggered a renewed interest in the study of mathematics, science, and engineering by American students.

The chapter concludes with several interesting speculations about the long-range impacts that future robot technology will have on the human race. The two areas presented here involve nanotechnology and self-replicating machines. Although the consequences of such future robotics technologies are hypothetical at the moment, no examination of the impact of the robot on the human race would be complete without some discussion of these intriguing technical possibilities.

THE ROBOT AS AN AGENT OF SOCIAL CHANGE

While commonly regarded as smart mechanical devices, industrial robots also serve as the agents for a sweeping wave of social change, shaping the world of the twenty-first century. The impact of industrial robots on the process of manufacturing is often measured in terms of economics, that is, the reduction in overall cost (or time) in producing some unit of manufactured goods, be it an automobile, dishwasher, or a customized computer. But the industrial robot is really an agent for social change, causing transformations in the workplace that are impacting many areas of human activity.

The most profound influence is that robots with their mechanical arms, sensors, and machine intelligence are taking over many tasks formerly performed by humans—task that are either extremely hazardous, boring, or simply onerous. In factories, for example, robots now perform many tasks with high precision and without the need for coffee breaks or vacations. Of course, the notion of "sick leave" has not been totally eliminated because robots require scheduled maintenance (well-care), as well as emergency repair (sick care).

Overall there is a general trend in the manufacturing industries (especially heavy industries, such as automobile manufacturing) in which less skilled blue-collar jobs are disappearing. With the arrival of modern industrial robots, the human workforce in developed (or industrialized) nations is experiencing a dramatic transformation, which is every bit as significant in its social consequences as the Agricultural Revolution and the First and Second Industrial Revolutions. Because of the rising number of robots now being used in manufacturing in factories of all sizes, there are many more desirable jobs in the information and services industry than in the heavy or light manufacturing industries. In some cases, robots are also penetrating the services industries, including medicine. For example, industrial robots have been modified for use in pharmacies and hospitals to distribute physician-prescribed medicines.

Some of these workforce changes are welcomed by the workers, as they are relieved from burdensome tasks and allowed to earn a living in a more personally fulfilling manner. In other instances, however, abrupt and poorly planned shifts in workforce demographics are causing enormous social upheaval, especially when there are inadequate steps taken to ease the psychological and economic distress of displaced human workers.

Many technical visionaries and futurists do not see the industrial robot as a villain, but rather as a hero or liberator, freeing human workers of dangerous, burdensome, or boring tasks, and allowing these workers to make better use of their talents and intellectual capabilities in jobs that have less vulnerability to physical injury or psychological stress. In perhaps what might be called a "utopian scenario," some robot system advocates suggest that the increased use of robots in modern factories will pave the way for shorter workweeks. A four-day (32-hour) workweek, for example, could greatly improve quality of life for factory workers. The social impact of a shortened workweek (with equivalent pay) offers the promise, at least in theory, of providing human workers with more time to spend with family, to enjoy leisurely physical activities, and to pursue mentally stimulating programs of self-improvement.

However, the reality of the this possible impact (based on late twentieth century labor trends in the United States), suggests that any financial benefits accruing from the introduction of robots in a factory will most likely go straight into corporate profits, with little, if any, benefit being passed on to the workers in terms of a shortened workweek for equivalent pay. This counterpoint, sometimes called the devil's advocate position, suggests that instead of enjoying a less stressful job environment, any surviving human workers will face additional stresses to upgrade their skills or "be replaced by a more competent machine."

Since the introduction of the Jacquard automated loom at the beginning of the nineteenth century (the start of the First Industrial Revolution), manufacturing technology has enjoyed a steady improvement with ever-more efficient and sophisticated machines continuously displacing less skilled human workers. The Second Industrial Revolution with its expanded use of electricity and electric motors further accelerated this labor transforming process in industrialized nations.

Henry Ford's innovative use of human-workers to operate a moving assembly line in the early part of the twentieth century to manufacture affordable personal automobiles transformed life in the United States. Ford's bold approach to high quality, mass manufacturing of a precision end product also set the stage

Figure 5-1 This picture shows the Autoscript III, a prescription filling robot at the National Naval Medical Center in Bethesda, Maryland, as it picks up a properly-coded bin of medication. The modified robot is one of two used in the hospital pharmacy. (Credit: Photograph courtesy of the U.S. Navy.)

for the rise of industrial robots at the end of the twentieth century. Today, on a global basis, the automobile manufacturing industry uses over 50 percent of the world's industrial robots. According to the *World Robotics 2005* report from the United Nations Economic Commission for Europe (UNECE), there will be over one million multipurpose industrial robot units installed and operated around the planet by the year 2008. Factories in Japan and the Republic of Korea will account for about 38 percent of the projected total, Europe approximately 33 percent, and North America (United States, Canada, and Mexico) some 14.9 percent. The rest of the world will account for the remaining 14.1 percent of the multipurpose industrial robot population.

Why are so many manufacturing industries in developed countries investing in industrial robots? The UNECE report suggests several reasons: cost savings, improvement in productivity, improvement in quality, to maintain a competitive edge in the global economy, and to improve the quality of life for factory workers by transferring dangerous or boring (repetitive) jobs from human workers to robots. A careful review of these reasons reveals the major impact industrial robots are having on the manufacturing industry throughout the world. And this trend is just beginning, because industrial robots are getting more skilled and (in many instances) less expensive.

Robot-enabled, automated fabrication represents a new industrial tool. Automated fabrication (or autofab) refers to a set of modern technologies that automate the processes of fabricating three-dimensional solid objects from raw materials. This manufacturing technology allows industrial engineers to transform digital designs into three-dimensional solid objects for production machine parts, prototypes, and molds. With the arrival of capable industrial robots, automated fabrication has progressed well beyond numerically controlled machining. In industrial engineering, fabrication basically means forming individual items out of raw materials. The age of mass production began at the turn of the twentieth century, when Henry Ford triggered a revolution in mass manufacturing by introducing the Model T automobile. To achieve his production objectives, Ford redesigned many manufacturing processes, including parts fabrication and assembly.

On Ford's moving assembly line, the assembly process was broken down into subprocesses, or tasks. An individual workstation was dedicated to each task. Ford also introduced another innovation, the decentralization of assembly plants. This allowed individual, specialized subassembly plants to become geographically separated. Ford's approach to mass production proved a great success, as evidenced by the dramatic price reductions in automobiles and other manufactured goods.

However, in the 1920s, General Motors overcame Ford's rigid stand on product customization (the Model T only came in one basic model and color, namely black) and began providing its automobile customers with product variety (that is a choice of a colors and styles). By combining the economies of scale with product customization, General Motors was able to expand its market share and from the 1930s through the 1970s held a dominant position in both the American and global automobile industry.

But the economic scene began to change in the 1970s, when Japanese automobile manufacturers embraced the use of robots and pursued a new philosophy of mass production with unlimited variety, or mass customization. The Japanese advancement of mass customization was greatly assisted by the development of new technologies, such as computer-aided design, computer-integrated manufacturing, automated fabrication, and robotics.

Today, the new word in the manufacturing industry is *flexible manufacturing*. More advanced industrial robots are found in a computer integrated manufacturing (CIM) facility. Once industrial engineers finish a product's design, using computer-assisted design (CAD) procedures, they then use a combination of computers, machines, and robots to bring that design to life in the form of a finished product. Industrial engineers suggest that the successful manufacturer of the twenty-first century will even invite the customer to participate in the process of product design and development. When this occurs on a larger scale, manufacturing technology will have advanced beyond the era of mass customization to a new manufacturing paradigm, called customer coconstruction. In customer coconstruction, the producer not only satisfies a variety of customer needs and wants, but also helps customers decide what it is they actually want and need. Industrial engineers believe that this process of customer coconstruction will encompass any or all stages of design, development, and production.

Figure 5-2 This photograph shows the author's own customized GRAVITON® space gyro created at the automated Toymaker 3000 computer-integrated-manufacturing (CIM) exhibit at Chicago's Museum of Science and Technology. The Toymaker 3000 is a state-of-the-art CIM automated assembly line that consists of eight interactive stations and 12 moving robots. The space gyro's parts are assembled, sonic welded where appropriate, laser inscribed, and inspected for quality—all automatically. (Credit: Photograph by author.)

The ultimate impact of the robot in the manufacturing industry will be a completely automated factory, which requires no human worker. A future, totally automated factory may retain a few human workers in a supervisory capacity. These people would turn the factory on or off and then watch for red lights on some control panel, indicating a serious malfunction. In the totally automated factory of the future, the human supervisor might even be located quite some distance away from the factory itself. When a red light blinks on his or her control panel, the human worker would simply decide which cadre of repair robots to dispatch to the area of the automated factory with the unusual problem. Otherwise, the robots themselves would perform routine maintenance and repairs.

Beyond the obvious customization in manufacturing, the fully automated factory has a very interesting social impact. Consider, for a moment, the first automated factory that has as its main product a variety of other industrial and mobile robots. At this point in the twenty-first century, with no or minimal human supervision, intelligent robots will start designing, producing, and even improving new robots—each generation becoming a little more capable than the previous generation of machines. This breakaway condition represents the critical point in history, at which the evolution of human intelligence has liberated (for the first time) consciousness from the confines of human biology. Sometime afterward, the ultimate step in the evolution of manufacturing technology takes place, when self-replicating space robots are unleashed into the solar system and begin to travel to the stars beyond. As discussed at the end of this chapter, the impact on the universe would be nothing short of enormous.

A REVOLUTION IN WARFARE AND MILITARY STRATEGY

Starting in about the middle of the twentieth century, military robots transformed warfare and in the process dramatically changed international politics and the course of human history. This section of the chapter examines the overall impact of three general classes of robotic military systems: guided missiles, military spacecraft, and mobile robots. The first two classes of robotic military systems exerted enormous changes on warfare and military strategies during the cold war, while the mobile military robots started to significantly influence military operations at the end of the twentieth century. During contemporary military operations in Bosnia, Afghanistan, and Iraq a variety of mobile robot systems (primarily aerial and ground) have played an increasing role in protecting American forces, while improving the American war fighter's ability to inflict damage on enemy forces, ranging from organized combatants to terrorists and insurgents.

The Ballistic Missile and a Revolution in Strategic Warfare

In the middle of the twentieth century, one robot weapon transformed the world more significantly than any other device ever created by man. The marriage of two powerful World War II-era weapon systems, the American atomic bomb and the technical descendents of the German V-2 ballistic missile, ultimately produced the single most influential weapon system in the twentieth century, if not all history, the intercontinental ballistic missile (ICBM). The ICBM and its technical sibling, the submarine-launched ballistic missile (SLBM), was the first robot weapon system designed to travel into and through space. The ICBM is capable of striking a target thousands of kilometers away with an accuracy and destructive capability previously unavailable throughout history. The arrival of the first generation of such robotic weapons (essentially program, shoot, and forget devices) in the late 1950s completely transformed the nature of strategic warfare.

The ICBM created a fundamental change in the national security policy of the United States. Before the ICBM, the chief purpose of the U.S. military establishment had been to fight and win wars. Once the operational nuclear weapon-equipped ICBM arrived, both the United States and the former Soviet Union possessed a weapon that could deliver megatons of destruction to distant points on the globe with little or no chance of being stopped. The nuclear-armed ICBM reduced political reaction time to minutes. For the first time in history, an international confrontation could escalate into a major strategic conflict that could then destroy both combatant states and their allies—all in a matter of hours.

From that moment on, the chief purpose of the U.S. military establishment became the avoidance of strategic nuclear warfare. A wholesale, unstoppable exchange of ballistic-missile-delivered nuclear weapons would destroy both adversaries and leave Earth's biosphere in total devastation. There would be no winners, only losers. With the development of the ICBM, for the first time in history, human beings possessed a weapon system that could end civilization in less than a few hours.

The existence of such powerful, quick-strike robot weapons made deterrence of nuclear war the centerpiece of national security policy during the cold war era—a policy appropriately called mutual assured destruction (MAD). Military leaders no longer focused on "winning" the next major war; rather, they created a variety of schemes and technologies to help them prevent any large-scale confrontation that could escalate to the use of nuclear weapons. The final course of action in this strategic plan was quite simple. If all else failed and a nuclear war started, each side would inflict lethal damage on the other.

Figure 5-3 A drawing showing the modern Minuteman III intercontinental ballistic missile (ICBM). Armed with one or several nuclear warheads, the modern ICBM represents a deadly and unstoppable weapon capable of inflicting massive damage on distant targets in less than 30 minutes after launch. (Credit: Illustration courtesy of the U.S. Air Force.)

It is interesting to note that the threat of nuclear Armageddon has helped restrain those nations with announced strategic nuclear weapons capabilities (such as the United States, the Russian Federation, the United Kingdom, France, and the People's Republic of China) from actually using such weapons in

Figure 5-4 An artist's rendering showing the post-boost vehicle of the Minuteman III (MM III) intercontinental ballistic missile (ICBM) streaking through outer space. As shown here, post-boost vehicle carries a payload of three W62/Mark 12 re-entry vehicles. The W62 nuclear weapon has a reported yield of 170-kilotons. First deployed in 1970, the W62 nuclear warhead is an example of the amazing engineering progress made in the development of robotic weapons-tipped with powerful nuclear weapons that took place during the cold war era. The Mark 12 (Mk 12) re-entry vehicle is only 1.8-meters long. (Credit: Illustration courtesy of the U.S. Air Force.)

resolving lower-scale, regional conflicts. Because of this standoff of unstoppable robot weapon (guided missile) versus unstoppable robot weapon, political scientists assert that the ICBM created a revolution in warfare and international politics—a revolution making nuclear warfare between rational actors (nations) impossible. (As discussed in Chapter 6, the issue of theater ballistic missiles armed with nuclear weapons is a growing global problem. A possible Pakistan–India regional nuclear conflict, an irrational nuclear-armed North Korea, and the rising nuclear-weapon ambitions of Iran currently top the list of concerns.)

To make conflict management even more important, land-based ICBM technology progressed from the relatively slow responding Atlas and Titan liquid-fueled missiles, to a family of quick-response, solid-fueled Minuteman missiles in hardened underground silos. Once launched, these "instant," solid-propellant-fueled missiles would streak towards enemy territory and their payload of nuclear warheads would detonate on various targets in about 30 minutes.

For almost five decades, ICBMs (including SLBMs) have served as the backbone of America's strategic nuclear deterrent forces. Throughout the cold war and up to the present, deterring nuclear war remains the top U.S. defense priority. Since 1959, strategic-force missileers have served around-the-clock on continuous alert. Buried in underground launch facilities, ICBMs are the most rapid-response strategic force available to the American president. The Minuteman III ICBM, for example, is capable of hitting targets more than 8,000 kilometers away within about 30 minutes with outstanding accuracy. In the post-cold war political environment of the early twenty-first century, military leaders still regard this ICBM force as America's most credible deterrent against nations that possess, or are in the process of developing, weapons of mass destruction (WMD). Submarine-launched ballistic missiles (SLBMs) represent a complementary (mobile) component of this long-standing nuclear deterrent policy.

Military Space Robots and the Information Revolution in National Security

Robot spacecraft, such as reconnaissance satellites, surveillance satellites, and other information-related Earth-orbiting military spacecraft changed the nature of military operations and national security planning forever. Space-based information collection produced enormous impacts on peacekeeping and war fighting.

Starting in the mid-twentieth century, the development and deployment of Earth-orbiting military spacecraft significantly transformed the practice of national security and the conduct of military operations. From the launching of the very first successful American reconnaissance satellite in 1960, "spying from space" produced an enormous change in how the United States government collected the essential information with which to conduct peacekeeping and war fighting. Recognizing the immense value of the unobstructed view of Earth provided by the high ground of outer space, defense leaders made space technology an integral part of projecting national power and protecting national assets.

Most of the early military space activities were conducted behind a veil of secrecy. So, generally, only civilian space accomplishments made the headlines in the 1960s and 1970s. Today, that veil of official secrecy cloaking some of the most important (but classified) military satellite programs has been partially removed by the United States government. So within the limits of newly available public information, this section briefly describes the very important role American space robots, that is military satellites, have played and continue to play in providing the information needed to stabilize a nuclear-armed world.

Reconnaissance satellites, surveillance satellites, and other information gathering and/or distributing space platforms dramatically changed the nature of military operations and also had an enormous impact on arms control verification and treaty monitoring activities. Once proven feasible in the early 1960s, military satellites became an essential part of the defense infrastructure of the United States, the former Soviet Union, and other nations. Today, an armada of American military space robots supplies information across the entire spectrum of national security needs from vigilant monitoring to the swift and successful conclusion of armed conflict. When armed conflict becomes necessary, a variety of military satellites support the efficient application of United States military power in any part of the globe.

The expanded collection and flow of information essential to national security by space-based military systems represents an essential component in the preservation of a stable global civilization. Rational leaders do not want political misunderstandings or the lack of vital information to lead to an armed conflict that could escalate to the level of strategic nuclear warfare. Nor does the family of nations want an accidental nuclear war to start between two states, like India and Pakistan, who share a long and bitter history of political animosity and now possess fledgling nuclear arsenals. Today, the large quantity of information collected by military satellites supports the use of common sense and diplomacy in the resolution of most modern international disagreements and conflicts.

However, when the use of common sense and diplomacy fails, battlespace information supremacy significantly enhances the application of force by the American military. This circumstance generally promotes a swifter conclusion of armed conflict against enemy military forces. Space robots, often in combination with unmanned aerial vehicles (UAVs) and unmanned ground vehicles (UGVs), are experiencing increased application in the international war against terrorism, as part of an American-led global military initiative given the code name, Operation Enduring Freedom (OEF). This operation started immediately after the terrorist attack on the World Trade Center in New York City on September 11, 2001.

But there is a significant paradox involving the creation of an essentially transparent battlespace by sensors on modern military spacecraft and UAVs, which are discussed in the next section. In traditional armed conflict situations, military satellites and UAVs can provide American forces unprecedented force-multiplying advantages. For example, satellites often provide near real-time strike reporting and damage assessment data. This timely information allows field commanders to quickly reprogram smart weapons and deploy them against functioning targets, thereby avoiding unnecessary strikes against targets that are already neutralized or destroyed. The avoidance of such unnecessary strikes also minimizes collateral damage and civilian casualties. However, in unconventional warfare situations, such as encountered when combating terrorists, the distinctive information advantage provided by military robots in space is often significantly reduced. Even the most sophisticated spy satellites can only go so far in providing useful information about terrorists who hide among civilian populations and then attack suddenly from the shadows in an indiscriminate fashion.

Here lies the paradox: the smaller the hostile group being fought, generally the less valuable the military advantage of "eyes in the sky." The following simple analogy summarizes the current global security circumstances reasonably well. Sophisticated military satellites allow defense officials to efficiently monitor, track, and (as necessary) contain or neutralize "rogue elephants" rampaging through the world's political jungle. However, data from these same military satellites provides little direct assistance against the pesky (but sometimes deadly) disease-bearing mosquitoes that lurk in the same global political jungle. That is why UAVs and UGVs (discussed in the next section) gained so much importance in the current conflicts in Afghanistan and Iraq.

This section of the chapter briefly describes the impact of the major types of military satellites developed by the United States. Because outer space is the modern equivalent to the high ground in classical defense thinking and is free from national jurisdiction, appropriately designed military space robots are also well suited to perform the following military activities: reconnaissance, surveillance, communications, and navigation.

Space-based reconnaissance involves the acquisition of detailed information of a specific type that supports either strategic or tactical intelligence needs. Surveillance involves the use of sensors on satellites to support some type of continuous monitoring activity. People sometimes have difficulty differentiating the specific meaning of each term, so the following analogy is provided. A reconnaissance satellite is much like a robot scout that travels through hostile territory in an effort to gather certain important pieces of information. Where is the enemy? How numerous are the hostile forces? What type of weapons do they have? And so forth.

A surveillance satellite is similar to a robot guard or sentinel—in this case, keeping watch from a tall tower and looking down across the hemisphere for signs of hostile activities. At the first sign of trouble, the space robot sentinel sounds the alarm, thereby giving the friendly forces time to take appropriate defensive actions. Modern military satellites perform surveillance in three general categories: early warning (especially against ballistic missile attack), nuclear detonation detection (especially in support of nuclear test ban treaties), and weather monitoring (especially at the tactical or regional level). There are also military communication satellites and navigation satellites, whose functions assist peacekeeping and combat operations.

During the cold war, military surveillance satellites provided an important level of sanity within a politically divided world that focused on mutual assured destruction (MAD). Surveillance satellites, especially the Defense Support Program (DSP) satellites, served as the cornerstone of the American early warning program and made feasible the national policy of strategic nuclear deterrence. These robot sentinels in geostationary orbit would immediately detect any enemy attempt to launch a surprise ICBM attack in a destructive first strike.

In the post-cold war era, surveillance satellites still stand guard, always ready to alert national authorities concerning a hostile ballistic missile attack. Now, their missile-surveillance mission has been expanded to include shorter-range missiles, launched by rogue nations during regional conflicts. By 1995, new techniques in processing DSP data provided U.S. theater-level forces improved warning of attack by short-range missiles.

In the twenty-first century, a new generation of space-based, infrared surveillance systems will continue to impact American military operations and force protection. These new space-based robot sentries, with greatly improved infrared "eyes," will support four critical defense missions: missile warning, missile defense, technical intelligence, and battlespace characterization (including timely battle-damage assessment).

The Impact of Mobile Military Robots

The recent conflict situations in Afghanistan and Iraq have demonstrated that modern, mobile robots—especially UAVs and UGVs—can significantly increase the operational capabilities of modern armed forces. Because of these

Figure 5-5 This artist's rendering shows a U.S. Air Force Defense Support Program (DSP) satellite in its role as a robotic orbiting sentry. Since 1970, these surveillance satellites have played a vital role in the defense of the United States by detecting and reporting missile launches. (Credit: Artist rendering courtesy of the U.S. Air Force and Northrop Grumman.)

recent successes and favorable impacts on combat operations, mobile military robots have now become an integral element of modern American war fighting capability.

UAVs—such as the U.S. Air Force's RQ/MQ-1 Predator and RQ-4 Global Hawk and the U.S. Army's RQ-7 Shadow 200—have shown that these systems support improved acquisition and rapid distribution of intelligence, surveillance, and reconnaissance (ISR) provided at the tactical (theater) level. UAVs now play a very important role in successful military operations against highly mobile targets and elusive adversaries. UAVs have earned star status in the war against terrorism and have become the most-requested capability among combatant commanders in Southwest Asia. What makes these UAVs so valuable is their inherent ability to loiter and beam real-time images to combat forces on the ground. This capability provides American forces "eyes in the sky" for extended periods of time and denies an elusive enemy sanctuary. By providing constant surveillance in the enemy's "backyard," terrorist forces cannot readily mass assets and strike in strength.

If the enemy is foolish enough to mass in strength, the UAVs have an additional capability—they can deliver a deadly attack. For example, early in Operation Enduring Freedom, an armed Predator UAV was credited with taking out one of al-Qaida's top lieutenants in Afghanistan with a Hellfire missile, and has since been widely used for offensive operations in Iraq. Although the Predator was not originally designed as a strike-platform, this flying-robot now combines an ability to provide continual surveillance and to respond quickly to on-the-ground threats. This quick reaction capability makes the Predator an especially valuable asset in the war on terror. Of special importance is that an armed surveillance aerial platform can take action very quickly in the cycle of enemy activity (as the hostile forces are just beginning to organize). In many cases, the UAV (under a missile release command from a distant human controller) then launches a Hellfire missile, which hits the mark with deadly accuracy and eliminates the threat entirely. But even the unarmed version of the Predator has an enormous impact in the war on terror. The Predator and other UAVs, with their low operating costs and generally extended loiter capabilities, scout suspected areas for signs of trouble and then identify targets, so other strike platforms, such as a U.S. Air Force's AC-130 gunship, can engage these elusive targets more quickly and effectively.

At present, the American military uses a wide variety of UAVs because no single system currently available can perform all surveillance and strike missions for all combat situations. At one end of the spectrum is the U.S. Air Force's long-term surveillance platform, called the Global Hawk. This supersophisticated UAV has joined the Predator in providing a special high-altitude surveillance capability. It is interesting to note that about a half dozen or so UAV systems, like the Predator and Global Hawk, are actually operated from locations within the United States as they fly in the skies over Iraq and Afghanistan, sometimes simultaneously. While these flying robots are based and serviced in or near the theater of operations, the humans who fly and monitor the systems can be located half-a-world away. It is important to realize that these surveillance UAVs are doing what people cannot, or (ideally) should not have to do. The large flying robots can provide comprehensive in-theater surveillance capabilities over extended regions of potentially hostile territory without tiring or losing concentration. Remember, the human operator is in a comfortable chair in a safe environment and, when fatigue sets in, shift-changes can be performed seamlessly without affecting the robot's mission. Human pilots would find it very difficult to operate for such extended periods, especially while operating over dangerous, high-stress environments.

At the other end of the UAV spectrum is the U.S. Marine Corp's Dragon Eye system. This small, hand-launched UAV gives squad- and company-level leaders, who are located and fighting in theater, a quick aerial snapshot of their immediate operating area. Small enough to break down into pieces that fit neatly into a marine's backpack, the Dragon Eye now supports numerous antiterrorist combat operations in Iraq.

U.S. Army troops use the Raven, another handheld UAV, to gather over-the-horizon views of trouble spots. Another tactical UAV, called the Shadow, is proving its value in Iraq, during improvised explosive device (IED) sweeps and reconnaissance missions. Small, tactical UAVs also provide situational awareness for

Figure 5-6 As part of Operation Enduring Freedom, U.S. Air Force personnel perform pre-flight checks on a MQ-1 Predator UAV, prior to the flying-robot's mission over Afghanistan on November 9, 2001. (Credit: Photograph courtesy of the U.S. Air Force.)

troops guarding garrisons and high-value targets, support mobile troops during scouting missions, and watch over convoy movements. The small flying robots give the troops a real combat advantage. If a convoy is going down the road and the lead truck observes something up ahead that looks suspicious or unusual, the convoy can quite literally stop, while the troops assemble and deploy a small UAV to see what is going on—without endangering the convoy. In other operations, the troops might use a "flock" of UAVs to provide aerial surveillance on all sides (left, right, front, and back), as military trucks rumble down a particularly dangerous stretch of road.

The age of the UAV in modern warfare has just begun and the demand for flying robot vehicles that can collect images, drop bombs (if necessary), or hover over targets without risking the lives of human pilots is growing exponentially. In December 2005, on at least five occasions, for example, armed Predators flown remotely by airmen sitting at consoles in a Nevada air base struck insurgent strongholds in western Anbar Province, Iraq. The Department of Defense budget request for 2007 contains a proposal to purchase six USAF Global Hawks, 26 USAF Predators, four U.S. Navy Fire Scouts, and 20 small UAVs for the U.S. Army. Furthermore, over the next five years, the Pentagon plans to purchase at least 219 Predators for the U.S. Air Force and special operations forces and 35 Global Hawks.

The U.S. Army and Marine Corps are exploring the great potential that UGVs have to lessen the risk to human beings and to improve fighting unit efficiencies during ground engagements and (for the marines) operations coming ashore, such as the reconnoitering of beach areas and landing zones prior to and during offensive operations. Troops want robots that can take the risk during operations that are dangerous, dirty, or simply dull (like guard duty). Robots that can look "around the corner" or investigate interior spaces during urban warfare operations would significantly reduce friendly casualties, while quickly pinpointing and assisting in the dispatch of hostile forces. Some ground robots are being equipped with weapons that can be fired by human soldiers from the vantage point of a relatively safer, remote location. Mobile robots are also being used to explore caves in search of terrorists.

Perhaps the most important current use of ground mobile military robots is to assist human soldiers as they search for and dispose of mines, booby traps, and the insidious IEDs that have been causing numerous casualties in nonconventional combat situations and insurgency activities currently taking place in Iraq and Afghanistan. A variety of versatile mobile robots with suitable manipulators, cameras, and sensors, have proven their great value in reducing the personal risk of EOD specialists, as these troops investigate and then (if appropriate) disarm or destroy potentially lethal devices intentionally placed by terrorists. Remote operation of such robots allows the EOD experts to do their job with a significant reduction in personal risk.

THE IMPORTANCE OF TELEOPERATORS AND NUCLEAR ROBOTS

Starting in the early 1940s and continuing up to the present, the development and use of progressively more capable teleoperators enabled many modern nuclear energy applications. Although their role is often overlooked, this human-machine partnership made possible the rapid growth of the American nuclear weapons program, opened the door to a variety of radioisotope uses (such as high intensity sources in medicine), and supported the rise and maintenance of civilian nuclear power generation around the globe.

The first really sophisticated teleoperators emerged in conjunction with the beginning of the American nuclear bomb project. As part of this extensive program, known as the Manhattan Project, the first large plutonium production reactors were constructed and then put into operation at the Hanford Complex in Washington State. The Hanford production reactors soon began producing large quantities of intensely radioactive materials. The success of the nuclear bomb program required that neutron-irradiated fuel rods (which now contained plutonium along with many other radioisotopes) be chemically processed so that the newly produced plutonium could be removed for use in weapons. Never before in history, however, had human beings had to handle such intensely radioactive materials. Despite the great urgency of this secret wartime program, workers at Hanford and elsewhere in the nuclear weapons complex had to be properly protected while the highly radioactive bomb-making materials were processed, handled, machined, and tested.

To protect themselves from intense nuclear radiation, the workers began using an innovative (though simple) system of mechanical manipulators, called teleoperators. With these action-at-a-distance devices, workers could safely handle the radioactive materials, which were kept on the other side of thickly shielded areas called hot cells. It was the timely development of these dexterous

Figure 5-7 A U.S. Marine Corps explosive ordnance disposal (EOD) technician prepares to deploy a remotely operated robot vehicle to detonate a buried improvised explosive device (IED) near Camp Fallujah in Iraq (November 27, 2005). The suspected IED was buried in a dirt mound on the side of the road next to an old IED crater. (Credit: Photograph courtesy of the U.S. Marine Corps.)

Figure 5-8 This photograph shows a Talon 3B robot as it recovers a stick of dynamite and other explosive devices at a military training range in Bahrain, Persian Gulf. The explosive devices used in this training exercise (June 2005) are commonly seen by American military personnel deployed in Iraq. Explosive ordnance disposal (EOD) technicians operate the robot from safe locations through the use of monitors and video equipment attached to the unit. The robot is designed for the search and destruction of improvised explosive devices (IEDs), as well as other forms of ordnance found in Iraq. (Credit: Photograph courtesy of the U.S. Navy.)

(but expendable) machines that allowed the nuclear workers to successfully and safely handle extremely hazardous materials. Once adequate quantities of plutonium were produced, the bomb material was shipped to Los Alamos National Laboratory where scientists molded, shaped, and machined the artificially produced metal into the precise components they needed to make the revolutionary nuclear fission bombs that operated on the implosion principle.

From a historic perspective, the first mechanical manipulators were essentially fancy metallic tongs, similar to the tools blacksmiths used to handle red-hot horseshoes, while they pounded and worked them into the proper shape. But, scientists working on the Manhattan Project needed something better and safer than tongs to handle intensely radioactive materials. So, engineers at the U.S. Atomic Energy Commission's (USAEC) Argonne National Laboratory (ANL) in Illinois designed the first generation of unilateral remote manipulators.

A unilateral manipulator is an electromechanical device (often with a small electric motor to operate the mechanical arms fingers or grippers) that does not provide force feedback to the human operator. The operator can see what the mechanical arms are doing but does not have a sense of touch or feel with respect to the on-going mechanical actions. Despite the lack of force feedback, a wide variety of well-engineered manipulators supported the development of the first nuclear reactors and first nuclear weapons—especially the processing of intensely radioactive materials in well-shielded facilities, called hot cells.

The unilateral manipulators of the Manhattan Project demonstrated two important principles of teleoperation: first, the mechanical arm/hand can be located a significant distance away from the human operator; second, the force exerted by the mechanical arm/hand can greatly exceed human capabilities.

In September 1944, the Manhattan Project's 100-B plutonium production reactor at Hanford achieved nuclear criticality and began operation. Two months later construction workers at Hanford completed the chemical separation plants in which human workers used a variety of teleoperated manipulator systems to process the highly radioactive irradiated fuel from the Hanford production reactors and provided the plutonium used in the implosion-design nuclear bombs. Although primitive by modern standards, the early manipulators systems did the job and nuclear weapons using plutonium became the cornerstone of the American nuclear weapons program.

At 05:29:45 A.M. (Mountain War Time) on July 16, 1945, the United States successfully detonated the world's first nuclear explosion in a remote portion of the southern New Mexican desert. Code-named Trinity, the bulky and primitive, spherical plutonium-implosion test device exploded with a yield of about 21 kilotons. The tremendous blast heralded the dawn of a new age in warfare—the age of nuclear weaponry. From that dramatic moment on, human beings possessed the means of swiftly bringing about their own destruction.

As mentioned previously, the development of nuclear weapons caused a fundamental change in the national security policy of the United States and of other nations throughout the world. For example, before the existence of the nuclear weapon, the primary purpose of the United States military establishment was to fight and win wars. Following World War II, however, American national security strategy experienced a series of dramatic transformations. Accelerating this process was the fact that American nuclear scientists, endowed with an increasing supply of plutonium, kept developing progressively more compact and more efficient nuclear weapons.

In 1949, Goertz and his coworkers at the ANL publicly demonstrated the first mechanical, bilateral master-slave manipulator device for the remote handling of hazardous materials. These hazardous materials included the highly radioactive materials associated with the rapidly expanding American civilian and military nuclear programs. Goertz's first bilateral master-slave manipulator had a crude sense of touch, which meant that when the mechanical fingers (grippers) of the slave manipulator arm closed on a glass beaker, the human operator handling the master manipulator arm could feel resistance of the beaker's glass wall to the pressure of the machine's mechanical fingers. This sense of touch (in reality a form of force feedback) greatly improved the deftness of the human-machine combination in teleoperation and also prevented the greater-than-human mechanical advantage of a machine manipulator from breaking delicate objects. Progress in teleoperation directly supported progress in nuclear technology applications.

After World War II, the primary focus of American national security strategy was to block further political expansion by the former Soviet Union. The American nuclear weapon emerged as a powerful instrument of geopolitics. The strategic transition took place primarily because the nuclear weapon was available, and as perceived by American national security strategists a relatively "inexpensive" way to keep the vastly superior Russian conventional forces at bay in postwar Western Europe and elsewhere in the world. In the tense days of the ideological conflict that characterized the early portion of the cold war, the United States parlayed the threat of using its nuclear weapons monopoly to devastate the Russian homeland to thwart any additional attempts by the former Soviet Union at territorial expansion. However, American strategists also recognized that the advantage of this nuclear monopoly would not last forever, because now that the physical principles of a nuclear fission weapon had been demonstrated, any industrialized nation could (in principle) construct a nuclear weapon. The spread of nuclear weapons was simply a matter of time and the commitment by other governments of large quantities of resources.

But these same analysts were quite surprised when the former Soviet Union broke the American nuclear weapon monopoly so quickly. The first Russian nuclear explosion took place in August 1949 and served as the spark that ignited an incredible arms race. The nuclear arms race dominated geopolitics and military strategies for the next four decades. The detonation of the first Russian nuclear device was also the world's first example of nuclear proliferation—the process by which a nation that did not formerly possess a nuclear weapon acquires and demonstrates it now has nuclear weapons capability.

Responding to the needs of the rapidly growing nuclear industry, in 1954, Goertz and his coworkers at the ANL further improved the art of teleoperation by applying the principles of cybernetics to manipulator design and constructed the first electric master-slave manipulator system. The new device represented another major milestone in teleoperation and robotics. Electric wires that carried control signals in one direction and force feedback in the other direction replaced the often cumbersome cables and metal tapes, which connected the master arms and hands to the slave counterparts. Now, when a human operator used his hand to close the grips on the master manipulator, the action sent electric signals to a servomotor in the remote slave manipulator. As a result of this breakthrough, the bilateral teleoperator, like its unilateral cousin, conquered distances with wires or radio frequency signals. Goertz's device established the principle of the teleoperation of machines (robots) at great distance—a principle with important applications well beyond the nuclear industry.

The act of nuclear proliferation by the former Soviet Union not only triggered the great nuclear arms race of the cold war, but it also created a permanent legacy of nuclear contamination in both nations. Today, environmental scientists in the United States and the Russian Federation are cooperatively searching for more efficient ways to cleanup the cold war's highly toxic and radioactive legacy. This massive environmental cleanup and waste disposal problem has several major characteristics, which make the application of modern nuclear robots an integral part of any effective solution. Many of the cold war era nuclear complex sites have hazardous radiation environments, chemical waste burdens, and abandoned facilities with physically restrictive passageways—conditions that make access by human workers totally unsuitable. The regulatory limits that establish allowable radiation exposure levels for nuclear industry workers create an additional, legal impediment on the use of human workers in large-scale cleanup activities. Unlike robots, human workers quickly exceed statutory radiation exposure limits (often in minutes or hours), even with the best protective equipment and well-planned operations. So the use of human workers becomes impractical, if not outright impossible. Remotely operated mobile robots are the only practical way of resolving the problem.

To make matters worse for human workers, many of these nuclear weapons complex sites are also loaded with harmful chemicals and now present hostile mechanical working environments, with numerous dangers posed by things like sharp metal objects. So it is only the use of specially designed nuclear robots that makes decontamination, waste removal, and decommissioning operations practical and (in many instances) possible. When properly designed, mobile nuclear robots can function in such harsh, cramped, and dangerous environments.

Today, the nuclear industry has adapted many mobile robots and portable teleoperator systems for applications in hazardous, highly radioactive environments. The mobile robots and teleoperators are used as substitutions for human presence in such hazardous work environments. Robot engineers have designed the systems to accommodate the terrain and navigate the narrow spaces occurring at various sites. Mobility is provided through multiple track systems or wheel systems, depending on the circumstances. Engineers enhance the nuclear robot's maneuverability with design features such as zero turning radius, skid steering, and the use of ultrasonic sensors or lasers for navigation.

Cleanup and environmental monitoring teams have used nuclear robots and teleoperators in a number of important applications. For example, the robot vehicles have been programmed to effectively collect, store, and analyze data obtained during the inventory and inspection of waste drums. Mobile vehicles have surveyed and mapped floors for radiation contamination and have decontaminated concrete surfaces using superheated water spray heads or scabbling equipment. Nuclear robots and portable teleoperators have also successfully excavated radioactive soil and cut and removed highly contaminated equipment in process areas.

A mobile robot named SWAMI (Stored Waste Autonomous Mobile Inspector) was developed by Department of Energy sponsored scientists at the Savannah River Technology Center to make the tedious and potentially dangerous job of drum inspection safer and more efficient. Department of Energy sites around the country now store tens of thousands of drums containing low-level radioactive, hazardous, and mixed wastes. Regulations require weekly inspection of these drums—a monotonous, time-consuming task that human workers must perform in hostile environments. Robotic systems, like SWAMI, have improved the efficiency, documentation, and accuracy of drum inspection and inventory activities, while greatly reducing the exposure of human personnel to hazardous materials. Operational experience with mobile inspection robots, like SWAMI, has also shown that a properly designed mobile robot can detect floor contamination (from leaky drums) more reliably and accurately. Looking beyond the cleanup of the cold war nuclear complex sites, as aging civilian nuclear facilities are closed and scheduled for decommissioning, these robotic systems will play an ever-expanding, crucial role in waste drum inspections and inventory operations.

In many industries, including the nuclear industry, inspection and maintenance tasks would be much easier if only human workers could have visual access to out-of-the-way places, such as the inside of tanks, ducts, and pipes, or safe access to dangerous environments. One solution is to deploy robot systems that carry viewing equipment (or other appropriate sensors) into remote areas. In that way, human operators can remotely perform the necessary inspection activities while stationed a safe distance from the hazardous environment operation. Robotic, remote viewing systems have been developed with a variety of equipment to enable video and audio monitoring of any area. The systems can use lights, video cameras (either color or black and white), radiation and chemical monitors, microphones, and speakers. The robot's cameras are often designed to zoom, pan, tilt, and rotate. Stereo viewing and the simultaneous observation of numerous locations are possible if engineers equip the robot with multiple camera systems. Depending on the needs of the particular inspection operation, the remote viewing robot may also make video recordings (in some standardized format) to permanently document its journey into a hazardous, inaccessible location.

In the late 1980s, Foster-Miller developed CECIL® a robotic maintenance and inspection system for nuclear steam generators. Sponsored by the Electric Power Research Institute (EPRI), Consolidated Edison, and Public Service Electric and Gas, CECIL® provides inspection and cleaning capability within the tube bundle of steam generators used in nuclear power plants. Using a flexible lance, the teleoperated nuclear robot allows access to remote steam generator locations. The system uses high-pressure water jets to remove hard and soft sludge deposits, while the human operator controls the operation by means of video feedback. The robot is installed through an inspection hand-hole in the steam generator unit and travels throughout the lower region of the tube bundle. In addition to cleaning, the CECIL® robot provides visual inspection, contaminant sampling, and foreign object retrieval. CECIL® is a mobile nuclear robot specifically designed to operate in restricted high-radiation environments under the remote supervision of human personnel in protective clothing. The CECIL® robot has been used in nuclear plants in the United States, Japan, Korea, France, and Canada.

In 1991, the U.S. Department of Energy started a large-scale demonstration decontamination and decommission (D&D) operation of the Chicago Pile-Five (CP-5) Research Reactor Facility at the ANL. Over its lifetime, the CP-5 Research Reactor Facility had been used as an intense neutron source, irradiating over 27,000 research specimens. The D&D operation ended in July 2000 and the facility was cleared of radiological contamination and formerly released as an "Industrial Use Area." During this large-scale D&D operation a variety of mobile robots and portable teleoperator systems played a significant role.

In summary, nuclear robots and teleoperators are an integral part of the nuclear industry. As a substitution for human presence, these well-demonstrated, robust systems have performed complex tasks in hazardous and difficult-to-access work locations. They take the punishment and radiation exposure, while their human partners observe and direct the tasks from safer, remote locations.

Figure 5-9 This picture shows a pipeline inspection mobile robot developed at the Intelligent Systems and Robotics Center of the Sandia National Laboratories in New Mexico. (Credit: Photograph courtesy of the U.S. Department of Energy and Sandia National Laboratories.)

MEETING THE UNIVERSE FACE-TO-FACE

Robot spacecraft have opened up the universe to exploration. Modern space robots are sophisticated exploring machines that have now visited all the eight major planets of the solar system. As a historic note, NASA's *New Horizons* spacecraft was successfully launched from Cape Canaveral, Florida, on January 19,

Figure 5-10 The picture provides a close-up view of a remotely operated field robot (built circa 1990) at the Oak Ridge National Laboratory (ORNL). The mobile nuclear robot is equipped with electronic systems, sensors, and computers. Designed to perform outdoor surveys of radioactive waste storage sites, the diesel-engine powered, eight-wheeled, all-terrain vehicle is driven across the field being inspected by a human teleoperator, who remains comfortably and safely seated at a remote control console. (Credit: Photograph courtesy of Department of Energy/Oak Ridge National Laboratory.)

2006. This robot probe is now traveling on its long one-way mission to conduct a scientific encounter with the dwarf planet Pluto and its system of three icy moons in July 2015.

Emerging out of the politically charged space race of the cold war, a progressively more capable family of robot spacecraft have dramatically changed what scientists know about the alien worlds that journey together with Earth around the Sun. In a little over four decades, scientists have learned more about these wandering lights, called πλαυετες (or planets) by the ancient Greek astronomers, than in all the previous centuries of astronomical observations. Thanks to space robots, each major planetary body and (where appropriate) its collection of companion moons has now become a much more familiar world.

Similarly, sophisticated robot astronomical observatories placed on strategically located platforms in space have allowed astronomers and astrophysicists to meet the universe face-to-face, across all the information-rich portions of the electromagnetic spectrum. No longer is the human view of the universe limited

Figure 5-11 This artist's rendering depicts NASA's *Cassini* spacecraft during the critical Saturn orbit insertion (SOI) maneuver, just after the main engines had begun firing on July 1, 2004. The smart exploring machine performed the SOI maneuver automatically. The successful maneuver reduced the robot spacecraft's speed, allowing *Cassini* to be captured by Saturn's gravity and enter orbit, beginning a planned four-year scientific investigation of the Saturn system. On December 25, 2004, *Cassini* successfully released its hitchhiking companion, the *Huygens* probe—sending the wok-shaped robot on its historic one-way journey into the atmosphere of Titan (Saturn's largest moon). (Credit: Artist's rendering courtesy of NASA/JPL.)

to a few narrow bands of radiation that trickle down to Earth's surface through an intervening atmosphere that is often murky and turbulent.

Space robots share certain common features with their mobile robot technical counterparts on Earth. However, space robots also require the blending of aerospace and computer technologies that are far more demanding, unusual, and sophisticated than generally needed for robots operating here on Earth. Therefore, space robots can be regarded as leading the robot technology parade—a major impact that sets the stage for incredible developments in robotic system technologies this century. For example, space robots need to be quite smart and "independent," since they generally have to work on their own in the harsh environment of outer space and sometimes on strange alien worlds, about which little is previously known.

Under certain circumstances, telepresence and virtual reality technologies allow a human being to form a real-time, interactive partnership with an advanced space robot, which serves as that person's dexterous mechanical surrogate capable of operating in a hazardous, alien world environment. This is true, perhaps, of an advanced, future space robot designed to explore remote regions of the Moon, while its human controller uses virtual reality technologies to make important new discoveries working in shirtsleeve comfort inside a permanent lunar surface base or even back on Earth. But as a space robot operates farther away, the round-trip communications distance with human controllers back on Earth must soon be measured not in thousands of kilometers, but rather in light-minutes. The great distances associated with deep space exploration make the real-time control of a robot spacecraft by human managers impractical, if not altogether impossible. So to survive and function around or on distant worlds, space robots need to be very smart, that is, they need to contain various levels of machine intelligence, or artificial intelligence (AI). As levels of machine intelligence continue to improve this century, truly autonomous space robots will become a reality.

Someday, human engineers will construct an especially intelligent robot that exhibits a cognitive "machine mind" of its own. AI experts suggest that smart exploring machines of the future will have (machine) intelligence capabilities sufficient to repair themselves, to avoid hazardous circumstances on alien worlds, and to recognize and report all the interesting objects or phenomena they encounter.

There is an interesting correlation between progress in space exploration by robots and parallel progress in computer technology and aerospace technology. To observe the connection, all a person has to do is take a brief look at some of the most interesting American space robots, as exemplified by NASA in Pioneer, Ranger, Mariner, Viking, and Voyager programs. As an integral part of the space age (1957), space robots emerged from simple, often unreliable, electromechanical exploring devices, into the fairly sophisticated scientific platforms that now extend human consciousness and intelligent inquiry to the edges of the solar system and far beyond.

One outstanding example of technical progress is the first intense search for life on Mars. This exciting effort started in 1975, when NASA launched the agency's Viking missions, consisting of two orbiter and two lander robot spacecraft. Development of this elaborate robot mission was divided between several NASA centers and private U.S. aerospace firms, with JPL building the Viking orbiter spacecraft, conducting mission communications, and eventually assuming management of the mission.

Credit for the single space robot mission that has visited the most planets goes to JPL's Voyager project. Launched in 1977, the twin *Voyager 1* and *Voyager 2* spacecraft flew by the planets Jupiter (1979) and Saturn (1980–1981). *Voyager 2* then went on to have an encounter with Uranus (1986) and with Neptune (1989). Both *Voyager 1* and *Voyager 2* are now traveling on different trajectories into interstellar space. In February 1998, *Voyager 1* passed the *Pioneer 10* spacecraft to become the most distant human-made object in space. The Voyager Interstellar Mission (VIM) should continue well into the second decade of the twenty-first century.

Millions of years from now—most likely when human civilization has completely disappeared from the surface of Earth—four robot spacecraft (*Pioneer 10* and *11*, *Voyager 1* and *2*) will continue to drift through the interstellar void. Each spacecraft serves as a legacy of human ingenuity and inquisitiveness. By carrying a special message from Earth, each far-traveling robot spacecraft also bears permanent testimony that at least for one moment in the often-bloodied history of the human species, a few people raised their foreheads to the sky and reached for the stars. Though primarily designed for scientific inquiry within the solar system, these four relatively simple robotic exploring machines are now a more enduring artifact of human civilization than any cave painting, great monument, giant palace, or high-rising city created here on Earth.

A new generation of more-sophisticated spacecraft appeared in the late 1980s and early 1990s. These spacecraft allowed NASA to conduct much more detailed scientific investigations of the planets and the Sun. Representative of the significant advances in sensor technology, computer technology, and aerospace engineering are the robot spacecraft used in the Galileo mission to Jupiter and the Cassini mission to Saturn.

The Galileo mission began on October 18, 1989, when the sophisticated spacecraft was carried into low Earth orbit by the space shuttle *Atlantis* and then started on its interplanetary journey by means of an inertial upper stage (IUS) rocket. Relying on gravity-assist flybys to reach Jupiter, the *Galileo* spacecraft flew past Venus once and Earth twice. As it traveled through interplanetary space beyond Mars on its way to Jupiter, *Galileo* encountered the asteroids Gaspra (October 1991) and Ida (August 1993). *Galileo*'s flyby of Gaspra on October 29, 1991, provided scientists their first-ever close-up look at a minor planet. On its final approach to Jupiter, *Galileo* observed the giant planet being bombarded by fragments of Comet Shoemaker-Levy-9, which had broken apart. On July 12, 1995, the *Galileo* mother spacecraft separated from its hitchhiking companion (an atmospheric probe) and the two robot spacecraft flew in formation to their final destination.

On December 7, 1995, *Galileo* fired its main engine to enter the orbit around Jupiter and gathered data transmitted from the atmospheric probe during that small robot's parachute-assisted descent into the Jovian atmosphere. During its two-year prime mission, the *Galileo* spacecraft performed ten targeted flybys of Jupiter's major moons. In December 1997, the sophisticated robot spacecraft began an extended scientific mission, which featured eight flybys of Jupiter's smooth, ice-covered moon Europa and two flybys of the pizza-colored, volcanic Jovian moon, Io.

Galileo started a second extended scientific mission in early 2000. This second extended mission included flybys of the Galilean moons Io, Ganymede, and Callisto, plus coordinated observations of Jupiter with the *Cassini* spacecraft. In December 2000 *Cassini* flew past the giant planet to receive a much-need gravity assist, which enabled the large spacecraft to eventually reach Saturn. *Galileo* conducted its final flyby of a Jovian moon in November 2002, when it zipped past the tiny inner moon, Amalthea.

The encounter with Amalthea left *Galileo* on a course that would lead to an intentional impact into Jupiter's atmosphere in September 2003. NASA mission controllers deliberately crashed the *Galileo* mother spacecraft into Jupiter at the end of the space robot's very productive scientific mission, because they wanted to avoid any possibility of contaminating Europa with terrestrial microorganisms. As an uncontrolled derelict, the *Galileo* spacecraft might have eventually crashed into Europa sometime within the next few decades. Many exobiologists suspect that Europa has a life-bearing, liquid-water ocean underneath its icy surface.

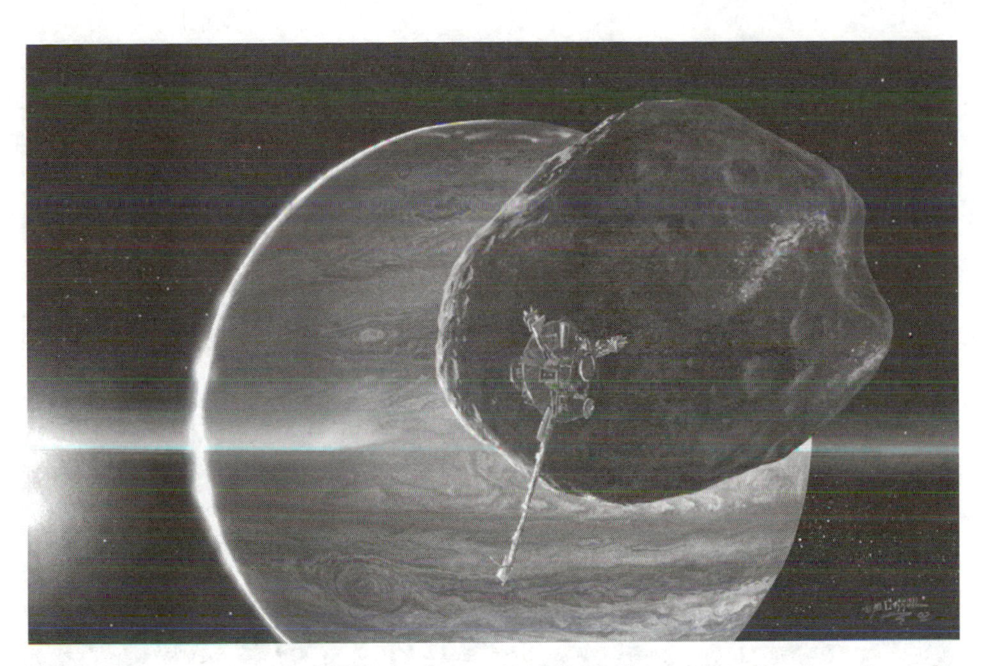

Figure 5-12 This artist's rendering depicts NASA's Galileo spacecraft as it performed a very close flyby of Jupiter's tiny inner moon Amalthea in November 2002. (Credit: Artist's rendering courtesy of NASA.)

NASA's sophisticated *Cassini* spacecraft, which is now exploring the Saturn system, and the robust *Spirit* and *Opportunity* Mars Exploration Rovers, which are now rolling across the surface of the Red Planet, are some of the latest examples of advanced robot spacecraft engineering.

Each portion of the electromagnetic spectrum (that is, radio waves, infrared radiation, visible light, ultraviolet radiation, X-rays, and gamma rays) brings astronomers and astrophysicists unique information about the universe and the objects within it. For example, certain radio frequency (RF) signals help scientists characterize cold molecular clouds. The cosmic microwave background (CMB) represents the fossil radiation from the big bang, the enormous ancient explosion considered by most scientists to have started the present universe about 15 billion years ago. The infrared (IR) portion of the spectrum provides signals that let astronomers observe nonvisible objects such as near-stars (brown dwarfs) and relatively cool stars. Infrared radiation also helps scientists peek inside dust-shrouded stellar nurseries (where new stars are forming) and unveil optically opaque regions at the core of the Milky Way Galaxy. Ultraviolet (UV) radiation provides astrophysicists special information about very hot stars and quasars, while visible light helps observational astronomers characterize planets, main sequence stars, nebulae, and galaxies. Finally, the collection of X-rays and gamma rays by space-based observatories brings scientists unique information about high-energy phenomena, such as supernovae, neutron stars, and black holes, whose presence is inferred by intensely energetic radiation emitted from extremely hot material as it swirls in an accretion disk before crossing the particular black hole's event horizon.

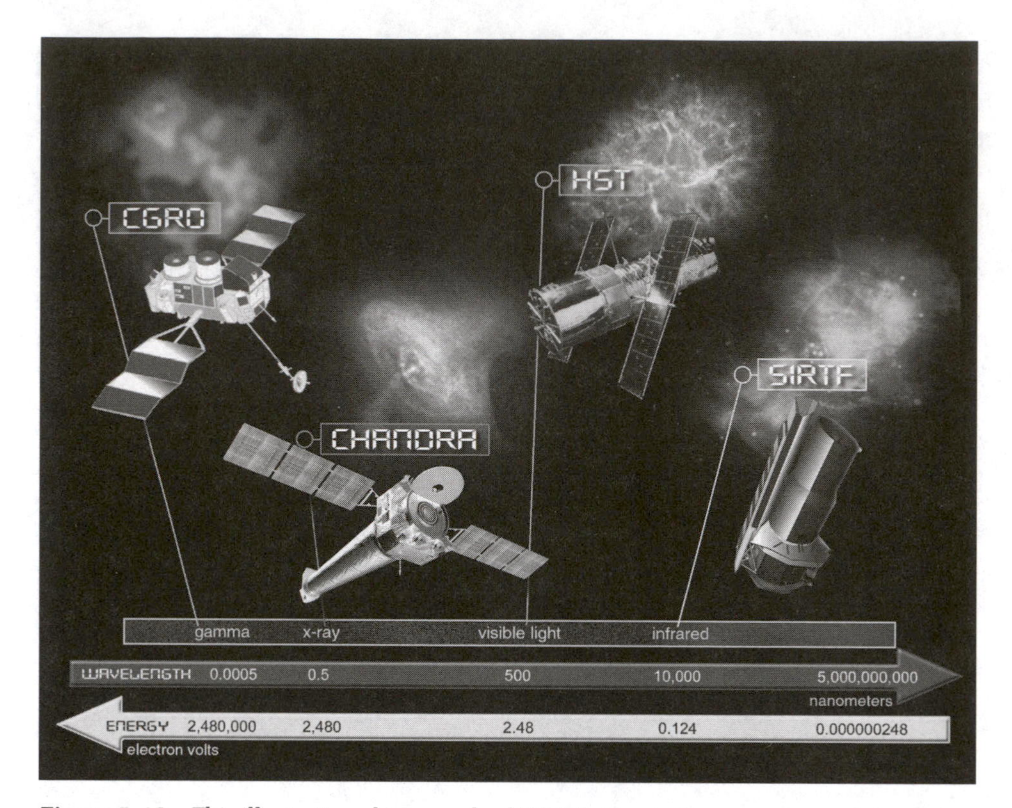

Figure 5-13 This illustration shows each of NASA's Great Observatories and the region of the electromagnetic spectrum from which the particular space robot (astronomy facility) collects scientific data. From left to right (in order of decreasing photon energy and increasing wavelength): the *Compton Gamma Ray Observatory* (CGRO); the *Chandra X-Ray Observatory* (CXO); the *Hubble Space Telescope* (HST); and the *Space Infrared Telescope Facility* (SIRTF), renamed by NASA as the *Spitzer Space Telescope* (SST). (Credit: Illustration courtesy of NASA.)

Scientists recognized that they could greatly improve their overall understanding of the universe, if they could observe all portions of the electromagnetic spectrum. As the technology for space-based astronomy matured toward the end of the twentieth century, NASA created the Great Observatories Program. This important program involved a series of four highly sophisticated space-based astronomical observatories—each carefully designed with state of the art equipment to gather "light" from a particular portion (or portions) of the electromagnetic spectrum. An observatory spacecraft is a robot spacecraft that does not have to travel to a celestial destination to explore it. Rather, the observatory spacecraft occupies a special orbit around Earth or the Sun, from which location the robot system can observe distant astronomical targets free of the obscuring and blurring effects of Earth's atmosphere. Infrared observatories must also operate in orbits that minimize interference from large background thermal radiation sources like Earth and the Sun.

NASA initially assigned each Great Observatory a development name and then renamed the orbiting astronomical facility to honor a famous scientist. The first Great Observatory was the *Space Telescope* (ST), which became the *Hubble Space Telescope* (HST). It was launched by the space shuttle in 1990 and then refurbished on-orbit through a series of subsequent shuttle missions. With constantly upgraded instruments and improved optics, this long-term space-based observatory is designed to gather light in the visible, ultraviolet, and near infrared portions of the spectrum. This spacecraft honors the American astronomer Edwin Powell Hubble (1889–1953). NASA is now examining plans for another (possibly robotic) refurbishment mission, which might keep the HST operating for several more years until being replaced by the *James Webb Space Telescope* (JWST) around 2011.

The second Great Observatory was the *Gamma Ray Observatory* (GRO), which NASA renamed the *Compton Gamma Ray Observatory* (CGRO), following its launch by the space shuttle in 1991. Designed to observe high-energy gamma rays, this robotic observatory collected valuable scientific information from 1991 to 1999 about some of the most violent processes in the universe. NASA renamed the observatory to honor the American physicist and Nobel laureate, Arthur Holly Compton (1892–1962). The CGRO's scientific mission officially ended in 1999. The following year, NASA mission managers commanded the massive spacecraft to perform a controlled deorbit burn. This operation resulted in a safe reentry in June 2000 and the harmless impact of surviving pieces in a remote portion of the Pacific Ocean. This action avoided the undesirable and adverse impact of having an abandoned space robot possibly crashing down on someone's house after it reentered Earth's atmosphere.

NASA originally called the third observatory in this series the *Advanced X-Ray Astrophysics Facility* (AXAF). NASA renamed this observatory the *Chandra X-Ray Observatory* (CXO) to honor the Indian-American astrophysicist and Nobel laureate, Subrahmanyan Chandrasekhar (aka Chandra) (1910–1995). The observatory spacecraft was placed into a highly elliptical orbit around Earth in 1999. The CXO examines X-ray emissions from a variety of energetic cosmic phenomena, including supernovas and the accretion disks around suspected black holes, and should operate until at least 2009.

The fourth and final member of NASA's Great Observatory Program is the *Space Infrared Telescope Facility* (SIRTF). NASA launched this observatory in 2003 and renamed it the *Spitzer Space Telescope* (SST) to honor the American astrophysicist, Lyman Spitzer, Jr. (1914–1997). The sophisticated infrared observatory provides scientists a fresh vantage point from which to study processes that have until now remained mostly in the dark, such as the formation of galaxies, stars, and planets. The SST also serves as an important technical bridge to NASA's Origins Program—an ongoing attempt to scientifically address such fundamental questions as "Where did we come from?" and "Are we alone?"

Today, incredibly complex robot exploring machines allow scientists on Earth to conduct detailed, firsthand investigations of alien worlds throughout the solar system and beyond. The robot–human partnership in space exploration makes the universe both a destination and a destiny. In its ultimate form, this partnership leads to the very exciting concept of the self-replicating system (SRS). (See Chapter 7 for a more detailed discussion of the self-replicating system concept.) If ever developed, the SRS unit would represent an extremely powerful tool for robot space exploration with ramifications on a cosmic scale. Using properly developed and controlled SRS technologies, a future generation of humans could set in motion a chain reaction that spreads organization, life, and conscious intelligence across the galaxy in an expanding wave-like bubble, limited in propagation velocity only by the speed of light itself.

ROBOTS AS TOOLS TO EXPLORE INNER SPACE

Underwater robots have made the oceans of Earth more readily accessible for scientific study, resource prospecting and harvesting, salvage and rescue operations, national defense activities, and archaeology. Underwater robots are mobile systems that can be divided into two general classes: the remotely operated vehicle (ROV) and the autonomous underwater vehicle (AUV). Each type of submersible robot system has been used in a variety of missions and each has several well-demonstrated applications in which sophisticated machines have successfully replaced the need for human divers. These robots can also work at depths in the ocean that exceed the reach of human divers and can operate for more extended times. For remotely operated systems, the human being is comfortably located at a workstation on a surface ship, ocean platform, or even ashore. Autonomous underwater vehicles are launched, operate independently of human interaction, and then (depending on the mission and circumstances) return to a designated location for recovery, refurbishment, and reuse.

The U.S. Navy pioneered development of underwater robots in the 1960s for a variety of national defense needs, including the ability to perform deep-sea rescue missions and to collect objects from the floor of the ocean at depths exceeding the capabilities of human divers or under conditions considered too hazardous for human divers.

One historic deep-water search mission is presented here to demonstrate one of the important roles that underwater robots play in national defense and the ocean sciences. The United States Navy has lost two nuclear submarines at sea: USS *Thresher* (lost during sea trials in 1963) and USS *Scorpion* (lost on patrol in 1968).

The USS *Scorpion* (SSN-589) was a 3,500-ton *Skipjack* class nuclear-powered attack submarine. After commissioning in 1960, the *Scorpion* made periodic deployments to the Mediterranean Sea and other ocean areas where the presence of a fast and stealthy submarine would be beneficial during the cold war. The *Scorpion* began another Mediterranean Cruise in February 1968. The following May, while homeward bound from that tour, the ship was lost along with its 99 officers and crewmen in the Atlantic Ocean about 640 kilometers southwest of the Azores. In late October 1968, the U.S. Navy located the wreckage of the USS *Scorpion* as it rested on the sea floor about 3,050 meters below the surface. Photographs taken by a towed deep-submergence vehicle deployed from the USNS *Mizar* showed that the *Scorpion's* hull had suffered fatal damage as the vessel submerged and that even more damage occurred as she sank.

Since the fatal accident, U.S. Navy personnel (assisted by advanced underwater robots) have made subsequent visits to the site and studied the wreck.

Figure 5-14 This picture shows a U.S. Navy sailor inspecting the Super Scorpio remotely operated vehicle (ROV) aboard the Military Sealift Command (MSC) special missions ship M/V *Kellie Chouest* on May 7, 2004. A navy salvage team on the ship is preparing to recover an F-14D Tomcat aircraft, which had crashed into the Pacific Ocean about three kilometers west of Point Loma, California, during a routine training mission from the USS *John C. Stennis* (CVN 74). The Super Scorpio ROV is equipped with remotely operated video cameras and teleoperated manipulators. (Credit: Photograph courtesy of the U.S. Navy.)

However, the cause of the initial damage to the USS *Scorpion* continues to generate controversy. The U.S. Navy also conducted environmental sampling of the wreckage site in 1968, 1979, and 1986. The results of sediment, water, and marine analyses indicated that the concentrations of total plutonium in these environmental samples were not significantly different than the background concentrations due to fallout from past atmospheric nuclear tests. Despite the tragic loss of the *Scorpion*, its reactor's inherent safety features against catastrophic contamination of the marine environment appear to be performing as designed and intended.

Industrial companies use ROVs and AUVs to perform inspections of underwater pipelines and the structural testing of offshore platforms. Scientists use underwater robots to perform studies of the ocean, its animals, and the conditions of the sea floor in ways never before achievable. Adventurers and historians use submersible robots to locate famous shipwrecks (such as the RMS *Titanic*) or to salvage ancient artifacts and treasures.

Underwater robots represent an exciting new tool for archaeologists. Today, with ROVs and AUVs, archaeologists can explore traces of human history now lost in ocean depths far beyond the 50-meter boundary of SCUBA diving. Underwater robots have dramatically increased the number of interesting underwater

Figure 5-15 The bow of the sunken nuclear submarine USS *Scorpion* (SSN-589) as photographed in October 1968 by a deep-submergence vessel deployed from the USNS *Mizar*. This image shows the top of the bow section from the vicinity of the sail (which has been torn off) at the left to the tip of the bow at the center of the image. The torpedo room hatch is visible about half way along the length of this hull section, with a lifeline track running aft from it. (Credit: Photograph courtesy of the U.S. Navy.)

sites now available for archaeological study. Over the past decade, successful projects in the Black and Mediterranean Seas have demonstrated the scientific merit of deep-water archaeology. Underwater robots equipped with chemical sensors, acoustic sensors, and high-definition camera systems provide data that often answers important archaeological questions. Similarly, ROVs equipped with manipulators and special tools enable archaeologists to make physical intrusions on otherwise inaccessible deep-water sites. Through the practice of teleoperation, archaeologists can now remotely collect artifacts on the sea floor or even excavate into the lower strata of the sea bottom—all from the comfort of a control console aboard a surface ship.

ROBOTS AS STIMULI FOR SCIENCE EDUCATION

For the past two decades, professional educators in the United States have repeatedly expressed their concerns about the general decline in interest in science, engineering, and mathematics by American students in upper elementary school through college. Business and political leaders have echoed similar concerns, because a decline in scientific and mathematical literacy among the younger portion of the population can easily translate into an overall loss of national competitiveness in an ever-more technology-based global economy.

But a ray of sunshine has suddenly appeared among all these gloomy projections. Interest in hobby robots, entertainment robots, and competitive robots

Figure 5-16 Personnel with the Naval Sea Systems Command (NAVSEA), work with another crew man to retrieve the Office of Naval Research (ONR)-funded Remote Environmental Monitoring Units (REMUS)—an autonomous underwater vehicle (AUV) developed by the Woods Hole Oceanographic Institution. During this operation off Cape Hatteras, North Carolina, in September 2005, scientists and historians were attempting to locate the *Alligator*—the U.S. Navy's first submarine. The *Alligator* was lost at sea in a fierce storm in 1863. The torpedo-like underwater robot performed an ocean floor survey, as part of an ongoing effort by the National Oceanographic and Atmospheric Administration (NOAA) to promote scientific and historic research, education, and ocean literacy. (Credit: Photograph courtesy of the U.S. Navy.)

(also known as battlebots) appears to be stimulating student interest in science, mathematics, and engineering. Students of all ages are participating in and enjoying organized robot competitions that represent sporting events (like robot soccer) or even gladiator competitions (essentially a remotely controlled demolition derby).

In one competition, sponsored by an organization called FIRST (For Inspiration and Recognition of Science and Technology), as the game begins four robots burst into a playing field and start scrambling to cross a bridge, collect balls, and score the maximum number of points within a fixed period of time. The match involves fast and furious action as each team's "homemade" machines compete with others under the watchful direction (radio-control) of anxious teams of students. These cheering students quickly learn that there is much more science and engineering involved in building a winning robot than simply inserting a battery pack and toggling a joystick.

Often adults from professional scientific and engineering organizations work with teams of students and their science teachers in designing, building, and testing a "winning bot." This growing interest in hobby-level robots stretches

Figure 5-17 The Bee Bots team (393) robot, named Dr. Beevil, scores points by gathering balls, during the FIRST Southeast Regional competition held at the Kennedy Space Center Visitor Complex, Florida, in March 2000. During this competition, teams of high school students tested the limits of their imagination, using robots they had designed with the support of business and engineering professionals and corporate sponsors. (FIRST is an organizational acronym meaning For Inspiration and Recognition of Science and Technology.) (Credit: Photograph courtesy of NASA/KSC.)

across international boundaries and a number of interesting sporting competitions (such as robot soccer leagues) have emerged. Many adults are also having fun with building and competing hobby or sports robots, as evidenced by the number of informal robot clubs and groups that have appeared in the last few years. (See Chapter 11 for some examples.)

Several years from now, educators, social scientists, and political leaders will be able to more accurately judge the true impact "hobby robots" are making on the overall technical and mathematical literacy of the American population. For now, the trend appears to be a giant leap in the right (more technical) direction, and one to be strongly encouraged by academic institutions, government agencies, and private firms.

6

Issues

There are many complex and interesting issues surrounding the use of modern robotic systems, including the tiny little devices promised by nanotechnology. Some of these issues, like the fear of robots destroying the human race, are both widespread and yet quite unfounded. Such anxieties originate primarily from science fiction stories and cinematic portrayals of robots as rogue androids and cyborgs. Unfortunately, these purely fictional portrayals have made dramatic (often subconscious) impressions on people, especially when experienced in childhood. Even when considered from a mature, adult perspective, many people find it quite difficult to shake the negative impressions. For example, once a person has encountered HAL 9000 misbehaving in the classic science fiction story, *2001: A Space Odyssey*, he or she usually finds it difficult to totally ignore the possibility of artificial intelligence gone wild. After all, if people can have mental breakdowns, why not machines?

There is hardly a person alive today who has not pounded on some malfunctioning machine, or yelled out at a "misbehaving" computer. These actions, when viewed rationally, under calm mental conditions, are actually quite senseless and rather foolish. The automobile that did not start or the computer that "glitched" and lost a file are inanimate objects, which are totally unaware of their existence and certainly quite incapable of making decisions specifically intended to thwart the desires of the human operator. Even so, people still pound on, kick, or scream at misbehaving machines with an uncanny regularity that reveals the presence of some basic antagonisms and fears that are deeply etched into the subconscious.

Similarly, the vast majority of moviegoers recognize that the relentless T-800 killer android in the movie *Terminator* is just a highly entertaining piece of cinematic fiction. But when the relentless terminator robot declares "I'll be back!" in this blockbuster science fiction motion picture, many unscientific and unrealistic impressions of robots as potential villains are reinforced. Once exposed to the misdeeds of a killer robot—even if only in fiction—the intriguing issue about dangerous, smart machines is reinforced. A commonly asked question today is: What happens if smart robots really become conscious; learn all they can from human beings, and then decide to compete with their human creators? With technical progress now taking place at an exponential rate, the boundary between science fiction and science fact is becoming blurred and lost in the fog of future-technology projections.

Several futurists have already suggested, quite seriously, that within several decades—possibly as early as 2030 or 2040—artificial intelligence will exceed human (biological) intelligence and that the long-term future of intelligent consciousness on Earth will involve a hybrid combination of man and machine. Improvements in prosthetics, the creation of artificial organs to replace most failed biological ones, the ability to perform atom-by-atom engineering through nanotechnology, and continued exponential improvements in artificial intelligence (resulting in smarter and smarter machines), would all be technological milestones along this very interesting pathway into the future. While not all scientists support this projection, clearly the concept raises some very significant technical, social, political, and even theological questions.

Machine phobias and fears are not always rational, but they nonetheless influence how people think. To further aggravate the issue, real industrial robots have killed real human workers. So much for Isaac Asimov's laws of robot behavior!

Long before twentieth-century science fiction stories began portraying machines (robots) as villainous, potential rivals of the human race, real world human beings developed and clearly demonstrated their fear and hatred of machines. In 1811, when faced with dire domestic economic conditions due to the Napoleonic Wars and threatened with the loss of jobs due to the automation of the textile factories, the weavers around Nottingham, England, formed something akin to a guerrilla army and began smashing the machines that threatened their livelihood. Called Luddites, after their legendary leader Ned Ludd, these rebellious workers continued to attack machines, destroy factories, and assault factory owners throughout the region, until about 1817 when the majority of the movement succumbed to persistent government legal pressure and force of arms.

To restore social order, the British Parliament dispatched over 10,000 troops to the region and had the leaders of the Luddite movement either executed or deported to Australia. Historians provide several somewhat conflicting interpretations of these events, including whether Ned Ludd actually existed as a person. Many British experts suggest that Ned Ludd was simply a fictional, heroic character, much like the legendary Robin Hood—who also battled government officials (such as Prince John) centuries earlier from the very same Sherwood Forest near Nottingham. Other historians think that Ned Ludd was just a bumbling worker, who decades before had accidentally destroyed several machines. Ludd's machine-bashing accident served as a convenient legend that grew to meet the needs of the politically tempestuous times. Whatever the actual case, the Luddites were real people, who fought and died because they felt their traditional way of life (as weavers) was being threatened by factory owners, who were automating the production of textiles.

As part of the First Industrial Revolution, British factory owners began introducing machines that could mass-produce textile products. These products were of an inferior quality, but could be produced at a much lower price. To aggravate the situation even further, relatively unskilled workers could operate the new machines. This transition was attacking the centuries-old textile industry and its apprenticeship program, which had provided the master weavers economic stability and job security. Many historians suggest that the Luddites may even have viewed the new technology (the automated power loom) as an evil instrument of oppression by which the factory owners could wrestle economic power and security away from the weavers. Prior to the arrival of the "accursed" machines, the now rebellious weavers had led relatively secure lives, producing high-quality lace and stockings. For many years prior, fine lace production around Nottingham had been an essentially family-owned and controlled cottage industry. Now, well-financed factory owners were taking control of their livelihood and future. The circumstances proved unacceptable and the famous social explosion resulted.

Whether the Luddites accepted the fact or not, the First Industrial Revolution was bringing about a major shift in the traditional British workforce. The Luddites were not the only labor group impacted by automation. A separate, but similar uprising by farm workers took place in the English countryside in 1830. This rebellion involved the destruction of several threshing machines by angry farm workers.

Today, the term Luddite is often used to describe a person who hates or resists new technology. However, in light of the previous discussion, it is appropriate to use a bit more precise definition—namely that of a person who is threatened with job loss by the arrival of a particular new technology.

Today, similar social issues—namely job loss and workforce displacement—accompany the introduction of industrial robots into manufacturing plants and automotive factories. Automation is necessary for economic competitiveness, but the displaced workers are human beings who cannot or should not be simply discarded like broken equipment. This chapter introduces several companion issues related to the use of industrial robots, such as the potential safety hazards that exist when a robot operates on the factory floor.

The use of military robots, such as unmanned aerial vehicles (UAVs) and unmanned ground vechicles (UGVs), raises several important technical issues and legal issues. Space robots with nuclear power supplies have raised some prickly political and environmental issues. Space robots designed to land on other worlds and bring soil samples back to Earth raise the issue of extraterrestrial contamination. There are also serious political and legal issues with any future decision to place robot weapons in space. The spread of robot weapons, especially the proliferation of nuclear-armed ballistic missiles to rogue nations and terrorist groups, remains a central global issue.

The chapter includes a summary of some of the major issues surrounding the development of nanotechnology. There is also a brief discussion of two more speculative issues related to future robotic systems. One issue involves the question of what happens when a very smart robot becomes self-aware (or conscious) of its existence. Another issue involves the ethics (on a cosmic scale) of sending self-replicating robot systems out into space, beyond the solar system.

ROBOTS AND THE WORKPLACE

This section presents several important issues dealing with the arrival of industrial robots in the workplace.

Job Displacement and the Unemployment Threat

From the arrival of the first automated looms in France and later Great Britain, workers have always been seriously concerned about loss of their jobs to a machine. Industrial automation is a marvelous application of technology, but there is a dark side. What does a society do with the displaced workers? The issue becomes especially acute when the rate of technology change occurs on the order of years versus decades, so human workers have a constant anxiety of becoming obsolete, no matter how well or how long they have performed their jobs.

A gradual workplace change that takes decades will generally correspond to a person's working lifetime, so the overall social impact is generally less severe. However, major workplace changes that occur in less than five or ten years produce far more dramatic economic and social consequences. One major consequence is widespread unemployment or displaced employment, as workers go from relatively high-paying jobs to minimum wage, so-called "burger-flipping." Some workers take advantage of their seniority and can find refuge in early retirement, but this alternative is generally accompanied by a sharply decreased income. Younger workers downsize into multiple minimum-wage jobs, just to survive.

In any case, the worker displacement can produce an enormous amount of social unrest at the local, regional, and possibly national level. Overstressed and unhappy parents often take their emotions out on their children, who then become unhappy or difficult children in school. Difficult or distracted students tend to underachieve in academics, thus perpetuating a dismal social scenario in which high-level aspirations for material prosperity become too stifled by a lifetime burden of low wage-earning prospects. While the highly automated factory down the road is using robots to produce better quality new cars, the displaced (that is, underemployed or unemployed) workers who used to work in that factory will no longer be able to afford one.

The increased use of robots in automobile manufacturing has displaced many less-skilled workers, but the severity of this adverse social impact is not uniformly felt throughout the world. Strong labor unions, as found in the American automobile industry, can blunt the impact of job loss by delaying the introduction of new automated equipment until workers with seniority are retired and other workers are retrained. The economic impact of robots as labor-saving technologies often varies with the rate of growth of the population and with the willingness and the ability of people to transition to different types of jobs.

No reasonable person would argue that the spray-painting robot is not a marvelous addition to automobile manufacturing. Human workers no longer have to be exposed to a very hazardous work environment. But what happened to the jobs of these particular automotive workers? In some companies the workers were retrained, in other companies management decided to close the entire less-efficient plant. The latter decision often placed entire communities (sometimes referred to as "company towns") at grave economic and social risk.

Keeping the manufacturing status quo at aging, human-worker oriented factories is not the solution either. Highly automated factories (many located in foreign countries) generally produce better quality manufactured products, which are also less expensive. Unable to compete in the global economy, the noncompetitive American factory closes and everybody loses—workers, managers, and investors.

At this point in the discussion it should become quite clear that the industrial robot is neither the villain nor the hero. These machines are the agents of a remarkable, ongoing social transformation, which is now taking place as developed nations journey into a postindustrialized world. (Economists and social scientists have suggested the word *superindustrialization* for this transformation.) In a postindustrial society, the manipulation of information becomes the key area of human activity, followed by the service industry, and then far more distantly by manufacturing and agriculture. Today, one to three persons on a highly mechanized American farm can produce food for 100 or so persons. Some economic experts suggest that a similar situation is taking place in industrialized countries regarding manufacturing. Soon, just five to ten persons will be needed to operate a highly automated factory versus the hundreds of workers previously required to produce the same quantity of goods at the same rate. Where do the other 90 to 95 workers go? Hopefully into better-paying jobs in the information or services sector of the economy.

Because of many not easily quantifiable factors—such as a worker's willingness and ability to be retrained and the terms of existing labor contracts (which strongly influence how rapidly management can introduce robot-assisted manufacturing)—it is quite difficult to predict precisely how many people will be affected when industrial robots arrive at a particular factory. All of the industrialized nations of the world face a precarious relationship between increased automation and "meaningful" human worker employment. One fact is very clear, however: Modern industrial robots are changing the work environment and redefining the very concept of the factory and the role of the so-called "blue-collar" worker.

One possible solution is that integrated employee groups will replace traditional job classifications. In a new, highly automated factory, human-machine production teams could share responsibilities for running the plant. New job functions, such as preventative-maintenance experts, robot technicians, and information experts, will appear. But can a person who has worked for many years spray-painting cars on an assembly line, for example, be easily transformed into the skilled technician who now services and repairs the spray-painting robot that took his job away? Obviously these changes will not be without social stress and economic difficulties.

Psychological Impact of Robots in Workplace

Imagine the psychological impact on a blue-collar worker, who has given twenty-five years of service performing arduous and dangerous assembly line work at a particular manufacturing plant, when he is suddenly told by management that the factory's new owners are "going-robot" and that his job will no longer exist. Psychologists list loss of a job as one the most severe stresses in a person's life, exceeded in physical and mental impact only by the death of a person's parents, the loss of a spouse (through death or divorce), and the loss of a child (through death or divorce). These human tragedies often leave a deep, indelible mark on the affected person's mind and cause a variety of physical ailments, including depression, addictions, and severe changes in personality.

Despite the often-used cliché about the indifference of blue-collar workers to their jobs, many workers (rightly or wrongly) associate what they do with who they are as a person. Most normal working people take some degree of pride in accomplishing their daily tasks well and enjoy receiving not only financial rewards for a job well done, but also some "psychic income" in the form of genuine praise from their supervisors. The need for psychic income starts in childhood with praise from nurturing parents and continues through a person's school years in the form of positive academic feedback, called "good grades." In an industrial setting, this psychic income continues in the form of praise, sometimes informally referred to as "at-a boys" or "warm fuzzies." There is also a reverse side to this psychic income coin. Any worker, who has ever incurred the wrath of a supervisor, is well aware of proverbial "cold prickly"—a biting, harsh comment that often cuts to the heart, as if by a blade of steel. In today's superindustrialized work environment, a laggard worker is more than likely to receive the cold-prickly statement "If you do not shape up and improve, we'll get a robot to replace you."

Perhaps the greatest psychological insult to a worker is to actually be replaced by a machine. Anyone who has experienced this ultimate on-the-job "insult," may start questioning their overall worth as a human being—especially if the worker regarded his job as his life. It is bad enough for an older worker to yield to the enthusiasm and ambition of a younger, more vibrant human worker. This has gone on since the formation of tribal groups in prehistoric times. But to have a machine take over his life's work might become too much to bear for a blue-collar worker.

Companies anticipating the integration of robots into an existing industrial workforce should do so in a way that does not adversely affect worker attitudes, the reward system (monetary and psychic), or general self-esteem. Properly integrated, industrial robots can raise employee self-esteem by providing the human worker with more meaningful and challenging work for the same or even better pay. In the automotive industry, for example, workers who earned a living by spray painting or spot welding generally would have little difficulty turning these hazardous jobs over to robots—if the human workers were retrained to function as robot systems engineers or technicians. The displaced workers then (in effect) would become partners with the new machines (robots) performing higher quality work more efficiently. Similar experiences take place when human assemblers are retrained as information technology (programming) personnel for the assembly robots, or when metalworking machine operators get "promoted" and become properly trained as robot maintenance personnel.

When improperly integrated, however, the sudden arrival of an industrial robot can demean a blue-collar worker's sense of worth. Soon, all sense of personal achievement, employee pride, and loyalty to the company vanish. The robot is viewed as a villain and tension between company management and labor grows. Lack of involvement by all the affected parties in an industrial setting is probably the single most significant factor contributing to workforce resistance.

The presence of robots in modern factories is a fact of economic life. Since the 1980s, the introduction of progressively more sophisticated robots in manufacturing facilities in the developed nations (many with strong labor union movements) has yielded an important lesson with respect to this issue. In the well-planned integration of robots, adverse psychological impact on displaced human workers is avoided or mitigated. Hasty, ill-planned injection of robots into the factory produces the opposite results. Worker resentment rises, the robots become villains, and openly (or subconsciously) all workers in the plant begin to participate in a subversive campaign to "make the machine fail." Stressed to the psychological limit, factory workers threatened with displacement will often "overlook" simple suggestions that would make the operation of the new robot a big success. Any worker who attempts to help management "win" with robot-assisted automation strategies at the expense of human jobs, is called a "scab" and is considered part of the "enemy camp" by his fellow blue-collar workers. This rising workplace tension erodes the company's production environment and turns any management success in bringing about more efficiency into a Pyrrhic victory. In some cases, this worker resentment helps make the installation of the robot fail. Once again everybody in the plant loses. First, the members of the plant's "white-collar" management team (the so-called "suits"), who pushed for the installation of the robot(s), are fired in disgrace. Then, the workers who enjoyed a brief, but hollow-victory, also lose their jobs because the factory can no longer compete and closes.

Displaced Worker Education and Training

Another major social issue in developed (industrialized) countries such as the United States and Great Britain, which arises from accelerated automation of the modern factory, involves education and training. As with the previous issue of job displacement, the education and training issue actually transcends the responsibility and jurisdiction of the employer. So, who is responsible to retrain a displaced worker? How is it to be achieved? When management at a manufacturing plant decides to go "all-robot," must they subsidize retraining of the displaced workers? And if the company pays for the retraining, are the workers obliged to work for the company for some fixed period of time in a more skilled (perhaps information technology) position? Or should any retraining package be viewed as just another form of severance pay? Should the local, state, or federal government be responsible for worker retraining? Anyone who had has to deal with a government agency (at any level) knows how a complex bureaucracy can squander financial resources intended for the displaced workers. For every hundred dollars budgeted to retrain displaced workers, a typical federal-state-local government combination could easily absorb about ninety dollars in administrative costs—leaving just ten dollars (per hundred) to pay for actual training of the workers. This is hardly a reasonable ratio. So government intervention may not significantly resolve the problem.

The issue of who retrains the blue-collar worker is an issue that must really be addressed by all affected social institutions, namely, manufacturing companies, labor unions, academic institutions (especially universities and community colleges), government agencies, and the most directly impacted local communities.

There is a short-term and longer-term component to this pressing issue. In the short term, management (industry) and labor might consider sharing the responsibility of retraining workers displaced when robots are inserted into the manufacturing complex. Depending on the age and seniority of the displaced worker, this training can take several forms. The results of U.S. Department of Labor studies conducted in the 1980s and 1990s suggest that younger factory workers (typically between the ages of 17 and 35) participate more eagerly and successfully in educational opportunities and job retraining programs. Older workers (those over 35 years of age) are far more resistive of change, feel alienated in academic environments, and generally do not perform well in retraining programs. Consequently, older industrial workers, especially those in semiskilled or skilled production line jobs, have a much greater risk of becoming "obsolete on the job," when the robots replace them as management embraces higher levels of automation.

This brings into focus the second, longer-term component of the education and training issue. Highly automated factories are needed for a company to remain competitive in today's global economy. When a manufacturing company fails to innovate and "go-robot," quite often that company loses a significant portion of its market share to foreign competition. In the extreme case, the company experiences a total failure. So, today's factory workers must anticipate a career that includes lifelong schooling—with much of this training aimed at just keeping current with the rapid changes taking place in manufacturing.

Taking a long-term view of the situation, American secondary schools, vocational schools, and institutions of higher learning must aggressively develop and provide appropriate curricula, which prepare their students for a lifetime of technical learning. Unfortunately, academic institutions, while centers of innovation in some instances, are more often bloated bureaucracies that propagate obsolete programs and curricula. Students passing through such programs are hardly prepared to pursue careers in the modern manufacturing industry or other rapidly emerging fields.

The problem becomes even more acute as less and less American students participate in science and technology courses while in high school or college. The absence of a sound secondary school foundation in mathematics, physics, and chemistry, for example, limits the efficacy of job-related training programs. It is very difficult for a young worker to train as a robot technician when he or she has had little formal technical education. Should industry be required to repair and rebuild years of inadequate high school or community college-level technical education? Any training program should also be transportable, that is, contain some degree of nationally recognized skill standardization. Transportable training would allow a displaced factory worker to achieve some limited amount of job security—either within his current company or perhaps in obtaining employment in a new industrial firm.

Where Will Future Robot Engineers and Technicians Come From?

Another issue closely related to technical education involves the source of future robot engineers and technicians. As the need for robot-savvy engineers, scientists, and technicians grows in the next two decades, where does a future employer search for the necessary workers? The future employer may be a federal agency, a national laboratory, a large industrial firm, or a small, innovative company.

Many young people get their first experience in building and operating robots as hobbyists in groups and clubs. Robot competitions involve students as early as elementary school and continue up through high school and college. For some, this entertaining "hands-on" experience also stimulates an interest in pursuing careers in engineering, typically mechanical, electrical, or computer system-related. Several American universities, such as Carnegie-Mellon University have active robotic research and development programs.

Military robot training programs provided by the American armed forces offer another source of robot system training—in this case "employer-sponsored" education. A young soldier, airman, or marine, for example, may be trained on how to maintain, repair, and operate the mobile robot used in explosive ordnance disposal (EOD) operations. Upon separation from the Armed Forces, that young person would probably have little difficulty seeking employment with a law enforcement agency or environmental cleanup company as a mobile robot field technician.

Community colleges, vocational schools, and high schools could develop robot-based, hands-on learning programs to stimulate interest and to provide practical experience in science, mathematics, and engineering. Such programs could provide a steady supply of qualified workers seeking future employment as robotic system technicians. Preparing the next generation of American workers for successful participation in a future world filled with robots starts in today's primary and secondary classrooms.

There are also a growing number of fascinating scientific and engineering positions becoming available in the nanotechnology field. Many of these career opportunities occur at the nanotechnology centers being set by the Office of Science within the U.S. Department of Energy (DOE) at various national laboratories. The nanotechnology research positions usually require an advanced degree in physics, chemistry, materials science, or engineering. Scientific positions, such as a micro/nano integration device scientist or an electron microscopy scientist, generally require that the candidate possess a doctoral degree in the physical sciences or related engineering area with two or more years of postdoctoral research experience. In addition, many scientific positions within the DOE national laboratory system require U.S. citizenship and the ability to receive a DOE-granted security clearance. The comments here are not intended to discourage, but to enlighten. The best way to prepare for a career at the frontiers of nanotechnology research is to know well ahead of time what formal academic training is being expected of successful job candidates. Technician positions at the national laboratories are equally demanding. A candidate must possess U.S. citizenship, be able to qualify for a security clearance, and have skills in performing "hands-on" science or engineering tasks that are usually computer-interactive.

Figure 6-1 Members of the Florida Space Coast FIRST Robotics Team, known also as the Pink Team, display their robot, called Roccobot, at the 2005 FIRST Robotics Regional Competition, which was held at the University of Central Florida in Orlando (March 10–12). The NASA-sponsored robot took first place in the regional competition. The Pink Team comprises students from Rockledge High School and Cocoa Beach Junior/Senior High School. (Credit: Photograph courtesy of NASA/KSC.)

Figure 6-2 At Camp Fallujah, Iraq (November 27, 2005), a U.S. Marine Corps explosive ordnance disposal (EOD) technician prepares to deploy a remotely operated vehicle (ROV) to neutralize a buried improvised explosive device (IED). The suspected IED was buried in a dirt mound on the side of the road next to an old IED crater. (Credit: Photograph courtesy of the U.S. Marine Corps.)

SAFETY IN THE USE OF ROBOT SYSTEMS

Robot Kills Worker (A True Case Story)

According to the Occupational Safety and Health Administration (OSHA) of the U.S. Department of Labor, on January 19, 2001, a 29-year-old male died from injuries sustained when he was struck on the head by a cycling, single-side gantry robot. The victim had recently performed a mold change on a 1,500-ton horizontal injection-molding machine (HIMM). Postaccident investigations revealed that he was apparently looking for tools that he may have left within the machine during the setup operation. The victim climbed on top of the purge guard and leaned over the top of the stationary platen of the HIMM in an attempt to see if the tools were left within the mold area. In the process, he placed his head beneath the robot's gantry frame. His actions placed his body between the robot's home position and the robot's support frame on the stationary platen.

While the worker tried to look inside the mold area, the robot cycled, moving from its home position to the mold area to retrieve a molded part. As the robot did so, it struck the victim on the right side of his head. The robot's movement crushed the victim's head between the robot arm and the vertical support for the robot's frame. Another employee noticed the victim lying on top of the HIMM and went to investigate. Upon seeing the victim's condition, he summoned other employees and they moved the victim to the floor. Emergency medical responders were also called. While awaiting the arrival of the emergency response team, the employees began chest compressions and tried other first aid procedures. Unfortunately, none of these efforts were successful and the worker was pronounced dead on arrival (DOA) at the local hospital.

Similar reports and studies from Sweden and Japan (countries with large populations of industrial robots) indicate that many of the reported robot accidents took place, not under normal operating conditions, but rather during programming, programming touch-up, maintenance, repair, testing, setup, or adjustment. During many of these "nonnormal" operations, the robot operator, programmer, or corrective maintenance worker may temporarily be within a robot's working envelope, where unintended operations could then result in injuries.

Potential Hazards with Robots in the Workplace

According to OSHA, the use of robots in the workplace can pose potential mechanical and human-error-induced hazards. These mechanical hazards might include workers colliding with equipment, being crushed or trapped by equipment, or being injured by falling components. A worker might collide with the robot's arm or peripheral equipment as a result of the machine's unanticipated movements, component malfunctions, or unpredicted changes in the robot's computer programming. In other mechanical hazard scenarios, a worker could be injured as a result of being trapped between the robot's arm and other peripheral equipment, or else by being crushed by peripheral equipment as a result of being impacted by the robot's arm into this auxiliary equipment.

Hazards can also result from the mechanical failure of the components associated with the robot or its power source, drive components, tooling or end-effector, or peripheral equipment. Examples of such mechanical hazards include the failure of gripper mechanisms with the resultant release of parts, or the failure of end effector power tools, such as grinding wheels, buffing wheels, deburring tools, and power screwdrivers.

Human errors create workplace hazards, both to personnel and equipment. Errors in programming the robot, interfacing peripheral equipment with the robot, or in connecting input/output sensors to the robot, can all result in unpredictable movement or unanticipated action by the robot. Such unanticipated action by the robot may then cause injury to any workers inside the robot's work envelope or result in equipment damage. Human errors in judgment often happen when programmers incorrectly activate the robot's teach pendant or control panel. Perhaps the greatest human judgment errors result from becoming familiar with a robot's redundant motions. Under such circumstances, human workers are too trusting in assuming the nature of the robot's motions and place themselves in hazardous or precarious positions while programming the robot or performing maintenance within the robot's work envelope.

Industrial robots in the workplace are usually associated with the facility's machine tools and processing equipment. When performing safety analyses, it is important to remember that robots are machines, and, as such, engineers must equip these systems with safeguards in ways similar to how they prevent any other hazardous remotely controlled machine from injuring human beings.

Movement enunciators (bells and sirens), flashing lights, and conveniently located "All Stop" buttons or switches are helpful complements to a properly designed robot workplace.

There are several techniques that safety engineers use to prevent (or at least minimize) human worker exposure to the hazards that robots can present in the workplace. The most common technique is to install perimeter guard devices (such as metal fences) with interlocked gates. A critical issue is precisely how the selected interlock mechanism works. Of major safety concern is whether the robot's computer program, control circuit, or primary power circuit is interrupted whenever an interlock is activated. At a minimum, the robot's primary motive power should be interrupted by activation of the interlock. This would prevent the robot from functioning whenever someone opens an interlocked gate and improperly enters the robot's work envelope.

With regard to safety concerning an industrial (usually fixed in place) robot, there are generally three circumstances when a human worker gets sufficiently close to the machine and is exposed to danger. These circumstances are: first, during the programming of the robot; second, when the robot is operating (and a human enters the work cell or workspace); and third, during maintenance of the robot.

The most obvious risk is that a human worker can experience physical injury, if an operating robot collides with or encounters the human being. Besides getting a potentially lethal slap by a massive and fast moving robot arm, the human worker could experience inadvertent spraying of paint or some other harmful substance, squeezing by a powerful hydraulic gripper, or intense heat from a welding device. There is also the possibility of electric shock, getting hit by a flying object (if a part or tool breaks off), or else having a heavy object (like the frame of an automobile) dropped on the worker by a pick-and-place robot.

The safety issues considered up to now have been primarily associated with fixed industrial robots, but some of the following comments are also appropriate for mobile robots—especially those that are massive or have potentially hazardous articulating arms with end effectors, and/or attached tools.

Safety engineers work with robot engineers to avoid almost all the potential hazards associated with the operation of a factory robot. Their combined efforts generally lead to a properly designed working envelope (or work cell) for the robot or even a collection of robots at a particular manufacturing station. One common sense engineering approach is to construct a physical barrier around the periphery of the robot's working envelope. This generally puts a cage or fence around the robot. A variety of intruder-detection devices placed along the perimeter of the working envelope, inside the working envelope, and even in close proximity to the robot supports a defense in-depth approach to safety. Emergency "all stop" buttons (for fixed robots) and "deadman switches" (for teleoperated mobile robots) would provide an additional level of safety. Should an unauthorized person (intruder) enter the work cell of an operating robot, an alarm would sound and the robot could reduce its speed to safe levels, or else completely stop. An obstacle-avoidance sensor could also be used, so a robot arm would not swing into any part of the working envelope in which an "intruder" (unauthorized object) was detected. A modern automatic garage door opener system is often equipped with a similar obstacle avoidance switch, and will stop closing if an object (like a small child) is detected in the path of the descending door.

Safety during maintenance of a robot is equally important, because human workers enter the working envelope and come in direct contact with the robot. Interlocks to guarantee that the robot's power is really off, or that the robot is locked in a safe, or quiet, mode, should be activated before workers attempt to enter the robot's working area. The same precautions are used by a professional electrician, when he makes sure the main circuit breaker is in an open (no electricity flow) condition before attempting to perform work involving the electric circuit in the home or office.

The "deadman switch" is useful during the programming of a factory robot. As the human worker "walks" an active robot through its required movements, the human worker must keep active pressure on the safety switch (or toggle) of the teach pendant. If something happens and the human worker drops the teach pendant or releases pressure on this switch, the robot stops immediately and does not function. The use of a deadman switch is a very important component in robot safety. One of the main reasons why industrial robots replace people in factories is to remove the human workers from operational hazards and dangerous tasks. It would be ironic to defeat this safety benefit by having the robot prove hazardous as a human was programming it or performing maintenance.

Similar safety steps must be taken to avoid potential hazards during operation of civilian mission mobile robots. Civilian activities involving mobile robots include: facility inspections and security patrols in hazardous areas at a rocket launch site, environmental monitoring and hazmat (hazardous materials) cleanup operations, urban search and rescue missions, and other law enforcement applications. A basic fail-safe design, an appropriately configured deadman switch to cover both maintenance and operational activities, and embedded interlocks for maintenance and repair should avoid the vast majority of potential human worker/controller hazards.

Military robots—either UGVs or UAVs—are generally employed in hostile, combat situations to reduce the risk to and loss of friendly forces. Since military robots must operate in such actively hostile environments (namely, enemy forces trying to harm friendly forces and the robot), the concept of "safety in the workplace" becomes rather moot. But the mobile military robot must be designed to alleviate, rather than contribute to, the hazard to friendly forces. This issue is especially significant when these military robots are armed and can intentionally inflict lethal damage within the battlespace. Operational procedures and safety devices must be included to avoid "friendly fire" casualties. While science fiction films, like *Westworld*, have provided delightful entertainment, an armed mobile robot that starts firing at its human operators is to be avoided at all costs. (In the 1974 movie *Westworld*, a gunfighter android, played by Yul Brynner, seriously malfunctions and begins stalking the high-technology resort's wealthy human guests.) When a remotely operated UGV is used to carry an explosive charge for placement and detonation at a enemy position (in urban warfare) or on a suspicious improvised explosive device (IED), design and operating procedures of the robot must ensure that the UGV will not prematurely drop the explosive near friendly troop positions. Furthermore, the soldier operating the robot must

Figure 6-3 Employees at NASA's Kennedy Space Center (KSC) learn about a mechanical robot during the 2004 Spaceport Super Safety and Health Day. Mobile robots, like their industrial robot counterparts, are machines that create safety issues and require employee diligence if accidents or injuries are to be avoided in the workplace. (Credit: Photograph courtesy of NASA/KSC.)

be provided a sufficient number of safety devices to avoid prematurely detonating the explosive charge while it is being carried to the enemy position or IED site. Premature detonation of the explosive charge (usually by encrypted radio signal) could injure friendly human troops, damage equipment, or destroy the military robot, which should only function as the messenger of destruction and not the machine-equivalent of a fanatical suicide-bomber.

PROLIFERATION OF NUCLEAR-ARMED BALLISTIC MISSILES

Nuclear Weapon Proliferation

Until the end of the cold war, a bipolar nuclear-deterrence-dominated world maintained a quasi-stable form of international security through a combined system of alliances, spheres of influence, and global and regional multilateral institutions (including the United Nations). However, following the disintegration of the former Soviet Union, a unipolar world emerged. The geopolitical landscape is now dominated by a technologically strong United States. Instead of facing and focusing on a singular superpower foe, the United States now faces

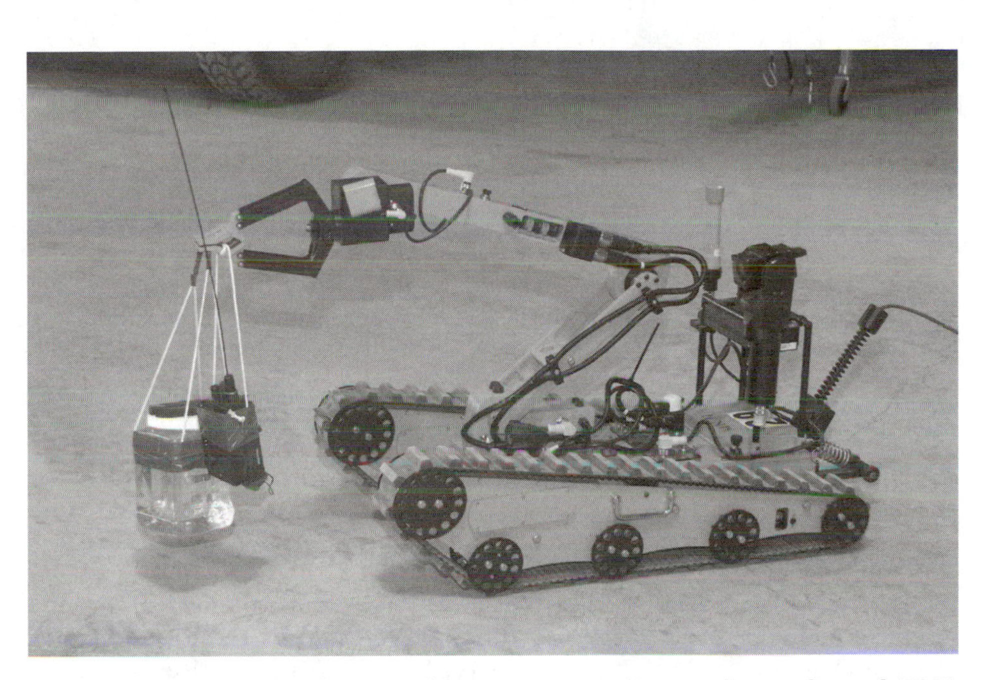

Figure 6-4 This mobile robot is used by U.S. Navy explosive ordnance disposal (EOD) technicians to carry an explosives deterrent device in its claw and properly place the device close to a suspected bomb threat. This type of military EOD robot is remotely operated as it places an explosive charge on the hostile location and then withdrawn to a safe distance—after which, the human operator sends an encrypted signal to detonate the robot-delivered explosive package. (Credit: Photograph courtesy of the U.S. Navy.)

numerous asymmetric threats in which state and nonstate adversaries try to avoid direct military engagements but devise strategies, tactics, and weapons to minimize the strengths of the U.S. military and exploit perceived weaknesses.

For example, there are strategic nuclear missile strike threats in which Russia, China, most likely North Korea, and probably Iran have the capability (now or by the year 2015) to hit targets in the United States. During the cold war, the United States and the former Soviet Union used a policy of mutually assured destruction (MAD). But the threat of nuclear destruction does not have the same deterrent potential against rogue regimes, like North Korea or Iran. Furthermore, nonstate actors (terrorist groups) may also acquire the ability to deliver a limited nuclear attack on American soil or against countries that are allies of the United States. The new global terrorism threat, highlighted by the September 11 attacks on New York City and Washington, DC, demonstrate the willingness of terrorists to sacrifice their own lives to achieve their evil political aims.

This section briefly presents some of the major negative global trends, concerning ballistic missile and nuclear weapon proliferation. Each trend involves complex political, social, and technical issues. In some areas around the world, such as South Asia and Northeast Asia, reliance on nuclear weapons and nuclear brinkmanship has dramatically increased since the end of the cold war. This increase might now promote an accidental nuclear war between India and Pakistan, could encourage attempts at nuclear blackmail by North Korea, or could inflame preexisting political tensions in a fragile region, such as the Middle East. For example, what would Israel's response be to Iran's development of a modest nuclear arsenal and the long-range ballistic missiles capable of delivering nuclear warheads? What would the U.S. response be to a thinly veiled technology blackmail threat by North Korea to strike at an American West Coast city with a nuclear-tipped long-range ballistic missile should certain political concessions with respect to a unified Korean peninsula not be met? These are very troublesome questions that did generally not arise during the cold war. But now, nuclear proliferation is producing several very difficult-to-handle asymmetric threats. The use of nuclear weapons in a regional nuclear war or in a well-organized urban attack by terrorists (possibly under the aegis of a rogue state) would have tremendous shock value and inflict massive civilian casualties.

Following the rapid dissolution of the former Soviet Union, there also remain serious nuclear threats and proliferation problems within the Russian nuclear arsenal. Russian insecurity and questionable weapons surety could lead to precipitous nuclear escalation or accidental/unauthorized nuclear release. The proliferation of ballistic missiles with longer and longer ranges is putting more U.S. overseas bases and allied countries at risk. The asymmetric nuclear-weapon strategies being pursued by such countries as North Korea and Iran are aimed at raising the costs of an American intervention in a regional crisis or conflict. Any miscalculation by India and/or Pakistan could easily lead to another border war that quickly escalates to the use of nuclear weapons. Poor command and control arrangements in new nuclear powers pose the major risk of either a serious "Broken Arrow" type nuclear incident or else the accidental or unauthorized detonation of a nuclear device.

Despite enormous efforts by the International Atomic Energy Agency (IAEA) to promote the peaceful uses of nuclear technology and to combat nuclear weapons proliferation, the nuclear weapon continues to serve as the coin of international power. Regional powers view nuclear weapons and long-range ballistic missiles as a way to deter the United States from intervening in a border crisis or local conflict. In many parts of the world, the possession of a nuclear weapons arsenal is seen as providing the nuclear-capable country with a substitute for alliances and external security guarantees. Possession of a nuclear weapon also ensures national/regime survival and provides despotic leaders with a high-stakes bargaining chip. A seldom discussed, but very real proliferation-related issue, is the fact that the sale of nuclear weapons technology, expertise, and/or materials is financially very lucrative. Organized crime and cash-strapped rogue states (like North Korea) would stand to profit substantially if they engaged in nuclear weapons brokering—regardless of the political or social consequences. In addition to the proliferation of nuclear weapons, illicit trafficking in radioactive sources poses a significant threat to modern civilization.

ISSUES RAISED IN THE USE OF SPACE ROBOTS

Three interesting and important issues are discussed in this section concerning the use of robots to explore outer space. The first addresses the thorny political issue of human explorers versus machine explorers. The second involves the concerns about the use of space nuclear power, especially plutonium-238 to provide electric power for robot spacecraft sent on deep space missions. The third issue involves the important planetary contamination issue as robots from one world visit another world in the solar system.

Should Robots or Astronauts Be Used to Explore Space?

One debate that has persisted in the American civil space program since 1958 involves the basic question: Should robots (machines) or astronauts (human beings) be used to explore the solar system? The ideal response, of course, is that *both* should be used in partnership. This approach was taken most effectively during the Apollo Project when robot spacecraft, like Ranger and Surveyor, served as precursors to the human landing missions.

However, following the remarkable success of the Apollo Project, many space experts began to reconsider the role of humans in space exploration. Healthy debates occurred throughout the 1970s and early 1980s, concerning future strategies for NASA's space exploration program. Some long-range planners concluded that most, if not all, future exploration goals could best be served by robot spacecraft—balancing: cost, risk, potential scientific return, and schedule. One major point brought up in favor of robot systems was that human space travelers require extensive and expensive life-support systems. In contrast, robots can survive long journeys into deep space and accomplish exploration goals just as well as humans. Furthermore, the loss of a robot spacecraft does not cause the same national numbness and paralyzing impact as when a human crew is lost during space exploration.

Other aerospace industry experts sharply disagreed during these debates post-Apollo Project debates. They argued that humans are and will remain very important in space exploration. These experts further proclaimed that robots and humans are not interchangeable. The proponents for human space flight also pointed out human beings are far more adaptable than robots and can react better to unexpected events. When things go wrong, human beings can use their creativity and intellect to make innovative repairs.

The tragic loss of the space shuttle *Challenger* and crew in January 1986 followed 17 years later by the loss of *Columbia* and crew in February 2003 have revived the robots versus humans for space exploration debate. Today, however, there is no real debate about using robot spacecraft to explore remote regions of interplanetary space—the regions beyond the main asteroid belt or the innermost portions of the solar system—namely Venus, Mercury, or near the Sun's corona. The present debate centers on the following specific question: Should a human expedition to Mars occur this century or should the detailed exploration of the Red Planet be assigned to a series of progressively more complex space robots. Flaming this debate are the obvious health and life risks to the astronaut crew on a three-year interplanetary mission, as well as its enormously large projected price tag. Some studies have estimated the total actual cost of a human expedition to Mars (with a crew of 10) would be about a trillion dollars. For comparison, the United States government spent a total of about 25 billion dollars to send human explorers to the Moon.

Responding to a directive from the White House regarding the exploration of space in the twenty-first century, NASA planners are now developing strategies for a return to the Moon and then human space flight to Mars. The goal is no longer humans or robots. It is humans *and* robots working together. Each brings complimentary capabilities that will support the detailed exploration and future settlement of these two worlds. A well-planned and organized robot–human partnership is essential for the successful return of humans to the Moon and for the construction of the first permanent lunar base. A dynamic, well-functioning human-robot partnership is just as crucial, if human explorers are to successfully travel to Mars later this century.

Robotic systems on the space shuttle and the *International Space Station* (ISS) provide a glimpse of how the robot–human partnership in space exploration and operations will grow over the next few decades. The space shuttle's remote manipulator system (RMS) is an excellent example of how this partnership should work. The 15-meter-long robot arm (also called the Canadarm because it was designed and constructed by Canada) is mounted near the forward end of the port side of the orbiter's payload bay. The device has seven degrees of freedom (DOF). Like a human arm, it has a shoulder joint that can move in two directions; an elbow joint; a wrist joint that can roll, pitch, and yaw; and a gripping device. The gripping device is called an end effector. The RMS's end effector is a snare device that closes around special posts, called grapple fixtures. The grapple fixtures are attached to the objects that the RMS is trying to grasp. Astronauts have made extensive use of the RMS during a wide variety of shuttle missions.

The ISS, currently under construction in Earth orbit, will have several robotic systems to help astronauts complete their tasks. The assembly and maintenance of the ISS relies heavily on the use of extravehicular robotic systems. When fully assembled, the ISS robotics complement will include three main manipulators, two small dexterous arms, and a mobile base and transporter system.

The most complex robotic system on the ISS is the mobile servicing system (MSS). Jointly developed by Canada and NASA, the MSS comprises five subsystems: the space station remote manipulator system (SSRMS), the mobile base system (MBS), mobile transporter (MT), the special purpose dexterous manipulator (SPDM), and the robotic workstation (RWS). The SSRMS is a 5.2-meter-long manipulator consisting of two booms, seven joints, and two latching end effectors. Astronauts can control and monitor the SSRMS from one of two modular workstations.

NASA strategic planners envision an expanded role for robots in the development and operation of a permanent lunar base and in assisting human explorers on Mars. Some of these future space robots will serve in precursor roles, such as performing focused exploration of candidate sites. Other space robots will be sent ahead to prepare a candidate site on the Moon (and eventually Mars) for the arrival of human beings. Finally, another group of space robots will work in direct partnership with astronauts, as they return to the Moon or explore Mars later this century. Space robot systems will display an entire spectrum of behavior characteristics from dexterous, teleoperated devices to fully autonomous machines, capable of performing their jobs without direct human supervision or guidance.

The Question of Aerospace Nuclear Safety

There is an ongoing debate concerning the use of nuclear energy in space. Current arguments involve the continued use of plutonium-fueled radioisotope thermoelectric generators (RTGs) and radioisotope heater units (RHUs) by the United States on deep space missions, such as the Cassini spacecraft to Saturn. However, rather than continuous this public debate on aerospace nuclear safety is quite cyclical. Vocal outcries spearheaded by various protest organizations generally rise to an audible level only months, weeks, or days before the scheduled launch of a spacecraft containing RTGs, RHUs, or both. From a scientific perspective, it is quite difficult to judge whether the magnitude and extent of each protest cycle is proportional to the amount of media coverage or vice versa.

The launch of NASA's nuclear-powered *Cassini* spacecraft in 1997 provides a vivid example of just how derisive and sharp-tongued these public debates and protests can be. Emotional arguments over safety raged and legal challenges continued right up to the very moment of launch. But presidential approval allowed the powerful Titan IV rocket to lift off from Cape Canaveral on October 15, 1997, and successfully send the *Cassini* spacecraft on its seven-year-long, gravity-assisted journey to Saturn. Somewhat ironically, the successful launch of NASA's RTG carrying the *New Horizons* spacecraft to Pluto on January 19, 2006, from Cape Canaveral, Florida, drew little adverse publicity and resulted in only minor public protests and demonstrations.

Why is there such great concern now about using an RTG to provide electric power for a scientific spacecraft? Radioisotope power systems are not a new part of the U.S. space program. In fact, they have enabled NASA's safe exploration of the solar system for many years. The Apollo landing missions to the surface of the Moon, as well as the later Pioneer, Viking, Voyager, Ulysses, Galileo, and Cassini robotic spacecraft missions all used RTGs. However, widely witnessed aerospace tragedies—namely, the space shuttle *Challenger* explosion in 1986 and more recently the *Columbia* reentry accident in 2003—have reinforced the uncontestable fact that rocketry and space travel remain inherently high-risk activities. Aerospace missions sometimes fail and accidents do happen. A launch pad explosion and the ensuing blazing inferno of chemical propellants is an indelible, culturally imbedded image of technology gone wrong—something like a very negative version of Carl Jung's collective unconscious. For many of these people, the perceived RTG risk and aerospace nuclear safety debate are also subconsciously influenced by a vague apprehension about nuclear energy. Without question, the Chernobyl nuclear plant disaster (April 1986)—an avoidable accident that led to numerous deaths and the radiological contamination of major portions of the former Soviet Union and Europe—continues to stimulate public anxiety and debate about the control and safe use nuclear technology.

Rightly or wrongly, some of the more vocal protestors during the Cassini launch debates referred to the spacecraft and its RTG power supply as a "flying Chernobyl." While technically incorrect, this inflammatory analogy helped emotion displace reason-based dialog. Consequently, the Cassini risk and safety debates often became public shouting contests between government officials from NASA and the DOE who sponsored and promoted the use of space nuclear power and various protestor groups whose members predicted (in the extreme) that Central Florida would soon become a nuclear wasteland. Even the legality (with respect to international space law) of the American government's use of nuclear power in space was questioned and challenged. It is beyond the scope of this book to address all the issues and arguments (pro and con) raised during these debates concerning the use of RTGs and RHUs in space missions. However, we will give some attention to the major aerospace nuclear safety issues that are most often subject to misunderstanding or misinterpretation.

It is very important to recognize that no amount of analysis, technical logic, or data can "prove" something is "safe" to an individual. The ultimate decision that some thing or action is safe involves *personal* choice and judgment. The scientist can only provide technical data and mathematical analyses to help an individual quantify the nature of a particular risk. Yet, even the most technically accurate and objectively presented risk analysis does not and cannot automatically invoke human acceptance. For example, transportation specialists and risk analysts can analytically demonstrate that travel on commercial jet aircraft is statistically safer (per kilometer traveled) than traveling in a motor vehicle. Yet, there are many people who continue to fear flying and will not accept this "safer" mode of transportation under any circumstances.

Within the international space community, especially through the work of the United Nations' Committee on the Peaceful Uses of Outer Space (COPUOS), it is recognized and accepted that RTGs "may be used for interplanetary missions and other missions leaving the gravity field of Earth. They may also be used in Earth orbit if, after conclusion of the operational part of their mission, they are stored in a high orbit." The international aerospace community further recommends the use of designs that "contain the radioisotope fuel under all operational and accidental circumstances." Therefore, the use of nuclear power systems to explore outer space is NOT prohibited by international law. What is prohibited by international law, however, is the testing of nuclear weapons in outer space and the deployment of nuclear weapons (that is, weapons of mass destruction) in orbit around Earth or on the surface of other planetary bodies. Failure to recognize this very important distinction often leads to much confusion and misunderstanding with respect to the use of RTGs and RHUs in space—since both of these devices contain nonweapons grade plutonium.

Another common misconception that fuels heated safety debates is the perception that RTGs and RHUs are inherently unsafe because they contain plutonium. Space nuclear power advocates point out that for decades NASA and DOE have placed the highest priority on assuring the safe use of radioisotope power systems on each space nuclear mission. For example, RTGs and RHUs use a ceramic form of plutonium-238 dioxide that has been designed specifically for safety. It is heat resistant and limits the rate of vaporization in fire or under reentry conditions. This ceramic material also has low solubility in water. Finally, by design, this material does not disperse or move easily through the environment. Upon impact, for example, it primarily fractures into large particles and chunks that cannot be inhaled as dust. This is an especially important safety feature. Plutonium-238 dioxide particles have to be of a sufficiently small size to be inhaled and deposited in lung tissue, where they lead to an increased lifelong chance of developing cancer. In the event of an accidental dispersal of plutonium dioxide into the terrestrial environment, other exposure pathways to human beings, such as ingestion, could occur—but such pathways would contribute far less to the potential development of cancer than inhalation.

Aerospace nuclear engineers use protective packaging and modularity of design to further reduce the likelihood that significant quantities of plutonium dioxide would be released during a launch or reentry accident. By design, the GPHS-RTG encloses its ceramic plutonium fuel in small, independent modular units—each with its own heat shield and impact shell. Radioisotope heater units (RHUs) enjoy a similar, multishell protective design to guard against the release of plutonium dioxide during a space mission accident.

Opponents of space nuclear power point out that accidents have already happened and that any such future accidents are unacceptable. Proponents for the use of RTGs and RHUs respond by stating that RTG safety features functioned as designed in all three American space nuclear accidents—thereby avoiding any serious environmental contamination or undo risk to the global population.

At this point, it is helpful to briefly review the three RTG accidents that have occurred in the United States space program. While RTGs have never been the cause of a spacecraft accident, they have been onboard spacecraft during three different space missions that did fail for other reasons. By way of a brief historic note for comparison, in January 1977, a Russian military satellite with an onboard nuclear reactor fell out of orbit and crashed into the Canadian wilderness near Great Slave Lake.

After four successful RTG launches, a U.S. Navy Transit-5-BN-3 navigational satellite with a SNAP-9A RTG (System for Nuclear Auxiliary Power) onboard failed to achieve orbit on April 21, 1964 due to a launch vehicle abort. Despite the ascent abort, however, the SNAP-9A RTG carried by the spacecraft performed as designed for a launch/mission abort and it burned up upon reentry into Earth's atmosphere somewhere over the Indian Ocean. The design of that particular RTG involved about 1 kilogram of plutonium-238 in metallic form and used high-altitude burn-up and atmospheric dispersion as a safety approach. During reentry over Madagascar, the plutonium metal completely burned up and dispersed at an altitude of between 45 and 60 kilometers over the West Indian Ocean. The United States government conducted airborne and surface sampling operations for months after this abort to ensure that the plutonium had burned up and dispersed in the stratosphere, as intended. Current American aerospace nuclear safety policy no longer uses atmospheric dispersion of an RTG's plutonium fuel as a means of avoiding surface contamination or direct hazards to people.

The second RTG accident in the United States space program occurred on May 18, 1968. This aerospace nuclear accident involved a SNAP-19B2 generator that was onboard NASA's Nimbus B-1 meteorological satellite. In this case, erratic behavior of the launch vehicle forced its intentional destruction by the U.S. Air Force Range Safety Officer at Vandenberg AFB in California. The launch vehicle was at an altitude of about 30 kilometers and traveling downrange from the launch site, when the safety officer destroyed the errant rocket along with its attached payload. Tracking data placed the impact point of the launch vehicle and spacecraft debris off the California coast in the Santa Barbara Channel about 5 kilometers north of San Miguel Island. Aerospace nuclear engineers had designed the SNAP-19B2 RTG for intact reentry and tested its ability to survive in a marine environment. Since recovery team data indicated that the radioisotope fuel capsules were still intact and that they posed no immediate environmental or health problem, officials felt there was no immediate urgency to recover them from the ocean floor. So, this SNAP-19B2 RTG, containing about 2 kilograms of plutonium-238, was recovered from the Pacific Ocean five months later. This incident provided verification to aerospace nuclear safety engineers that a properly designed RTG could remain in a marine environment for long periods of time following a launch/mission abort without concern for radioisotope fuel release. Postaccident examination of the plutonium fuel capsules indicated that they had experienced no harmful effects from either the destruction of the launch vehicle, free fall impact into the ocean, or nearly five months residency on the floor of the Pacific Ocean.

The third major American RTG accident involved the aborted Apollo 13 mission to the Moon in April 1970. In this incident, a SNAP-27 RTG, destined for placement on the lunar surface, reentered Earth's atmosphere along with the Aquarius lunar excursion module (LEM) that had served as a translunar trajectory lifeboat for the three in-flight stranded Apollo 13 astronauts. When it reentered Earth's atmosphere, the SNAP-27 RTG resided in a graphite fuel cask attached to the LEM. Both objects reentered at approximately 122 kilometers altitude above the South Pacific Ocean near the Fiji Islands. High and low altitude atmospheric sampling in the area indicated that there was no release of plutonium by the SNAP-27 RTG during its reentry and plunge into the ocean. As a result safety officials assumed that the SNAP-27 RTG, which contained 44,500 curies of plutonium-238 in the form of oxide microspheres, functioned as designed and impacted intact in the deep ocean south of the Fiji Islands. It now resides on the ocean bottom near the Tonga Trench in some 6 to 9 kilometers of water.

Prior to each RTG mission by the United States, federal agencies (such as NASA, DOD, and DOE) jointly conduct extensive safety reviews supported by safety testing and analysis. In addition, an adhoc Interagency Nuclear Safety Review Panel (INSRP) performs an independent safety evaluation of the mission as part of the overall presidential nuclear safety launch approval process. Based upon recommendations by DOE and other agencies and the INSRP evaluation, NASA may then submit a request for nuclear safety launch approval to the White House Office of Science and Technology Policy (OSTP). The OSTP director (that is, the president's science advisor) may make the decision or refer the matter directly to the president. In either case, the normal process for launch cannot proceed until nuclear safety approval has been granted.

However, despite international acceptance and decades of prudent use of RTGs on numerous space missions, advocates of space nuclear power still find it difficult to develop a popular consensus as to whether the benefits offered by the use of RTGs and/or RHUs outweigh their risks. For purposes of risk assessment, NASA and DOE have suggested that the risk of using plutonium-fueled RTGs can be defined as the probability (per unit radiation dose) of producing, in an individual or population, a radiation-induced detrimental health effect, such as cancer. In risk assessment, risk is mathematically defined as the probability of an undesirable event taking place times the magnitude of the consequence of that event (often expressed as the number of fatalities within some affected population). Government risk analyses performed prior to the Cassini mission concluded that an early launch accident with plutonium dioxide release had a probability of occurrence of 1 in 1,400 and would cause 0.1 fatalities in the affected population. In the language of risk assessment, this represents an overall risk factor of 0.00007. Government risk assessments also suggested that an accident with plutonium dioxide release occurring later in the launch profile or during spacecraft reentry had a probability of 1 in 476 and would cause an estimated 0.04 fatalities in the affected population. In the mathematical language of risk analysis, this scenario represented an overall risk factor of 0.00008.

But not everyone agrees with the risk-assessment approach. Opponents often maintain that government-agency-conducted risk assessment studies are, by virtue of their sponsorship, untrustworthy because they are biased toward a pronuclear technology outcome. But, no matter how unbiased and objective a particular risk assessment study is, it can only provide a numerical expression of risk for a particular scenario that is based upon a certain set of assumptions. The acceptability of a particular risk (no matter how numerically insignificant it might appear on various comparative scales) is still a very personal, subjective judgment that can never be forced by mathematical arguments alone. This clearly happened during the relatively heated debates prior to the Cassini launch. So, space nuclear power advocates should anticipate similar human behavior and responses during the safety debates that should arise when future RTG space missions approach their scheduled launch dates.

The Issue of Extraterrestrial Contamination

Sending robot space probes to other worlds raises an interesting "space age" issue not faced by previous generations of human beings, namely, the problem of extraterrestrial contamination. In general, extraterrestrial contamination is the accidental contamination of one world by life-forms, especially microorganisms, from another world. Using the Earth and its biosphere as a reference, scientists refer to this planetary contamination process as *forward contamination*, if an extraterrestrial sample or the alien world itself is contaminated by contact with terrestrial organisms, and *back contamination* if alien organisms are released into the Earth's biosphere.

An alien species will usually not survive when introduced into a new ecological system, because it is unable to compete with native species that are better adapted to the environment. Once in a while, however, alien species actually thrive, because the new environment is very suitable and indigenous life-forms are unable to successfully defend themselves against these alien invaders. But if this "war of biological worlds" ever occurs, the result could be a permanent disruption of the host ecosphere, with severe biological, environmental, and possibly economic consequences.

Frequently, alien organisms that destroy resident species are microbiological life-forms. Such microorganisms may have been nonfatal in their native habitat, but once released in the new ecosystem, they become unrelenting killers of native life-forms that are not resistant to them. In past centuries on Earth, entire human societies fell victim to alien organisms against which they were defenseless, as, for example, the rapid spread of diseases that were transmitted to native Polynesians and American Indians by European explorers.

But an alien organism does not have to directly infect humans to be devastating. The consequences of the potato blight fungus that swept through Europe and the British Isles in the nineteenth century caused a million people to starve to death in Ireland alone.

In the space age it is obviously of extreme importance to recognize the potential hazard of extraterrestrial contamination (forward or back). Before any species is intentionally introduced into another planet's environment, scientists must carefully determine not only whether the organism is pathogenic (disease-causing) to any indigenous species but also whether the new organism will be able to force out native species—with destructive impact on the original ecosystem. The introduction of rabbits into the Australian continent is a classic terrestrial example of a nonpathogenic life-form creating immense problems when introduced into a new ecosystem. The rabbit population in Australia simply exploded in size because of their high reproduction rate, which was essentially unchecked by native predators. As discussed elsewhere in the chapter, before engineered nanotechnology devices are released into the terrestrial environment or in the human body in medical procedures, similar concerns of uncontrolled behavior, possibly changes in structure leading to unchecked growth and behavior, must also be resolved.

At the start of the space age, scientists were already aware of the potential extraterrestrial-contamination problem in either direction. Quarantine protocols (procedures) were established to avoid the forward contamination of alien worlds by outbound unmanned spacecraft, as well as the back contamination of the terrestrial biosphere when lunar samples were returned to Earth as part of the Apollo program. For example, the United States is a signatory to a 1967 international agreement, monitored by the Committee on Space Research (COSPAR) of the International Council of Scientific Unions, which establishes the requirement to avoid forward and back contamination of planetary bodies during space exploration.

Quarantine is basically a forced isolation to prevent the movement or spread of a contagious disease. Historically, quarantine was the period during which ships suspected of carrying persons or cargo (for example, produce or livestock) infected with contagious diseases were detained at their port of arrival. The length of the quarantine, generally 40 days, was considered sufficient to cover the incubation period of most highly infectious terrestrial diseases. If no symptoms appeared at the end of the quarantine, then the travelers were permitted to disembark. In modern times, the term *quarantine* has obtained a new meaning, namely, that of holding a suspect organism or infected person in strict isolation until it is no longer capable of transmitting the disease. With the Apollo Project and the advent of the lunar quarantine, the term now has elements of both meanings. Of special interest in future space missions to the planets and their major moons is how scientists avoid the potential hazard of back contamination of Earth's environment when robot spacecraft and (later possibly human explorers) bring back samples for more detailed examination in laboratories on Earth.

NASA started a planetary quarantine program in the late 1950s at the beginning of the U.S. civilian space program. This quarantine program, conducted with international cooperation, was intended to prevent, or at least minimize, the possibility of contamination of alien worlds by early space probes. At that time, scientists were concerned with forward contamination. In this type of extraterrestrial contamination, terrestrial microorganisms, "hitchhiking" on initial planetary probes and landers, would spread throughout another world, destroying any native life-forms, life-precursors, or perhaps even remnants of past life-forms. If forward contamination occurred, it would compromise future scientific attempts to search for and identify extraterrestrial life-forms that had arisen independently of the Earth's biosphere.

A planetary quarantine protocol was therefore established. This protocol required that outbound unmanned planetary missions be designed and configured to minimize the probability of alien-world contamination by terrestrial life-forms. As a design goal, these spacecraft and probes had a probability of 1 in 1,000 (1×10^{-3}) or less that they could contaminate the target celestial body with terrestrial microorganisms. Decontamination, physical isolation (for example, prelaunch quarantine) and spacecraft design techniques have all been employed to support adherence to this protocol.

Of course, today, as scientists keep learning more about the environments on other worlds in the solar system, they can keep refining their probability estimates. Just how well terrestrial life-forms grow on Mars, Venus, Europa, Titan, Enceladus, and other interesting celestial bodies will be the subject of future in situ (on site) laboratory experiments performed by robot spacecraft acting as surrogates for Earthbound exobiologists.

As a reference point of aerospace technical history, the early U.S. Mars flyby missions (for example, *Mariner 4*, launched on November 28, 1964, and *Mariner 6*, launched on February 24, 1969) had probability values ranging from 4.5×10^{-5} to 3.0×10^{-5}. These missions achieved successful flybys of the Red Planet on July 14, 1965, and July 31, 1969, respectively. Postflight calculations indicated that there was no probability of planetary contamination as a result of these successful precursor missions.

The human-crewed U.S. Apollo Project missions to the Moon (1969–1972) also stimulated a great deal of debate about forward and back contamination. Early in the 1960s, scientists began speculating in earnest: Is there life on the Moon? Some of the most bitter technical exchanges during the Apollo Project concerned this particular question. If there was life, no matter how primitive or microscopic, we would want to examine it carefully and compare it with life-forms of terrestrial origin. This careful search for microscopic lunar life would, however, be very difficult and expensive because of the forward-contamination problem. For example, all equipment and materials landed on the Moon would need rigorous sterilization and decontamination procedures. There was also the glaring uncertainty about back contamination. If microscopic life did indeed exist on the Moon, did it represent a serious hazard to the terrestrial biosphere? Because of the potential extraterrestrial-contamination problem, some members of the scientific community urged time-consuming and expensive quarantine procedures.

On the other side of this early 1960s contamination argument were those exobiologists who emphasized the suspected extremely harsh lunar conditions: virtually no atmosphere; probably no water; extremes of temperature ranging from 120°C at lunar noon to −150°C during the frigid lunar night; and unrelenting exposure to lethal doses of ultraviolet, charged particle and X-ray radiations from the Sun. No life-form, it was argued, could possibly exist under such extremely hostile conditions.

This line of reasoning was countered by other exobiologists who hypothesized that trapped water and moderate temperatures below the lunar surface could sustain very primitive life-forms. And so the great extraterrestrial-contamination debate raged back and forth, until finally the *Apollo* 11 expedition departed on the first lunar-landing mission. As a compromise, the *Apollo* 11 mission flew to the Moon with careful precautions against back contamination but with only a very limited effort to protect the Moon from forward contamination by terrestrial organisms.

The Lunar Receiving Laboratory (LRL) at the Johnson Space Center in Houston, Texas, provided quarantine facilities for two years after the first lunar landing. What scientists learned during its operation serves as a useful starting point for planning new quarantine facilities, Earth-based or space-based. In the future, these quarantine facilities will be needed to accept, handle, and test extraterrestrial materials from Mars and other solar-system bodies of interest in our search for alien life-forms (present or past).

During the Apollo Project, no evidence was discovered that native alien life was then present or had ever existed on the Moon. Scientists at the Lunar Receiving Laboratory performed a careful search for carbon, since terrestrial life is carbon-based. One hundred to 200 parts per million of carbon were found in the lunar samples. Of this amount, only a few tens of parts per million are considered indigenous to the lunar material, while the bulk amount of carbon has been deposited by the solar wind. Exobiologists and lunar scientists have concluded that none of this carbon appears derived from biological activity. In fact, after the first few Apollo expeditions to the Moon, even back-contamination quarantine procedures of isolating the Apollo astronauts for a period of time were dropped.

There are three fundamental approaches toward handling extraterrestrial samples to avoid back contamination. First, scientists could sterilize a sample while it is enroute to Earth from its native world. Second, they could place it in quarantine in a remotely located, maximum-confinement facility on Earth while scientists examine it closely. Finally, they could also perform a preliminary hazard analysis (called the extraterrestrial protocol tests) on the alien sample in an orbiting quarantine facility before allowing the sample to enter the terrestrial biosphere. To be adequate, a quarantine facility must be capable of (1) containing all alien organisms present in a sample of extraterrestrial material, (2) detecting these alien organisms during protocol testing, and (3) controlling these organisms after detection until scientists could dispose of them in a safe manner.

One way to bring back an extraterrestrial sample that is free of potentially harmful alien microorganisms is to sterilize the material during its flight to Earth. However, the sterilization treatment used must be intense enough to guarantee that no life-forms as scientists currently know or understand them could survive. An important concern here is also the impact the sterilization treatment might have on the scientific value of the alien world sample. For example, use of chemical sterilants would most likely result in contamination of the sample, preventing the measurement of certain soil properties. Heat could trigger violent chemical reactions within the soil sample, resulting in significant changes and the loss of important planetary geological data. Finally, sterilization would also greatly reduce the biochemical information content of the sample. It is even questionable as to whether any significant exobiology data can be obtained by analyzing a heat-sterilized alien material sample.

If scientists do not sterilize the alien samples en route to Earth, they have only two general ways of avoiding possible back-contamination problems. They can place the unsterilized sample of alien material in a maximum quarantine facility on Earth and then conduct detailed scientific investigations, or intercept and inspect the sample at an orbiting quarantine facility before allowing the material to enter Earth's biosphere.

The technology and procedures for hazardous-material containment have been employed on Earth in the development of highly toxic chemical and biological warfare agents and in conducting research involving highly infectious diseases. A critical question for any quarantine system is whether the containment measures are adequate to hold known or suspected pathogens while experimentation is in progress. Since the characteristics of potential alien organisms are not presently known, scientists must assume that the hazard they could represent is at least equal to that of terrestrial Class IV pathogens. (A terrestrial Class IV pathogen is an organism capable of being spread very rapidly among humans; no vaccine exists to check its spread; no cure has been developed for it; and the organism produces high mortality rates in infected persons.) The alternative to this potentially explosive controversy is quite obvious: locate the quarantine facility in outer space.

A space-based facility, possibly teleoperated by scientists on Earth, provides several distinct advantages. First, it eliminates the possibility of a sample return spacecraft's crashing and accidentally releasing its deadly cargo of alien microorganisms. Second, it guarantees that any alien organisms that might escape from confinement facilities within the orbiting complex cannot immediately enter Earth's biosphere. Third, since the facility is in-orbit around Earth, teleoperation during protocol testing becomes practical by scientist on Earth. They can make natural movements and measurements, without the added difficulty waiting of many minutes during each step, because of the physical distances to other alien worlds and the finite time it takes radio waves (and light) to travel back and forth.

As scientists expand the human sphere of influence into heliocentric space by means of robot spacecraft from Earth, they must also remain conscious of the potential hazards of extraterrestrial contamination. One solution to this issue is a properly designed and operated robotic orbiting quarantine facility. At this Earth-orbiting automated facility, alien-world materials can be tested for potential hazards. Three hypothetical results of such protocol testing are: (1) no replicating alien organisms are discovered; (2) replicating alien organisms are discovered, but they are also found not to be a threat to terrestrial life-forms; or (3) hazardous replicating alien life-forms are discovered. If potentially harmful replicating alien organisms were discovered during these protocol tests, then the robotic facility would either render the sample harmless (for example, through heat- and chemical-sterilization procedures); retain it under very carefully controlled conditions in the orbiting complex and perform more detailed analyses on the alien life-forms; or properly dispose of the sample before the alien life-forms could enter Earth's biosphere and infect terrestrial life-forms.

Increasing interest in Mars exploration has also prompted a new look at the planetary protection requirements for forward contamination. In 1992, the Space Studies Board of the U.S. National Academy of Sciences recommended changes in the requirements for Mars landers that significantly alleviated the burden of planetary protection implementation for these missions. The board's recommendations were published in the document, "Biological Contamination of Mars: Issues and Recommendations," and presented at the 29th COSPAR Assembly which was held in 1992 in Washington, DC. In 1994, a resolution addressing these recommendations was adopted by COSPAR at the thirtieth Assembly; it has been incorporated into NASA's planetary protection policy. Of course, as scientists learn more about Mars, planetary protection requirements may change again to reflect current scientific knowledge.

These new recommendations recognize the very low probability of growth of (terrestrial) microorganisms on the Martian surface. With this assumption in mind, the forward contamination protection policy shifts from probability of growth considerations to a more direct and determinable assessment of the number of microorganisms with any landing event. For landers that do not have life-detection instrumentation, the level of biological cleanliness required is that of the *Viking* spacecraft prior to heat sterilization. Class 100,000 clean-room assembly and component testing can accomplish this level of biological cleanliness. This is considered a very conservative approach that minimizes the chance of compromising future exploration. Landers with life-detection instruments would be required to meet *Viking* spacecraft poststerilization levels of biological cleanliness or levels driven by the search-for-life experiment itself. Scientists recognize that the sensitivity of a life-detection instrument may impose the more severe biological cleanliness constraint on a Mars lander mission.

Included in recent changes to COSPAR's planetary protection policy is the option that an orbiter spacecraft is not required to remain in orbit around Mars for an extended time, if it can meet the biological cleanliness standards of a lander without life-detection experiments. In addition, the probability of inadvertent early entry (into the Martian atmosphere) has been relaxed compared to previous requirements.

The present policy for samples returned to Earth remains directed toward containing potentially hazardous Martian material. Concerns still include a difficult-to-control pathogen capable of directly infecting human hosts (currently considered extremely unlikely) or a life-form capable of upsetting the current ecosystem. Therefore, for a future Mars Sample Return Mission (MSRM), the following backward-contamination policy now applies: All samples would be enclosed in a hermetically sealed container; the contact chain between the return space vehicle and the surface of Mars must be broken in order to prevent the transfer of potentially contaminated surface material by means of the return spacecraft's exterior; the sample would be subjected to a comprehensive quarantine protocol to investigate whether or not harmful constituents are present. It should also be recognized that even if the sample return mission has no specific exobiological goals, the mission would still be required to meet the planetary protection sample return procedures as well as the life-detection protocols for forward-contamination protection. This policy not only mitigates concern of potential contamination (forward or back), but it also prevents a hardy terrestrial microorganism "hitchhiker" from masquerading as a Martian life-form.

Figure 6-5 This is an artist's rendering of a Mars Sample Return Mission (MSRM). The sample-return spacecraft is shown departing the surface of Mars after soil and rock samples, previously gathered by robot rovers, have been stored on board in a specially sealed capsule. To support planetary protection protocols, once in rendezvous orbit around Mars, the sample return spacecraft would use a mechanical device to transfer the sealed capsule of soil samples to an orbiting Earth-return mother spacecraft. This craft would then take samples back to Earth for detailed study by scientists. (Credit: Artist's rendering courtesy of NASA/JPL; artist Pat Rawlings.)

OVER THE TECHNOLOGY HORIZON ISSUES

Two extremely hypothetical issues are presented here as related to long-term developments in artificial intelligence and robot technology. The issues have been around for some time and no discussion of smart, advanced robotic systems would be complete without including them. The first issue involves the nature of a thinking machine that achieves self-consciousness. The second question involves the concept of perhaps the ultimate robot, the self-replicating system (SRS).

Can Machines Think Like the Human Mind?

Perhaps the most intriguing philosophical question involved with advanced robot systems is the question of machine intelligence and machine consciousness. Simply stated: Is a machine that thinks, conscious and aware of its existence? The first great modern philosopher René Descartes believed that the bodies of humans and animals are complex automata. In his treatise *Discourse on Method*, published in 1637, Descartes discusses how humans, who have the power of reason, and animals, which cannot reason, can be distinguished from one another and machines. His most famous quote (as found in *Discourse on Method*) is: "Cogito, ergo sum" (which means, "I think, therefore I am"). This statement highlights some of the deep philosophical arguments Descartes raised in developing his mind-body dualism. The nature of consciousness and the mind is an issue that has intrigued philosophers for ages. The issue arises again from an interesting new perspective as robot specialists speculate about endowing very smart machines with a sense of consciousness and cognition. At what point does a so-called "thinking machine" become truly conscious?

At the dawn of the Age of Science, René Descartes began revisiting the concept of mind as it had wandered down in Western civilization from the ancient Greek philosophers, like Plato and Aristotle, and the great Medieval Christian theologians, like Thomas Aquinas. In his *Principles of Philosophy*, Descartes proposed the philosophical concept that mind (soul) and body (matter) are separate and distinct entities. His postulation represents the birth of modern *dualism* and the start of the famous mind/body problem. For Descartes, the rational mind (soul) was an entity (substance) distinct from matter (the body). Within his model of mind, there were two very different kinds of substance: an invisible, unextended thinking substance (which he called the *res cogitans*) and a physical, extended substance (labeled the *res extensa*) that could be measured and divided. According to Cartesian dualism, the human mind (soul) was responsible for such invisible activities as thinking, willing, desiring, and so forth. It represented the *res cogitans* (the thinking substance) of a human being. In contrast, the human body (including the brain and the entire nervous system) was a physical, extended substance (that is, the *res extensa*).

At death, the soul (mind) would leave the body (which subsequently decays) and then continues to exist in some transformed (invisible) state of consciousness. Within the context of Christian theology, Descartes' dualism further suggested that the soul (as the immortal, spiritual seat of human consciousness) experiences an *afterlife*—a state of continual happiness (heaven) or perpetual pain (hell).

From at least as far back in human history as wandering Neanderthal tribes and their primitive burial ceremonies, human beings in almost every civilization and culture have expressed anticipation of some kind of life after death. The survival of personal human consciousness has been and still remains a pressing question in philosophy and theology. No study of mind, human or artificial, is complete without exploring this issue. The following statement introduces a major milestone on the journey through these discussions: *the human mind (as a conscious personal entity) either survives the death and destruction of the body, or it doesn't*. More bluntly stated, a person's mind either knows who the person is after death, or else that person simply no longer personally exists as that particular individual.

If personal consciousness survives the biological death, then where does it "go," what does it "do," when it gets there? Perhaps the most interesting question of all is: Can the human mind still interact on some level with the physical world and normal (living) human beings who reside there? On the other hand, if personal consciousness terminates with biological death, then a person's "mind" is no more. Now what about the "mind" of a conscious machine? Can it die? Or, once created or arisen, does conscious machine intelligence find a way to "become immortal"? The simple human response is, pull the machine's electric plug and that will be that. Perhaps. But when a person powers down his or her computer in the evening and then turns the power on the next day, programs and data "come back to life" (in a manner of speaking). Would an advanced form of adaptive machine intelligence exhibit similar survival traits in the absence of an active supply of electricity?

Descartes himself recognized many of the philosophical difficulties he created in trying to explain how an invisible (spiritual) mind could influence physical matter (the body) to perform voluntary physical actions and how a distinctly separate body could affect the mind through such conscious sensations as pain and pleasure. Yet, following in the philosophical footsteps of Plato and Aristotle, Descartes vigorously rationalized his own existence as a *thinking being*. As previously mentioned, this important connection between mind (consciousness) and existence he eloquently summarized in his famous quotation: "Cogito, ergo sum" ("I think, therefore I am."). Descartes's dualistic model of the mind, presented during the great scientific revolution of the 17th century, greatly influenced subsequent philosophers, and the debate about mind-matter interactions continues to the present day.

Today, neuropsychologists and other "mind" scientists, recommend the acceptance of a monist versus dualist model of mind. This modern position, often referred to as *emergent materialism* (and sometimes as *emergent psychoneural monism* or *monistic materialism*), rejects Descartes' hypothesis that the mind and body are different substances and proposes, instead, that all mental activities and states are actually the result of collective processes occurring within the (physical) brain. Under the concept of emergent materialism, consciousness and mental states exist, but do so as an interactive, integral part of the brain and not as a separate, invisible entity. However, proponents of this model also point out that mind is not just a simple result of the brain's complex composition of cells, but rather mind comes from a special collection and association of emergent biophysical activities. Neuropsychologists suggest that functions like thinking, perceiving, feeling, and willing arise from a currently unexplained collective ("emergent") property of the brain's overall physical structure and not just the electrochemical or mechanical responses of brain cells to stimulations by the body's nervous system. In other words, within this model, a mind is definitely much greater than the sum of its numerous biological parts. This particular collection of living tissues, cells, and energy gives rise to a very special biophysical property: intelligent consciousness.

But exactly where in the brain does this consciousness reside? Unfortunately, even with all the tools and skills of modern science, no one can now say for sure. Does this elusive intelligent consciousness, this "mind," survive and transcend the physical death of the body? If "mind" is just an emergent property of the brain, and the brain needs a living body to survive, then the logical answer is: no! But this represents a most uncomfortable conclusion that flies in the face of millennia of collective human thinking and belief. How can scientists hope to reconcile such neuroscientific models of mind (as centered in the brain) with philosophical and theological models (which treat mind and consciousness as manifestations of an eternal human soul)? The creation of smart machines that achieve some level of consciousness, only amplifies this already complicated philosophical issue.

In 1950, the British mathematician and computer science pioneer Alan Mathison Turing raised a similar question in his intriguing paper, "Computing Machines and Intelligence." As part of his pioneering discussion on artificial intelligence, Turing gave the world a test, now called the Turing test, for judging whether a machine is successfully simulating the thought processes of the human mind.

In 1949, Turing became the deputy director of the University of Manchester's computing laboratory. His duties included developing software for the Manchester Mark I machine. During this period, he did pioneering work in the field of artificial intelligence. His 1950 paper, "Computing Machines and Intelligence," raised the interesting question of machine intelligence and consciousness. He introduced the Turing test in an attempt to create a standard for determining when a machine is conscious or like the human mind in its thinking behavior. Today, computer scientists everywhere recognize the Turing test as a simple, yet clever, procedure, which examines whether a computing machine is capable of thinking like a human being.

It its simplest form, the Turing test involves a human being (called the interrogator) sitting at a teletype machine, connected to but isolated from two other correspondents. One of these correspondents is a human being, while the other is a "thinking" computer. By asking questions and examining the responses, the interrogator tries to determine which one of the correspondents is the computer and which one is the human being. The advanced computer is programmed to give delayed answers and even deceptive answers, mimicking how the human mind would respond to questions. If it is impossible for the human interrogator to determine which correspondent is a machine and which one is a human being, then the computer has passed the Turing test and is considered capable of humanlike thought. Many computer scientists regard Turing's 1950 paper as the start of the field of artificial intelligence.

The Issue of Controlling a Self-Replicating System

Whenever engineers discuss the technology and role of self-replicating systems (SRS), their conversations inevitably turn to the interesting question: What happens if an SRS gets out of control? Before human beings seed the solar system or interstellar space with even a single SRS unit, engineers and mission planners should know how to pull an SRS unit's plug if things get out of control. Some engineers and scientists have already raised this very legitimate concern about SRS technology. Another question that robot engineers often encounter concerning SRS technology is whether smart machines represent a long-range threat to human life. In particular, will machines evolve with such advanced levels of artificial intelligence that they become the main resource competitors and adversaries of human beings—whether the ultra-smart machines can replicate or not? Even in the absence of advanced levels of machine intelligence that mimic human intelligence, the SRS might represent a threat just through its potential for uncontrollable exponential growth.

These questions can no longer remain entirely in the realm of science fiction. Robot engineers must start examining the technical and social implications of developing advanced machine intelligences and SRS *before* they bring such systems into existence. Fully engaging in such prudent and reasonable forethoughts will avoid a future situation (now very popular in science fiction) in which human beings find themselves in a mortal conflict over planetary (or solar system) resources with their own intelligent machine creations.

Of course, human beings definitely need smart machines to improve life on Earth, to explore the solar system, to create a solar system civilization, and to probe the neighboring stars. So robot engineers and scientists should proceed with the development of smart machines, but temper these efforts with safeguards to avoid the ultimate undesirable future situation, in which the machines turn against their human masters and eventually enslave or exterminate them. In 1942, the science fact/fiction writer Isaac Asimov suggested a set of rules for robot behavior in his story "Runaround," which appeared in *Astounding* magazine.

Over the years, Asimov's "laws" have become part of the cult and culture of modern robotics. They are: (Asimov's First Law of Robotics) "A robot may not injure a human being, or, through inaction, allow a human being to come to harm;" (Asimov's Second Law of Robotics) "A robot must obey the orders given it by human beings except where such orders would conflict with the first law;" and (Asimov's Third Law) "A robot must protect its own existence as long as such protection does not conflict with the first or second law." The message within these so-called laws represents a good starting point in developing benevolent, people-safe, smart machines.

However, any machine sophisticated enough to survive and reproduce in largely unstructured environments would probably also be capable of performing a certain degree of self-reprogramming or automatic improvement (that is, have the machine behavior of evolution). An intelligent SRS unit might eventually be able to program itself around any rules of behavior that were stored in its memory by its human creators. As it learns more about its environment, the smart SRS unit might decide to modify its behavior patterns to better suit its own needs. If this very smart SRS unit really "enjoys" being a machine and making (and perhaps improving) other machines, then when faced with a situation in which it must save a human master's life at the cost of its own, the smart machine may decide to simply shut down instead of performing the life-saving task it was preprogrammed to do. Thus, while it does not harm the endangered human being, it may not help the person out of danger either.

Science fiction contains many interesting stories about robots, androids, and even computers turning on their human builders. The conflict between the human astronaut crew and the interplanetary spaceship's feisty computer, HAL, in Arthur C. Clarke and Stanley Kubrick's cinematic masterpiece *2001: A Space Odyssey* is an incomparable example. The purpose of this brief discussion is not to invoke a Luddite-type response against the development of very smart robots; only to suggest that such exciting research and engineering activities be tempered by some forethought concerning the potential technical and social impact of these developments.

As previously mentioned in this chapter, early in the Industrial Revolution, a group of British workers, ostensibly influenced by someone called Ned Ludd, rioted and destroyed newly installed textile machinery that was taking their jobs away. The term Luddite now generally refers to a person, who exhibits a very strong fear or hatred of technology—that is, a person who is an extreme technophobe. This term is often encountered during discussions about the social impact of robots here on Earth.

One or all of the following techniques might control an SRS population in space. First, the human builders could implant machine-genetic instructions (deeply embedded computer code) that contained a hidden or secret cutoff command. This cutoff command would be automatically activated after the SRS units had undergone a predetermined number of replications. For example, after each machine replica is made, one regeneration command could be deleted—until, at last, the entire replication process is terminated with the construction of the last (predetermined) replica. A very simple example, which illustrates the principle behind an embedded reproduction limit code, is that of a motion picture rented on a disposable DVD. After two or three plays, the disposable DVD disables (or erases) itself and the motion picture on the DVD can no longer be viewed.

Second, a special signal from Earth at some predetermined emergency frequency might be used to shutdown individuals, selected groups, or all SRS units at any time. This approach is like having an emergency stop button, which when pressed by a human being causes the affected SRS units to cease all activities and go immediately into a safe, hibernation posture. Many modern machines have either an emergency stop button, flow cutoff valve, heat limit switch, or master circuit breaker. The signal activated "all-stop" button on an SRS unit would just be a more sophisticated version of this engineered safety device.

For low-mass SRS units (perhaps in the 100 kilograms to 4,500 kilograms class) population control might prove more difficult because of the shorter replication times, when compared to much larger mass SRS factory units. To keep these mechanical critters in line, robot engineers might decide to use a predator robot.

The predator robot would be programmed to attack and destroy only the type of SRS units, whose populations were out of control due to some malfunction or other. Robot engineers have also considered SRS unit population control through the use of a universal destructor (UD). This machine would be capable of taking apart any other machine it encountered. The universal destructor would recover any information found in the prey robot's memory prior to recycling the prey machine's parts. Wildlife managers use (biological) predator species on Earth today to keep animal populations in balance. Similarly, robot managers in the future could use a linear supply of nonreplicating machine predators to control an exponentially growing population of misbehaving SRS units.

Robot engineers might also design the initial SRS units to be sensitive to population density. Whenever the smart robots sensed overcrowding or overpopulation, the machines could lose their ability to replicate (that is, become infertile), stop their operations, and go into a hibernation state, or even (like lemmings on Earth) report to a central facility for disassembly. Unfortunately, SRS units might mimic the behavior patterns of their human creators too closely. So, without preprogrammed behavior safeguards, overcrowding could force such intelligent machines to compete among themselves for dwindling supplies of resources (terrestrial or extraterrestrial). Dueling, mechanical cannibalism, or even some highly organized form of robot versus robot conflict might result.

Hopefully, future human engineers and scientists will create smart machines that only mimic the best characteristics of the human mind. For it is only in partnership with very smart and well-behaved SRSs that the human race can some day hope to send a wave of life, conscious intelligence, and organization through the Milky Way Galaxy.

In the very long term, there appear to be two general pathways for the human species: either human beings are a very important biological stage in the overall evolutionary scheme of the universe; or else humans are an evolutionary dead end. If the human race decides to limit itself to just one planet (Earth), a natural disaster or humankind's own foolhardiness will almost certainly terminate the species—perhaps in just a few centuries or a few millennia from now. Excluding such unpleasant natural or human-caused catastrophes, without an extraterrestrial frontier, a planetary society will simply stagnate due to isolation, while other alien civilizations (should such exist) flourish and populate the galaxy.

Replicating robot system technology offers the human race very interesting options for continued evolution beyond the boundaries of Earth. Future generations of human beings might decide to create autonomous, interstellar self-replicating robot probes (von Neumann probes), and send these systems across the interstellar void on missions of exploration. Or, future generations of human beings could elect to develop a closely knit (symbiotic) human-machine system—a highly automated interstellar ark—that is capable of crossing interstellar regions and then replicating itself when it encounters star systems with suitable planets and resources.

According to some scientists, any intelligent civilization that desires to explore a portion of the galaxy more than 100 light years from their parent star would probably find it more efficient to use self-replicating robot probes. This galactic exploration strategy would produce the largest amount of directly sampled data about another star system for a given period of exploration. One estimate suggests that the entire galaxy could be explored in about one million years, assuming the replicating interstellar probes could achieve speeds of at least one-tenth the speed of light. If other alien civilizations (should such exist) follow this approach, then the most probable initial contact between extraterrestrial civilizations would involve a self-replicating robot probe from one civilization encountering a self-replicating probe from another civilization.

If these encounters are friendly, the probes could exchange a wealth of information about their respective parent civilizations and any other civilizations previously encountered in their journeys through the galaxy. The closest terrestrial analogy would be a message placed in a very smart bottle that is then tossed into the ocean. If the smart bottle encounters another smart bottle, the two bump gently and provide each other a copy of their entire content of messages. One day, a beachcomber finds a smart bottle and discovers the entire collection of messages that has accumulated within.

If the interstellar probes have a hostile, belligerent encounter, they will most likely severely damage or destroy each other. In this case, the journey through the galaxy ceases for both probes and the wealth of information about alien civilizations, existent or extinct, vanishes. Returning to the simple message in smart bottle analogy here on Earth, a hostile encounter damages both bottles, they sink to the bottom of the ocean, and their respective information contents are lost forever. No beachcomber will ever discover either bottle and so he will never have the chance of reading the messages contained within.

One very distinct advantage of using interstellar robot probes in the search for other intelligent civilizations is the fact that these probes could also serve as a cosmic safety deposit box, carrying information about the technical, social, and cultural aspects of a particular civilization through the galaxy long after the parent civilization has vanished. The gold-anodized records of NASA engineers included on the *Voyager 1* and *2* spacecraft and the special plaques they placed on the *Pioneer 10* and *11* spacecraft are humans' first attempts at achieving a tiny degree of cultural immortality in the cosmos.

Star-faring self-replicating machines should be able to keep running for a long time. One speculative estimate by exobiologists suggests that there may exist at present only 10 percent of all alien civilizations that ever arose in the Milky Way Galaxy—the other 90 percent having perished. If this estimate is correct then, on a simple statistical basis, nine out of every 10 robotic star probes within the galaxy could be the only surviving artifacts from long-dead civilizations. These self-replicating star probes would serve as emissaries across interstellar space and through eons of time. Here on Earth, the discovery and excavation of ancient tombs and other archaeological sites provides a similar contact through time with long vanished peoples.

Perhaps later this century, human space explorers and/or their machine surrogates will discover a derelict alien robot probe, or recover an artifact the origins of which are clearly not from Earth. If terrestrial scientists and cryptologists are able to decipher any language or message contained on the derelict probe (or recovered artifact), humans may eventually learn about at least one other ancient alien society. The discovery of a functioning or derelict robot probe from an extinct alien civilization may also lead human investigators to many other alien societies. In a sense, by encountering and successfully interrogating an alien robot star probe, the human team of investigators may actually be treated to a delightful edition of the proverbial *Encyclopedia Galactica*—a literal compendium of the technical, cultural, and social heritage of thousands of extraterrestrial civilizations within the galaxy (most of which are probably now extinct).

There are a number of interesting ethical questions concerning the use of interstellar self-replicating probes. Is it morally right, or even equitable, for a self-replicating machine to enter an alien star system and harvest a portion of that star system's mass and energy to satisfy its own mission objectives? Does an intelligent species legally "own" its parent star, home planet, and any material or energy resources residing on other celestial objects within its star system? Does it make a difference whether the star system is inhabited by intelligent beings? Or, is there some lower threshold of galactic intelligence quotient (GIQ) below which star-faring races may ethically (on their own value scales) invade an alien star system and appropriate the resources needed to continue on their mission through the galaxy? If an alien robot probe enters a star system to extract resources, by what criteria does it judge the intelligence level of an indigenous life form—perhaps in an effort not to severely disturb existing life-bearing ecospheres? Further discussion about and speculative responses to such intriguing SRS-related questions extends far beyond the scope of this book. However, the brief line of inquiry introduced here cannot end without at least mention of the most important question in cosmic ethics: Now that the human species has developed space technology, are humans and their solar system above (or below) any galactic appropriations threshold?

7

The Future of Robot Technology

Attempting to "project" (*not* predict) the future of robot technology is somewhat like looking at the Wright Brothers' first aircraft in 1903 and coming up with the Boeing 777, or like looking at the gigantic ENIAC computer in 1946 and coming up with any of the numerous, powerful laptop computers with flat panel screens now being sold commercially. From an historic perspective, under support from the U.S. Army, John Presper Eckert and John W. Mauchy completed the ENIAC (Electronic Numerical Integrator And Calculator) at University of Pennsylvania in 1946. The ENIAC is considered by science historians as the world's first electronic digital computer, and at the time of its completion, was the world's most complex electronic machine. ENIAC was a massive, room-sized machine containing over 18,000 vacuum tubes. But the device could only handle numbers.

The business of futurists is to make technical projections. Such technical projections are useful, because they provide some indication of future possibilities—technical, social, economic, and political—available to the human race. But, in all likelihood, many such projections will be hopelessly off the mark. Some will be far too conservative, falling victim to unanticipated technology breakthroughs. The discovery of the transistor in 1947, for example, changed the course of human history. The transistor greatly miniaturized electronics and made the modern digital computer a practical reality. Some projections will be too optimistic and fail to include preliminary assessment of the possible impacts, consequences, and so-called "downsides" in a new technology. The promise of "meter-less electricity for everyone" at the dawn of the civilian nuclear power industry in the mid-1950s is an example. Yes, civilian nuclear power now generates a significant portion of the world's electricity, but it is hardly free and there are some long-term social obligations inherent in the use of this technology. These obligations include the requirement and expense of decommissioning the nuclear power plant (at the end of its useful operational life) and the need to manage high-level nuclear wastes (both defense and civilian) for essentially the entire future history of the human race. These long-term commitments were never prominently displayed during the initial hype surrounding the use of nuclear energy for civilian applications.

The point here is that projections about the future of robot technology will follow a similar pattern. Some will be on the mark, some will be wildly optimistic, and some will be unnecessarily pessimistic—perhaps hampered by the inability to anticipate a "wild card" breakthrough in technology. Armed with such caveats, what is the best way to look at the future of robot technology?

One reasonable approach at *robot futurism* is to simply "follow the money." An examination of the technology areas into which significant amounts of the current federal research and development budget are being invested should provide some insight into where the breakthroughs may appear. This approach yields three areas worthy of special attention in this chapter: space robots, autonomous military vehicles (aerial and ground), and nanotechnology. Each of these technology areas is currently enjoying significant levels of funding with the Department of Defense (DOD), the Department of Energy (DOE), and the National Aeronautics and Space Administration (NASA). There are also interesting collaborations taking place that could promote the occurrence of an important technical nonlinearity, or wild card event. For example, NASA research on an autonomous robot airplane for use in the exploration of Mars may provide the technology key to the development of an unmanned combat aerial vehicle (UCAV) by the DOD, or vice versa.

SOME FUTURE SPACE ROBOTS

NASA engineers are planning to add a strong dose of artificial intelligence (AI) to planetary orbiters, landers, and rovers to make these robot spacecraft much more self-reliant and capable of making basic decisions during a mission without human control or supervision. In the past, robot rovers contained very simple AI systems, which allowed them to make a limited number of basic, noncomplicated decisions. However, in the future, mobile robots will possess much higher levels of AI or machine intelligence and be able to make decisions now being made by human mission controllers on Earth.

One of the technical challenges that robot engineers face is how to encapsulate the process by which human beings make decisions in response to changes in their surroundings into a robot rover or complex lander spacecraft sitting on a planet millions of kilometers away. To make the detailed exploration of the Moon and Mars by mobile robots practical over the next two decades, future robot rovers will have to be intelligent enough to navigate the surface of the Moon or Mars without a continuous stream of detailed instructions from and decision making by scientists on Earth.

Large teams of human beings on Earth are needed to direct the Mars Exploration Rovers (MER) *Spirit* and *Opportunity* as the two robot rovers roll across the terrain of Mars looking for evidence of water. In a very slow and deliberate process, it takes human-robot teams on two worlds millions of kilometers apart several days to achieve each of many individual mission milestones and objectives. Specifically, it takes about three (Earth) days for the *Spirit* or *Opportunity* robot rover to visualize a nearby target, get to the target, and do some contact science. Mission controllers currently measure a great day of robot exploring on Mars in terms of travel up to 100 meters per Martian day (sol) across the surface of the planet. (A sol is a Martian day and is about 24 hours, 37 minutes, 23 seconds in duration using Earth-based time units.) Imagine trying to explore an entire continent here on Earth using a system that travels a maximum distance each day equivalent to the length of just one football or soccer field.

This section examines how future advanced space robots (especially mobile robots with more onboard machine intelligence [AI]) will gather data about their environment and then make on the spot evaluations of appropriate tasks and actions without being dependent on decisions made by humans. Advanced AI systems onboard such smart future mobile robots will eventually allow them to mimic human thought processes and perform tasks a human explorer would want done. For example, such smart rovers might pause to make an on the spot soil analysis of an interesting sample, communicate with an orbiting robot spacecraft for additional data about the immediate location, or even signal with other robot rovers to gather (swarm) at the location to perform a collective (or group) evaluation of the unusual discovery.

Within the next two decades, teams of smart robots, interacting with each other, should be able to map and evaluate large tracts on the surface of the Moon or Mars. An interactive team of smart robot rovers would provide much better coverage of a large area of land, possess redundancy, and perhaps even exhibit a level of collective intelligence while performing tasks too difficult or complex for a single robot system. With a team of robots, the mission objectives can be accomplished, even if one robot fails to perform or is severely damaged in an accident.

Using Smart Robots to Prospect for Lunar Water on the Moon

The Moon is nearby and accessible, so it is a great place to try out many of the new space technologies, including the advanced space robots that are also critical in the detailed scientific study and eventual human exploration of more distant alien worlds, such as Mars. Whether a permanent lunar base turns out to be feasible hinges on the issue of logistics, especially the availability of water in the form of water ice. The logistics problem is quite simple. Water is dense and rather heavy, so shipping large amounts of water from Earth's surface to sustain a permanent human presence on the Moon this century could be prohibitively expensive. Establishing a permanent human base on the Moon becomes much easier and far more practical if large amounts of water (frozen in water ice deposits) are already there.

This unusual resource condition is possible, because scientists now hypothesize that comets and asteroids smashing into the lunar surface eons ago left behind some water. Of course, water on the Moon's surface does not last very long. It evaporates in the intense sunlight and quickly departs this airless world by drifting off into space. Only in the frigid recesses of permanently shadowed craters do scientists expect to find any of the water that might have been carried to the Moon and scattered across the lunar surface by ancient comet or asteroid impacts. In the 1990s, two spacecraft, *Clementine* and the *Lunar Prospector*, collected tantalizing data suggesting that the shadowed craters at the lunar poles may contain significant quantities of water ice.

NASA plans to resolve this very important question by using smart robots as scouts. First into action will be the *Lunar Reconnaissance Orbiter* (LRO)—a robot spacecraft mission planned for launch by late 2008. The LRO mission emphasizes the overall objective of collecting science data that will facilitate a human return to the Moon. As part of NASA's strategic plan for solar system exploration, a return to the Moon by human beings is considered a critical step in field testing the equipment necessary for a successful human expedition to Mars later this century.

The LRO will orbit the Moon for at least one year using a 30- to 50-kilometer altitude, polar orbit to map the lunar environment in greater detail than ever before. The six instruments planned for the LRO will do many things. First, the instruments will map and photograph the Moon in great detail, paying special attention to the permanently shadowed polar regions. The LRO's instruments will also measure the Moon's ionizing radiation environment and conduct a very detailed search for signs of water ice deposits. No single spacecraft-borne instrument can provide definitive evidence of ice on the Moon. However, scientists feel that, if all the data from the LRO's collection of water-hunting instruments point to suspected ice in the same area, then those data would be most compelling and warrant further investigation.

Within NASA's current strategic vision for robot–human partnership in space exploration, the LRO is just the first in a string of smart robots with missions to the Moon over the next two decades. Once compelling evidence for the presence of water ice is obtained by the LRO, then the next logical step is to send a smart scout robot to that location to scratch and sniff the site and to perform on the spot (in situ) analyses. The rover robot's detailed investigations will confirm the existence of any water ice. The semiautonomous mobile robot may expand investigations of the area to provide a first order estimate of the total quantity of the water available.

Finally, if suitable water resources are located and inventoried, teams of smart robot prospectors would be sent to the Moon to harvest the particular site or sites in preparation for the return of human beings to the lunar surface. Supervised and teleoperated by humans from Earth, a team of semiautonomous water-harvesting robots would make the construction and operation of a permanent human base practical (from a logistics perspective) and prepare the way for an eventual human expedition to Mars.

Smarter Robots to the Red Planet

NASA's *Phoenix Mars Scout*, currently in development, will land in icy soils near the north polar permanent ice cap of the Red Planet and explore the history of water in these soils and any associated rocks. This sophisticated space robot serves as NASA's first exploration of a potential modern habitat on Mars and open the door to a renewed search for carbon-bearing compounds, last attempted with the *Viking 1* and *2* lander spacecraft missions in the 1970s.

The *Phoenix* spacecraft is currently being constructed and should launch in August 2007. The robot explorer will land in May 2008 at a candidate site in the

Figure 7-1 This is an artist's concept of an advanced, semi-autonomous robot rover making remote sample collections at the Moon's south pole. With minimal supervision and teleoperation by controllers on Earth, this type of advanced robot sample collector would help validate the presence of water ice and quantify any promising resource data collected by lunar-orbiting resource reconnaissance spacecraft. The presence of ample quantities of water ice in the permanently shadowed polar regions of the Moon would be a major stimulus in the development of permanent human bases. Robot-assisted lunar ice mining could become the major industry on the Moon later this century. (Credit: Artist's concept courtesy of NASA/Johnson Space Center.)

Martian polar region previously identified by the *Mars Odyssey* orbiter spacecraft as having high concentrations of ice just beneath the top layer of soil. *Phoenix* is a fixed lander spacecraft, which means it cannot move from place to place on the surface of Mars. Rather, once the spacecraft has safely landed on the surface, it will use its robotic arm to dig the ice layer and bring samples to its suite of on-deck science instruments. These instruments will analyze samples directly on the Martian surface, sending science data back to Earth via radio signals, which will be collected by NASA's Deep Space Network.

The *Phoenix* spacecraft's stereo color camera and a weather station will study the surrounding environment, while its other instruments check excavated soil samples for water, organic chemicals, and conditions that could indicate whether the site was ever hospitable to life. Of special interest to exobiologists, the spacecraft's microscopes would reveal features as small as one one-thousandth the width of a human hair.

The *Phoenix Lander's* science goals of learning about ice history and climate cycles on Mars complements the robot spacecraft's most exciting task—to evaluate whether an environment hospitable to microbial life may exist at the ice–soil boundary. One tantalizing question is whether cycles on Mars, either short term or long term, can produce conditions in which even small amounts of near surface water might stay melted. As studies of arctic environments on Earth have indicated, if water remains liquid—even just for short periods during long intervals—life can persist, if other factors are right.

Figure 7-2 This is an artist's concept of a robot field geologist, called the TeleProspector. This advanced mobile robot would be capable of allowing human geologists comfortably located at a permanent lunar base (or back on Earth) to extend their visual and tactile senses to a remote location on the Moon through telepresence and virtual reality technologies. Enabled by the robot's stereovision, motion sensors, and its ability to duplicate human movements and provide tactile sensations, the human operator is surrounded with a virtual experience that mimics much of the environment the robot is physically experiencing in the field. Here, for example, both the robot and the human geologist (through virtual reality and telepresence) have just discovered a cluster of interesting crystals carried up to the Moon's surface from many miles below by an ancient lava flow. (Credit: Courtesy of NASA/Johnson Space Center; artist Pat Rawlings.)

Figure 7-3 This artist's concept shows NASA's planned *Phoenix* robot lander spacecraft deployed on the surface of Mars (circa 2008). The lander would use its robotic arm to dig into a spot in the water-ice rich northern polar region of Mars for clues concerning the Red Planet's history of water. The robot explorer would also search for environments suitable for microscopic organisms (microbes). (Credit: Artist's concept courtesy of NASA.)

Building upon the success of the two Mars Exploration Rover (MER) spacecraft, *Spirit* and *Opportunity*, which arrived on the surface of the Red Planet in January 2004, NASA's next mobile rover mission to Mars is being planned for arrival on the planet in late 2010. Called the *Mars Science Laboratory* (MSL), this mobile robot will be twice as long and three times as massive as either *Spirit* or *Opportunity*. The MSL will collect Martian soil samples and rock cores and analyze them on the spot for organic compounds and environmental conditions that could have supported microbial life in the past, or possibly even now in the present.

Recent advances in microelectronic technology and mobile robotics have made it possible for engineers to consider the creation and use of extremely small automated or remote-controlled vehicles in planetary surface exploration missions. For convenience, engineers often define a nanorover as a robot system with a mass of between 10 grams and 50 grams. Sometimes the terms millibot and microbot are also encountered in the technical literature. A millibot is a small, semiautonomous to fully autonomous robot deployed from a larger robot—the parent or marsupial robot. Microbot is a mobile robot rover with a

Figure 7-4 This artist's concept shows NASA's planned *Mars Science Laboratory* (MSL) on the Red Planet (circa 2010) with a robot arm extended to the front of the rover. With a greater range than any previous robot rover used on Mars, the MSL will be able to analyze dozens of samples scooped up from the soil and cored from rocks at scientifically interesting locations on the planet. One of the primary objectives of this sophisticated robot explorer is to investigate the past or present ability of Mars to support life. This artist's rendering shows the MSL. (Credit: Artist's concept courtesy of NASA/JPL-Caltech.)

microcontroller onboard. Some robot designers use the prefix *micro* in the term microbot to emphasize the fact that the robot is controlled by an elegant, onboard microprocessor. Other robot engineers use the same prefix (micro) to emphasize that the mobile robot under discussion is very, very small. Although mobile robot terminology is not standardized, the message is quite clear—these mobile robots are very, very small. One or several of these tiny robots could be used to survey areas around a lander and to look for a particular substance, such as water ice or microfossils. The nanorover would then communicate its scientific findings back to Earth via the lander spacecraft, possibly in conjunction with an orbiting mother spacecraft or communications hub, such as the proposed *Mars Telecommunications Orbiter* (MTO).

A cluster of nanorovers endowed with some degree of collective intelligence could perform detailed analysis of an interesting Martian surface or subsurface site suspected of harboring microbial life. How did the nanorovers get to that interesting site? In one possible exploration scenario, a larger surface rover (like the planned MSL) serves as a mother spacecraft and mobile base camp. The larger rover might carry several populations of such nanorovers, releasing or injecting them as part of its own test protocol in the search for suspected life sites (extinct or existent) on the Red Planet.

NASA engineers expect to launch the *Mars Telecommunications Orbiter* (MTO) in September 2009, have the spacecraft arrive at Mars in August 2010, and then start performing its mission for six to ten years from a high altitude orbit around the Red Planet. This future spacecraft's mission is to serve as the Mars hub for interplanetary telecommunications. By providing reliable and more available communications channels to Earth for rovers and stationary landers working on the surface of Mars, the MTO greatly increases the overall information payoff from all future robot missions.

Eventually, mobile space robots will achieve higher levels of artificial intelligence, autonomy, and dexterity, so that servicing and exploration operations will become less and less dependent on a human operator being present in the control loop. These robots would be capable of interpreting very high-level command structures and executing commands without human intervention. Erroneous command structures, incomplete task operations, and the resolution of differences between the robot's built-in "world model" and the real-world environment it is encountering would be handled autonomously. This level of intelligence will be even more important when future advanced robots are sent deeper into the outer solar system and telecommunications time delays of minutes become hours.

Collective intelligence is another interesting concept for future robots. Just as human beings can self-organize into groups or teams to achieve complicated goals, collections of smart robots will learn to self-organize into teams (or swarms) to perform more complicated missions. For example, a team of robot rovers could gather at a particularly interesting surface site to harvest all the science data available; or else several mobile robots might rush to the assistance of a stranded robot. Such collective actions and group behavior will allow teams of future space robots to exceed the performance capabilities and artificial intelligence levels of any individual machine. Collective machine intelligence would open up entirely new avenues for the use of robot systems on Mars and elsewhere in the solar system.

The development of higher levels of autonomy and the demonstration of collective machine intelligence by teams of robots are very important technology milestones in robotics. Once attained, these capabilities will also result in the effective use of robots in the construction and operation of permanent lunar or Martian surface bases. In such complex undertakings on alien worlds, teams of smart machines will serve as scouts, mobile science platforms, and eventually construction workers, who set about their tasks with little or no direct human supervision. As future space robots learn to think a little more like humans, these machines will anticipate the needs of their human partners in space exploration, and simply perform the necessary tasks with little or no human supervision. If a human explorer shows strong interest in a particular outcropping on Mars, his or her mobile robot companion will also focus its sensors and attention on the site. When an astronaut drops a tool on Mars during the construction of a surface base, his or her companion construction robot will immediately fetch the tool with its mechanical arm and "hand" it back to the astronaut, without blinking an electronic eye.

Anticipating this future scenario, NASA engineers have already laid the foundation for more productive human-robot partnerships in space exploration. Robonaut is a humanoid robot designed by the Robot Systems Technology Branch at the Johnson Space Center in Houston, Texas, in a collaborative effort with DARPA. The project seeks to develop and demonstrate a humanoid robot system that can function as an extravehicular activity (EVA) astronaut. One of

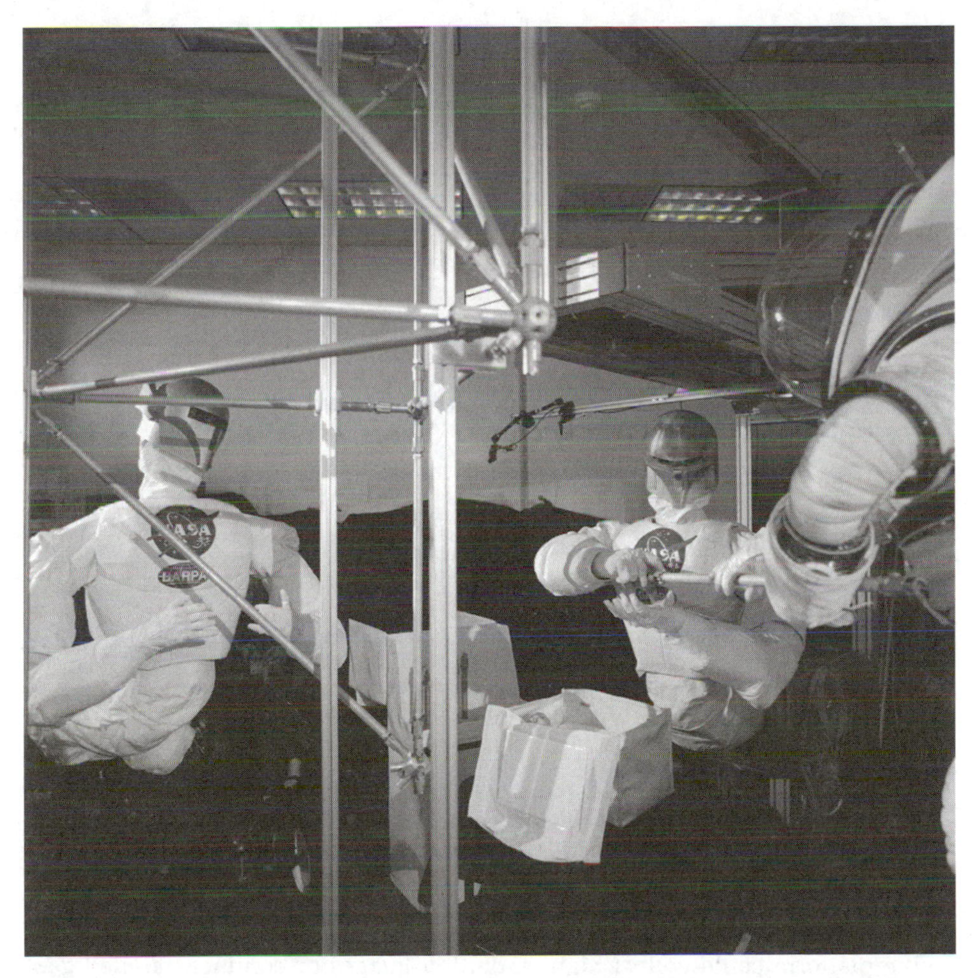

Figure 7-5 A space-suited human astronaut (on right) practices an assembly and construction task together with two Robonauts (humanoid robots), during a simulated extravehicular activity at the Johnson Space Center in 2003. Second- and third-generation designs of such humanoid robots would serve as valuable machine partners, helping human beings permanently settle the Moon and (later) explore Mars. (Credit: Photograph courtesy of the NASA/Johnson Space Center.)

the major goals of the Robonaut program is to construct a machine (humanoid robot) with a dexterity that exceeds the dexterity of a space-suited astronaut.

Mars Airplane

The Mars airplane is a conceptual, low-mass, autonomous, unmanned aerial vehicle (UAV) that can deploy experiment packages or conduct detailed reconnaissance operations on Mars. In some mission scenarios, the Mars airplane would be used to deploy a network of science stations, such as seismometers or meteorology stations, at selected Martian sites with an accuracy of a kilometer.

Figure 7-6 An artist's rendering of one candidate concept for an autonomous umanned aerial vehicle for use in scientific scout missions on Mars. This concept, developed by NASA Langley Research Center (LaRC), is called ARES (Aerial Regional-scale Environmental Survey of Mars). (Credit: Artist's rendering courtesy of NASA/LaRC.)

When designed with a payload capacity of about 50 kilograms, this robotic flying platform could collect high-resolution images or conduct detailed geochemical surveys of candidate surface sites of great interest in exobiology. The ultra-light aerial robot would be capable of flying at altitudes between 500 and 15,000 meters, with corresponding ranges of 25 to 6,700 kilometers. Scientists might deploy a robot airplane on Mars to perform aerial reconnaissance up long valleys and canyons. Flying in a giant canyon, the robot airplane would cover a large amount of interesting territory and gather very high-resolution images. Such scouting missions would identify specific sites worthy of more detailed study by surface rovers and/or human explorers.

NASA strategic planners have entertained two basic design approaches for a Mars airplane. In the first approach, the airplane is designed as a one-way, disposable aerial platform. After descending into the thin Martian atmosphere from a mother spacecraft, the robot airplane automatically deploys its large wings and performs aerial surveys, atmospheric soundings, and other scientific investigations, finally crashing when its hydrazine fuel supply is exhausted.

In the second scenario, engineers have equipped the Mars airplane with a small, variable-thrust rocket motor and land gear, so that it can make a soft (survivable) landing on the surface of the Red Planet, conduct some scientific investigation, and then take off. Because the Martian atmosphere is so thin, taking off from the ground requires an aircraft with very big wings and a power plant that supports a very fast takeoff. A rocket-assisted takeoff represents one viable engineering approach. This type of robot aircraft would have the ability to make in situ measurements and to gather samples at several widely separated sites on the Red Planet. The soil specimens can be examined on the spot or else delivered to a lander/ascent vehicle robot spacecraft, as part of a Mars sample return mission.

Mars mission planners recognize that a fleet of robot aircraft would provide a great deal of exploration flexibility and support to a human expedition to Mars. These aerial platforms could help the astronauts evaluate candidate-landing sites, deploy special sensors in support of network science projects, or collect soil and rock specimens from remote locations. Should several of the astronauts get stranded or lost while exploring the surface, Mars airplanes could effectively perform wide-area search operations. Finally, a Mars airplane, equipped with radio frequency transmitter/receiver hardware, could loiter in a fixed high-altitude holding pattern and serve as a temporary telecommunications relay station between astronaut explorers and their base camp or between astronauts at the base camp and a team of robot rovers, automated science stations, or other robot aircraft.

The First Interstellar Probe

An interstellar probe is a highly automated robot spacecraft sent from this solar system to explore another star system. Most likely this type of probe would make use of very smart machine systems capable of operating autonomously for decades or centuries.

Once the robot probe arrives at a new star system, it would begin a detailed exploration procedure. The target star system is scanned for possible life-bearing planets, and if any are detected, they become the object of more intense scientific investigations. Data collected by the mother spacecraft probe and any miniprobes (deployed to explore individual objects of interest within the new star system) are transmitted back to Earth. There, after light-years of travel, the signals are intercepted and analyzed by scientists, and interesting discoveries and information are used to enrich human knowledge and understanding about the galaxy and, by extrapolation, about the universe.

The robot interstellar probe could also be designed to carry a payload of specially engineered microorganisms, spores, and bacteria or even a "seed population" of tiny, self-replicating machines—the product of late twenty-first-century efforts in nanotechnology. If the robot probe encounters ecologically suitable planets on which life has not yet evolved, then it could make the decision to "seed" such barren, but potentially fertile worlds, with primitive life-forms or at least life precursors. A swarm of nanomachines might be deposited to make that world potentially more habitable. In that way, human beings (in partnership with their smart machines) would not only be exploring neighboring star systems, but also be participating in the spreading of life itself through nearby portions of the Milky Way Galaxy.

NASA's long-range strategic planners have examined some of the engineering and operational requirements for the first interstellar robot probe, as might

Figure 7-7 This artist's concept shows the human race's first interstellar robot probe departing the solar system (circa 2075) on an epic journey of scientific exploration. (Credit: Artist's concept courtesy of NASA.)

be launched at the end of this century to a nearby (within 10 light-years distance or less) star system. Some of these challenging requirements (all of which exceed current levels of technology by at least one or two orders of magnitude) are briefly mentioned here. The interstellar probe must be capable of sustained, autonomous operation for more than 100 years. The robot spacecraft must be able to manage its own health, that is, being able to anticipate or predict a potential problem, detect an emerging abnormality, and then prevent or correct the situation. For example, if a subsystem is about to overheat (but has not yet exceeded thermal design limits), the smart robot probe might redirect operations and adjust the thermal control system to avoid the potentially serious overheating condition.

The first interstellar robot probe requires a very high-level of artificial intelligence. The space robot must be able to perform fault management through repair, redundancy, and workarounds without any human guidance or assistance. The totally autonomous smart robot must also be able to carefully manage its onboard resources, supervising the generation and distribution of electric power, allocating the use of consumables, deciding when and where to commit emergency reserves and the limited supply of spare parts and components. The main onboard computer (or "machine brain") of the probe must exercise data management skills and be capable of an inductive response to unknown or unanticipated environmental changes and circumstances. When faced with unknown difficulties or opportunities, the robot probe must be able to modify the mission plan (established by its human creators decades before) and generate new tasks.

For example, during the mission, long-range sensors onboard the probe might discover that a Jupiter-sized extrasolar planet lies within the target star system and that this planet has a large (previously unknown) moon with an atmosphere and a liquid water ocean. Instead of sending one of its dwindling supply of miniprobes ahead to investigate the Jupiter-sized planet, the smart robot mother spacecraft makes a decision to release its miniprobe to make close-up measurements on this interesting moon. Since the mother spacecraft will probably be over eight light-years from Earth when the (hypothetical) discovery is made, the decision to change the mission plan must be made exclusively by the robot spacecraft, which is less than a few *light-days* away from the encounter. Sending a message back to Earth and asking for instructions would take over 16 years (for round-trip communications) and by then the interstellar probe would have completely passed through the target star system and disappeared into the interstellar void.

Similarly, instruments onboard the interstellar robot (regarded here as the mother spacecraft) and its supporting cadre of miniprobes must be capable of deductive and inductive learning, so as to adjust to how measurements are taken in response to unfolding opportunities, feedback, and unanticipated values (high and low). Some of the greatest scientific discoveries on Earth happened because of an accidental measurement or unanticipated reading.

For example, while investigating the energy content of sunlight with the help of a thermometer and a prism, the German-born British astronomer Sir (Frederick) William Herschel (1738–1822) slowly ran his thermometer across the visible portion of the solar spectrum. As he pushed the thermometer past red light into a dark (black to the human eye) region, Herschel was amazed that his thermometer suddenly indicated a higher temperature reading in a region of apparent darkness beyond the visible spectrum. He had accidentally discovered the infrared portion of the electromagnetic spectrum, which though invisible to the human eye, certainly has measurable energy content.

The instruments onboard the robot probe must be capable of exercising a similar level of curious inquiry and then be able to respond to unanticipated, but quite significant, new findings. The robot probe must have a level of artificial intelligence capable of "knowing" when new information is especially significant. This is a difficult task for human scientists, who often overlook the most significant pieces of data in an experiment or observation. To ask a robot's machine brain to respond "eureka" ("I've found it") at the moment of a great discovery is pushing artificial intelligence well beyond the technical horizon projected for the next few decades in this field. Yet, if the human race is going to make significant discoveries with robot interstellar probes that is precisely what these advanced exploring machines must be capable of doing.

From a spacecraft engineering perspective, the interstellar robot probe should consist of low-density, high-strength materials to minimize propulsion requirements. Remember, to keep a mission to the nearby stars within 100 years or so duration, the robot spacecraft should be capable of cruising at about one-tenth the speed of light (or more). Any less than that would take a star probe mission to even the nearest stars several centuries to achieve. How would a future society keep the great, great, great grandchildren of the probe engineers interested in receiving the signals from the all-but-forgotten space robot? Consequently, the first interstellar probe mission (using advanced but nonreplicating technology) should involve a 100-year or less journey to the target star.

The materials used on the outside of the robot probe must maintain their integrity for over a century or longer, even when subjected to hostile deep space environmental conditions, such as ionizing radiation, cold, vacuum, and interstellar dust. The structure of the robot spacecraft should be capable of autonomous reconfiguration. The power system must be able to provide reliable base power (typically at a level of 100 kilowatts-electric up to possibly one megawatt-electric) on an autonomous and self-maintaining basis for over 100 years. Finally, the star probe must be capable of autonomous data collection, assessment, storage, and communications (back to Earth) from a wide variety of scientific instruments and onboard spacecraft state-of-health sensors.

Some of the intriguing challenges in information technology include the proper calibration of instruments and collection of data over a period of years after decades of sensor dormancy. The robot probe must be able to transmit data back to Earth over distances ranging from 4.5 to 8.0 light-years. Finally, after decades of handling modest levels of data, the spacecraft's information systems must be capable of handling a gigantic burst of incoming data as the robot probe and its miniprobes encounter the target star system.

The Theory and Operation of Self-Replicating Systems

The brilliant Hungarian-American mathematician John von Neumann was the first person to seriously consider the problem of self-replicating systems. His book on the subject, *Theory of Self-reproducing Automata*, was edited by a colleague, Arthur W. Burks (b.1915) and published posthumously in 1966—almost a decade after von Neumann's untimely death due to cancer in 1957.

Von Neumann became interested in the study of automatic replication as part of his wide-ranging interests in complicated machines. His work during the World War II Manhattan project (the top secret American atomic bomb project) led him into automatic computing. Through this association, he became fascinated with the idea of large complex computing machines. In fact, he invented the scheme used today in the great majority of general-purpose digital computers—the von Neumann concept of serial processing stored-program—which is also referred to as the von Neumann machine.

Following his pioneering work in computer science—a field of which he is one of the founding fathers—von Neumann decided to tackle the larger problem of developing a self-replicating machine. The theory of automata provided him a convenient synthesis of his early efforts in logic and proof theory and his more recent efforts (during and after World War II) on large-scale electronic computers. Von Neumann continued to work on the intriguing idea of a self-replicating machine and its implications until his death.

Von Neumann actually conceived of several types of self-replicating systems, which he called the kinetic machine, the cellular machine, the neuron-type machine, the continuous machine, and the probabilistic machine. Unfortunately, he was only able to develop a very informal description of the kinetic machine before his death in 1957.

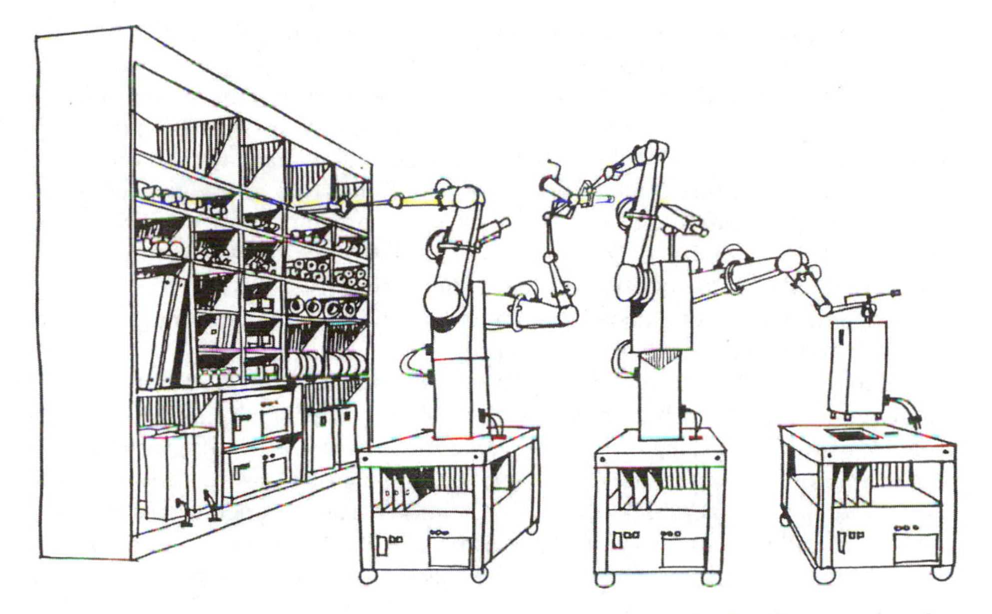

Figure 7-8 This drawing illustrates the kinematic machine, which is the most often discussed of von Neumann-type self-replicating systems (SRSs). (Credit: Drawing courtesy of NASA.)

The kinematic machine is the most often discussed of the von Neumann-type self-replicating systems. For this type of SRS, von Neumann envisioned a machine residing in a "sea of spare parts." The kinematic machine would have a memory tape that instructed the device to go through certain mechanical procedures. Using manipulator arms and its ability to move around, this type of SRS would gather and assemble parts. The stored computer program would instruct the machine to reach out and pick up a certain part, and then go through an identification and evaluation routine to determine whether the part selected was or was not called for by the master tape. (Note: in von Neumann's day microprocessors, minicomputers, floppy disks, CD ROMs, and multi-gigabyte capacity hard drives did not exist.) If the component picked up by the manipulator arm did not meet the selection criteria, it was tossed back into the parts bin (that is, back into the "sea of parts"). The process would continue until the required part was found and then an assembly operation would be performed. In this way, von Neumann's kinematic SRS would eventually make a complete replica of itself—without, however, understanding what it was doing. When the duplicate was physically completed, the parent machine would make a copy of its own memory tape on the (initially) blank tape of its offspring. The last instruction on the parent's machine tape would be to activate the tape of its mechanical progeny. The offspring kinematic SRS could then start searching the "sea of parts" for components to build yet another generation of SRS units.

In dealing with his self-replicating system concepts, von Neumann concluded that these machines should include the following characteristics and capabilities: (1) logical universality; (2) construction capability; (3) constructional universality; and (4) self-replication. Logical universality is simply the device's ability to function as a general-purpose computer. To be able to make copies of itself, a machine must be capable of manipulating information, energy, and materials. This is what is meant by the term construction capability. The closely related term constructional universality is a characteristic that implies the machine's ability to manufacture any of the finite-sized machines that can be built from a finite number of different parts, which are available from an indefinitely large supply. The characteristic of self-replication means that the original machine, given a sufficient number of component parts (of which it is made) and sufficient instructions, can make additional replicas of itself.

One characteristic of SRS devices that von Neumann did not address, but that has been addressed by subsequent investigators, is the concept of evolution. In a long sequence of machines making machines like themselves, can successive robot generations learn how to make themselves better machines? Robot engineers and artificial intelligence experts are exploring this intriguing issue as part of the larger question of thinking machines that are self-aware. Can robots be made smart and alert enough to learn from the experiences encountered in daily operations and thus improve their performance? If so, will such improvements simply reflect a primitive level of machine learning? Or, will the smart machines somehow begin to develop an internal sense of "knowing" that they know. If and when this ever occurs, the smart robot will begin to mimic the consciousness of its human creators. Some AI researchers like to boldly speculate that an advanced "thinking" robot in the distant future could be capable of formulating famous philosophical postulate of René Descartes: "Cogito, ergo sum" (I think, therefore I am). An SRS unit exhibiting the behavior of evolution might certainly be capable of achieving some form of machine self-awareness. From von Neumann's work and the more recent work of other investigators, five broad classes of SRS behavior have been suggested:

1. *Production*. The generation of useful output from useful input. In the production process, the unit machine remains unchanged. Production is a simple behavior demonstrated by all working machines, including SRS devices.
2. *Replication*. The complete manufacture of a physical copy of the original machine unit by the machine unit itself.
3. *Growth*. An increase in the mass of the original machine unit by its own actions, while still retaining the integrity of its original design. For example, the machine might add an additional set of storage compartments in which to keep a larger supply of parts or constituent materials.
4. *Evolution*. An increase in the complexity of the unit machine's function or structure. This is accomplished by additions or deletions to existing subsystems, or by changing the characteristics of these subsystems.
5. *Repair*. Any operation performed by a unit machine on itself that helps reconstruct, reconfigure, or replace existing subsystems, but does not change the SRS unit population, the original unit mass, or its functional complexity.

In theory, replicating systems can be designed to exhibit any or all of these machine behaviors. When such machines are actually built, however, a particular SRS unit will most likely emphasize just one or several kinds of machine behavior, even if it were capable of exhibiting all of them. For example, the fully

Figure 7-9 An artist's rendering of the general components of a (conceptual) self-replicating lunar factory. (Credit: Artists rendering courtesy of NASA/MSFC.)

autonomous, general-purpose self-replicating lunar factory, proposed in 1980 by Georg von Tiesenhausen and Wesley A. Darbo of the Marshall Space Flight Center (MSFC), is an SRS design concept that is intended for unit replication. There are four major subsystems that make up this proposed SRS unit. First, a materials processing subsystem gathers raw materials from its extraterrestrial environment (the lunar surface) and prepares industrial feedstock. Next, a parts production subsystem uses this feedstock to manufacture other parts or entire machines.

At this point, the conceptual SRS unit has two basic outputs. Parts may flow to the universal constructor (UC) subsystem where they are used to make a new SRS unit (this is replication); or else, parts may flow to a production facility subsystem where they are made into commercially useful products. This self-replicating lunar factory has other secondary subsystems, such as a materials depot, parts depot, power supply, and command and control center.

The universal constructor (UC) manufactures complete SRS units that are exact replicas of the original SRS unit. Each replica can then make additional replicas of itself until a preselected SRS unit population is achieved. The universal constructor would retain overall command and control (C&C) responsibilities for its own SRS unit as well as for its mechanical progeny—until, at least, the C&C functions themselves have been duplicated and transferred to the new units.

To avoid cases of uncontrollable exponential growth of such SRS units in some planetary resource environment, the human masters of these devices may reserve the final step of the C&C transfer function to themselves or so design the SRS units such that the final C&C transfer function from machine to machine can be overridden by external human commands.

AUTONOMOUS MOBILE ROBOTS FOR THE MILITARY

The successful combat experiences in Afghanistan and Iraq, as part of Operation Enduring Freedom (OEF) and Operation Iraqi Freedom (OIF), by American forces using remotely operated unmanned ground vehicles (UGVs) and unmanned aerial vehicles (UAVs) has encouraged military leaders in the Department of Defense to aggressively pursue improvements in military robot systems. Plans now include designs that increase the levels of autonomy for both UGVs and UAVs. For example, the success of the Predator UAV, carrying Hellfire missiles, on armed reconnaissance and surveillance missions, has expanded research and development efforts on unmanned combat aerial vehicles (UCAVs) within the United States Air Force and Navy. Similarly, the U.S. Army is actively investing research and development resources in the development of UGVs for a variety of important combat missions, including armed reconnaissance, search and destroy operations, and active perimeter defense of key bases, installations, and logistics depots. The vast majority of these efforts has received initial sponsorship or continued cooperative assistance from the military robot and artificial intelligence programs in the Defense Advanced Research Projects Agency (DARPA).

Concerning on-going robotic technology efforts by and for the U.S. armed services, three types of autonomous systems (either ground, aerial, or underwater) are generally considered. Autonomous military robots may be scripted, supervised, or intelligent. A *scripted autonomous robot system* uses a preplanned script (program) with embedded physical models to carry out its intended mission or objectives. A guided missile and a guided (smart) bomb are examples. Military analysts and engineers describe scripted autonomous robot systems as "point (program), fire (launch), and forget" systems. This is because once deployed the scripted autonomous robot system does not need or have any further human interaction. The autonomous guided missile finds its way to the target; once the scripted autonomous underwater vehicle (AUV) gets placed into the water, it submerges, runs its mission, and returns to the designated recovery station—all on its own.

A *supervised autonomous robot system* has some or all of its planning, sensing, monitoring, and networking functions automated, that is, performed by the robot. However, a distant human operator uses a communications link to provide the robot with cognitive (thinking) abilities, such as decision making, the fusion and perception of sensor data, the diagnosis of anomalies and problems, and any intended collaboration with other systems, either manned or robotic. Most "conventional" autonomous military robots being considered for the near future would fall into this category. For example, an unmanned combat aerial vehicle (UCAV) would take off on its own, fly itself over a particular course, and modify the flight path to avoid other aerial vehicles, adverse weather, and

Figure 7-10 A U.S soldier deploys a remotely controlled explosive ordnance disposal (EOD) robot to detonate a possible improvised explosive device (IED) in Al Iskandariyah, Iraq on February 27, 2005. (Credit: Photograph courtesy of the U.S. Army.)

hostile threats. At the end of the mission it would return to the home base, landing itself. However, as the UCAV travels around collecting data and sensing the environment, it might encounter and identify a hostile enemy target. It is at this point that the human operator is immediately alerted so he or she can make a quick decision, as to whether to engage and destroy the target or else hold fire.

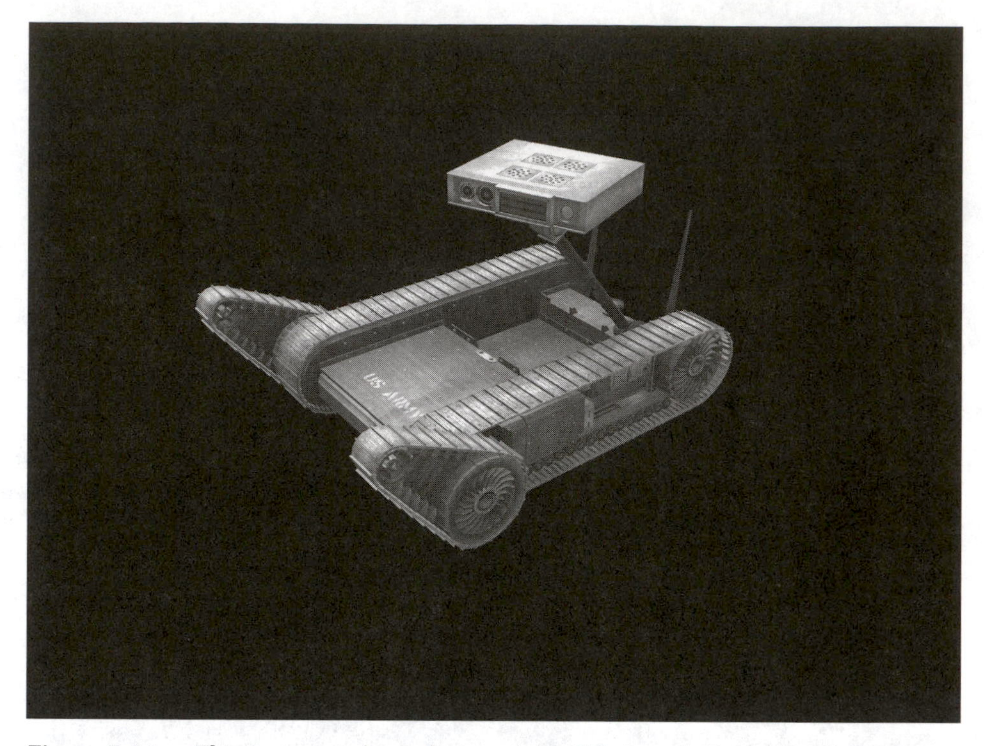

Figure 7-11 This is an artist's rendering of the U.S. Army's Future Combat Systems (FCS) small unmanned ground vehicle (UGV). (Credit: Artist's rendering courtesy of the U.S. Army.)

Keeping the human in the loop with respect to armed robot systems is very important. First, the human soldier, airman, marine, or sailor quite literally keeps his finger on the trigger and that responsibility is not assigned to an automated machine. Second, the human being can make a quick judgment concerning friendly or hostile forces, or even about any possible noncombatants in the area. The robot takes the high risk and potentially lethal exposure in the hot (or combat) zone, the human from a remote, presumably safer location, provides all the cognitive abilities necessary to attack or to avoid friendly casualties and excessive collateral damage. The armed flying robot or ground robot brings the fight directly and surgically to the enemy, compounding his loses and minimizing the risk to friendly forces or to noncombatants.

The third type of autonomous military robot system approaches the realm of science fiction in concept and operation. Called an *intelligent autonomous robot system* this type of military robot would have machine intelligence that controls all levels of operation of the autonomous vehicle, whether armed or not. This system's AI capability would allow the robot to make decisions (perhaps based on a set of planned options). The intelligent autonomous military robot must be capable of perceiving and interpreting the meaning of data sensed that it collects from the environment, as well as actions by hostile and friendly forces. This intelligent robot would also be able to diagnose itself, detect pending or actual subsystem failures, and (within the limits of its design) take corrective actions. These capabilities would allow the robot to continue to perform its mission despite encountered difficulties and to achieve the military objectives for which it was deployed. The intelligent autonomous military robot would also be able to collaborate with other robot systems, using various protocols (such as the current rules of engagement) and communications networks linked to other machines and supervising humans.

Imagine it is now 2020 and a squad of twelve well-armed, intelligent autonomous ground vehicles is advancing on a fortified enemy position. As they cross open ground, the autonomous military robots evasively maneuver and "talk" to each other. The robots continue to probe for soft spots in the enemy's defense. All the details of the unfolding battle are relayed back to a human supervisor, who is monitoring the performance of his squad of twelve "mechanical super-grunts" as they encircle the enemy stronghold. Even though several robots get hit, the others maneuver in for the kill. Suddenly, the enemy soldiers raise a white flag. What do these intelligent autonomous robots do? Is their AI software sufficiently advanced that they recognize a possible surrender situation? Do they suspect a trick and keep attacking? Or do they stop and await orders from the human supervisor?

Robot engineers define autonomy as "the ability to make decisions without human intervention." Does the intelligent autonomous robot of the mid-twenty-first century begin to resemble the behavior of the relentless T-800 robot in *The Terminator* motion picture?

THE PROMISE OF NANOTECHNOLOGY

What would the future be like, if scientists and engineers developed very tiny devices that could manipulate and reconstruct matter one atom or molecule at a time? That is precisely the overarching promise held out by the advocates of nanotechnology. The term is quite elusive and has taken on various meanings in the technical and popular literature. Basically, nanotechnology refers to materials and devices that exist and operate on the nanoscale. To provide some perspective, a nanometer is one-billionth (10^{-9}) of a meter. So, when scientists and engineers talk about nanoparticles, nanoelectronics, nanomachines, and nanmedicine, they are speaking about very tiny realms, typically between one and 100 nanometers in dimension. By way of comparison, a human red blood cell is over 2,000 nanometers long or about two micrometers (μm) in length. On the periphery of this revolutionary frontier area of science and engineering, incredibly small devices, called microelectromechanical systems (or MEMS), have already been constructed. Working in many of the U.S. government's national laboratories and federal research centers, scientists and engineers have developed a variety of very delicate and precise techniques to create MEMS devices and are now beginning to push the manipulation of materials on a regular and repeatable basis into the realm of nanotechnology.

Large sums of federal research money are now being invested into various areas of nanotechnology, including nanomedicine and nanoelectronics. At this point, no one can say with any degree of certainty whether these efforts will really be successful, and if so, where will such research successes lead. This section identifies some of the challenges and promises of nanotechnology, as identified within the U.S. Department of Energy's expanding research efforts at several national laboratories. According to DOE scientists and engineers, nanotechnology promises to make major contributions toward solving some of today's more serious problems, such as the control of diseases, handling the adverse aspects of global change, fighting industrial pollution, cleaning up toxic waste sites, and improving food production. There is also the promised revolution in electronics and computer technology.

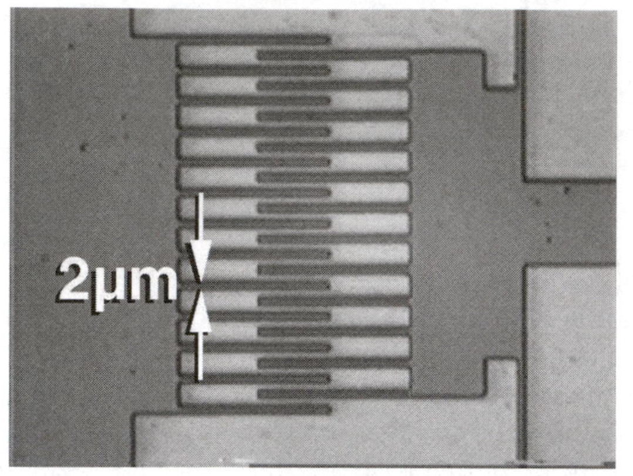

Figure 7-12 In 2000, Sandia National Laboratories researchers created the world's first diamond micromachine that drives a tiny diamond piston in the comb drive (shown here). The drive is powered by a tiny alternating electric current. As the two sets of comb teeth repel and attract each other, one comb slides back and forth. The dimension noted by arrows in the image is two micrometers, or two-millionths of one meter. (Credit: Picture courtesy of the U.S. Department of Energy/Sandia National Laboratories.)

One of the major technical areas facing great potential change is that of materials and manufacturing. Nanotechnology promises to fundamentally change the way materials and devices will be produced in the future. Nanostructures, polymers, metals, ceramics, and other materials could have greatly improved mechanical and physical properties. The ability to build things one atom or one molecule at a time offers to create entire new classes of incredible structural materials. Nanotechnology has the potential of making products lighter, smarter, stronger, cleaner, with higher precision, and perhaps, best of all, less expensive.

The ability to synthesize nanoscale building blocks, which have precisely controlled dimensions and composition, and then to assemble these building blocks into larger structures with unique properties and functions will revolutionize large segments of the materials and manufacturing industry. As envisioned by scientists in the DOE's national laboratories, nanostructuring is expected to promote lighter, stronger, and programmable materials. Visionaries anticipate the rise of molecular manufacturing leading to innovative devices based on new principles and architectures of material assembly. The so-called buckyball is just the beginning. Molecular (or cluster) manufacturing will take advantage of assembly at the nanoscale for a given purpose. Material scientists suggest that structures not previously observed in nature will also be developed.

Nanotechnology promises to revolutionize medicine by providing previously unthinkable tools and procedures for curing some of humankind's most notorious health problems. Imagine what all a modern physician could do if she could perform molecular-scale surgery to repair or rearrange individual cells in a patient's body. Nanotechnology advocates suggest that since disease is the result of physical disorder, misarranged molecules, and cells, through nanomedicine a physician should be able to "reach in" unobtrusively and cure most diseases. Mutations in DNA could be repaired and cancer cells, toxic chemicals, and viruses might be destroyed through the use of nanoscale medical devices. Medical researchers would be able to use nanoscale material manipulation techniques and devices to probe and characterize living cells. Research access to biological materials at this level promises to create an incredible revolution in the life sciences.

Lifetimes could be greatly extended as nanotechnology supports the development of more durable, rejection-resistant artificial tissues and organs. Therapeutic medicine would be based on precisely delivering new types of medicine to individual cells in all parts of the body, including those previously inaccessible. Nanoscale sensor systems (injected, inhaled, or ingested by a person) would shift the focus of patient care from disease treatment to early detection and prevention.

In the field of nanoelectronics and computer technology, engineers envision the ability to construct incredibly small, nanoscale circuits and computers. Smaller circuits run faster and this, in turn, enables far greater computing speeds. Nanotechnology offers the promise of making crystalline materials of ultrapurity and with better thermal conductivity and longer life. The nanostructured microprocessor device would continue the contemporary trend in declining energy use and cost per gate. Projections of this trend (due to nanoelectronics) suggest the efficiency of future computers will be improved by a factor of millions (that is six orders of magnitude or more).

The nanotechnology revolution would also influence national defense and space exploration, especially in the areas of improved materials, better sensors, incredibly more powerful computers, and a variety of interesting, very tiny little matter-manipulators (nano-sized robots), which could start terraforming a planet or else bring a rogue nation back into peaceful harmony with the world community.

There are, of course, enormous technical challenges facing the men and women who labor at the frontiers of nanotechnology in laboratories throughout the United States and the world. There are also social, ethical, and political questions associated with the rise of nanotechnology. Can, or should, self-replicating nanoscale devices be released into a person's body or into the general environment? How do scientists and engineers intend to control nanoscale devices that can replicate or even "mutate" into less desirable matter manipulators?

For the person, who really wishes to look far into the future, imagine a self-replicating robot, carrying a payload of nanotechnology devices. Sometime in the early twenty-second century, the human race might send this self-replicating system out of our transformed and fully engineered solar system on an incredible interstellar voyage. The overall mission of this self-replicating space robot is to trigger a wave of life and consciousness in other star systems. This wave (or more correctly, exponentially growing bubble) of replicating robot starships eventually expands out across the Milky Way. The name of the original self-replicating space robot is *Let There Be Life*.

8

Glossary of Terms Used in Robot Technology

absolute temperature. Temperature value relative to absolute zero, which corresponds to 0 K, or −273.15°C (after the Swedish astronomer Anders Celsius [1701–1744]). In SI units, the absolute temperature values are expressed in kelvins (K), a unit named in honor of the Scottish physicist Baron William Thomson Kelvin (1827–1907). In the traditional engineering unit system, absolute temperature values are expressed in degrees Rankine (°R), named after the Scottish engineer William Rankine (1820–1872).

absolute zero. The temperature at which molecular motion vanishes and an object has no thermal energy (or heat). From thermodynamics, absolute zero is the lowest possible temperature, namely zero degrees kelvin (0 K).

accelerated life test(s). The series of test procedures for a robot spacecraft that approximate in a relatively short period of time the deteriorating effects and possible failures that might be encountered under normal, long-term space mission conditions. Accelerated life tests help engineers detect critical design flaws and material incompatibilities (for example, excessive wear or friction) that eventually might affect the performance of a spacecraft component or subsystem over its anticipated operational lifetime.

acceleration (symbol: a). The rate at which the velocity of an object changes with time. Acceleration is a vector quantity and has the physical dimensions of length per unit time to the second power (for example, meters per second per second, or m/s^2).

accelerometer. An instrument that measures acceleration (time rate of change of velocity) or gravitational forces capable of imparting acceleration. Often used as a sensor on terrestrial robots to detect and measure the rate of change of velocity in a specific direction. Such data help determine the overall position of the robot and/or the position of the robot's arm, manipulator, or end effector with respect to some referenced (initial) position. Engineers also use accelerometers on robot spacecraft to assist in guidance and navigation and on planetary probes to support scientific data collection.

acceptance test(s). In the robot industry, the required formal tests conducted to demonstrate the acceptability of a unit, component, or complete robot system for delivery. These tests demonstrate performance to purchase specification requirements and serve as quality-control screens to detect deficiencies of workmanship and materials.

accumulator. A device or mechanism that stores up or accumulates something. For example, in hydraulics, an accumulator stores fluid under pressure. In computer engineering, an accumulator stores a number and then replaces that number with the new number that results when the originally stored number is operated upon by the computer system—for example, added, subtracted, multiplied, divided, squared, etc.

accuracy. In robotics, the degree to which the actual position of a robot (especially a robot's arm/end-effector) corresponds to the desired or commanded position.

acronym. A word formed from the first letters of a name, such as MER, which means Mars Exploration Rover or USA for the United States of America. An acronym is also a word formed by combining the initial parts of a series of words, such as lidar, which means light detection and ranging. Acronyms are frequently used in the robotics, computer science, and information technology.

activation mechanism. The situation required to invoke a procedure—usually a match of the system state to the preconditions required to exercise a production rule.

active control. The automatic activation of various control functions and equipment onboard unmanned aerial vehicle (UAV) or robot spacecraft. For example, to achieve active attitude control a UAV's current attitude is measured automatically and compared with a reference or desired value. Any significant difference between the flying robot's current attitude and the reference or desired attitude produces an error signal, which is then used to initiate appropriate corrective maneuvers by onboard actuators. Since both the measurements and the automatically imposed corrective maneuvers will not be perfect, the active control cycle usually continues through a number of iterations until the difference between the UAV's actual and desired attitude is within preselected, tolerable limits.

active homing guidance. A navigation and guidance system wherein a robot carries within itself both the source for illuminating the target (destination) and the receiver for detecting the signal reflected by the target (destination). This type of system can help an autonomous mobile robot navigate through a maze or a collection of obstacles. Active homing guidance systems may be used to assist a team of smart robots as they rendezvous at a specific location for cooperative (joint) operations.

active sensor. A sensor that illuminates a target, producing a return signal in the form of secondary radiation, which is then detected for the purpose of tracking and possibly identifying the target. A lidar is an example of an active sensor.

actuator. A device (usually electromechanical) that translates energy into motion and/or the application of a force. In an automated system, the actuator is responsible for a specific action or sequence of actions. Actuators are used to move a robot's manipulator joints. Three basic types of actuators currently are used in contemporary robots: pneumatic, hydraulic, and electrical. Pneumatic actuators employ a pressurized gas to move the manipulator joint. When the gas is propelled by a pump through a tube to a particular joint, it triggers or actuates movement. Pneumatic actuators are inexpensive and simple, but their movement is not precise. So, this kind of actuator usually is found in nonservo or pick-and-place robots. Hydraulic actuators are quite common and capable of producing a large amount of power. The main disadvantages of hydraulic actuators are their accompanying apparatus (pumps and storage tanks) and problems with fluid leaks. Electrical actuators provide smoother movements, can be controlled very accurately, and are very reliable. However, these actuators cannot deliver as much power as hydraulic actuators of comparable mass. Nevertheless, for modest power actuator functions, electrical actuators often are preferred.

adapter. Any device used or designed primarily to fit or adjust one component to another; for example, a fitting to join two pipes that have different threads or different diameters.

adaptation. Modification of an organism or its parts that makes it more fit for existence under the conditions of its environment.

adaptive control system. A control system that continuously monitors the dynamic response of the system being controlled and then automatically adjusts critical system parameters to satisfy preassigned response criteria, thereby producing the same response over a wide range of environmental conditions.

aerial survey. The search for sources of nuclear radiation (ionizing radiation) using sensitive instruments mounted in a helicopter, airplane, or UAV. Generally, the instrumentation records the type (through spectral analyses), intensity, and location of the radiation sources.

aero-. A prefix that means of or pertaining to the air, the atmosphere, aircraft, or flight through the atmosphere of a planet.

aerobot. An autonomous robotic aerovehicle (such as, a free-flying balloon or a specially designed extraterrestrial airplane) that is capable of flying in the atmospheres of Venus, Mars, Titan, or the outer planets. For Martian or Venusian aerobots, the balloon system would be capable of one or more of the following activities: autonomous state determination; periodic altitude variations; altitude control and the ability to follow a designated flight path within a planetary atmosphere using prevailing planetary winds; and landing at a designated surface location. The Mars airplane would be a low mass, unpiloted (robot) aircraft that carries experiment packages or performs detailed reconnaissance operations on the Red Planet.

AI. *See* **artificial intelligence**.

algorithm. A prescribed set of well-defined rules, processes, or mathematical equations for solving a problem in a finite number of steps.

alien life form (ALF). A general expression used by exobiologists to describe extraterrestrial life at any level of development from simple microscopic organisms to intelligent technically advanced beings.

alphanumeric. (alphabet plus numeric) A term that includes letters and numerical digits, such as, JEN42675WZ18.

ampere (symbol: A). The SI unit of electric current, defined as the constant current that, if maintained in two straight parallel conductors of infinite length, of negligible circular cross sections, and placed 1 meter apart in a vacuum, would produce a force between these conductors equal to 2×10^{-7} newtons per meter of length. The unit is named after the French physicist André M. Ampère (1775–1836).

amplifier. A device capable of reproducing an input electrical or electromagnetic radiation signal with increased intensity or gain. The energy required to increase the intensity of the input signal is drawn from an external source. If the output signal is a linear function of the input signal, the device is called a linear amplifier; otherwise, it is called a nonlinear amplifier.

amplitude modulation (AM). In telemetry and communications, a form of modulation in which the amplitude of the carrier wave is varied, or "modulated," about its unmodulated value. The amount of modulation is proportional to the amplitude of the signal wave. The frequency of the carrier wave is kept constant.

analog. Information represented by a quantity that can change continuously with time, as opposed to a quantity represented by a discrete (digital) set of incrementally changing values. An analog device depicts values by a continuously variable physical property, such as voltage, pressure, or position. An analog representation of a signal can vary continuously over a range, while a digital representation of the same signal is restricted to a discrete set of numbers. A mercury-filled thermometer is an example of a simple analog device. As the temperature varies, the mercury moves within the device continuously to indicate all new temperature values (within the range of temperatures for the particular thermometer).

analog computer. A computing device that processes continuously variable physical (analog) data, such as voltage or pressure variations. Specialized analog computers are used in selected scientific and industrial applications. However, the vast majority of modern computers are digital devices that process discrete data, such as binary numbers. A slide rule is an example of a simple analog computer. *See also* **digital computer.**

analog-to-digital converter (ADC). A device that transforms continuously variable analog data or signals into discrete, digitized signals. In the ADC device, the incoming analog signal is sampled, digitized, and encoded.

androgynous interface. A nonpolar interface; one that physically can join with another of the same design; literally, having both male and female characteristics.

android. A term found in science fiction, which describes a robot with near-human form or features; a synthetic man or woman constructed with artificial materials that simulate natural biological materials.

anechoic chamber. A test enclosure especially designed for experiments in acoustics. The interior walls of the chamber are covered with special materials (typically sound-absorbing, pyramid-shaped surfaces) that absorb sufficiently well the sound incident upon the walls, thereby creating an essentially sound-free condition in the frequency range(s) of interest.

angle. The inclination of two intersecting lines to each other, measured by the arc of a circle intercepted between the two lines forming the angle. There are many types of angles. An acute angle is less than 90°; a right angle is precisely 90°; an obtuse angle is greater than 90° but less that 180°; and a straight angle is 180°.

angle of attack. The angle (commonly used symbol: α) between a reference line fixed with respect to the airframe of an UAV and a line in the direction of movement of the UAV.

angstrom (symbol: Å). A unit of length used to indicate the wavelength of electromagnetic radiation in the visible, near-infrared, and near-ultraviolet portions of the electromagnetic spectrum. Named after the Swedish physicist Anders Jonas Ångström (1814–1874), who quantitatively described the Sun's spectrum in the year 1868. One angstrom equals 0.1 nanometer (10^{-10}meters). Although this unit is sometimes encountered in microrobotics, nanotechnology, and microelectronics, the use of the nanometer is preferred.

angular acceleration (symbol: α). The time rate of change of angular velocity (ω).

angular frequency (symbol: ω). The frequency of a periodic quantity expressed as angular velocity in radians per second. It is equal to the frequency (in hertz or cycles per second) times 2π radians per cycle.

angular momentum (symbol: L). A measure of an object's tendency to continue rotating at a particular rate around a certain axis. It is defined as the product of the angular velocity (ω) of the object and its moment of inertia (I) about the axis of rotation; that is, $L = I\,\omega$.

angular velocity (symbol: ω). The change of angle per unit time; usually expressed in radians per second.

anisotropic. Exhibiting different properties along axes in different directions; an anisotropic radiator would, for example, emit different amounts of radiation in different directions as compared to an isotropic radiator, which would emit radiation uniformly in all directions.

anode. The positive electrode in a battery, fuel cell, or electrolytic cell. *Compare with* **cathode.**

anomaly. A deviation from the normal or anticipated result.

antecedent. The left-hand side of a production rule. The pattern needed to make the rule applicable. *See also* **consequent.**

antenna. A device used to detect, collect, or transmit radio waves. A radio telescope is a large receiving antenna, while many robot spacecraft have both a directional antenna and an omnidirectional antenna to transmit (downlink) telemetry and to receive (uplink) instructions.

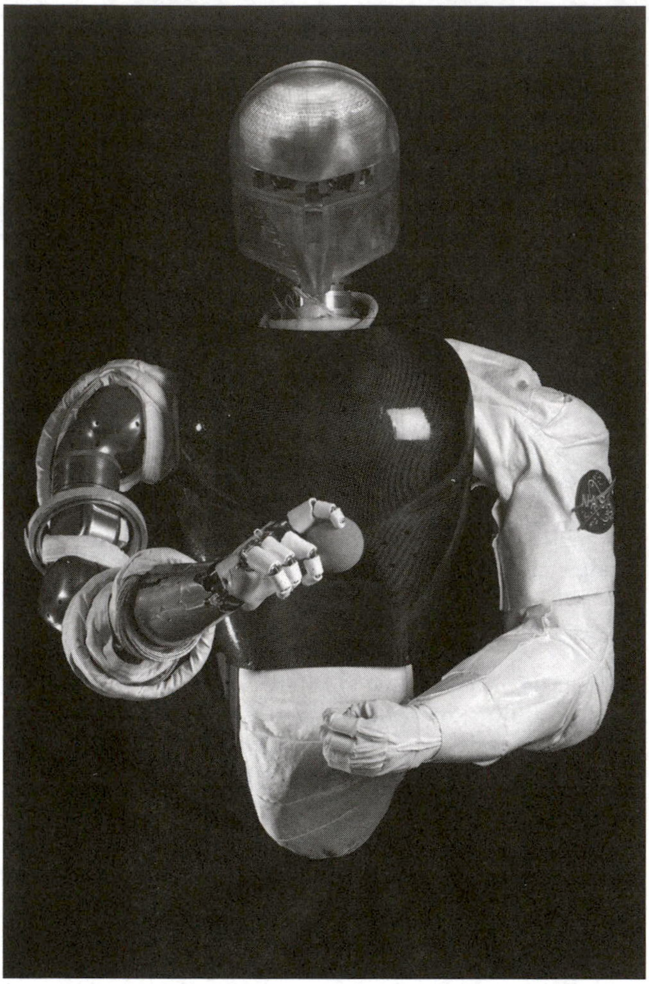

antiextrusion ring. Ring installed on the low-pressure side of a seal or packing to prevent extrusion of the sealing material; sometimes called a *backup ring*.

antirotation device. Mechanical device (such as a key) used in rotating machinery to prevent rotation of one component relative to an adjacent component.

antisatellite (ASAT) spacecraft. A robot spacecraft designed to destroy other satellites in space. An ASAT spacecraft could be deployed in space disguised as a peaceful satellite that quietly lurks as a secret hunter/killer satellite, awaiting instructions to track and attack its prey.

***Aqua* spacecraft.** An advanced Earth-observing satellite placed into polar orbit by NASA on May 4, 2002. The primary role of *Aqua*, as its name implies (Latin for "water"), is to gather information about changes in ocean circulation and how clouds and surface water processes affect Earth's climate. Equipped with six state-of-the-art instruments, this robot spacecraft is collecting data on global precipitation, evaporation, and the cycling of water on a planetary basis.

arc. 1. In mathematics, a part of a curved line, such as a portion of a circle. 2. In physics, a luminous glow that appears when an electric current passes through ionized air or gas.

architecture. The logical or physical structure of a manufacturing process, a computer, or a computer-based system.

ARPANET. A network of computers and computational resources used by the United States artificial intelligence (AI) and computer science community and sponsored by the Defense Advanced Research Projects Agency (DARPA).

artificial intelligence (AI). The discipline within the fields of information technology and computer science in which scientists attempt to give smart machines and advanced computers reasoning powers that resemble and approach logical operations of the human brain. This term is often taken to mean the study of thinking and perceiving as general information processing functions by machines. Also called *machine intelligence* and *heuristic programming*.

articulated. Segmented or jointed and thereby able to accommodate motion.

Figure 8-1 This is NASA's Robonaut A, an android-like robot system being developed at the Johnson Space Center (JSC) in cooperation with the Defense Advanced Research Project Agency. Robonaut A is helping demonstrate how astronaut-friendly robots can assist human beings during hazardous extravehicular activities. (Credit: Courtesy of NASA/JSC.)

assembler. In nanotechnology, a postulated tiny construction machine that would be capable of manipulating individual atoms and molecules and building a variety of important micromachines and nanostructures. Extrapolating this concept to the extreme, a legion of programmable self-replicating assemblers would function like the matter compilers speculated upon in the science fiction literature. Given energy and raw materials these programmable self-replicating assemblers would set about making any one of a wide variety of macroscopic-sized objects. *See also* **nanotechnology**; **Santa Claus machine**.

assembly robot. A computerized industrial robot, generally with sensors, which is designed for assembly line and manufacturing tasks.

astronaut. The name used by the United States for its human space travelers; comparable to the Russian term **cosmonaut**.

axis of motion. The separate motion (or degrees of freedom) that a robot has in its manipulator, wrist, and base. The four most common types of motion are: Cartesian (rectangular) coordinate motion, cylindrical coordinate motion, jointed spherical coordinate motion, and spherical coordinate motion.

astronomical unit (AU). A convenient unit of distance defined as the semi-major axis of Earth's orbit around the Sun. One AU, the average distance between Earth and the Sun, is equal to approximately 149.6 10^6 kilometers (approximately 92.9 10^6 miles), or 499.01 light-seconds.

asymmetric. Lacking a mirror-image construction on both sides of a dividing line.

atmosphere. The gaseous envelope of a celestial body, such as a planet or a large moon.

atmospheric probe. A space robot (usually released by a mother spacecraft) that contains a special collection of scientific instruments for determining the pressure, composition, and temperature of a planet's atmosphere at different altitudes. An example is the probe released by NASA's *Galileo* spacecraft in December 1995. As it plunged into Jupiter's atmosphere, the robot probe successfully transmitted its scientific data to the mother spacecraft (the *Galileo* robot spacecraft) for about 58 minutes.

atom. A tiny particle of matter (the smallest part of an element) indivisible by chemical means. It is the fundamental building block of the chemical elements. The elements, such as hydrogen (H), helium (He), carbon (C), iron (Fe), lead (Pb), and uranium (U), differ from one another because they consist of different types of atoms. According to (much simplified) modern atomic theory, an atom consists of a dense inner core (the nucleus) that contains protons and neutrons and a cloud of orbiting electrons. Atoms are electrically neutral, with the number of (positively charged) protons being equal to the number of (negatively charged) electrons.

atomic clock. A precise device for measuring or standardizing time that is based on periodic vibrations of certain atoms (cesium) or molecules (ammonia). Widely used in military and civilian robot spacecraft, as, for example, the Global Positioning System (GPS).

atomic mass. The mass of a neutral atom of a particular nuclide usually expressed in atomic mass units (amu). *See also* **mass number**.

atomic mass unit (amu). One-twelfth (1/12) the mass of a neutral atom of carbon-12, the most abundant isotope of carbon.

atomic number (symbol: Z). The number of protons in the nucleus of an atom and also its positive charge. Each chemical element has its characteristic atomic number. For example, the atomic number for carbon is 6, while the atomic number for uranium is 92.

atomic weight. The mass of an atom relative to other atoms. At present, the most abundant isotope of the element carbon, namely carbon-12, is assigned an atomic weight of exactly 12. As a result, 1/12 the mass of a carbon-12 atom is called one atomic mass unit, which is approximately the mass of one proton or one neutron. Also called *relative atomic mass*.

atomize. To divide a liquid into extremely minute particles, either by impact with a jet of steam or compressed air or by passage through some mechanical device, such as an injector. A spray-painting robot usually has a mechanical device to dispense paint in a very fine mist uniformly on the surface of the work piece. Air saturated with atomized paint would be very hazardous to a human worker, but does not cause any problems for the industrial robot.

attitude. The position of an object as defined by the inclination of its axes with respect to a frame of reference. The orientation of a robotic system that is either in motion or at rest, as established by the relationship between the system's axes and a reference line or plane. Attitude is often expressed in terms of pitch, roll, and yaw. *See also* **axis.**

attitude control system. The onboard system of computers, low-thrust rockets (*thrusters*), and mechanical devices (such as a momentum wheel) used to keep a robot spacecraft stabilized during flight and to precisely point its instruments in some desired direction. Stabilization is achieved by spinning the spacecraft or by using a three-axis active approach that maintains the spacecraft in a fixed, reference attitude by firing a selected combination of thrusters when necessary.

***Aura* spacecraft.** A NASA Earth-observing spacecraft designed is to study ozone, air quality, and climate. This advanced robot spacecraft was launched into polar orbit on July 15, 2004, and is now gathering data that help scientists study the environment and climate change.

automated approach. The automated (that is, without human supervision) maneuvers of a robot spacecraft from its normal orbital position (station-keeping position) toward another orbiting spacecraft for the purpose of conducting rendezvous and docking operations.

automatic pilot. Equipment that automatically stabilizes the attitude of an aerial vehicle about its pitch, roll, and yaw axes. Also called *autopilot*; sometimes referred as "George"—as in "let George (the autopilot) fly the vehicle."

automaton (plural: automata). A self-operating machine, such as a mechanical cuckoo clock. The term usually describes a purely mechanical device, as opposed to a robotic system, which can contain electromechanical and electronic components. The eighteenth-century French engineer Jacques de Vaucanson constructed several popular automatons, including a famous mechanical duck.

autonomous robot. A robot capable of independent action; one that operates without preprogrammed behaviors and without direct supervision from human beings.

axis (plural: axes). Straight line about which a body rotates (axis of rotation) or along which its center of gravity moves (*axis of translation*). Also, one of a set of reference lines for a coordinate system, such as the *x*-axis, *y*-axis, and *z*-axis in the Cartesian coordinate system.

backtracking. Returning (usually due to depth-first search failure) to an earlier point in a search space. Also the name given to *depth-first backward reasoning*.

backup. A unit or item kept available to replace one that fails to perform satisfactorily.

backward chaining. A form of reasoning in artificial intelligence that starts with a goal and recursively chains backward to its antecedent goals or states by applying applicable operators until an appropriate earlier state is reached or the system backtracks. This process is a form of depth-first search. When the application of operators changes a single goal or state into multiple goals or states, the approach is referred to as problem reduction.

bang-bang robot. *See* **pick-and-place robot.**

batch manufacturing. A manufacturing process in which parts are produced or materials processed in discrete batches or runs, as opposed to in a continuous operation.

Between each batch or run, the equipment may be used to support other production operations involving different parts or materials.

battery. An electrochemical energy storage device that serves as a source of direct current or voltage, usually consisting of two or more electrolytic cells that are joined together and function as a single unit.

baud (rate). A unit of signaling speed. The baud rate is the number of electronic signal changes or data symbols that can be transmitted by a communications channel per second. Named after J. M. Baudot (1845–1903), a French telegraph engineer.

bel (symbol: B). A logarithmic unit (n) used to express the ratio of two power levels, P_1 and P_2. Therefore, n (bels) $= \log_{10} (P_2/P_1)$. The decibel (symbol: dB) is encountered more frequently in physics, acoustics, telecommunications, and electronics, where 10 decibels = 1 bel. This unit honors the American inventor Alexander Graham Bell (1847–1922). Compare with **neper.**

bent-pipe communications. An engineering expression (jargon) for the use of relay stations to achieve nonline of sight (LOS) transmission links.

berthing. The joining of two orbiting spacecraft using a manipulator or other mechanical device to move one into contact (or very close proximity) with the other at a selected interface. For example, NASA astronauts have used the space shuttle's remote manipulator system (long robotic arm) to carefully berth a large free-flying spacecraft (like the *Hubble Space Telescope*) onto a special support fixture located in the orbiter's payload bay during an on-orbit servicing and repair mission.

binary digit (bit). Only two possible values (or digits) in the binary number system, namely 0 or 1. Binary notation is a common telemetry (information) encoding scheme that uses binary digits to represent numbers and symbols. For example, digital computers use a sequence of bits, such as an eight-bit-long byte (binary digit eight), to create a more complex unit of information.

binary notation. A numeric system that uses only two different characters, usually 0 and 1. Because the numbers 0 and 1 can be represented easily by the "Off" and "On" conditions of an electric circuit, binary notation is widely used in digital computers.

biomimetic system. A human-made system that can mimic a natural biological system. *See also* **insect robot.**

biotechnology. Any technique that uses living organisms, or parts of organisms, to make or modify products, improve plants and animals, or to develop microorganisms for specific uses.

bit. A binary digit, the basic unit of information (either 0 or 1) in binary notation.

blackboard approach. A problem-solving approach in artificial intelligence whereby the various system elements communicate with each other by means of a common working data storage called the blackboard.

black box. A unit or subsystem (often involving an electronic device) of a robot that is considered only with respect to its input and output characteristics, without any specification of its internal elements. If engineers design robots with embedded diagnostic sensors, which support maintenance and/or repair, then repairing a disabled robot in the factory or in the field may only require replacing the failed black box with a properly functioning one. A human technician or possibly another robot can perform the black box swap.

blind search. An ordered approach in artificial intelligence that does not rely on knowledge for searching for a solution.

blocks world. A small artificial world, consisting of blocks and pyramids, used to develop ideas in robotics, computer vision, and natural language interfaces.

bottom-up control structure. A problem-solving approach in artificial intelligence that employs forward reasoning from current or initial conditions. Also called an *event-driven* or *data-driven control structure.*

breadboard. An assembly of preliminary circuits or parts used to prove the feasibility of a device, circuit, system, or principle without regard to the final configuration or packaging of the parts.

breadth-first search. An artificial intelligence approach in which, starting with the root node, the nodes in the search tree are generated and examined level by level (before proceeding deeper). This approach is guaranteed to find an optimal solution, if it exists.

burst disk. Passive physical barrier in a fluid system that blocks the flow of fluid until ruptured by (excessive) fluid pressure.

byte (binary digit eight). A basic unit of information or data consisting of eight binary digits (bits). The information storage capacity of a computer system often is defined in terms of kilobytes (kb), megabytes (Mb), and even gigabytes (Gb). One kilobyte corresponds to 2^{10}, or 1,024 bytes, while 1 megabyte corresponds to 2^{20}, or 1,048,576 bytes.

cannibalize. The process of taking functioning parts from a nonoperating robot and installing these salvaged parts in another (usually similar) robot in order to make the latter operational.

capacitor. Passive circuit element that stores electrical charge, creating a voltage differential. Capacitors can be fabricated within integrated circuits, as well as in the form of discrete components.

capture. The event when the end effector of a robot's mechanical arm makes contact with and firmly grasps the targeted object. Engineers sometimes design target objects with a special grappling fixture to facilitate the capture process. For example, orbiting spacecraft captured by the space shuttle's remote manipulator system (RMS) (large robotic arm) usually have a special external grappling fixture assist in the capture process.

carrier wave (CW). In telecommunications, an electromagnetic wave intended for modulation. This wave is transmitted at a specified frequency and amplitude. Information then is superimposed on this carrier wave by making small changes in (i.e., modulating) either its frequency or its amplitude.

Cartesian coordinate motion. A robotic system in which all robot motions travel in right angle lines (perpendicular) to each other. There are no radial motions and, therefore, the robot's work envelop has a rectangular shape. Also called *rectangular coordinate motion.*

Cartesian coordinate system. A coordinate system, developed by the French mathematician Renè Descartes (1596–1650), in which locations of points in space are expressed by reference to three mutually perpendicular planes, called coordinate planes. The three planes intersect in straight lines called the coordinate axes. The distances and the axes are usually marked (x, y, z) and the origin is the (zero) point at which the three axes intersect.

Cartesian robot. A robot that has its tooling mounted to an arm, which travels with Cartesian coordinate motion—that is, along the x-, y-, and z-axes. Unlike other types of industrial robots, the Cartesian coordinate robot does not revolve around a stationary rotary axis. This type of robot tends to have greater accuracy and repeatability than other types of industrial robots, especially for heavy loads. Also called a *rectangular coordinate robot* or *Cartesian coordinate robot.*

cathode. The negative electrode in a battery, fuel cell, electrolytic cell. *Compare with* **anode.**

center of gravity. That point in a rigid body at which all the external forces appear to act.

center of mass. The point at which the entire mass of a body (or system of bodies) appears to be concentrated. For a body (or system of bodies) in a uniform gravitational field, the center of mass coincides with the center of gravity.

central force. A force that for the purposes of computation can be assumed to be concentrated at one central point with its intensity at any other point being a function of the distance from the central point. For example, gravitation is considered as a central force in orbital mechanics.

central processing unit (CPU). The computational and control unit of a computer—the device that functions as the brain of a computer system. The CPU interprets and executes instructions and transfers information within the computer. Microprocessors, which have made possible the personal computer revolution, contain single-chip CPUs, while the CPUs in large mainframe computers and many early minicomputers contain numerous circuit boards (each packed full of integrated circuits).

centrifugal force. A reaction force that is directed opposite to a centripetal force, such that it points out along the radius of curvature away from the center of curvature.

centripetal force. The central (inward-acting) force on a body that causes it to move in a curved (circular) path. Consider a person carefully whirling a stone secured by a strong (but lightweight) string in a circular path at a constant speed. The string exerts a radial tug on the stone, which is called the centripetal force. Now as the stone keeps moving in a circle at constant speed, the stone also exerts a reaction force on the string, which is called the centrifugal force. It is equal in magnitude but opposite in direction to the centripetal force exerted by the string on the stone.

***Chandra X-ray Observatory* (CXO).** One of NASA's major robot astronomical observatories. The spacecraft was launched in July 1999 and named after the Indian-American physicist, Subrahmanyan Chandrasekar (aka: Chandra) (1910–1995). CXO studies some of the most interesting and puzzling X-ray sources in the universe.

charge coupled device (CCD). An electronic (solid state) device containing a regular array of sensor elements that are sensitive to various types of electromagnetic radiation (e.g., light) and emit electrons when exposed to such radiation. The emitted electrons are collected and the resulting charge analyzed. CCDs are used as the light-detecting component in modern television cameras, telescopes, and advanced robot vision systems.

checkout. The sequence of actions (such as functional, operational, and calibration tests) performed to determine the readiness of a robot to perform its intended tasks.

circuit board. A card or board of insulating material on which components such as a semiconductor devices, capacitors, and switches are installed.

circumferential seal. Seal composed of a continuous ring or of one or more segmented rings whose sealing surface is parallel to the centerline of the flow passage. Also called *radial seal.*

clean room. A controlled work environment in which dust, temperature, and humidity are carefully controlled during the fabrication, assembly, and/or testing of critical components. Engineers often assemble space robots and nanorobots in clean rooms. Specially designed, clean room certified robots assist in the automated manufacture of microelectronic components.

clevis. A fitting with a U-shaped end for attachment to the end of a pipe or rod.

clock. An electronic circuit, often an integrated circuit, which produces high-frequency timing signals. A common application is synchronization of the operations performed by a computer or microprocessor-based system. Typical clock speeds (rates) in microprocessor circuits are in the megahertz range with some clocks reaching almost the gigahertz range, where one megahertz (MHz) equals 10^6 cycles per second and one gigahertz (GHz) equals 10^9 cycles per second.

closed loop. Term applied to an electrical or mechanical system in which the output is compared with the input (command) signal, and any discrepancy between the two results in corrective action by the system elements.

closed system. In thermodynamics, a system for which only energy (but not matter) can cross the boundaries.

closing rate. The speed at which two objects approach each other.

cognition. An intellectual process by which knowledge is gained about perceptions or ideas.

color. A quality of light that depends on its wavelength. The spectral color of emitted light corresponds to its place in the spectrum of the rainbow. Visual light, or perceived color, is the quality of light emission as recognized by the human eye. Simply stated, the human eye contains three basic types of light-sensitive cells that respond in various combinations to incoming spectral colors. For example, the color brown occurs when the eye responds to a particular combination of blue, yellow, and red light. Violet light has the shortest wavelength, while red light has the longest wavelength. All the other colors have wavelengths that lie in between. In order to imitate how the human eye responds to color, the robot vision system in an advanced crop-harvesting robot should be to differentiate at some distance between a ripe red tomato ready for picking and a green one best left growing on the vine.

combinatorial explosion. The rapid growth of possibilities as the search space expands. If each branch point (decision point) has an average of n branches, the search space tends to expand as n^d, as the depth of search (d) increases.

command. A signal that initiates or triggers an action in the device that receives the signal. In the operation of robot systems, also called *an instruction.*

common sense. The ability of a human being to act appropriately in everyday situations based on the person's lifetime accumulation of experiential knowledge.

common sense reasoning. Low level reasoning based on a wealth of experience.

communication. The successful transmission of information through a common system of symbols, signs, behavior, speech, writing, or signals.

compact disk (CD). An optical storage medium used to store computer data, digitized images, music, and other types of information.

compile. The act of translating a computer program written in a high-level language into the machine language that controls the basic operations of the computer.

composite materials. Structural materials of metals, ceramics, or plastics with built-in strengthening agents that may be in the form of filaments, foils, powders, or flakes of a different compatible material.

***Compton Gamma Ray Observatory* (CGRO).** A major NASA robot astrophysical observatory dedicated to gamma ray astronomy. The CGRO was placed in orbit around Earth in April 1991. At the end of its useful scientific mission, flight controllers intentionally commanded the massive (16,300 kilograms) spacecraft to perform a deorbit burn. This caused it to reenter and safely crash in June 2000 in a remote region of the Pacific Ocean. The spacecraft was named in honor of the American physicist Arthur Holly Compton (1892–1962).

computer architecture. The manner in which various computational elements are interconnected to achieve a computational function.

computer graphics. Visual representations generated by a computer—usually observed on a monitoring screen.

computerized robot. A servo-controlled robot run by a computer. This type of robot is also called a smart robot because the controller for such machine devices can learn new instructions through electronic signal transmissions. The electronic signal teaching process is far simpler than the traditional industrial robot teaching method, during which a human being leads the robot's gripper-arm through whatever series of motions, the robot is expected to perform.

computer network. An interconnected set of communicating computers.

computer vision. Perception by a computer (often within a robot) in which a symbolic description of a scene depicted in an image derived from visual sensory input is developed. Computer vision is frequently a knowledge-based, expectation-guided process that uses models to interpret sensory data. Also called *machine vision.*

conceptual dependency. An approach to natural language understanding in which sentences are translated into basic concepts that are expressed as a small set of semantic primitives.

conduction (thermal). The transport of heat (thermal energy) through an object by means of a temperature difference from a region of higher temperature to a region of lower temperature. For solids and liquid metals, thermal conduction is accomplished by the migration of fast-moving electrons, while atomic and molecular collisions support thermal conduction in gases and other liquids. *Compare with* **convection.**

conflict resolution. Selecting a procedure from a conflict set of applicable competing procedures or rules.

conflict set. The set of rules that matches some data or pattern in a global data base.

conjunct. One of several subproblems; each of the component formulas in a logical conjunction.

conjunction. A problem composed of several subproblems. A logical formula built by connecting other formulas using logical ANDs.

connectives. In artificial intelligence, operators (such as AND, OR, etc.) connecting statements in logic so that the truth-value of the composite is determined by the truth-value of the components.

consequent. The right-hand side of a production rule.

conservation of angular momentum. The principle of physics that states that absolute angular momentum is a property which cannot be created or destroyed but can only be transferred from one physical system to another through the action of a net torque on the system. As a consequence, the total angular momentum of an isolated system remains constant.

conservation of energy. The principle of physics that states that the total energy of an isolated system remains constant if no interconversion of mass and energy takes place within the system. Also called the *first law of thermodynamics.*

conservation of mass and energy. From special relativity and Albert Einstein's famous mass-energy equivalence formula ($E = \Delta mc^2$), this conservation principle states that for an isolated system, the sum of the mass and energy remains constant, although interconversion of mass and energy can occur within the system.

conservation of momentum. The principle of physics that states that in the absence of external forces, absolute momentum is a property that cannot be created or destroyed. Consequently, the total momentum of an isolated system remains constant. *See also* **Newton's laws of motion.**

console. A desk-like array of controls, indicators, and video display devices for the monitoring and controlling of space robot operations or the flight of a UAV in Earth's atmosphere. During the critical phases of a space robot's mission or a UAV's atmospheric flight, the console becomes the central place from which to issue commands to or at which to display information concerning the space robot or UAV. For complex space robot projects or long-duration UAV reconnaissance operations, the mission control center will often contain a cluster of consoles—each assigned to specific monitoring and control tasks.

constellation (aerospace). A term used to collectively describe the number and orbital disposition of a set of satellites, such as the constellation of Global Positioning Satellites.

constraint propagation. A method in artificial intelligence for limiting search by requiring that certain constraints be satisfied. Constraint propagation may also be viewed as a mechanism for moving information between subprograms.

context. The set of circumstances or facts that define a particular situation, event, etc. The portion of the situation, which remains the same, when an operator is applied in a problem-solving situation.

continuous path robot. One of two basic types of servo-controlled robots. To teach a continuous path robot its path, a human being physically moves the robot's manipulator arm through whatever series of motions it is expected to perform. These learned or rehearsed motions are then stored in the robot's computer for future recall.

controller. 1. A device that converts an input signal from a controlled variable (such as temperature, pressure, fluid level, or fluid flow rate) into a valve actuator (pneumatic, hydraulic, electrical, or mechanical) input signal to vary the valve position so as to provide the required correction of the controlled variable. 2. A robot's computer "brain." In the case of an industrial robot, the controller stores data and directs the movement of the robot's manipulator. Robot controllers can be relatively simple devices or quite complex; a typical controller usually permits storage and execution of more than one program.

control structure. In artificial intelligence, the strategy for manipulating the domain knowledge to arrive at a problem solution. Also called *reasoning strategy*.

convection. A fundamental form of heat transfer characterized by mass motions within a fluid resulting in the transport and mixing of the properties of that fluid. The up-and-down drafts in a fluid heated from below in a gravitational environment. Because the density of the heated fluid is lowered, the warmer fluid rises (natural convection); after cooling, the density of the fluid increases, and it tends to sink. *Compare with* **conduction** and **radiation**.

cooperative target. A three-axis stabilized orbiting object that has signaling devices to support automated rendezvous and docking/capture operations by a (robot) chaser spacecraft.

coorbital. Sharing the same or very similar orbit.

***Copernicus* (spacecraft).** An astronomical observatory launched by NASA on August 21, 1972. This robot spacecraft was the third in the Orbiting Astronomical Observatory (OAO) program and the second successful spacecraft to observe the celestial sphere from above Earth's atmosphere. An ultraviolet (UV) telescope with a spectrometer measured high-resolution spectra of stars, galaxies, and planets with the main emphasis being placed on the determination of interstellar absorption lines. Named in honor of the famous Polish astronomer Nicholas Copernicus (1473–1543), whose advocacy of heliocentric cosmology stimulated the Scientific Revolution.

computerized robot. A servo robot that is run by computer. This kind of industrial robot is programmed by instructions fed into the controller electronically. Some versions of these smart robots even have the ability to improve upon their basic work instructions. *See also* **industrial robot**.

cosmonaut. The name used by Russia (formerly the Soviet Union) for its human space travelers; comparable to the American term **astronaut**.

Cosmos spacecraft. The general name given to a large number of Soviet and later Russian robot spacecraft, ranging from military satellites to scientific platforms investigating near-Earth space. *Cosmos 1* was launched in March 1962; since then well over 2,000 Cosmos spacecraft have been sent into outer space. Also called *Kosmos*.

coulomb (symbol: C). The SI unit of electric charge. The quantity of electric charge transported in one second by a current of one ampere. Named after the French physicist Charles de Coulomb (1736–1806).

creep. The slow (but continuous) permanent deformation of material caused by a constant tensile or compressive load that is less than the load necessary for the material to give way under pressure (i.e., to yield); some time is required to induce creep, and the process is accelerated at elevated temperatures.

cruise missile. A guided missile traveling within the atmosphere at aircraft speeds and, usually, low altitude whose trajectory is either preprogrammed or updated while enroute to the target. The modern cruise missile is a smart, flying robotic weapon capable

Figure 8-2 This is an artist's concept of a cryobot—a conceptual robot probe that NASA will use to penetrate the icy surface of a planet (Mars polar region) or moon (Europa). The cryobot moves through ice by melting the surface directly in front of it, while allowing liquid to flow around the torpedo-shaped robot probe and refreeze behind it. (Credit: Courtesy of NASA/JPL.)

of achieving high accuracy in striking a distant target. It is maneuverable during flight, is constantly propelled, and, therefore, does not follow a ballistic trajectory. Modern cruise missiles may be armed with nuclear weapons or with conventional warheads (i.e., high explosives).

cruise phase. For a robot spacecraft on an interplanetary scientific mission, the part of the mission (usually months or even years in duration) following launch and prior to planetary encounter.

cryobot. A planned NASA robot probe for penetrating into the icy surface of a planet or moon. The cryobot moves through ice by melting the surface directly in front of it, while allowing liquid to flow around the torpedo-shaped robot probe and refreeze behind it. As it makes its mole-like passage into an alien world, the cryobot's instruments take measurements of the encountered environment and send collected data back to the surface lander. On Mars, it appears that a communications cable could be used for penetration of shallower depths. On Europa, the thicker ice would require use of a network of miniradio wave transceiver relays embedded in the ice. The use of semiautonomous steering and levels of artificial intelligence that promote fault management will help reduce the risk of the robot probe getting trapped by subsurface obstructions, such as large rocks.

current (symbol: I). The flow of electric charge through a conductor. The ampere (symbol: A) is the SI unit of electric current.

cycle. Any repetitive series of operations or events; the complete sequence of values of a periodic quantity that occur during a period—for example, one complete wave. The period is the duration of one cycle, while the frequency is the rate of repetition of a cycle. The hertz (Hz) is the SI unit of frequency; and one hertz equals one cycle per second.

cycle life. The number of times a component or unit may be operated (for example, opened and closed) and still perform within acceptable limits. Engineers try to design robots such that all mechanical or electromechanical components have compatible cycle lives during routine operation. Parts, components, or subsystems that have a tendency to wear out more quickly are usually identified for inspection and (if necessary) replacement during routine maintenance. Engineered with this design approach, a robot system will function properly over its anticipated operating lifetime.

cycle time. The period of time from starting one machine operation to starting another (in a pattern of continuous repetition).

cylindrical coordinate robot. A robot that has a horizontal shaft that goes in and out , and rides up and down on a vertical shaft, which (shaft) also rotates about the base.

cylindrical coordinates. A system of curvilinear coordinates in which the position of a point in space is determined by (a) its perpendicular distance from a given line; (b) its distance from a selected reference plane perpendicular to this line; and (c) its angular distance from a selected reference line when projected onto this plane. The coordinates thus form the elements of a cylinder and by convention are written r, θ, and z, where r is the radial distance from the cylinder's axis z and θ is the angular position from a reference line in a cylindrical cross section normal to z. Also called *polar coordinates*. The relationships between the cylindrical coordinates and the rectangular Cartesian coordinates (x, y, z) are: $x = r \cos \theta$; $y = r \sin \theta$; and $z = z$.

damping. The suppression of oscillations usually because energy is being expended by the oscillating system to overcome friction or other types of resistive forces.

database. A collection of interrelated or independent data items stored together to serve one or more applications.

database management system. A computer system for the storage and retrieval of information about some domain.

data-driven. A forward reasoning, bottom-up problem solving approach.

data fusion. The technique in which multivariate data from multiple sources are retrieved and processed as a single, unified entity. A significant set of a priori databases is crucial to the effective functioning of the data fusion process. For example, this technique might be used in supporting the development of an evolving world model for a robotic lunar rover, which is teleoperated from a control station on Earth. Inputs from the robot vehicle's sensors would be blended, or fused, with existing lunar environment databases to support a more intelligent exploration strategy and more efficient field operations.

data handling subsystem. The onboard computer responsible for the overall management of a spacecraft's activity is usually the same computer that maintains timing; interprets commands from Earth; collects, processes, and formats the telemetry data that is to be returned to Earth; and manages high-level fault protection and saving routines. This spacecraft computer often is referred to as the command and data handling subsystem.

data link. 1. The means of connecting one location to another for the purpose of transmitting and receiving data. 2. A communications link suitable for the transmission of data. 3. Any communications channel used to transmit data from a sensor to a computer, a readout device, or a storage device.

data processing (DP). A general term describing the systematic application of procedures—electrical, mechanical, optical, computational, and so on—whereby data are changed from one form to another. Sometimes used to describe the overall work performed by computers. Also called automatic data processing (ADP) and electronic data processing (EDP).

data reduction. The transformation of raw or observed data into more compact, ordered, or useful information.

data smoothing. The mathematical process of fitting a smooth curve to a dispersed set of data points.

data structure. The form in which data are stored in a computer.

datum. 1. A single unit of information; the singular of data. 2. Any numerical or geometrical quantity or set of such quantities that may serve as reference or base for other quantities. Where the concept is geometric, the preferred plural form is datums.

deadband. In general, an intentional feature in the guidance and control system of a mobile robot or a UAV that prevents a path error from being corrected until that error exceeds a specified magnitude.

dead spot. In a robot's control system, a region about the neutral control position where small movements of the actuator do not produce any response in the system.

debug (debugging). To isolate and remove malfunctions from a hardware system, subsystem, or component; to correct mistakes in computer software.

decade. 1. A group or series of ten; for example, a period of 10 years. 2. The interval between any two quantities having the ratio 10:1.

deceleration. The act or process of moving, or causing to move, with decreasing speed; negative acceleration.

decrement. The decrease in the value of a variable.

dedicated. Serving a single function; for example, a dedicated battery is a power source serving a single load, such as a special ordnance circuit.

deduction. A process of reasoning in which the conclusion follows from the premises given.

deep space. By an arbitrary but widely used definition in the American space program, the region of outer space at altitudes greater than 5,600 kilometers above Earth's surface.

Deep Space Network (DSN). NASA's global network of antennas that serve as the radio wave communications link to distant interplanetary spacecraft and probes, transmitting instructions to them and receiving data from them. Large radio antennas of the DSN's three Deep Space Communications Complexes (DSCCs) are located in Goldstone, California; near Madrid, Spain; and near Canberra, Australia—providing almost continuous contact with a robot spacecraft in deep space as Earth rotates on its axis.

deep space probe. A robot spacecraft designed for exploring deep space, especially to the vicinity of the Moon and beyond. This includes lunar probes, Mars probes, outer planet probes, solar probes, and so on.

default value. A value to be used when the actual value is unknown.

Defense Support Program (DSP). The family of missile surveillance satellites operated by the U.S. Air Force since the early 1970s. Placed in geostationary orbit around Earth, these military surveillance satellites serve as robot sentinels and can detect missile launches, space launches, and nuclear detonations occurring around the world.

degree (usual symbol: °). A term that has commonly been used to express units of certain physical quantities, such as angles and temperatures. The ancient Babylonians are believed to be the first people to have subdivided the circle into 360 parts, or degrees, thereby establishing the use of the degree in mathematics as a unit of angular measurement.

degrees of freedom (DOF). A mode of motion, either angular or linear, motion with respect to a coordinate system, independent of any other mode. A body in motion has six possible degrees of freedom, three linear (sometimes called *x*-, *y*-, and *z*-motion with reference to linear [axial] movements in the Cartesian coordinate system) and three angular (sometimes called: pitch, yaw, and roll with reference to angular movements). For example, each joint in a serial robot represents a degree of freedom.

delta-V (symbol: ΔV) Velocity change; a numerical index of the maneuverability of a robot spacecraft. This term often represents the maximum change in velocity that a robot spacecraft's propulsion system can provide. Typically described in terms of kilometers per second (km/sec) or meters per second (m/sec).

demodulation. The process of recovering the modulating wave from a modulated carrier.

density (usual symbol: ρ). The mass of a substance per unit volume at a specified temperature.

deposition. The operation by which a (thin) film is placed on the surface of an object.

depth-first search. In artificial intelligence, a search that proceeds from the root node to one of the successor nodes and then to one of that node's successor nodes, etc., until a solution is reached or the search is forced to backtrack.

design. An iterative decision-making process that produces plans by which resources are converted into products or systems to meet human needs or wants or to solve problems.

design principle. Design rules regarding rhythm, balance, proportion, variety, emphasis, and harmony, used to evaluate existing designs and guide the design process.

detent. A releasable element used to restrain a part before or after its motion. For technician safety during maintenance or repair operations, detents can be used to temporarily secure potentially hazardous moving mechanisms on a robot (such as a large mechanical arm).

dexterity. Of a human, skill in using the hands or body; of a robot, flexibility in the use of a manipulator arm or its end effector.

diaphragm. A thin membrane that can be used as a seal to prevent fluid leakage or as an actuator to transform an applied pressure into a linear force.

difference reduction. An approach to problem solving in artificial intelligence that tries to solve a problem by iteratively applying operators, which will reduce the difference between the current state and the goal state. Also called *"means-ends" analysis.*

diffuser. A specially designed duct in a pneumatic system, sometimes equipped with stationary guide vanes, that decreases the velocity of a gaseous working fluid (such as air or nitrogen) and also increases the fluid's pressure.

digit. A single character or symbol in a number system. For example, the binary system has two digits, 0 and 1; while the decimal system has ten digits: 0, 1, 2, 3, 4, 5, 6, 7, 8, and 9.

digital computer. The most common type of computer in use today; one that processes data that have been converted into binary notation. *Compare with* **analog computer**.

digital image processing. Computer processing of the digital number (DN) values assigned to each pixel in an image. For example, all pixels in a particular image with a digital number value within a certain range might be assigned a special color or might be changed in value some arbitrary amount to ease the process of image interpretation by a human analyst. Furthermore, two images of the same scene taken at different times or at different wavelengths might have the digital number values of corresponding pixels computer manipulated (e.g., subtracted) to bring out some special features.

This is a digital image processing technique called differencing or change detection. *See also* **machine vision; remote sensing.**

digital transmission. A technique in telecommunications that sends the signal in the form of one of a discrete number of codes (for example, in binary code as either 0 or 1). The information content of the signal is concerned with discrete states of this signal, such as the presence or absence of voltage or a contact in a closed or open position.

digitize. To express an analog measurement in discrete units, such as a series of binary digits. For example, an image or photograph can be scanned and digitized by converting lines and shading (or color) into combinations of appropriate digital values for each pixel in the image or photograph.

dipole antenna. A half-wave (dipole) antenna typically consists of two straight, conducting metal rods each one-quarter of a wavelength long that are connected to an alternating voltage source. The electric field lines associated with this antenna configuration resemble those of an electric dipole. (An electric dipole consists of a pair of opposite electric charges [+q and −q] separated by a distance [*d*].) The dipole antenna is commonly used to transmit (or receive) radio-frequency signals below 30 megahertz (MHz).

directional antenna. An antenna that radiates or receives radio frequency (RF) signals more efficiently in some directions than in others. A collection of antennas arranged and selectively pointed for this purpose is called a directional antenna array.

direct readout. The information technology capability that allows ground stations on Earth to collect and interpret the data messages (telemetry) being transmitted from satellites.

diplexer. A device that permits an antenna to be used simultaneously or separately by two transmitters.

disconnect. Short for quick-disconnect—a separable connector characterized by two separable halves, an interface seal, and, usually, a latch-release locking mechanism; it can be separated without the use of tools in a very short time.

discrete. Composed of distinct elements.

distortion. 1. In general, the failure of a system (typically optical or electronic) to transmit or reproduce the characteristics of an input signal with exactness in the output signal. 2. An undesired changed in the dimensions or shape of a structure; for example, the distortion of a robot's manipulator arm after experiencing a large temperature gradient.

docking. The act of physically joining two orbiting robot spacecraft. Usually accomplished by independently maneuvering one spacecraft (the chaser spacecraft) into contact with the other (the target spacecraft) at a chosen physical interface. Once contact is made, a variety of automated mechanical devices, such as latches, grippers, and even small robotic arms, can be activated to adjust and secure the physical connection between the two spacecraft.

docking interface. The area of contact between two docking mechanisms.

docking mechanism. A mechanism that performs appropriate physical (mechanical) functions to connect one robot spacecraft to another during an automated docking operation.

domain. The problem area of interest.

downlink. The telemetry signal received at a ground station from a robot spacecraft or space probe.

downtime. A period during which a robot is not operating; this outage can be due to planned maintenance or unplanned component or subsystem failure.

drone. An unpiloted air vehicle that is operated by remote control. Drones often are used as test targets for missile and fighter aircraft weapon systems. *See also* **unmanned aerial vehicle (UAV).**

duct. A tube or passage that confines and conducts the flow of a fluid.

dud. A munition that has not been armed as intended or that has failed to explode after being armed. Explosive ordnance disposal (EOD) robots are often used to find, inspect, and remove or render safe in place duds found on a weapons test range or in a combat zone.

duplexer. A device that permits a single antenna system to be used for both transmitting and receiving. (Should not be confused with diplexer—a device that permits an antenna to be used simultaneously or separately by two transmitters.)

dust detector. A direct-sensing science instrument that measures the velocity, mass (typical range 10^{-16} g to 10^{-6} g), flight direction, charge (if any), and number of dust particles striking the instrument, carried by some robot spacecraft on interplanetary missions, especially missions that encounter comets.

dynamics. The branch of mechanics that studies the motion of rigid bodies under the influence of external forces.

dyne (symbol: d). A unit of force in the centimeter-gram-second (cgs) system; equal to the force required to accelerate a one gram mass one centimeter per second per second; that is, 1 dyne = 1 gm-cm/sec^2. Compare **newton.**

early warning satellite. A military spacecraft whose primary mission is the detection and notification of the launch of an enemy ballistic missile attack. This type of *surveillance satellite* is essentially an around-the-clock robot sentinel that uses its sensitive infrared (IR) radiation sensors to detect the heat released when a missile is launched. *See also* **Defense Support Program (DSP).**

Earth-observing spacecraft. A sophisticated robot spacecraft in orbit around Earth that has a specialized collection of sensors capable of monitoring important environmental variables. Data from such satellites help support Earth system science. Also called an *environmental satellite* or a *green satellite.*

editor. A software tool to aid in modifying a software program.

elasticity. The ability of a body that has been deformed by an applied force to return to its original shape when the force is removed.

electric circuit. The complete path of an electric current including usually the source of the electric energy.

electric current. A flow of electric charge. *See also* **ampere.**

electricity. Flow of energy due to the motion of electric charges; any physical effect that results from the existence of moving or stationary electric charges (e.g., static electricity, lightning).

electric potential (symbol: V). The work done in moving a unit of positive charge from infinity to the point in an electric field whose potential is being specified. The unit of electric potential is the volt. For example, if 1 joule is required to transfer a charge of 1 coulomb, then the electric potential is 1 volt.

electric robot. A robot that uses electrical energy (electricity) to actuate its manipulator arm, including the wrist and end effector. *See also* **industrial robot.**

electrode. A conductor (terminal) at which electricity passes from one medium into another. The positive electrode is called the anode; the negative electrode is called the cathode. In semiconductor devices, an element that performs one or more of the functions of emitting or collecting electrons or holes, or of controlling their movements by an electric field. In electron tubes, a conducting element that performs one or more of the functions of emitting, collecting, or controlling the movements of electrons or ions usually by means of an electromagnetic field.

electroexplosive device (EED). A pyrotechnic device in which electrically insulated terminals are in contact with, or adjacent to, a material mixture that reacts chemically (often explosively) when the required electrical energy level is discharged through the terminals. An EOD robot will often attach an EED to a suspicious package, withdraw to a safe distance, and then have its human operator send the proper signal to the EED. This action is taken to neutralize or destroy a suspected bomb (or dud) in place, when it appears that the device is too unstable to be moved or to be rendered safe by a human bomb disposal expert.

electromagnetic. Having both electric and magnetic properties; pertaining to magnetism produced or associated with electricity.

electromagnetic (EM) communications. The technology involving the development and production of a variety of telecommunication equipment used for electromagnetic transmission of information over any media. The information may be analog or digital, ranging in bandwidth from a single voice or data channel to video or multiplexed channels occupying hundreds of megahertz. When used to interact with UAVs and robot spacecraft, includes communications equipment and laser communications techniques capable of automatically acquiring and tracking signals and maintaining communications through (as appropriate) atmospheric media and interplanetary space.

electromagnetic radiation (EMR). Radiation made up of oscillating electric and magnetic fields and propagated with the speed of light. Includes (in order of decreasing frequency) gamma radiation, X-rays, ultraviolet, visible, and infrared (IR) radiation, and radar and radio waves. EM radiation travels at the speed of light (about 300,000 kilometers per second and is the basic mechanism for energy transfer through the vacuum of outer space.

electromotive force (emf). The characteristic of an electrical energy source that enables a current to flow in a circuit. It is the sum (algebraic) of the potential differences acting in an electric circuit. The emf (typical unit: volts) is measured by the energy liberated when a unit electric charge passes completely around the circuit.

electron (symbol: e). A stable elementary particle with a unit negative electrical charge (1.602×10^{-19} coulomb) and a rest mass (m_e) of approximately 1/1837 that of a proton (namely, 9.109×10^{-31} kilogram). Electrons surround the positively charged nucleus and determine the chemical properties of the atom. Positively charged electrons, or positrons, also exist. Electrons were first discovered in the late 1890s by the British scientist Sir Joseph John Thomson (1856–1940).

electronic. Of or pertaining to any of the large number and wide variety of devices that involve the generation, transmission, use, or control of electricity.

electronics. The branches of physics and engineering that deal with the understanding, design, and application of devices based on the conduction of electricity through a vacuum, gas, or semiconductor. Although this term originated with vacuum (electron) tube applications, modern electronics is concerned primarily with semiconductor devices and solid-state physics. Microelectronics involves extremely small electronic components.

electron volt (eV). A unit of energy equivalent to the energy gained by an electron when it experiences a potential difference of one volt. Larger multiple units of the electron volt are encountered frequently—as, for example: keV for thousand (or kilo-) electron volts (10^3 eV); MeV for million (or mega-) electron volts (10^6 eV); and GeV for billion (or giga-) electron volts (10^9 eV). One electron volt is equal to 1.602×10^{-19} joule.

embed. To write a computer language on top of (embedded in) another computer language.

encounter. The close flyby or rendezvous of a spacecraft with a target body. The target of an encounter can be a natural celestial body (such as a planet, asteroid, or comet) or a human-made object (such as another spacecraft).

end effector. A robot's end effector (hand or gripping device) generally is attached to the end of the manipulator arm. Typical functions of the end effector include grasping, pushing and pulling, twisting, using tools, performing insertions, and various types of assembly activities. End effectors can be mechanical, vacuum, or magnetically operated, can use a snare device or have some other unusual design feature. The final design of the end effector is determined by the shapes of the objects that the robot must grasp. Usually most end effectors are some type of gripping or clamping device.

end-point robot. *See* **pick-and-place robot.**

energy (symbol: *E*). The capacity to do work. Energy appears in many different forms such as mechanical, thermal, electrical, chemical, and nuclear. According to the first law of thermodynamics, energy can neither be created nor destroyed, but simply changes form (including mass–energy transformations).

engineer. A person who is trained in and uses technological and scientific knowledge to solve practical problems and to apply physical principles for specific objectives, such as transportation, power generation, lighting, or manufacturing.

engineering. The profession of or work performed by an engineer. Engineering involves the knowledge of the mathematical and natural sciences (biological and physical) gained by study, experience, and practice that are applied with judgment and creativity to develop ways to utilize the materials and natural principles for the benefit of humankind. For example, a mechanical engineer uses thermodynamics to design efficient heat engines for transportation or the generation of electric power.

engineering design. The systematic and creative application of scientific and mathematical principles to practical ends such as design, manufacture, and operation of efficient and economical structures, machines, processes, and systems.

entertainment robot. A robot system constructed not to perform work, but simply to provide amusement and recreation to its human owner or to (fee-paying) human audiences, who enjoy observing its interactions and performance. Entertainment robots trace their heritage back to the puppets and movable statues of antiquity and the elegant mechanical automatons fashioned in Western Europe during the seventeenth and the eighteenth centuries. In twentieth century, modern theme parks used a variety of robotic animals (such as dinosaurs), cinematic robots, and humanoid robots (typically moving and speaking historic figures, such as American presidents) to amuse guests. With continued improvement in microelectronics, a variety of relatively sophisticated robot toys appeared in the marketplace since the 1970s. These robot toys include an amusing collection of robot pets, such as Sony's robot dog called *Aibo*.

entropy (symbol: *S*). A measure of the extent to which the energy of a system is unavailable; as entropy increases, energy becomes less available to perform useful work.

escape velocity (common symbol: V_e). The minimum velocity that an object must acquire to overcome the gravitational attraction of a celestial body. The escape velocity for an object launched from the surface of Earth is approximately 11.2 kilometers per second (km/s), while the escape velocity from the surface of Mars is 5.0 kilometers per second.

Europa. The smooth, ice-covered moon of Jupiter, discovered by Galileo Galilei in 1610 and currently thought to have a liquid water ocean beneath its frozen surface.

European Space Agency (ESA). An international organization that promotes the peaceful use of outer space and cooperation among the European member states in space research and applications.

evaluation function. A function (usually heuristic) used to evaluate the merit of the various paths emanating from a node in a search tree.

event-driven. In artificial intelligence, a forward-chaining problem-solving approach based on the current problem status.

expectation-driven. In artificial intelligence, processing approaches that proceed by trying to confirm models, situations, states, or concepts anticipated by the system.

expert system. A computer program that uses knowledge and reasoning techniques to solve problems normally requiring the abilities of human experts.

Figure 8-3 A U.S. Army explosive ordnance disposal (EOD) robot, called i-Robot, pulls the wire of an alleged improvised explosive device (IED) found by Iraqi police on November 3, 2004. The EOD robot (shown here) was operated by a team of American soldiers, who examined the suspicious device from a safe distance, using the robot's two onboard cameras and mechanical arm. In this instance, the suspicious device proved to be a decoy, set in place by terrorists to harass Iraqi policemen and the U.S. soldiers assisting them. (Credit: U.S. Navy New Photo.)

***Explorer 1*.** The first American satellite to successfully orbit around Earth. Launched from Cape Canaveral on January 31, 1958, by a Juno I four-stage rocket vehicle, this satellite involved a quickly assembled team from the U.S. Army (under the direction of Wernher von Braun) and Caltech's Jet Propulsion Laboratory (JPL). Dr. James van Allen (State University of Iowa) provided the satellite's instruments that discovered a portion of Earth's trapped radiation belts, which were subsequently named after him.

Explorer spacecraft. The large family of NASA scientific spacecraft, starting in 1958, that have investigated astronomical and astrophysical phenomena, the properties and structure of Earth's magnetosphere and atmosphere, and our planet's precise shape and geophysical surface features.

explosive ordnance disposal (EOD) robot. A mobile (field) military robot that is teleoperated by a team of soldiers (or law enforcement officers), who can inspect suspicious packages or known unexploded bombs from a safe distance, using the robot's onboard camera system, mechanical arms, and other threat evaluation sensors. Some EOD robots are also designed to carry a small high explosive charge (like C-4) and place this charge near the unexploded ordnance. When the charge has been set and the robot has withdrawn to a safe distance, the human controller detonates the intentionally placed charge, thereby destroying the unexploded bomb.

extraterrestrial contamination. The contamination of one world by life-forms, especially microorganisms, from another world. Taking Earth's biosphere as the reference, planetary contamination is called forward contamination, when an alien world is contaminated by contact with terrestrial organisms, and back contamination, when alien organisms are released into Earth's biosphere.

Figure 8-4 (EVA robot) Looking like he is playing with a high tech soccer ball, astronaut Winston Scott reaches out and retrieves the free-flying Autonomous EVA Robotic Camera (AERCam), during Winston's space walk in the payload bay of the space shuttle *Columbia*. These interesting astronaut-robot spacecraft interactions took place in low Earth orbit, during the STS-87 shuttle mission in December 1997. (Credit: Courtesy of NASA/Johnson Space Center.)

extravehicular activity (EVA) robot. A space-qualified robot system designed to assist astronauts and cosmonauts as they work *outside* the pressurized volume of a large, orbiting space system, like the *International Space Station* (ISS). NASA is now investigating the development of a spherical, soccer-ball sized, autonomous space robot and an android-like robot (called robonaut) to inspect the outside portions of a space vehicle, while space-suited astronauts work alongside. In time, future EVA robots will be operated by astronauts who remain in the shirtsleeve comfort of the space station's pressurized volume. As the EVA robots become more capable and reliable, these systems could perform routine inspections on, maintenance of, and (when necessary) make local minor repairs to, the outside of a large orbiting facility like the space station with little or no supervision from the astronaut crew.

farad (symbol: F). The SI unit of electrical capacitance. It is defined as the capacitance of a capacitor whose plates have a potential difference of one volt when charged by a quantity of electricity equal to one coulomb. This unit is named after the nineteenth-century British scientist Michael Faraday (1791–1867), who was a pioneer in the field of electromagnetism. Since the farad is too large a unit for typical applications, submultiples, such as the microfarad (10^{-6} F), the nanofarad (10^{-9}F), and the picofarad (10^{-12}F), are encountered frequently.

fatigue. In engineering, a weakening or deterioration of metal or other material occurring under load, especially under repeated cyclic or continued loading. Self-explanatory compounds of this term include fatigue crack, fatigue failure, fatigue load, fatigue resistance, and fatigue test.

fault diagnosis. Determining the source of trouble in an electromechanical system.

fault tolerance. The capability of a robot system to function despite one or more critical failures; usually achieved by the use of redundant circuits or functions and/or reconfigurable components.

feedback. The return of a portion of the output of a device to the input. Positive feedback adds to the input; negative feedback subtracts from the input.

femto- (symbol: f). The SI prefix for 10^{-15}. This prefix is used to designate very small quantities, such as a femtosecond (fs), which corresponds to 10^{-15} second—a very brief flash of time.

field of view (FOV). The area or solid angle than can be viewed through or scanned by a remote-sensing (optical) instrument.

field robot. A mobile robot that operates in unpredictable, unstructured environments, typically outdoors (on Earth) and often operates autonomously or by teleoperation over a large workspace—generally a square kilometer or more. For example, in surveying a potentially dangerous site, a human operator will stay at a safe distance away in a protected work environment and control (by cable or radio frequency link) the field robot, which then actually operates in the hazardous environment. These terrestrial field robots are technical first cousins to the more sophisticated, teleoperated robot rovers that have roamed on the surface of the Moon and Mars and will continue to do so in future missions throughout this century. *See also* **explosive ordnance disposal (EOD) robot; military robot; space robot.**

fifth generation computer. A non-von Neumann, intelligent, parallel processing form of computer.

first order predicate logic. A popular form of logic used by the artificial intelligence community to represent knowledge and to perform logical inference. By using first order predicate logic, computer scientists can make assertions about variables in a proposition.

fluid mechanics. The major branch of science that deals with the behavior of fluids (both gases and liquids) at rest (fluid statics) and in motion (fluid dynamics). This scientific field has many subbranches and important applications, including aerodynamics (the motion of gases, including air), hydrostatics (liquids at rest), and hydrodynamics (the motion of liquids, including water).

flyby. An interplanetary or deep space mission in which the flyby robot spacecraft passes close to its target celestial body (e.g., a distant planet, moon, asteroid, or comet), but does not impact the target or go into orbit around it. *See also* **space robot.**

flywheel. A massive, rotating wheel that can store energy as kinetic (or motion) energy as its rate of rotation is increased; energy then can be removed from this system by decreasing the rate of rotation of the wheel.

force (symbol F). The cause of the acceleration of material objects as measured by the rate of change of momentum produced on a free body. Force is a vector quantity, mathematically expressed by Newton's second law of motion: Force = mass × acceleration.

forward chaining. Event-driven or data-driven reasoning.

frame. A data structure for representing stereotyped objects or situations. A frame has slots to be filled for objects and relations appropriate to the situation.

frequency (common symbol: *f* or *ν*). In general, the rate of repetition of a recurring or regular event; for example, the number of vibrations of a system per second or the number of cycles of a wave per second. For electromagnetic radiation, the frequency (ν) of a quantum packet of energy (i.e., a photon) is given by: $\nu = E/h$, where E is the photon energy and h is the Planck constant. The SI unit of frequency is the hertz (Hz), which is defined as 1 cycle per second.

frequency modulation (FM). An information transfer technique used in telecommunications in which the frequency of the carrier wave is modulated (i.e., increased or

Figure 8-5 This is NASA's Field Integrated Design and Operations (FIDO) rover robot being used in field tests to simulate driving conditions on Mars (circa April 1999). FIDO is at a geologically interesting site in central Nevada, while being controlled by human beings at the Jet Propulsion Laboratory (JPL) in Pasadena, California. The robot rover is about the size of a coffee table and has a mass of about 70 kilograms. FIDO used its articulate, mechanical arm to manipulate science instruments. During these field tests FIDO was powered by both the solar panels that cover the top of the rover and by replaceable, rechargeable batteries. (Credit: Courtesy of NASA/JPL.)

decreased) as the signal (to be transferred) increases or decreases in value but the amplitude of the carrier wave remains constant. Specifically, angle modulation of a sine carrier wave in which the instantaneous frequency of the modulated wave differs from the carrier frequency by an amount proportional to the instantaneous value of the modulating wave.

fuel cell. A direct conversion device that transforms chemical energy directly into electrical energy by reacting continuously supplied chemicals. In a modern fuel cell, an electrochemical catalyst (like platinum) promotes a noncombustible reaction between a fuel (such as hydrogen) and an oxidant (such as oxygen).

functional application. The generic task or function performed in an application.

fuzzy set. A generalization of set theory that allows for various degrees of set membership, rather than all or none.

***g*.** The symbol used for the acceleration due to gravity. At sea level on Earth, *g* is approximately 9.8 meters per second-squared (m/s^2). This term is used as a unit of stress for bodies experiencing acceleration.

Galileo Project. NASA's highly successful scientific mission to Jupiter launched in October 1989. With electricity supplied by two radioisotope-thermoelectric generator (RTG) units, the Galileo spacecraft has extensively studied the Jovian system since December 1995. Upon arrival, it also released an probe into the upper portions of Jupiter's atmosphere.

gamma rays (symbol: γ). High-energy, very-short-wavelength packets or quanta of electromagnetic radiation. Gamma ray photons are similar to X-rays, except that they are usually more energetic and originate from processes and transitions within the atomic nucleus. Gamma rays typically have energies between 10,000 electron volts and 10 million electron volts (i.e., between 10 keV and 10 MeV) with correspondingly short wavelengths and high frequencies. The processes associated with gamma ray emissions in astrophysical phenomena include: (1) the decay of radioactive nuclei, (2) cosmic ray interactions, (3) curvature radiation in extremely strong magnetic fields, and (4) matter–antimatter annihilation. Gamma rays are very penetrating and are best stopped or shielded against by dense materials, such as lead or tungsten. Sometimes called *gamma radiation*.

Ganymede. With a diameter of 5,262 kilometers, the largest moon of Jupiter and in the Solar System. Discovered by Galileo in 1610.

generate and test. In artificial intelligence, a common form of state space search based on reasoning by elimination. The system generates possible solutions and the tester prunes those solutions that fail to meet appropriate criteria.

geographic information system (GIS). A computer-assisted system that acquires, stores, manipulates, compares, and displays geographic data, often including multispectral sensing data sets from Earth-observing satellites and UAVs.

giga- (symbol: G). A prefix meaning multiplied by 10^9.

***Giotto* spacecraft.** Scientific spacecraft launched by the European Space Agency (ESA) in July 1985 that successfully encountered the nucleus of Comet Halley in mid-March 1986 at a distance of about 600 kilometers.

global data base. Complete database describing the specific problem, its status, and that of the solution process.

Global Positioning System (GPS). The constellation of over 20 U.S. Air Force satellites in circular 20,350 kilometers altitude orbits around Earth that provide accurate navigation data to military and civilian users on a global basis.

goal-driven. A problem-solving approach in artificial intelligence that works backward from the goal.

goal regression. A technique in artificial intelligence for constructing a plan by solving one conjunctive subgoal at a time—checking to see that each solution does not interfere with the other subgoals, which have already been achieved. If interference occurs, the offending subgoal is moved to an earlier noninterfering point in the sequence of subgoal accomplishments.

graph. A set of nodes connected by arcs.

gravity assist. The change is a spacecraft's direction and speed achieved by a carefully calculated flyby through a planet's gravitational field. This change in spacecraft velocity occurs without the use of supplementary propulsive energy.

gripper. *See* **end effector**.

gyroscope. A device that uses the angular momentum of a spinning mass (rotor) to sense angular motion of its base about one or two axes orthogonal (mutually perpendicular) to the spin axis. Also called a *gyro*.

half-life 1. (radioactive). The time in which half the atoms of a particular radioactive isotope disintegrate to another nuclear form. Measured half-lives vary from millionths of a second to billions of years. The half-life ($T_{1/2}$) is given by the expression:

$$T_{1/2} = (\ln 2)/\lambda = 0.69315/\lambda$$

where λ is the decay constant for the particular radioactive isotope and ln 2 is the natural (Napierian) logarithm of the number 2 with a numerical value of approximately 0.69315.

halo orbit. A circular or elliptical orbit in which a spacecraft remains in the vicinity of a Lagrangian libration point.

hard landing. A relatively high velocity impact of a lander spacecraft or probe on a solid planetary surface. The impact usually destroys all equipment, except perhaps a very rugged instrument package or payload container.

heat. Energy transferred by a thermal process. Heat (or thermal energy) can be measured in terms of the mechanical units of energy, such as the joule (J), or in terms of the amount of energy required to produce a definite thermal change in some substance, as, for example, the energy required to raise the temperature of a unit mass of water at some initial temperature (e.g., calorie). 1 joule = 0.239 calorie.

heat engine. A thermodynamic system that receives energy in the form of heat and that, in the performance of energy transformation on a working fluid, does work. Heat engines function in cycles. An ideal heat engine works in accordance with the Carnot cycle, while practical heat engines use thermodynamic cycles such as Brayton, Rankine, and Stirling. The steam engine, which helped create the Industrial Revolution, is a heat engine. Gas turbines and automobile engines are also heat engines.

henry (symbol: H). The SI unit of inductance (L). Inductance relates to the production of an electromotive force (E) in a conductor when there is a change in the magnetic flux (j) in that conductor. The induced electromotive force (E) is proportional to the time rate of change of the current (dI/dt), namely $E = -L\,(dI/dt)$, where the inductance (L) serves as a proportionality constant and depends on the geometric design of the circuit. One henry (H) is defined as the inductance occurring in a closed electric circuit in which an electromotive force (or emf) of 1 volt is produced when the current (I) in the circuit is varied uniformly at the rate of 1 ampere per second. This unit has been named in honor of the American physicist Joseph Henry (1797–1878).

hertz (symbol: Hz). The SI unit of frequency. One hertz is equal to 1 cycle per second. Named in honor of the German physicist Heinrich R. Hertz (1857–1894).

heuristics. Rules of thumb or empirical knowledge used to help guide a problem solution.

heuristic search techniques. Graph searching methods that use heuristic knowledge about the domain to help focus the search. These techniques operate by generating and testing intermediate states along potential solution paths.

hierarchial planning. A planning approach in which first a high-level plan is formulated considering only the important (or major) aspects. Then, the major steps of the initial high-level plan are refined into more detailed subplans.

hierarchy. A system of things ranked one above another.

higher order language (HOL). A computer language (such as FORTRAN), which requires fewer statements than machine language and which is usually substantially easier to use and read.

High Energy Astronomy Observatory (HEAO). A series of three NASA observatory spacecraft placed in Earth orbit (HEAO-1 launched in August 1977; HEAO-2 in November 1978; and HEAO-3 in September 1979) to support X-ray astronomy and gamma ray astronomy. After launch, NASA renamed HEAO-2 the *Einstein Observatory* in honor of the famous physicist Albert Einstein.

"housekeeping" (spacecraft). The collection of routine tasks that must be performed to keep an spacecraft functioning properly during an orbital flight or interplanetary mission.

***Hubble Space Telescope* (HST).** A cooperative European Space Agency (ESA) and NASA program to operate a long-lived space-based optical observatory. Launched on April 25, 1990, by NASA's Space Shuttle *Discovery* (STS-31 mission), subsequent on-orbit repair and refurbishment missions have allowed this powerful Earth-orbiting optical observatory to revolutionize our knowledge of the size, structure, and makeup of the universe. This robot spacecraft is named after the American astronomer Edwin Powell P. Hubble (1889–1953).

***Huygens* probe.** A scientific probe sponsored by the European Space Agency (ESA) and named after Christiaan Huygens. After being carried and released by NASA's *Cassini* spacecraft, the *Huygens* probe descended into the atmosphere of Saturn's moon Titan and landed on the moon's frozen surface on January 14, 2005.

humanoid robot. A sophisticated robot system constructed with some resemblance to a human being, such as arms, legs, a torso, and a head, but which still retains sufficient mechanical characteristics so that the system is not an android—the fully autonomous humanlike machine so often hypothesized in the science fiction literature.

hydraulic. Operated, moved, or affected by liquid used to transmit energy.

hydraulic robot. An industrial robot that uses hydraulic power to move its arm, wrist, and end effector. The hydraulic power supply is often located some distance away from the robot's work site and generally consists of a motor-driven pump, reservoir for the hydraulic fluid, a filter, heat exchanger, and pipes to deliver the pressurized hydraulic fluid to the robot. High-pressure fluid leaks are a major problem with hydraulic robots. *See also* **industrial robot**.

hyperbolic orbit. An orbit in the shape of a hyperbola; all interplanetary, flyby spacecraft follow hyperbolic orbits, both for Earth departure and again upon arrival at the target planet.

hypothesis. A scientific theory proposed to explain a set of data or observations; can be used as basis for further investigation and testing.

ideal gas. The pressure (p), volume (V), and temperature (T) behavior of many gases at low pressures and moderate temperatures is approximated quite well by the ideal (or perfect) gas equation of state, which is

$$pV = NR_uT$$

where N is the number of moles of gas and R_u is the universal gas constant.

$$R_u = 8314.5 \text{ joules/kg-mole-K}$$

This very useful relationship is based on the experimental work originally conducted by Robert Boyle (1627–1691) (*Boyle's Law*), Jacques Charles (1746–1823) (*Charles's*

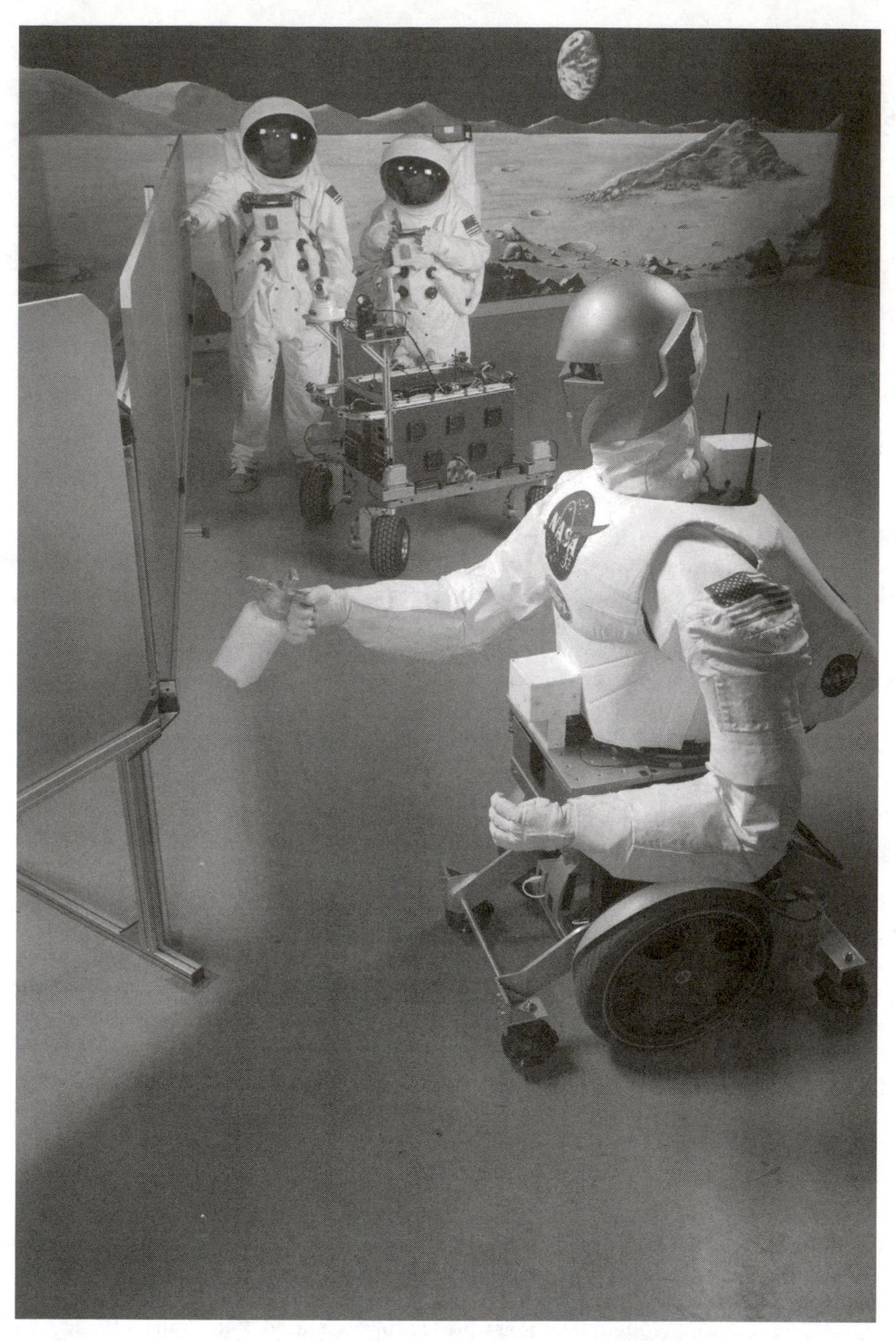

Figure 8-6 This picture provides a close-up look at NASA's humanoid robot, called Robonaut B, as it performs a mock weld. In the background, two space-suited human astronauts and another robot system (called the K10 robot) perform an inspection of a previously welded seam. The sophisticated simulation of human beings interacting with robot assistants during the assembly and construction of a (mock) lunar base took place at the Johnson Space Center (JSC) in November 2005. Robonaut B has a distinctive anthropomorphic design, but the humanoid robot's mobility system is clearly an electromechanical construct. (Credit: Courtesy of NASA/JSC.)

Law), and Joseph Louis Gay-Lussac (1778–1850) (*Gay-Lussac's Law*). In the ideal gas approximation, scientists assume that there are no forces exerted between the molecules of the gas and that these molecules occupy negligible space in the containing region. The ideal gas equation above and its many equivalent forms enjoy widespread application in physics and engineering.

identity. Two propositions (in logic) that have the same truth value.

image understanding (IU). Visual perception by a computer (or robot) employing geometric modeling and the artificial intelligence techniques of knowledge representation and cognitive processing to develop scene interpretations from image data. IU generally deals with three-dimensional (3D) objects.

impedance (symbol: Z). A quantity describing the total opposition to current flow (both resistance R and reactance X) in an alternating current (AC) circuit. For an AC circuit, the impedance can be expressed as:

$$Z^2 = R^2 + X^2$$

where Z is the impedance, R is the resistance, and X is the reactance, all expressed in ohms (Ω).

impeller. A device that imparts motion to a fluid; for example, in a centrifugal compressor, the impeller is a rotary disk that, faced on one or both sides with radial vanes, accelerates the incoming fluid outward into a diffuser.

impulse (symbol: I). In general, a mechanical "jolt" delivered to an object that represents the total change in momentum the object experiences. Physically, the thrust force (F) integrated over the period of time (t_1 to t_2) it is applied.

impulse intensity. Mechanical impulse per unit area. The SI unit of impulse intensity is the pascal-second (Pa-s). A conventionally used unit of impulse intensity is the tap, which is 1 dyne-second per square centimeter. 1 tap = 0.1 Pa-s.

incompressible fluid. A fluid for which the density ρ is assumed constant. The fluids used in hydraulic robots are often treated as incompressible fluids to a good engineering first approximation.

increment. An increase in the value of a variable.

independent system. In robot engineering, a system not influenced by other systems. For example, an independent circuit would not require other circuits to be functioning properly in order to perform its task. This "independent" characteristic requires that the circuit to be powered from an independent power supply, be controlled from a single source, and (in terms of redundancy) be tolerant of all credible failure modes in the corresponding redundant circuit or system. To make a system truly independent, the engineer also must exclude interface items such as mounting brackets and connectors that can create a common failure.

induction heating. The heating of an electrically conducting material using a varying electromagnetic field to induce eddy currents within the material. This may be an undesirable effect in electric power generation and distribution equipment but a desirable effect in materials processing (e.g., an induction heating furnace).

industrial robot. A robot designed to support manufacturing, to perform hazardous assembly or processing operations, to handle materials and products in a reliable, repetitive manner, or to accomplish similar industrial tasks. A typical industrial robot consists of one or more manipulators (arms), end effectors (hands), a controller, a power supply, and possibly an array of sensors to provide information about the manufacturing or warehouse environment in which the robot must operate. There are many types of industrial robots, so engineers have developed general classifications based on the basic functions each type of robot performs. The general classes of industrial robot are: nonservo (or pick-and-place), servo, programmable, computerized, sensory, and assembly robots. Of course, there are always a few robot systems that defy simple categorization. Is a small mobile, pipeline inspection robot that crawls inside a new oil pipeline checking the integrity of welds before the flow of oil commences an industrial robot, an environmental security and surveillance robot, or neither? Similarly, is a multiarmed robot system, which the works along the outside of a dry docked naval warship, scraping off encrusted barnacles and other unwanted materials, a military robot or an industrial robot, or both?

Figure 8-7 An arc-welding (industrial) robot in action, complete with an optical scanning and vision system that examines each weld as it is being made. (Credit: Courtesy of the U.S. Department of Energy and the National Engineering and Environmental Laboratory [INEEL].)

inertia. The resistance of a body to a change in its state of motion. Mass is an inherent property of a body that helps scientists to quantify inertia. *See also* **Newton's laws of motion**.

inference. The process of reaching a conclusion based on an initial set of propositions, the truths of which are known or assumed.

inference engine. The control structure of an artificial intelligence problem solver in which the control is separate from the knowledge.

infinity (symbol: ∞). A quantity beyond measurable limits.

infrared (IR) radiation. That portion of the electromagnetic (EM) spectrum lying between the optical (visible) and radio wavelengths. It is generally considered to span three decades of the EM spectrum, from 1 micrometer (μm) to 1,000 micrometers (μm) wavelength. The British-German astronomer Sir William Herschel (1738–1822) is credited with the discovery of infrared radiation.

inhibit device. A electromechanical device that prevents a hazardous event from occurring. This device has direct control, that is, it is not simply a device monitoring a potentially hazardous situation, nor is it in indirect control of some device experiencing the hazardous circumstances. All inhibit devices are independent from each other and are verifiable. A temperature limit switch that shuts down a device or system before a potentially hazardous temperature condition is reached (i.e., at some safe preset temperature limit) is an example of an inhibit device.

insect robot. A small, agile robot that functions by imitating the simple, repetitive biological processes of insects rather than trying to mimic complex, human-like operations.

intelligence. The degree to which an individual can successfully respond to new situations or problems. Intelligence is based on an individual's knowledge level and the ability of the person to appropriately manipulate and reformulate that knowledge (as well as incoming data) as required by the situation or problem.

intelligent assistant. An artificial intelligence computer program (usually an expert system) that helps a person in the performance of a task.

intelligent robot. *See* **smart robot**.

interactive environment. A computational system in which the user interacts (dialogues) with the system (in real time) during the process of developing or running a computer program.

interface. In general, the junction between two components. Specifically, the system by which the user interacts with the computer or robot.

Figure 8-8 Called insect robots or BEAM (biology, electronics, aesthetics, and mechanics) robots, the small agile robots shown here were designed and built at the Los Alamos National Laboratory (LANL) in New Mexico. Scientists at LANL decided to model the simple, repetitive biological processes of insects rather than trying to mimic complex, human-like operations, which generally has been the approach of traditional robotic technology programs that focused on android or humanoid robot systems. (Credit: Courtesy of the U.S. Department of Energy and the Los Alamos National Laboratory.)

integrated circuit (IC). Electronic circuits, including transistors, resistors, capacitors, and their interconnections, fabricated on a single small piece of semiconductor material (chip). Categories of integrated circuits such as LSI (large-scale integration) and VLSI (very large-scale integration) refer to the level of integration, which denotes the number of transistors on a chip.

Internet. An enormous global computer network that links many government agencies, research laboratories, universities, private companies, and individuals. This worldwide computer network has its origins in the "ARPA-net"—a small experimental computer network established by the Advanced Research Project Agency (ARPA) of the U.S. Department of Defense in the 1970s to permit rapid communication among universities, laboratories, and military project offices.

interstellar probe. A conceptual, highly automated, robotic interstellar spacecraft launched by the people of Earth (or some other advanced alien civilization) in the mid-twenty-first century to explore other star systems.

intravehicular activity (IVA) robot. A space-qualified robot system designed to assist astronauts and cosmonauts as they work *inside* the pressurized volumes of large orbiting space systems, like the *International Space Station* (ISS). *See also* **extravehicular activity (EVA) robot; humanoid robot.**

inviscid fluid. A hypothesized "perfect" fluid that has zero coefficient of viscosity. Physically this means that shear stresses are absent despite the occurrence of shearing deformations in the fluid. The inviscid fluid also glides past solid boundaries without sticking. No real fluids are inviscid. In fact, since real fluids are viscous, they stick to solid boundaries during flow processes creating thin boundary layers where shear forces are significant. However, the inviscid (perfect) fluid approximation provides a useful model that approximates the behavior of real fluids in many flow situations.

ion. An atom or molecule that has lost or (more rarely) gained one or more electrons. By this ionization process, the atom or molecule becomes electrically charged.

isolated system. A thermodynamic system that does not experience either matter or energy transfer across its boundaries.

isothermal process. In thermodynamics, any process or change of state of a system that takes place at constant temperature.

isotope. One of two or more atoms with the same atomic number (Z) (i.e., the same chemical element) but with different atomic weights. An equivalent statement is that the nuclei of isotopes have the same number of protons but different numbers of neutrons. Therefore, carbon-12 ($^{12}{}_{6}C$), carbon-13 ($^{13}{}_{6}C$), and carbon-14 ($^{14}{}_{6}C$), are all isotopes of the element carbon. The subscripts denote their common atomic number (i.e., $Z = 6$), while the superscripts denote their differing atomic mass numbers (i.e., $A =$ 12, 13, and 14, respectively), or approximate atomic weights. Isotopes usually have very nearly the same chemical properties but different physical and nuclear properties. For example, the isotope carbon-14 is radioactive, while the isotopes carbon-12 and carbon-13 are both stable (nonradioactive).

jansky (symbol: Jy). A unit used to describe the strength of an incoming electromagnetic wave signal. The jansky frequently is used in radio and infrared astronomy. It is named after the American radio engineer Karl Guthe Jansky (1905–1950), who discovered extraterrestrial radio wave sources in the 1930s—a discovery generally regarded as the birth of radio astronomy. 1 jansky (Jy) = 10^{-26} watts per meter-squared per hertz [W/(m^2-Hz)].

jointed arm. A robot arm that can perform actions similar to those performed by the shoulder, arm, and elbow arrangement of a human being.

joule (symbol: J). The SI unit of energy or work. One joule is the work done by a force of one newton moving through a distance of one meter. Named after James Prescott Joule.

kelvin (symbol: K). The SI unit of absolute thermodynamic temperature, honoring Lord Kelvin. By international agreement, one degree kelvin represents the fraction 1/273.16 of the thermodynamic temperature of the triple point of water.

kilo- (symbol: k). An SI unit system prefix meaning that a basic space/time/mass unit is multiplied by one thousand (1,000); as for example, a kilogram (kg) or a kilometer (km). Note, however, that in computer technology and digital data processing, kilo designated with a capital "K" refers to a precise value of 1,024 (which corresponds in binary notation to 2^{10}). Therefore, a kilobyte (abbreviated KB or K-byte) is actually 1,024 bytes and similarly a kilobit (Kb or K-bit) stands for 1,024 bits.

kilogram (symbol: kg). The fundamental unit of mass in the International Unit System (SI). 1 kilogram (kg) = 1,000 grams = 2.205 pounds-mass (lb_{mass}) (approximately).

kinetic energy (common symbols: KE or E_{KE}). The energy an object possesses as a result of its motion. In Newtonian (nonrelativistic) mechanics, kinetic energy is one-half the product of mass (m) and the square of its velocity (v), that is $E_{KE} = 1/2\ mv^2$.

knowledge base. Artificial intelligence data bases that are not merely files of uniform content, but are collections of facts, inferences, and procedures, which correspond to the types of information needed for problem solution.

knowledge base management. Management of a knowledge base in terms of storing, accessing, reasoning with the knowledge.

knowledge engineering. The approach within the field of artificial intelligence that focuses on the use of knowledge (as found, for example, in an expert system) to solve problems.

knowledge representation (KR). The form of the data structure used to organize the knowledge required for a problem.

knowledge source. An expert system component that deals with a specific area or activity.

Lagrangian libration point. The five points in outer space (called L_1, L_2, L_3, L_4, and L_5) where a small object can experience a stable orbit in spite of the force of gravity exerted by two much more massive celestial bodies when they orbit about a common center of mass. Joseph Louis Lagrange calculated the existence and location of these points in 1772.

lambert (symbol: L). A unit of luminance defined as equal to 1 lumen of (light) flux emitted per square centimeter of a perfectly diffused surface. This unit is named in honor of the German mathematician Johann H. Lambert (1728–1777).

lander (spacecraft). A robot spacecraft designed to safely reach the surface of a planet or moon and survive long enough on the planetary body to collect useful scientific data that it sends back to Earth by telemetry. *See also* **space robot.**

LANDSAT. The family of versatile, NASA-developed, Earth-observing spacecraft that have demonstrated numerous applications of space-based multispectral sensing since 1972. *See also* **remote sensing.**

latch. A device that fastens one object or part to another but is subject to ready release on demand so the objects or parts can be separated. For example, a sounding rocket can be held on its launcher by a latch or several latches and then quickly released after ignition and proper thrust development.

leaf. A terminal node in a tree representation. *See also* **tree structure.**

least commitment. A technique in artificial intelligence for coordinating decision making with the availability of information, so that problem-solving decisions are not made arbitrarily or prematurely, but are postponed until there is enough information.

light time. The amount of time it takes for light or radio wave signals to travel a certain distance at optical velocity ($c = 299{,}792.5$ km/s). For example, one light-second corresponds to a distance of approximately 300,000 kilometers.

light-year (symbol: ly). The distance light (or other forms of electromagnetic radiation) can travel in one year. One light-year equals a distance of approximately 9.46×10^{12} kilometers or 63,240 astronomical units (AU).

limited sequence robot. *See* **pick-and-place robot.**

limit switch. A mechanical device that can be used to determine the physical position of equipment. For example, an extension on a valve shaft mechanically trips a limit switch as it moves from open to shut or shut to open. The limit switch gives "ON/OFF" output that corresponds to valve position. Normally, limit switches are used to provide full open or full shut indications. Many limit switches are the push-button variety. When the valve extension comes in contact with the limit switch, the switch depresses to complete, or turn on, the electrical circuit. As the valve extension moves away from the limit switch, spring pressure opens the switch, turning off the circuit.

line of force. A line indicating the direction that a force acts.

line of sight (LOS). The straight line between a sensor or the eye of an observer and the object or point being observed. Sometimes called the *optical path.*

link. In telecommunications, a general term used to indicate the existence of communications pathways and/or facilities between two points. In referring to communications between a ground station and a spacecraft or satellite, the term "uplink" describes communications from the ground site to the spacecraft, while the term "downlink" describes communications from the spacecraft to the ground site.

list processing language (LIST). A fundamental programming language in the field of artificial intelligence.

liter (symbol: l or L). A unit of volume in the metric (SI) system. Defined as the volume of 1 kilogram of pure water at standard (atmospheric) pressure and a temperature of 4° Celsius. Also spelled *litre*. 1 liter = 0.2642 gallons = 1.000028 cubic decimeters (dm^3).

logarithm. The power (*p*) to which a fixed number (*b*), called the base (usually 10 or *e* [2.71828. . .]), must be raised to produce the number (*n*) to which the logarithm corresponds. Any number (n) can be written in the form: $n = b^p$. The term "*p*" is then the logarithm to the base "*b*" of the number "*n*"; that is, $p = \log_b$ n. *Common logarithms* have 10 as the base and usually are written: log or $\log_{10}$. *Natural logarithms* (also called Napierian logarithms) have e as the base and often are denoted: ln or $\log_e$. The irrational number *e* is defined as the limit as x tends to infinity (∞) of the expression $[1 + 1/x]^x$. An antilogarithm (or inverse logarithm) is the value of the number corresponding to a given logarithm. Using the previous nomenclature, if "*p*" is the logarithm, then "*n*" is the antilogarithm.

logarithmic scale. A scale in which the line segments that are of equal length are those representing multiples of 10.

logical operation. Execution of a single computer instruction.

logical representation. Knowledge representation by a collection of logical formulas (usually in first order predicate logic) that provide a partial description of the world.

longitudinal axis. The fore-and-aft line through the center of gravity of a robotic craft.

lumen (symbol: lm). The SI unit of luminous flux. It is defined as the luminous flux emitted by a uniform point source with an intensity of one candela in a solid angle of one steradian.

lumped mass. Concept in engineering analysis wherein a mass is treated as if it were concentrated at a point.

Luna. A series of Russian robot spacecraft sent to explore the Moon in the 1960s and 1970s.

lunar orbiter. A spacecraft placed in orbit around the Moon; specifically, the series of five *Lunar Orbiter* robot spacecraft NASA used from 1966 to 1967 to precisely photograph the Moon's surface in support of the Apollo Project.

lunar probe. A planetary probe for exploring and reporting conditions on or about the Moon. *See also* **Luna and Ranger Project**.

Lunar Prospector. A NASA orbiter spacecraft that circled the Moon from 1998 to1999, searching for mineral resources. Data suggest the possible presence of lunar (water) ice deposits in permanently shadowed polar regions.

lunar rover. Human-crewed or automated (robot) rover vehicles used to explore the Moon's surface. NASA's lunar rover vehicle (LRV) served as a "Moon car" for Apollo Project astronauts during the Apollo 15, 16, and 17 expeditions. Russian *Lunokhod 1* and *2* robot rovers were operated on Moon from Earth between 1970 and 1973.

Lunokhod. A Russian eight-wheeled robot vehicle, controlled by radio wave signals from Earth and used to perform lunar surface exploration during the *Luna 17* (1970) and *Luna 21* (1973) missions to the Moon.

lux (symbol: lx). The SI unit of illuminance. It is defined as 1 lumen per square meter.

machine. The engineer defines a machine as a device with fixed and moving parts that modifies mechanical energy in order to do work. The physicist defines a machine as a device or structure designed to transmit and modify forces. Both definitions are useful. Several simple machines (or basic components of most machines) are the hammer, lever, pulley, wheel and axel, the gear, the clutch, the wedge, the shaft, and the screw. Whether a machine is an extremely simple tool or complex, complicated mechanism (like an industrial robot), the main purpose for its existence remains the same—namely, to transform input forces into output forces. The mechanical efficiency of a machine (η) is defined as the output work divided by the input work. A hypothetical ideal machine has a mechanical efficiency of 100 percent, or $\eta = 1$.

machine intelligence (MI). *See* **artificial intelligence.**

magnetism. A class of physical phenomena that include the attraction for iron observed in lodestone and a magnet, are inseparably associated with moving electricity, are exhibited by both magnets and electric currents, and are characterized by fields of force.

manipulator. A mechanical devices used for handling objects; frequently involving remote operations (i.e., teleoperation) and/or hazardous substances or environmental conditions. That portion of a robot system, which is capable of grasping or handling. A robot's manipulator is often designed to mimic the movement of the human shoulder, arm, wrist, hand, and fingers. A manipulator generally has a versatile end effector (i.e., the special tool or "grasping element" installed at the end of the manipulator) that can respond to a variety of different handling requirements. For example, the U.S. space shuttle has a very useful manipulator called the remote manipulator system (RMS).

man-machine interface. The boundary where human and machine characteristics and capabilities are joined in order to obtain optimum operating conditions and maximum efficiency of the combined man-machine system. A joystick and a control panel are examples of man-machine interfaces. Also called *human-machine interface.*

manufacturing. The process of transforming a raw material into a finished product; especially in large quantities.

manufacturing cell. The overall system of hardware and software need to accomplish the total automation of a particular manufacturing task.

Mariner. A series of NASA planetary exploration spacecraft that performed flyby and orbital missions to Mercury, Mars, and Venus in the 1960s and 1970s.

Mars Global Surveyor (MGS). A NASA orbiter spacecraft launched in November 1996 that has been performing detailed studies of the Martian surface and atmosphere since March 1999.

Mars Odyssey. Launched from Cape Canaveral in April 2001, the *2001 Mars Odyssey* is NASA's latest orbiter spacecraft to explore Mars, specifically searching for geological features that could indicate the presence of water—past or present (subsurface).

Mars Pathfinder. An innovative NASA mission that successfully landed a Mars surface rover—a small robot called *Sojourner*—in the Ares Vallis region of the Red Planet in July 1997. For over 80 days, personnel on Earth used teleoperation and telepresence to cautiously drive the six-wheeled minirover to interesting locations on the Martian surface.

Mars surface rovers. Automated robot rovers and human-crewed mobility systems used to satisfy a number of surface exploration objectives on Mars in the twenty-first century.

marsupial robot. A robot that carries one or more smaller robots within itself and then releases these smaller, minirobots at a special area of interest for scientific inquiry, national defense, or environmental monitoring.

mass (symbol: m). Mass describes "how much" material makes up an object and gives rise to its inertia. The SI unit for mass is the kilogram (kg). An object that has one kilogram of mass on Earth will also have one kilogram of mass on the surface of Mars, or anywhere else in the universe.

mass number (symbol: A). The number of nucleons (i.e., the number of protons and neutrons) in an atomic nucleus. It is the nearest whole number to an atom's atomic weight. For example, the mass number of the isotope uranium-235 is 235.

mass spectrometer. An instrument used to measure the relative atomic masses and relative abundances of isotopes. A sample (usually gaseous) is ionized, and the resultant stream of charged particles is accelerated into a high vacuum region where electric and magnetic fields deflect the particles and focus them on a detector. A *mass spectrum* (i.e., a series of lines related to mass/charge values) then is created. This characteristic pattern of lines helps scientists identify different molecules.

Figure 8-9 These nuclear industry workers are using sophisticated master–slave manipulators to safely handle highly radioactive materials. The technician controls the master manipulator, while on the other side of the shielded wall (which includes special leaded-glass windows) the slave manipulator performs the operations that technician wants to perform involving intensively radioactive materials. This type of hot cell facility is the only practical way that such operations can take occur without exposing the workers to dangerous doses of nuclear radiation. (Credit: Courtesy of the U.S. Department of Energy.)

master–slave manipulator. A class of teleoperator that contains isomorphic "master and slave" arms. The master manipulator is held and positioned by a technician (human being) and the slave manipulator duplicates the motions, sometimes with a change in scale in displacement (providing either exaggerated or reduced movements) or force (providing either more fore than a human hand can exert or at other times being limited to a present "gentle squeeze" independent of the tactile force applied by the human operator.) The hot cell facilities used by the U.S. Department of Energy to handle highly radioactive materials employ some of the most sophisticated master/slave manipulators ever developed.

material. The tangible substance (chemical, biological, or mixed) that goes into the makeup of a physical object. One of the basic resources used in a technological system.

means–ends analysis. A problem-solving approach in artificial intelligence in which problem-solving operators are chosen in an iterative fashion to reduce the difference between the current problem-solving state and the goal state.

mechanical efficiency (η). The mechanical efficiency of a machine is defined as the output work divided by the input work. A hypothetical ideal machine has a mechanical efficiency of 100 percent, or $\eta = 1$.

mega- (symbol: M). A prefix in the SI unit system meaning multiplied by 1 million (10^6), as, for example, megahertz (MHz), meaning 1 million hertz.

melting. In thermodynamics, the transition of a material from the solid phase to the liquid phase, usually as a result of heating.

meta rule. In artificial intelligence, a higher-level rule used to reason about lower-level rules.

meter (symbol: m). The fundamental SI unit of length. 1 meter = 3.281 feet. Also spelled *metre* (British spelling).

metric system. The international system (SI) of weights and measures based on the meter as the fundamental unit of length, the kilogram as the fundamental unit of mass, and the second as the fundamental unit of time. Also called the *mks system.*

metrology. The science of dimensional measurement; sometimes includes the science of weighing.

MeV. An abbreviation for 1 million electron volts, a common energy unit encountered in the study of nuclear reactions. (1 MeV = 10^6 eV).

micro- (symbol: μ). A prefix in the SI unit system meaning divided by 1 million; for example, a micrometer (mm) is 10^{-6} meter. The term also is used as a prefix to indicate something is very small, as in *micrometeoroid* or *micromachine.*

microcode. A computer program at the basic machine level.

micrometer. 1. An SI unit of length equal to one-millionth (10^{-6}) of a meter; also called a micron. 1 μm = 10^{-6} m. 2. An instrument or gauge for making very precise linear measurements (e.g., thicknesses and small diameters) in which the displacements measured correspond to the travel of a screw of accurately known pitch.

micron (symbol: μm). An SI unit of length equal to one-millionth (10^{-6}) of a meter. Also called a *micrometer.*

microorganism. A tiny plant or animal, especially a protozoan or a bacterium.

microsecond (symbol: μs). A unit of time equal to one-millionth (10^{-6}) of a second.

microwave (radiation). A comparatively short-wavelength electromagnetic (EM) wave in the radio-frequency portion of the EM spectrum. The term "microwave" usually is applied to those EM wavelengths that are measured in centimeters, approximately 30 centimeters to 1 millimeter (with corresponding frequencies of 1 gigahertz [GHz] to 300 gigahertz [GHz]).

milestone. An important event or decision point in a program or plan. The term originates from the use of stone markers set up on roadsides to indicate the distance in miles to a given point. Milestone charts are used extensively in aerospace programs and planning activities.

military robot. Any one of a large number of robot systems designed, developed, and operated in support of national security goals and missions. Includes UAVs like the Predator, numerous UGVs, and a variety remotely operated vehicles (ROVs) for undersea activities.

military satellite (MILSAT). A robot spacecraft satellite used for military or defense purposes such as missile surveillance, navigation, and intelligence gathering.

milli- (symbol: m). The SI unit system prefix meaning multiplied by 1/1000 (10^{-3}). For example a millivolt (mV) is 0.001 volt; a millimeter (mm) is 0.001 meter; and a millisecond (msec) is 0.001 second.

millibar (symbol: mbar or mb). A unit of pressure equal to 0.001 bar (i.e., 10^{-3} bar) or 1,000 dynes per square centimeter. The millibar is used as a unit of measure of atmospheric pressure, with a standard atmosphere being equal to about 1,013 millibars or 29.92 inches (760 millimeters) of mercury 1 mbar = 100 newtons/m^2 = 1,000 dynes/cm^2.

millimeter (symbol: mm). One-thousandth (1/1000;10^{-3}) of a meter .1 mm = 0.001 m = 0.1 cm = 0.03937 in.

millisecond(symbol: msec or ms). One-thousandth (1/1000, 10^{-3}) of a second. 1 msec = 0.001 sec.

mini-. An abbreviation for "miniature."

minute. 1. A unit of time equal to the 60th part of an hour; that is, 60 minutes = 1 hour. 2. A unit of angular measurement such that 60 minutes (60′) equal 1 degree (1°) of arc.

mock-up. A full-size replica or dummy of something, such as a spacecraft, often made of some substitute material, such as wood, and sometimes incorporating actual functioning pieces of equipment, such as engines or power supplies. Mock-ups are used to study construction procedures, to examine equipment interfaces, or to train personnel.

modeling. A scientific investigative technique that uses a mathematical or physical representation of a system or theory. This representation, or "model," accounts for all, or at least some, of the known properties of the system or the characteristics of the theory. Models are used frequently to test the effects of changes of system components on the overall performance of the system or the effects of variation of critical parameters on the behavior of the theory.

model-driven. A top-down approach to problem solving in artificial intelligence in which the inferences to be verified are based on the domain model used by the problem solver.

modulation. The process of modifying a radio frequency (RF) signal by shifting its phase, frequency, or amplitude to carry information. The respective processes are called phase modulation (PM), frequency modulation (FM), and amplitude modulation (AM).

modus ponens. A mathematical form of argument in deductive logic. It has the form:

If A is true, then B is true.
A is true.
Therefore B is true.

mole (symbol: mol). The SI unit of the amount of substance. It is defined as the amount of substance that contains as many elementary units as there are atoms in 0.012 kilograms of carbon-12, a quantity known as Avogadro's number (N_A). (The Avogadro number, N_A, has a value of about 6.022×10^{23} molecules/mole.)

molecule. A group of atoms held together by chemical forces. The atoms in the molecule may be identical, as in hydrogen (H_2), or different, as in water (H_2O) and carbon dioxide (CO_2). A molecule is the smallest unit of matter that can exist by itself and retain all its chemical properties.

moment of inertia (symbol: I). For a massive body made up of many particles or "point masses" (m_i), the moment of inertia (I) about an axis is defined as the sum (Σ) of all the products formed by multiplying each point mass of particle (m_i) by the square of its distance $(r_i)^2$ from the line or axis of rotation; that is: $I = \Sigma_i \, m_i \, (r_i)^2$. The moment of inertia can be considered as the analog in rotational dynamics of mass in linear dynamics.

momentum (linear). The linear momentum (p) of a particle is the product of the particle's mass (m) and its velocity (v). Newton's second law of motion tells us that the time rate of change of momentum of a particle is equal to the resultant force (F) on the particle. *See also* **Newton's laws of motion**.

mother spacecraft. A main exploration spacecraft that also carries and deploys one or several atmospheric probes and rover or lander spacecraft, when arriving at a target planet. The mother spacecraft then relays their data back to Earth and may orbit the planet to perform its own scientific mission. NASA's *Galileo* spacecraft to Jupiter and *Cassini* spacecraft to Saturn are examples.

multispectral sensing. The remote-sensing method of simultaneously collecting several different bands (wavelength regions) of electromagnetic radiation (such as the visible, the near-infrared, and the thermal infrared bands) when observing a target.

nano-(symbol: n) . A prefix in the SI unit system meaning multiplied by 10^{-9}.

nanometer (nm). A billionth of a meter (i.e., 10^{-9} meter).

Figure 8-10 This prototype nanorover is only 20 centimeters long. One possible space application of this type of miniature robot explorer is to send back information about the surface of an asteroid to an orbiting mother spacecraft. The rover's camera can be focused to take panoramic shots as well as microscopic images. Engineers can place solar cells on all sides of the operational nanorover, so even if it flips over due to an asteroid's low-surface gravity, the tiny robot will always have enough power to activate the appropriate motors to right itself. (Credit: Courtesy of NASA.)

nanorover. A tiny robotic vehicle, usually with a total mass of between 10 and 50 grams. In space exploration applications, one or several of these tiny robots could be used to survey areas around a lander spacecraft and to look for a particular substance, such as water ice or microfossils. The nanorover would then communicate its scientific findings back to Earth via the lander spacecraft—possibly with an orbiting mother spacecraft serving as the communications relay.

nanosecond (ns). A billionth of a second (i.e., 10^{-9} second).

nanotechnology. A general term that describes the manufacture and application of microminiature machines, electronic devices, and chemical and biological sensors all of which have characteristic dimensions on the order of a micron (10^{-6} meter) or less.

NASA. The National Aeronautics and Space Administration, the civilian space agency of the United States. Created in 1958 by an act of Congress, NASA's overall mission is to plan, direct, and conduct civilian (including scientific) aeronautical and space activities for peaceful purposes.

National Aeronautics and Space Administration (NASA). *See* **NASA.**

natural deduction. Informal reasoning.

natural language interface (NLI). A system for communicating with a computer by using a natural language.

natural language processing (NLP). Processing of a natural language (such as English) by a computer to facilitate communication with the computer, or for other purposes such as language translation.

natural language understanding (NLU). Response by a computer based on the meaning of a natural language input.

natural material. Material found in nature, such as wood, stone, gases, and clay.

negate. In artificial intelligence to change a proposition into its opposite.

neper (symbol: N or N_p) . A natural logarithmic unit (x) used to express the ratio of two power levels, P_1(input) and P_2(output), such that x (nepers) = 1/2 ln (P_1/P_2). The unit is named after John Napier (1550–1617), the Scottish mathematician who developed natural logarithms (symbol: ln). This unit is often encountered in telecommunications engineering. 1 neper = 8.686 decibels.

neutron (symbol: n). An uncharged elementary particle with a mass slightly greater than that of the proton. It is found in the nucleus of every atom heavier than ordinary hydrogen. A free neutron is unstable, with a half-life of about 10 minutes, and decays into an electron, a proton, and a neutrino. Neutrons sustain the fission chain reaction in a nuclear reactor and support the supercritical reaction in a fission-based nuclear weapon.

newton (symbol: N). The SI unit of force, named after the British mathematician and physicist, Sir Isaac Newton (1642–1727). One newton is the amount of force that gives a one kilogram mass an acceleration of one meter per second per second.

Newton's law of gravitation. The physical law proposed in 1687 by Sir Isaac Newton (1642–1727), stating that every particle of matter in the universe attracts every other particle with the force of gravitational attraction (F_G) acting along the line joining the two particles and being proportional to the product of the particle masses (m_1 and m_2), and inversely proportional to the square of the distance (*r*) between the particles. This law expressed as an equation is, $F_G = [Gm_1m_2] / r^2$, where *G* is the universal gravitational constant [approximately 6.6732 (±0.003) 10^{-11} N-m^2/kg^2 in SI units].

Newton's laws of motion. The three postulates of motion formulated by Sir Isaac Newton (1642–1727) in about 1685. His first law (the conservation of momentum) states that a body continues in a state of uniform motion (or rest) unless acted upon by an external force. The second law states that the rate of change of momentum of a body is proportional to the force acting upon the body and occurs in the direction of the applied force. The third law (the action and reaction principle) states that for every force acting upon a body, there is a corresponding force of the same magnitude exerted by the body in the opposite direction. The third law is the basic principle by which every rocket operates. These important physical principles form the basis of classical mechanics.

node. A point (representing an object or the state of a system) in a graph connected to other points in the graph by arcs, which usually represent relationships.

nonmonotonic logic. A logic in which the results are subject to revision as more information is gathered.

nondestructive testing. Testing to detect internal and concealed defects in materials and components using techniques that do not damage or destroy the items being tested. Aerospace engineers and technicians frequently use X-rays, gamma rays, and neutron irradiation, as well as ultrasonics, to accomplish nondestructive testing.

nonservo robot. The simplest type of industrial robot. This type of robot picks up an object and places it at another location. The robot's freedom of movement usually is limited to two or three directions. *See also*: **industrial robot; robot.**

nuclear-electric propulsion (NEP). A space-deployed propulsion system that uses a space-qualified, compact nuclear reactor to produce the electricity needed to operate a space robot's electric propulsion engine(s).

nuclear radiation. Ionizing radiation consisting of particles (such as alpha particles, beta particles, and neutrons) and very energetic electromagnetic radiation (i.e., gamma rays). Atomic nuclei emit this type of radiation during a variety of energetic nuclear reaction processes, including radioactive decay, fission, and fusion.

object-oriented programming. A computer programming approach that focuses on objects that communicate by message passing. An object is considered to be a package of information and descriptions of procedures that can manipulate that information.

observatory. The place (or facility) from which astronomical observations are made. For example, the Keck Observatory is a ground-based observatory, while *Hubble Space Telescope* is a robot space-based observatory in orbit around Earth.

oersted (symbol: Oe). The unit of magnetic field strength in the centimeter-gram-second (cgs.) system of units. (1 oersted = 79.58 amperes/meter) This unit is named in honor of Hans Christian Oersted (1777–1851), a Danish physicist who was first to demonstrate the relationship between electricity and magnetism.

ohm (symbol: Ω). The SI unit of electrical resistance. It is defined as the resistance (R) between two points on a conductor produced by a current flow (I) of one ampere when there is a constant voltage difference (potential) (V) of one volt between these points. From Ohm's law, the resistance in a conductor is related to the voltage and the current by the equation: $R = V/I$, so that, 1 ohm (W) of resistance = 1 volt per ampere. The unit and physical law are named in honor of the German physicist George Simon Ohm (1787–1854).

one-way communications. Communications mode consisting only of downlink received from a robot spacecraft.

open loop. A control system operating without feedback or perhaps with only partial feedback. An electrical or mechanical system in which the response of the output to an input is preset; there is no feedback of the output for comparison and corrective adjustment.

open system. A thermodynamic system that can experience both matter and energy transfer across its boundaries.

operating life. The maximum operating time (or number of cycles) that an item can accrue before replacement or refurbishment without risk of degradation of performance beyond acceptable limits.

operators. Procedures or generalized actions that can be used for changing situations.

optoelectronic device. A device that combines optical (light) and electronic technologies, such as a fiber optics communications system.

orbit. A path described by one body in its revolution about another (as by Earth about the Sun or a human-made spacecraft around Earth).

orbiter (spacecraft). A spacecraft especially designed to travel through interplanetary space, achieve a stable orbit around the target planet (or other celestial body), and conduct a program of detailed scientific investigation.

Orbiting Astronomical Observatory (OAO). A series of large, Earth-orbiting robot astronomical observatories developed by NASA in the 1960s to broaden scientific understanding of the universe—especially as related to ultraviolet astronomy.

Orbiting Quarantine Facility (OQF). A proposed Earth-orbiting laboratory in which soil and rock samples from Mars and other worlds could first be tested for potentially harmful alien microorganisms before such extraterrestrial materials are allowed to enter Earth's biosphere. Robot systems will play a major role in the handling and analysis of alien soil and rock samples.

organism (biological). An individual life form, such as a plant, animal, bacterium, virus, or fungus; a body made up of organs, organelles, or other parts that work together to carry out the various processes of life.

orthogonal. At right angles; pertaining to or composed of right angles.

outgassing. Release of gas from a material when it is exposed to an ambient pressure lower than the vapor pressure of the gas. Generally refers to the gradual release of gas from enclosed surfaces when an enclosure is vacuum pumped or to the gradual release of gas from a robot spacecraft's surfaces and components when they are first exposed to the vacuum conditions of outer space following launch.

parallel processing. Simultaneous processing, as opposed to the sequential processing in a conventional (von Neumann) type of computer architecture.

parking orbit. The temporary (but stable) orbit of a spacecraft around a celestial body, used for assembly and/or transfer of equipment or to wait for conditions favorable for departure from that orbit.

pascal (symbol: Pa). The SI unit of pressure. It is defined as the pressure that results from a force of one newton (N) acting uniformly over an area of 1 square meter (1 pascal [Pa] $= 1\ N/m^2$). This unit is named after the French scientist Blaise Pascal (1623–1662).

passive. Containing no power sources to augment output power or signal, such as a passive electrical network or a passive reflector. Applied to a device that draws all its power from the input signal. A dormant device or system, that is, one that is not active.

passive sensor. A sensor that detects radiation naturally emitted (e.g., infrared) by or reflected (e.g., sunlight) from a target.

path. A particular track through a state graph.

pattern directed invocation. The activation of procedures by matching their antecedent parts to patterns present in the global database (the system status).

pattern matching. Matching patterns in a statement or image against patterns in a global database, templates, or models.

pattern recognition. The process of classifying data into predetermined categories.

payload. With respect to a robot's arm, the maximum mass (on Earth "weight") the arm can carry and still perform properly. An articulating arm usually has less of a payload capacity when fully extended than when operating in a folded or nonextended condition. For a space robot, the scientific payload is the amount of mass set aside for the science instruments and experiments.

perfect fluid. In simplifying assumptions made as part of preliminary engineering analyses, a fluid chiefly characterized by a lack of viscosity and, usually, by incompressibility. Also called an *ideal fluid* or an *inviscid fluid*.

perfect gas. A gas that obeys the following equation of state: $pv = RT$, where p is pressure, v is specific volume, T is absolute temperature, and R is the gas constant.

perception. An active process in which hypotheses are formed about the nature of the environment, or sensory information is sought to confirm or refute hypotheses.

personal AI computer. Small, interactive, stand-alone computers used by computer scientists and AI researchers in developing AI programs. Usually such computers are specifically designed to run an AI language.

phase modulation (PM). A type of modulation in which the relative phase of the carrier wave is modified or varied in accordance to the amplitude of the signal. Specifically, a form of angle modulation in which the angle of a sine-wave carrier is caused to depart from the carrier angle by an amount proportional to the instantaneous value of the modulating wave.

photometer. An instrument that measures light intensity and the brightness of objects.

photon. According to quantum theory, the elementary bundle or packet of electromagnetic radiation, such as a photon of light. Photons have no mass and travel at the speed of light. The energy (E) of the photon is equal to the product of the frequency (ν) of the electromagnetic radiation and Planck's constant (h): $E = h\nu$, where h is equal to 6.626×10^{-34} joule-sec, and ν is the frequency (hertz).

photovoltaic conversion. A form of direct conversion in which a photovoltaic material converts incoming photons of visible light directly into electricity. The solar cell is an example.

pick-and-place robot. One of the two basic types of industrial robot (the other being the servo robot). This type of robot has direction-control stops or valves, which are either fully opened or closed, thereby limiting positioning capability and program capacity. Also called a *bang-bang robot*, an *end-point robot*, a *limited-sequence robot*, or a *nonservo-controlled robot*.

***Pioneer 10, 11* spacecraft.** NASA's twin exploration robot spacecraft that were the first to navigate the main asteroid belt, the first to visit Jupiter (1973 and 1974), the first to visit Saturn (*Pioneer 11* in 1979), and the first human-made objects to leave the solar system (*Pioneer 10* in 1983). Each spacecraft is now on a different trajectory to the stars, carrying a special message (the "Pioneer plaque") for any intelligent alien civilization that might find it millions of years from now.

Pioneer Venus mission. Two robot spacecraft launched by NASA to Venus in 1978. *Pioneer 12* was an orbiter spacecraft that gathered data from 1978 to 1992. The *Pioneer Venus Multiprobe* served as a mother spacecraft, launching one large and three identical small planetary probes into the Venusian atmosphere (December 1978).

pitch. The rotation (angular motion) of a robot or robot's arm about its lateral axis. The wrist at the end of a robot's arm typically has three basic motions, which engineers describe as pitch, roll, and yaw. *See also* **roll; yaw**.

pixel. Contraction for picture element; the smallest unit of information on a screen or in an image; the more pixels, the higher the potential resolution of the video screen or image.

plan. A sequence of actions to transform an initial situation into a situation that satisfies the goal conditions.

plasticity. The tendency of a loaded body to assume a (deformed) state other than its original state, when the load is removed.

pneumatic. Operated, moved, or effected by a pressurized gas (typically air) that is used to transmit energy.

pneumatic robot. An industrial robot that is pneumatically actuated. The power for pneumatic actuation is usually provided by a remote compressor, which may provide pressurized working fluid (e.g., compressed air) to other equipment at the industrial facility. Sometimes called an *air-logic robot*.

point-to-point robot. An industrial robot, which represents one of the two basic types of servo robots. The expression "point-to-point" refers to the fact that this type of robot must be taught to perform its assigned task one step (or point) at a time. The human robot technician positions the robot's arm (especially the end effector or hand) at a particular point in space and then instructs (programs) the robot to store that particular position in its computer memory. The technician repeats this procedure on a point by point basis, until the robot has stored in its memory the complete sequence of motions and actions it is expected to perform. *See also* **industrial robot**.

poise (symbol: P). Unit of dynamic viscosity in the centimeter-gram-second (cgs) unit system. It is defined as the tangential force per unit area (dynes/cm^2) required to maintain a unit difference in velocity (1 cm/sec) between two parallel plates in a liquid that are separated by a unit distance (1 cm).

1 poise = (1 dyne-second)/(centimeter)^2 = 0.1 (newton-second)/(meter)^2. The unit is named after the French scientist Jean Louis Poiseuille (1799–1869). The centipoise, or 0.01 poise, is encountered often. For example, the dynamic viscosity of water at 20°C is approximately 1 centipoise.

polar coordinate system. A coordinate system in which a point (P) that is defined as P(x, y) in two-dimensional Cartesian coordinates is now represented as P(r, θ), where $x = r \cos \theta$ and $y = r \sin \theta$. Physically, in the polar coordinate system: r is the radial distance in the x–y plane from the origin (O) to point P and θ is the angle formed between the x-axis and the radial vector (i.e., the line of length r from the origin O to point P).

In three dimensions, the Cartesian coordinate system point P(x, y, z) becomes P(r, θ, z) in cylindrical polar coordinates—where the point P(r, θ, z) is now regarded as lying on the surface of a cylinder. The terms r and θ are as previously defined, while z represents the height above (or below) the x–y plane.

power. In general, the source of energy over time that actuates the robot's manipulator and other systems. Industrial robots have three basic power supplies: electric (usually wall plug provided), hydraulic, and pneumatic. As an historic note, many early automatons used hydraulic or pneumatic power. In a modern industrial facility, the pressurized liquid (hydraulic power) or gas (pneumatic power) is often created away from the robot's immediate work site and made available to the robot through pressurized lines. Power is a very important issue with mobile/field robots, whether they operate on Earth or in outer space. Batteries are a common source of electric power, sometimes in combination with solar cells. Large UGVs and UAVs often use gasoline or diesel engines for motive power and can use a simple generator to provide electricity during the mission. Space robots use solar cells, rechargeable batteries, or radioisotope thermoelectric generators (RTGs) for their electricity. Fuel cells are also an option. On certain occasions, teleoperated, mobile robots operating on Earth can be supplied electric power through a tethered cable.

portability. In computer science, the ease with which a computer program developed in one programming environment can be transferred to another programming environment.

predicate. That part of a proposition, which makes an assertion (for example, states a relation or attribute) about individuals.

predicate logic. A modification of prepositional logic to allow the use of variables and functions of variables.

premise. A first proposition upon which subsequent reasoning rests.

pressure (symbol: p). A thermodynamic property that two systems have in common when they are in mechanical equilibrium. Pressure is defined as the normal component of force per unit area exerted by a fluid on a boundary. The SI unit of pressure is the pascal (Pa).

probe. An instrumented robot spacecraft moving through the upper atmosphere or outer space or landing on another celestial body in order to obtain information about the specific environment, as, for example, a deep-space probe, a lunar probe, a Jovian atmosphere probe. *See also* **space robot**.

problem reduction. A problem-solving approach in AI in which operators are used to change a single problem into several subproblems that are usually much easier to solve.

problem-solving. In the field of AI, a procedure that uses a control strategy to apply operators to a situation in an attempt to achieve a goal.

problem state. The condition of a problem at a particular state.

procedural knowledge representation. A representation of knowledge about the world by a set of procedures—small computer programs that know how to do specific things or how to proceed in well-specified situations.

process. The collection or set of human activities used to create, invent, design, transform, produce, control, maintain, and use products or systems. A systematic sequence of actions that combines resources to produce an output.

production rule. A modular knowledge structure representing a single chunk of knowledge—usually expressed in If–Then or Antecedent–Consequent form. Often found in expert systems.

programmable robot. A type of industrial robot that is essentially a servo robot, which is driven by a programmable controller. The controller memorizes (stores) a sequence of movements and then repeats these movements and actions continuously. Often, engineers program this type of robot by "walking" the manipulator and end effector through the desired movement.

programming environment. The total programming setup, including the interface, the languages, the editors, and other programming tools.

Progress. A robot Russian supply spacecraft configured to perform automated rendezvous and docking operations with space stations and other orbiting spacecraft.

property list. A knowledge representation technique by which the state of the world is described by objects in the world by means of lists of their pertinent properties.

proposition. A statement (in logic) that can be true or false.

propositional logic. An elementary logic that uses argument forms to reduce the truth or falsehood of a new proposition from known propositions.

proton (symbol: p). A stable elementary nuclear particle with a single positive charge and a rest mass of about 1.672×10^{-27} kilograms, which is about 1,837 times the mass of an electron. A single proton makes up the nucleus of an ordinary (or) light hydrogen atom. Protons are also constituents of all other nuclei. The atomic number (Z) of an atom is equal to the number of protons in its nucleus.

prototype. A full-scale working model used to test a design concept by making actual observations and necessary adjustments. The prototype often serves as a base for constructing future models or systems.

pseudo-reduction. An approach to solving the difficult problem case in AI where multiple goals must be satisfied simultaneously. Plans are found to achieve each goal independently and then integrated using knowledge of how plan segments can be intertwined without destroying their important effects.

pulse code modulation (PCM). The transmission of information by controlling the amplitude, position, or duration of a series of pulses. An analog signal (e.g., an image, music, a voice, etc.) is broken up into a digital signal (i.e., binary code) and then transmitted in this series of pulses via telephone line or radio waves.

pump. A machine for transferring mechanical energy from an external source to the fluid flowing through it. The increased energy is used to lift the fluid, to increase its pressure, or to increase its rate of flow.

pyrometer. An instrument for the remote (noncontact) measurement of temperatures. This term is generally applied to instruments that measure temperatures above 600° Celsius.

quantization. The fact that electromagnetic radiation (including light) and matter behave in a discontinuous manner, and manifest themselves in the form of tiny "packets" of energy called *quanta* (singular: quantum).

quantum (*plural:* quanta). In modern physics, a discrete bundle of energy possessed by a photon.

quantum mechanics. The physical theory that emerged from Max Planck's original quantum theory and developed into wave mechanics, matrix mechanics, and relativistic quantum mechanics in the 1920s and 1930s. Within the realm of quantum mechanics, the Heisenberg uncertainty principle and the Pauli exclusion principle provide a framework that dictates how particles behave at the atomic and subatomic levels.

radar. An active form of remote sensing generally used to detect objects in the atmosphere and space by transmitting electromagnetic waves (such as, radio or

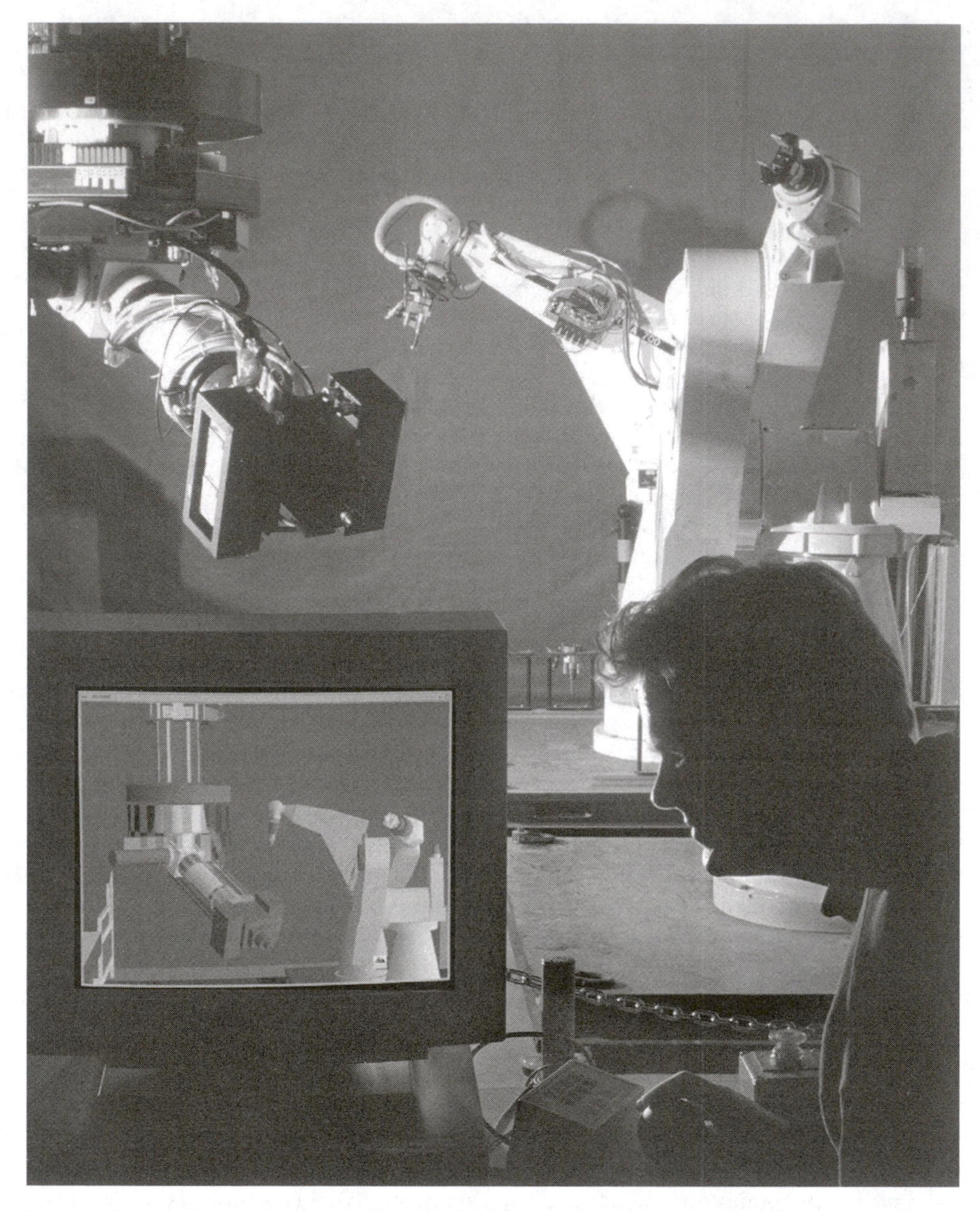

Figure 8-11 At technician at the Sandia National Laboratories in Albuquerque, New Mexico, is shown here programming three large manipulator robots in Sandia's manufacturing science and engineering laboratory. The technician is also using the laboratory's graphical programming system, which allows a human operator to test robot motions on a computer and then download and execute the motions on the actual manipulator robots. (Credit: Courtesy of the U.S. Department of Energy and Sandia National Laboratories, Albuquerque.)

microwaves) and sensing the waves reflected by the object. The reflected waves (called returns or echoes) provide information on the distance to the object and the velocity of the object (if it is moving). The reflected waves also can provide information about the shape of the object, its mass, and whether it is rotating or maintaining a fixed orientation or attitude. The term radar is actually an acronym for radio detection and ranging.

radar altimeter. An active instrument, carried onboard an UAV or a robot spacecraft, used for measuring the distance (or altitude) of the vehicle or craft above the surface of a planet. An accurate determination of altitude is obtained by carefully timing the travel of a radar pulse down to the surface and back.

radian. A unit of angle. One radian is the angle subtended at the center of a circle by an arc equal in length to a radius of the circle (that is, 1 radian $= 360^\circ/(2\pi) = 57.2958^\circ$).

radiant heat transfer. The transfer of thermal energy (heat) by electromagnetic radiation that arises due to the temperature of a body. Most energy transfer of this type is in the infrared portion of the electromagnetic spectrum. However, if the emitting object has a high enough temperature, it also will radiate in the visible spectrum and beyond. The term thermal radiation often is used to distinguish this form of electromagnetic radiation from other forms, such as radio waves, light, X-rays, and gamma rays. Unlike convection and conduction, radiant heat transfer takes place in and through a vacuum.

radiation. The propagation of energy by electromagnetic waves (photons) or streams of energetic nuclear particles is called radiation. Nuclear radiation generally is emitted from atomic nuclei (as a result of various nuclear reactions) in the form of alpha particles, beta particles, neutrons, protons, and/or gamma rays.

radio frequency (RF). In general, a frequency at which electromagnetic radiation is useful for communication purposes; specifically, a frequency above 10,000 hertz and below 3×10^{11} hertz. One hertz is defined as 1 cycle per second.

radioisotope thermoelectric generator (RTG). A portable electric power system in which thermal energy (heat) deposited by the absorption of alpha particles from a radioisotope source (generally plutonium-238) is converted directly into electricity. Radioisotope thermoelectric generators, or RTGs, have been used in American space missions where long life, high reliability, operation independent of the distance or orientation to the Sun, and operation in severe environments (such as, lunar night, Martian dust storms, Jupiter's intense radiation belts) are critical.

radiometer. An instrument for detecting and measuring radiant energy, especially infrared radiation.

radionuclide. A radioactive isotope characterized according to its atomic mass (A) and atomic number (Z). Radionuclides experience spontaneous decays in accordance with their characteristic half-life and can be either naturally occurring or human-made.

radio waves. Electromagnetic waves of wavelength between about 1 millimeter (0.001 meter) and several thousand kilometers; and corresponding frequencies between 300 gigahertz and a few kilohertz. The higher frequencies are used for telecommunications with robot spacecraft.

Ranger Project. The Ranger spacecraft were the first U.S. robot spacecraft sent toward the Moon in the early 1960s to pave the way for the Apollo Project's human landings at the end of that decade. The Rangers were a series of fully attitude-controlled space robots designed to photograph the lunar surface at close range before impacting. *Ranger 1* was launched on August 23, 1961, and set the stage for the rest of the Ranger missions, by testing spacecraft navigational performance. *Ranger 2* through *9* were launched from November 1961 through March 1965. All of the early Ranger missions (*Ranger 1* through *6*) suffered setbacks of one type or another. Finally, *Ranger 7*, *8*, and *9* succeeded with flights that returned many thousands of images (before impact).

raw data. Data that have not been reduced or processed.

reaction engine. An engine that develops thrust by its physical reaction to the ejection of a substance (including possibly photons and nuclear radiations) from it; commonly, the reaction engine ejects a stream of hot gases created by combusting a propellant within the engine. A reaction engine operates in accordance with Sir Isaac Newton's third law of motion (i.e., the action–reaction principle). Both rocket engines and jet engines are reaction engines. Sometimes called a *reaction motor*.

readout 1. (verb) The action of a UAV or robot spacecraft's transmitter sending data that are either being instantaneously acquired or else extracted from storage (often by playing back a magnetic tape upon which the data have been recorded previously). 2. (noun) The data transmitted by the action described in sense 1. 3. (verb) In computer operations, to extract information from storage.

readout station. A recording (or receiving) station at which the data-carrying radio-frequency signals transmitted by a UAV or a robot spacecraft are acquired and initially processed.

real-time. Time in which reporting on or recording events is simultaneous with the events; essentially, "as it happens."

real-time data. Data presented in usable form at essentially the same time the event occurs.

reconnaissance satellite. A robot military spacecraft in orbit around Earth that performs a reconnaissance mission (such as gathering images or collecting radio frequency emissions) against enemy nations and potential adversaries.

rectangular coordinate motion. *See* **Cartesian coordinate motion.**

rectangular coordinate robot. *See* **Cartesian coordinate robot.**

rectifier. A device for converting alternating current (AC) into direct current (DC); usually accomplished by permitting current flow in one direction only.

recursive operations. Operations in AI that are defined in terms of themselves. In mathematics, a recursion is an expression (such as a polynomial), each term of which is determined by applying a formula to the preceding terms.

redline. Term denoting a critical value for a parameter or a condition that, if exceeded, threatens the integrity of a system, the performance of a robot, or the success of its activity or mission.

regenerator. A device used in a thermodynamic process for capturing and returning to the process thermal energy (heat) that otherwise would be lost. The use of a regenerator helps increase the thermodynamic efficiency of a heat engine cycle.

regulator. Flow-control device that adjusts the pressure and controls the flow of fluid to meet the demands of a robot system, which has fluid flow devices within its subsystems and assemblies.

relative atomic mass (symbol: A). The total number of nucleons (that is, both protons and neutrons) in the nucleus of an atom. Also called the atomic mass or sometimes *atomic mass number*. For example, the relative atomic mass of the isotope carbon-12 is 12.

relative state. In a operation involving several robot systems, the position and motion of one robot relative to another.

relaxation approach. An iterative problem-solving approach in which initial conditions are propagated using constraints until all goal conditions are adequately satisfied.

reliability. The probability of specified performance of a piece of equipment or system under stated conditions for a given period of time.

remote control. Control of an operation from a distance, especially by means of telemetry and electronics; a controlling switch, level, or other device used in this type of control, as in remote-control arming switch. *See also* **teleoperation.**

remotely piloted vehicle (RPV). An aerial vehicle whose pilot does not fly onboard but rather controls it at a distance (i.e., remotely) using a telecommunications link from a crewed aircraft or ground station. RPVs are often used on extremely hazardous missions or on long-duration missions involving extended loitering and surveillance activities. *See also* **unmanned aerial vehicle (UAV).**

remote manipulator system (RMS). The Canadian-built, 15.2-meter-long articulating robot arm that is remotely controlled from the aft flight deck of NASA's space shuttle orbiter. The elbow and waist movements of the RMS permit payloads to be grappled for deployment out of the cargo bay or to be retrieved and secured in the cargo bay for on-orbit servicing or return to Earth. There is a similar system on the *International Space Station* (ISS).

remote sensing. The sensing of an object, event, or phenomenon without having the sensor in direct contact with the object being studied. Information transfer from the object to the sensor is accomplished through the use of the electromagnetic spectrum.

resilience. The property of a material that enables it to return to its original shape and size after deformation. For example, the resilience of a sealing material is the property that makes it possible for a seal to maintain sealing pressure despite wear, misalignment, or out-of-round conditions. This term is also applied in aerospace operations to describe the relative hardiness or "robustness" of a robot spacecraft or crew-occupied space vehicle that can suffer several significant component or subsystem degradations or failures and still function (through hardware and software "work-arounds" and automated fault isolation procedures) at a performance level sufficient to continue and/or complete the mission.

resistance (symbol: R). 1. Electrical resistance (R or Ω) is defined as the ratio of the voltage (or potential difference) (V) across a conductor to the current (I) flowing through it. In accordance with Ohm's law, $R = V/I$. The SI unit of resistance is the ohm (Ω), where 1 ohm = 1 volt per ampere. 2. Mechanical resistance is the opposition by frictional effects to forces tending to produce motion. 3. Biological resistance is the ability of plants and animals to withstand poor environmental conditions and/or attacks by chemicals or disease. This ability may be inborn or developed—as, for example, through the application of nanotechnology in genetic engineering.

resolution. 1. Generally, a measurement of the smallest detail that can be distinguished by a sensor system under specific conditions. 2. The degree to which fine details in an image or photograph can be seen as separated or resolved. *Spatial resolution* often is expressed in terms of the most closely spaced line-pairs per unit distance that can be distinguished. For example, when the resolution is said to be 10 line-pairs per millimeter, this means that a standard pattern of black-and-white lines whose line plus space width is 0.1 millimeter is barely resolved by an optical system, finer patterns are not resolved, and coarser patterns are more clearly resolved. *Spectral resolution* involves how finely the lines in a particular spectrum can be resolved and studied as a function of wavelength or energy level.

resource. In a technological system, the basic technological resources are energy, capital, information, machines and tools, materials, people, and time.

resolving power. In general, the finest detail an optical instrument can provide. With respect to remote sensing, a measure of the ability of individual components or the entire remote-sensing system, to define (and therefore "resolve") closely space targets.

retrofit. The modification of or addition to a robot system after it has become operational.

retroreflector. A mirror-like instrument, usually a corner reflector design, that returns light or other electromagnetic radiation (e.g., an infrared laser beam) in the direction from which it comes. Robots with machine vision systems are often assisted by the

Figure 8-12 In the grasp of the shuttle's 15.2-meter-long articulating robot arm, called the remote manipulator system (RMS), the U.S. Laboratory (*Destiny*) is carefully moved from its stowage position in the cargo bay of the space shuttle *Atlantis*. The RMS then helped astronauts connect the *Destiny* laboratory to the *International Space Station* (ISS). This photograph was taken during the STS-98 mission (February 2001) by astronaut Thomas D. Jones during his extravehicular activity (EVA). (Credit: Photograph courtesy of NASA/MSFC.)

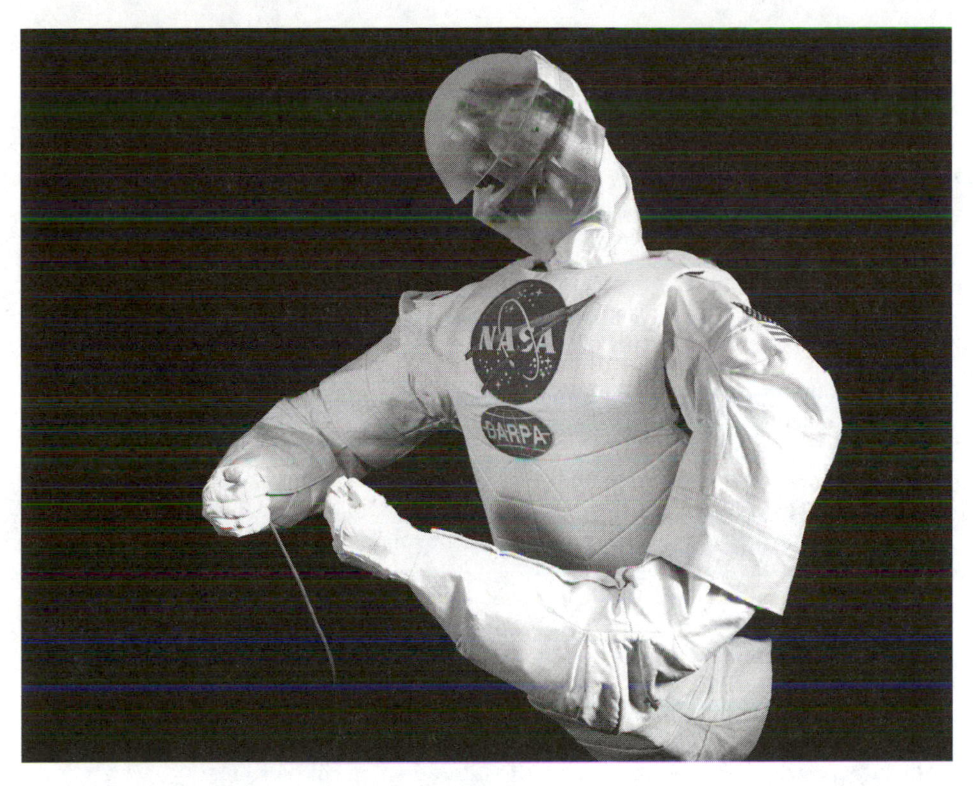

Figure 8-13 Robonaut is a humanoid robot designed by NASA at the Johnson Space Center (JSC) in a collaborative effort with the Defense Advanced Research Projects Agency (DARPA). The Robonaut project seeks to develop and demonstrate a robot system that can function as a human astronaut during extravehicular activity (EVA). This picture shows Robonaut B, the newer of two NASA robots used in hand-in-hand testing with human astronauts. (Credit: Courtesy of NASA/JSC.)

placement of retroreflectors on target objects, along assigned routes, or at strategic locations in the anticipated area of operation (for the purposes of range and location determination).

Robonaut. A humanoid robot designed by NASA's Johnson Space Center (JSC) in a collaborative effort with the Defense Advanced Research Projects Agency (DARPA). Once fully demonstrated, Robonaut leads to a humanoid robot that can function like a human astronaut during hazardous extravehicular activities (EVAs) in outer space and on the lunar and Martian surfaces.

robot. A smart machine that does routine, repetitive, hazardous mechanical tasks, or performs other operations either under direct human command and control or on its own, using a computer with embedded software (which contains previously loaded commands and instructions) or with an advanced level of machine (artificial) intelligence (which bases decisions and actions on data gathered by the robot about its current environment.) The word robot is derived from *robata*, a Czech word meaning "compulsory labor" or "servitude." The expression "robot" is attributed to Czech writer Karel Capek, who wrote the play *R.U.R. (Rossum's Universal Robots)*. Robots can be fixed in a place or mobile; they can be odd-looking machines or human-like systems. As computer technology and materials science continue to improve, the variety, characteristics, capabilities, and intelligence of robots will likewise expand this century. Contemporary robots include entire technical series and families of: industrial robots, environmental monitoring and cleanup robots, military and national security robots, and space exploration robots. Other contemporary robot systems support research (e.g., laboratory robots), the practice of medicine (telemedicine and robot-assisted surgery), the leisure time and entertainment industry (pet robot animals), archaeology and marine salvage activities (underwater ROVs), and law enforcement functions (such as, bomb disposal robots).

robotics. The science and technology of designing, building, and programming robots. Robotic devices, or robots as they are usually called, are primarily smart machines with manipulators or other devices that can be programmed to do a variety of manual or human labor tasks automatically.

robot rover. *See* **rover.**

robot spacecraft. A semiautomated or fully automated spacecraft capable of executing its primary exploration mission with minimal or no human supervision.

roll. The rotational or oscillatory movement of robotic vehicle about its longitudinal (lengthwise) axis. *See also* **pitch; yaw.**

root node. The initial (apex) node in a tree representation.

rover. A robot vehicle that can travel across the surface of Earth (such as members of the family of military robots designated as UGVs) or is used to study the surface of another world. The rover can be totally dependent on human controllers for instruction, semiautonomous, or fully autonomous—depending on the goals of its mission, the environmental conditions encountered, and its level of machine intelligence (AI).

rule-interpreter. The control structure for a production rule system.

Santa Claus machine. A postulated supermachine that can manufacture and assemble any object or structure from an initial supply of elemental materials, energy, and information. The American physicist Theodore (Ted) Taylor (1925–2004) introduced the concept for this type of incredible machine in 1978 and gave the hypothesized device its quaint name.

safety analysis. The determination of potential sources of danger and recommended resolutions in a timely manner. A safety analysis addresses those conditions found in either the hardware/software systems, the human-machine interface, or the human/environment relationship (or combinations thereof) that could cause the injury or death of supporting personnel, damage or loss of the robot system, injury or loss of life to the public, or harm to the environment.

safety device. A device that prevents unintentional functioning of a robot system or one or more of its potentially hazardous subsystems, such as a gripper or a long, massive mechanical arm.

scalar. Any physical quantity whose field can be described by a single numerical value at each point in space. A scalar quantity is distinguished from a vector quantity by the fact that a scalar quantity possesses only magnitude, while a vector quantity possesses both magnitude and direction. *Compare with* **vector.**

scheduling. The development of a time sequence of things to be done.

scripts. Frame-like structures for representing sequences of events.

science payload. The complement of scientific instruments on a robot spacecraft, including both remote-sensing and direct-sensing devices that together cover large portions of the electromagnetic spectrum, large ranges in particle energies, or a detailed set of environmental measurements.

scientific notation. A method of expressing powers of 10 that greatly simplifies writing large numbers. In scientific notation, a number expressed in a positive power of 10 means the decimal point moves to the right (e.g., $3 \times 10^6 = 3{,}000{,}000$); a number expressed in a negative power of 10 means that the decimal moves to the left (e.g., $3 \times 10^{-6} = 0.000003$).

sealant. Liquid/solid mixture installed at joints and junctions of components to prevent leakage of fluid (especially gas) from the joint or junction.

search space. The implicit graph representing all possible states of the system, which may have to be searched to find a solution. In many cases the search space is infinite. The term search space is also used for non-state-space representations.

second. 1. The SI unit of time (symbol: s) is now defined as the duration of 9,192,631,770 periods of radiation corresponding to the transition between two hyperfine levels of the ground state of the cesium-133 atom. Previously, this unit of time had been based on astronomical observations. 2. A unit of angle (symbol: ″) called the arc second, equal to 1/3600 of a degree of angle or 1/60 of an arc minute.

self-replicating system (SRS). An advanced space robot system, first postulated by John von Neumann (1903–1957), that would be capable of gathering materials, maintaining itself, manufacturing desired products, and even making copies of itself ("self-replication").

semantic network. A knowledge representation for describing the properties and relations of objects, events, concepts, situations, or actions, by a directed graph consisting of nodes and labeled edges (arcs connecting nodes).

semantic primitives. Basic conceptual units in which concepts, idea, or events can be represented.

sensor. In general, a device that detects and/or measures certain types of physically observable phenomena. More specifically, that part of an electronic instrument that detects electromagnetic radiations (or other characteristic emissions, such as nuclear particles) from a target or object at some distance away and then converts these incident radiations (or particles) into a quantity (i.e., an internal electronic signal) that is amplified, measured (quantified), displayed, and/or recorded by another part of the instrument. A *passive sensor* uses characteristic emissions from the object or target as its input signal. In contrast, an *active sensor* (like a radar) places a burst of electromagnetic energy on the object or target being observed and then uses the reflected signal as its input.

sensory robot. A computerized robot with one or more artificial senses to observe and record its environment and to feed information back to the controller. The artificial senses most frequently employed are sight (robot or computer vision) and touch (tactile sensors). The sensory robot can be an advanced industrial robot or a sophisticated mobile robot that supports such missions, as national defense activities, search and rescue operations, environmental monitoring and cleanup activities, and civilian law enforcement.

sentry robot. A mobile robot designed to perform security activities, such as surveillance operations, random-walk patrols, and intruder detection. The sentry robot can be integrated into the normal security operations of a protected facility (such as a weapons storage site), or be set up as a rapidly emplaced (or emergency) surveillance system that supplements the human guard force maintaining a security perimeter around some important item, facility, or geographic location, which requires a high-level of intrusion protection on a temporary basis. Sentry robots are not normally designed to carry weapons that would inflict lethal injuries upon intruders. However, some sentry robots are equipped with self-defense systems that can incapacitate intruders. When disturbed or attacked by an intruder, a sentry robot will sound an alarm, up to an including the point of "die screaming." In a well-designed security system the collection of sentry robots are in constant communications, as they wander around the facility or survey the scene of assigned positions. A disruption of the communications link automatically notifies the human command post that there is a breach in security.

Figure 8-14 A prototype (early-1980s) sentry robot, called ROBART I. (Credit: Photograph courtesy of the U.S. Navy.)

serial robot. A robot that is a single chain of joints connected by links.

servo. A device that helps control (usually by hydraulic means) a large moment of inertia by the application of a relatively small moment of inertia. Also called a *servomechanism.*

servo robot. The servo robot (or servo-controlled robot) represents several categories of industrial robots, including the point-to-point robot and the continuous path robot. This type of robot has servomechanisms for the manipulator and end effector to enable it to change direction in midair (or mid stroke) without having to strip or trigger a mechanical limit switch. Five to seven directions of motion are common, depending on the number of joints in the manipulator. The servo robot is the most common industrial robot in use today. *See also* **industrial robot**.

shaft. A bar or rod (almost always cylindrical) used to support rotating pieces or to transmit power or motion by rotation.

shake table. Device for subjecting components or assemblies to vibration in order to reveal vibrational mode patterns.

shear strength. In materials science, the stress required to produce fracture in the plane of cross section, the conditions of loading being such that the directions of force and of resistance are parallel and opposite, although their paths are offset a specified minimum amount.

shelf life. Storage time during which an item, such as a battery, remains serviceable—that is, will operate satisfactorily when put to use.

shutoff valve. Valve that terminates the flow of fluid; usually a two-way valve that is either fully open or fully closed.

siemens (symbol: S). The SI unit of electrical conductance. It is defined as the conductance of an electrical circuit or element that has a resistance of one ohm. In the past this unit sometimes was called the mho or reciprocal ohm. The unit is named in honor of the German scientist Ernst Werner von Siemens (1816–1892).

signal. 1. Information to be transmitted over a communications system. 2. A visible, audible, or other indication used to convey information. 3. Any carrier of information; as opposed to noise. 4. In electronics, any transmitted electrical impulse. The variation of amplitude, frequency, and waveform are used to convey information.

signal-to-noise ratio (SNR). The ratio of the amplitude of the desired signal to the amplitude of noise signals at a given point in time. The higher the signal-to-noise ratio, the less interference with reception of the desired signal.

simulation. 1. The art of replicating relevant portions of the real-world environment to test equipment, train mission personnel, and prepare for emergencies. Simulations can involve the use of physical mass and energy replicants, high-fidelity (that is, reasonably close to the original) hardware, and supporting software. With the incredible growth in computer and display technologies, computer-based simulations are assuming an ever-increasing role as robot system design tools and human controller training aids. Engineers often refer to top quality computer-based simulations as "reality in a box."

SI unit(s). The Systéme International d'Unités (International System of Units, or SI units) is the internationally agreed-upon system of coherent units that is now in use throughout the world for scientific and engineering purposes. The metric (or SI) system was developed originally in France in the late eighteenth century. Its basic units for length, mass, and time—the meter (m), the kilogram (kg), and the second (s)—were based on natural standards. The modern SI units are still based on natural standards and international agreement, but these standards now are ones that can be measured with greater precision than the previous natural standards.

skin. The outer covering of a robot, especially a mobile robot.

slew. To change the position of a sensor or an antenna assembly by injecting a signal into the positioning servo-mechanism.

slip flow. Flow in the transition regime of gas dynamics, wherein the mean free path of the gas molecules is of the same order of magnitude as the thickness of the boundary layer. The gas in contact with a body surface immersed in the flow is no longer at rest with respect to the surface.

slipstream. The flow of fluid around a structure that is moving through the fluid.

slot. An element in a frame representation to be filled with designated information about the particular situation.

smart robot. A computer-enriched robot that can be programmed to make performance choices, which are contingent upon the input from the various sensors (such as vision, tactile, environmental, radiation, etc). For example, a smart mobile robot with a machine-vision system would be capable of detecting and then navigating around most obstacles without direct human supervision. A teleoperated, smart environmental monitoring robot might stop in its tracks when its radiation sensors encounter a certain level of ionizing radiation (preprogrammed as unacceptable) at a nuclear waste cleanup site. The robot's human controllers could then note the precise location of the radiation hot spot, before directing the robot to continue with its surveillance task.

snubber. A device used to increase the stiffness of an elastic system (usually by a large factor), whenever the displacement of the system becomes larger than a specified amount.

soft landing. The act of landing on the surface of a planet without damaging any portion of a robot spacecraft or its payload, except possibly an expendable landing gear structure. *Compare with* **hard landing.**

software. The programs (that is, sets of instructions and algorithms) and data used to operate a digital computer. *Compare with* **hardware.**

solar cell. A direct energy conversion device that has been used for over four decades to provide electric power for robot spacecraft. A solar cell or photovoltaic system turns sunlight directly into electricity. The solar cell has no moving parts to wear out and produces no noise, fumes, or other polluting waste products. However, the space environment, especially trapped radiation belts and the energetic particles released in solar flares, can damage solar cells used on spacecraft and reduce their useful lifetime.

solar panel. A wing-like set of solar cells used by a spacecraft to convert sunlight directly into electric power; also called a solar array.

solar photovoltaic conversion. The direct conversion of sunlight (solar energy) into electrical energy by means of the photovoltaic effect. A single photovoltaic (PV) converter cell is called a solar cell, while a combination of cells, designed to increase the electric power output, is called a solar array or a solar panel.

solenoid. Helical coil of insulated wire that, when conducting electricity, generates a magnetic field that actuates a movable core.

solid. A state of matter characterized by a three-dimensional regularity of structure. When a solid substance is heated beyond a certain temperature, called the melting point, the forces between its atoms or molecules can no longer support the characteristic lattice structure, causing it to break down as the solid material transforms into a liquid or (more rarely) transforms directly into a vapor (sublimation).

solid angle (Symbol: Ω). Three-dimensional angle formed by the vertex of a cone; that portion of the whole of space about a given point, bounded by a conical surface with its vertex at that point and measured by the area cut by the bounding surface from the surface of a sphere of unit radius centered at that point. The *steradian* (sr) is the SI unit of solid angle.

solid lubricant. A dry film lubricant.

solid-state device. A device that uses the electric, magnetic, and photonic properties of solid materials, mainly semiconductors. It contains no moving parts and depends on the internal movement of charge carriers (i.e., electrons and "positive" holes) for its operation.

solution path. In AI, a successful path through a search space.

sorbent. A substance or material that takes up gas by absorption, adsorption, chemisorption, or any combination of these processes.

sound. In physics and engineering, a vibration in an elastic medium that is at a frequency and intensity, which can be heard by a human being. The normal human ear can respond to sounds in the frequency range from approximately 20 to 20,000 hertz

Figure 8-15 This unusual photograph shows NASA's *Surveyor 3* robot spacecraft located on the Moon's surface in the Ocean of Storms. The picture was taken in November 1969 by the Apollo 12 astronauts (Charles Conrad, Jr. and Alan L. Bean), who visited the robot spacecraft during their second extravehicular activity (EVA) on the surface. The lunar excursion module (LEM), which carried the two astronauts to the Moon's surface, can be seen in the background on the horizon. (Credit: Photograph courtesy of NASA.)

(Hz). Vibrations lower than this frequency range are called *infrasounds*, while those above this frequency range are called *ultrasounds*.

spacecraft. A platform that can function, move, and operate in outer space or on a planetary surface. Spacecraft can be a human-occupied or robotic. Some spacecraft operate in orbit around Earth or in interplanetary space. Other spacecraft orbit around another celestial body, plunge through its atmosphere, or land on its surface. Robots spacecraft used for exploration are often categorized as either: flyby, orbiter, atmospheric probe, lander, or rover spacecraft.

spacecraft clock. Generally, the timing component within a robot spacecraft's command and data-handling subsystem. This important device meters the passing time during the life of the space robot and regulates nearly all activity within the system.

space robot. Any one of a wide variety of robot spacecraft used to explore interplanetary space and other planetary systems. NASA uses the following general categories to describe the agency's large family of free flying space robots: flyby, orbiter, atmospheric probe, lander, and rover spacecraft. Space robots also include the versatile, well-articulated manipulator systems on the space shuttle and the *International Space*

Station (IVA), as well as astronaut-interactive robot assistants, such as an android-like system, called robonaut, and a soccer-ball sized freely robot.

spalling. Flaking off of particles and chunks from the surface of a material as a result of localized stresses.

spare. An individual part, subassembly, or assembly supplied for the maintenance or repair of robot systems or support equipment. Also a complete, functioning robot system kept in a dormant, nonoperating mode, which serves as an immediate replacement of a similar or identical functioning unit.

specific volume (symbol: *v*). Volume per unit mass of a substance; the reciprocal of density.

spectroscopy. The study of spectral lines from different atoms and molecules. Scientists use emission spectroscopy to infer the material composition of the objects that emitted the light and absorption spectroscopy to infer the composition of the intervening medium.

speech recognition. Recognition by a computer (primarily by pattern matching) of spoken words or sentences.

speech synthesis. Developing spoken speech from text or other representations.

speech understanding. Speech perception by a computer or a robot that possesses a certain level of AI.

speed of light (symbol: c). The speed of propagation of electromagnetic radiation (including light) through a perfect vacuum. Scientists consider the speed of light as a universal constant equal to 299,792.458 kilometers per second.

speed of sound. The speed at which sound travels in a given medium under specified conditions. The speed of sound at sea level in the International Standard Atmosphere is 1,215 kilometers per hour. Sometimes called the *acoustic velocity*.

spherical coordinate robot. An industrial robot that has a configuration similar to a tank turret. The robot's arm can move (or slide) in and out. The arm can also be raised and lowered in an arc (much like a tank's cannon can be raised or lowered in adjust the firing elevation). Finally, the robot's arm can rotate about the base (much like a tank's turret rotates about the weapon system's treaded chassis.) *See also* **industrial robot**.

spin stabilization. Directional stability of a robot spacecraft obtained by the action of gyroscopic forces that result from spinning the body about its axis of symmetry.

star probe. A conceptual NASA robot scientific spacecraft, capable of approaching within one million kilometers of the Sun's surface (photosphere) and providing the first in situ measurements of its corona (outer atmosphere).

starship. A conceptual, very advanced space vehicle capable of traveling the great distances between star systems within decades or less. The term "starship" is generally reserved for vehicles that could carry intelligent beings, while interstellar probe applies to an advanced robot spacecraft capable of interstellar travel at 10 percent or more of the speed of light.

state graph. A graph in which the nodes represent the system state and the connecting arcs represent the operators that can be used to transform the state from which the arcs emanate to the state at which the arcs arrive.

state of the art (SOA). The level to which technology and science have been developed in any given discipline or industry at some designated cutoff time.

statics. The branch of mechanics that studies stationary (nonmoving) rigid bodies under the influence of external forces.

static testing. The testing of a robot system in a stationary or hold-down position, either to verify structural design criteria, structural integrity, hardware/software compatibility, the continuity of electric circuitry, leak integrity for subsystems containing fluids, and similar engineering evaluations. For a robot system under development, static testing helps avoid costly, compounded errors and catastrophic failures during dynamic, full-scale testing activities. For an operational robot system, static testing is similar in scope and purpose to the preflight check and pilot "walkaround" performed on an aircraft before the vehicle is committed to flight.

stationkeeping. The sequence of maneuvers (usually performed automatically) that maintains a robot spacecraft in a predetermined orbit or on a desired trajectory.

steady state. Condition of a physical system in which parameters of importance (fluid velocity, temperature, pressure, etc.) do not vary significantly with time.

stochastic process. A statistical process; a process in which there is a random variable.

strain. In engineering, the change in the shape or volume of an object due to applied forces. There are three basic types of strain: longitudinal, volume, and shear. Longitudinal (or tensile) strain is the change in length per unit length, as occurs, for example, with the stretching of a wire. Volume (or bulk) strain involves a change in volume per unit volume, as occurs, for example, when an object is totally immersed in a liquid and experiences a hydrostatic pressure. Finally, shear strain is the angular deformation of an object without a change in its volume. Shear strain occurs, for example, when a rectangular block of metal is strained or distorted in such a way that two opposite faces become parallelograms, while the other two opposite do not change their shape.

stress. In engineering, a force per unit area on an object that causes it to deform (that is, experience strain). Stress can be viewed as either the system of external forces applied to deform an object or the system of internal "opposite" forces (a function of the material composition of the object) by which the object resists this deformation. The three basic types of stress are: compressive (or tensile) stress, hydrostatic pressure, and shear stress.

stretchout. An action whereby the time for completing a robot system development project is extended beyond the time originally programmed or contracted for. Cost overruns, unanticipated technical delays, and budget cuts are frequent reasons why the stretchout is needed during the development of a new robot system. Often occurs when engineers attempt to blend complex new technologies together for the first time.

subassembly. A portion of an assembly, consisting of two or more parts, that can be provisioned and replaced as an entity.

subgoals. Goals that must be achieved in order to achieve the original goal.

sublimation. In thermodynamics, the direct transition of a material from the solid phase to the vapor phase, and vice versa, without passing through the liquid phase.

subplan. A plan to solve a portion of the problem.

subproblems. The set of secondary problems that must be solved in order to solve the original problem.

subsonic. Of or pertaining to speeds less than the speed of sound. *See also* **speed of sound**.

superluminal. With a (hypothetical) speed greater than the speed of light.

supersonic. Of or pertaining to speed in excess of the speed of sound. *See also* **speed of sound**.

surface penetrator spacecraft. A robot space probe designed to enter the surface of a celestial body, such as a comet or asteroid. The penetrator is capable of surviving a high-velocity impact and then making in situ measurements of the penetrated surface. Data are sent back to a mother spacecraft for retransmission to scientists on Earth.

surface rover spacecraft. An electrically powered robot vehicle designed to explore a planetary surface. Depending on the size of the rover and its level of sophistication, this type of mobile craft is capable of semiautonomous to fully autonomous operation. The rover can perform a wide variety of exploratory functions, including the acquisition of multispectral imagery, soil sampling and analysis, and rock inspection and collection. Data are transmitted back to Earth either directly by the rover vehicle or via a lander spacecraft or orbiting mother spacecraft.

surface tension. The tendency of a liquid that has a large cohesive force to keep its surface as small as possible, forming spherical drops. Surface tension arises from intermolecular forces and is manifested in such phenomena as the absorption of liquids by porous surfaces, the rise of water (and other fluids) in a capillary tube, and the ability of liquids to wet a surface. Surface tension is related to capillary forces and the movement of fluids in wicks.

surveillance satellite. A robot military spacecraft that orbits Earth and watches regions of the planet for hostile military activities, such as ballistic missile launches and nuclear weapon detonations.

Surveyor Project. NASA's Moon exploration effort in which five robot lander spacecraft softly touched down on the lunar surface between 1966 and 1968. These advanced (for the time) robot spacecraft served as technical precursors for the Apollo Project human expeditions.

survivability. The capability of a robot system to avoid or withstand hostile environments without suffering irreversible impairment of its ability to accomplish its designated mission.

syllogism. A deductive argument in logic whose conclusion is supported by two premises.

symbolic. Relating to the substitution of abstract representations (symbols) for physical, real world objects.

syntax. The grammar of a language; how a language is ordered or arranged.

synthetic aperture radar (SAR). A radar system that correlates the echoes of signals emitted at different points along a UAV's flight path or robot spacecraft's orbit. The SAR system generally illuminates its target to the side of its direction of movement and travels a known distance along its planned flight path (orbit) while the reflected, phase shift-coded pulses are returned and collected. With extensive computer processing, this procedure provides the basis for synthesizing an antenna (aperture) on the order of kilometers in size. The highest resolution achievable by such a system is theoretically equivalent to that of a single large antenna as wide as the distance between the most widely spaced points along the flight path (orbit) that are used for transmitting positions. Unlike visible spectrum images, images collected by a SAR system are independent of time of day or cloud cover.

synthetic material. Material that is not found in nature, such as concrete or various types of plastics.

system. A group of interacting, interrelated, or independent elements or parts that function together as a whole to accomplish a goal.

system integration process. The process of uniting the parts (components, assemblies, and subsystems) into a complete and functioning robot system.

tap. A unit of impulse intensity, defined as one dyne-second per square centimeter. 1 tap = 0.1 pascal-second.

technology. Human innovation in action that involves the generation of knowledge and processes to develop systems that solve problems and extend human capabilities. It can also be defined as the innovation, change, or modification of the natural environment to satisfy perceived human needs and wants.

telecommunications. The transmission of information over great distances using radio waves or other portions of the electromagnetic spectrum.

telemetry. The process of making measurements at one point and transmitting the information via radio waves over some distance to another location for evaluation and use. Telemetered data on a robot spacecraft's communications downlink often include scientific data, as well as spacecraft state-of-health (SOH) data.

teleoperation. The technique by which a human controller operates a versatile robot system that is at a distant, often hazardous, location. High-resolution vision and tactile sensors on the robot, reliable telecommunications links, and computer-generated virtual reality displays enable the human worker to experience telepresence.

telepresence. The process, supported by an information-rich control station environment, that enables a human controller to manipulate a distant robot through teleoperation and almost feel physically present in the robot's remote location.

telescience. A mode of scientific activity in which a distributed set of users (investigators) can interact directly with their instruments, whether in space or remote facilities on Earth, with databases, data handling and processing facilities, and with each other.

telescope. A device that collects electromagnetic radiation from a distant object so as to form an image of the object or to permit the radiation to be analyzed. Optical (or astronomical) telescopes are divided into two main classes: refracting telescopes and reflecting telescopes.

temperature (symbol: T). A thermodynamic property that determines the direction of heat (thermal energy) flow. From the laws of thermodynamics, when two objects or systems are brought together, heat naturally will flow from regions of higher temperature to regions of lower temperature.

tera- (symbol: T). A prefix in the SI units meaning multiplied by 10^{12}.

terminal node. The final node emanating from a branch in a tree or graph representation. Also called *leaf node*.

terraforming. The proposed, highly automated, large-scale modification or manipulation of the environment of a planet, such as Mars or Venus, to make that planet more suitable for human habitation. Also called *planetary engineering*.

tesla (symbol: T). The SI unit of magnetic flux density. It is defined as one weber of magnetic flux per square meter. The unit is named in honor of Nikola Tesla (1870–1943), a Croatian-American electrical engineer and inventor. [1 tesla $= 1$ weber/(meter)$^2 = 10^4$ gauss.]

test. In engineering and science, a procedure taken to determine, under simulated or real conditions, the capabilities, limitations, characteristics, effectiveness, reliability, or suitability of a material, device, system, or method. For example, engineers test a mobile robot to determine whether the robot (as designed) can successfully move across or navigate around certain terrain features or surface conditions.

test bed. A base, mount, or frame within or upon which a piece of equipment, such as an experimental robot vision system, is secured for testing.

theorem. A proposition or statement to be proved based on a given set of premises.

theorem-proving. A problem-solving approach in AI in which a hypothesized conclusion (theorem) is validated by deductive reasoning.

thermal. Of or pertaining to heat or temperature.

thermal conductivity (symbol: k). An intrinsic physical property of a substance, describing its ability to conduct heat (thermal energy) as a consequence molecular motion. Typical units for thermal conductivity are joules/(second-meter-kelvin).

thermal control. Regulation of the temperature of a robot system, with special attention being given to make sure that no component or subsystem experiences temperatures that exceed or fall below specified (safe) limits. For space robots, temperature control is a very complex problem due to the extreme temperature conditions encountered in outer space, where radiation heat transfer provides the only natural way to transfer thermal energy (heat) into or out of the system.

thermal cycling. Exposure of a component to alternating levels of relatively high and low temperatures.

thermal efficiency (symbol: η_{th}). *See* **thermodynamic efficiency**.

thermal equilibrium. A condition that exists when energy transfer as heat between two thermodynamic systems (for example, System A and System B) is possible but none occurs. Scientists say that System A and System B are in thermal equilibrium and that they have the same temperature.

thermistor. A semiconductor electronic device that uses the temperature-dependent change of resistivity of the substance. The thermistor has a very large negative temperature coefficient of resistance; that is, the electrical resistance decreases as the temperature increases. This device is generally used for temperature measurements or electronic circuit control. If a component or subsystem inside a robot overheats, a thermistor in a safety subsystem can detect the growing temperature problem and provide a signal that causes a temporary (safe) shutdown of the entire system. This is an engineered safety feature, which avoids permanent thermal damage in the offending component, subsystem, or overall robot system.

thermocouple. A device consisting of two conductors made of different metals, joined at both ends, producing a loop in which an electric current will flow when there is a difference in temperature between the two junctions. The amount of current that will flow in an attached circuit is dependent on: the temperature difference between the measurement (hot) and reference (cold) junction; the characteristics of the two different metals used; and the characteristics of the attached circuit. Depending on the different metals chosen, a thermocouple can be used as a thermometer over a certain temperature range.

thermodynamic efficiency (symbol: η_{th}). In thermodynamics, the ratio of work done by a heat engine (W_{out}) to the total heat supplied by the thermal energy source (Q_{in}). Also called *thermal efficiency* and *Carnot efficiency*.

thermodynamics. The branch of science that treats the relationships between thermal energy (heat) and mechanical energy. Within physics and engineering, thermodynamics involves the study of systems. A thermodynamic system is a collection of matter and space with its boundaries defined in such a way that energy transfer (as work and heat) across the boundaries can be identified and understood easily. The surroundings represent everything else that is not included in the thermodynamic system under study. Engineers usually place thermodynamic systems within one of three distinct groups: closed systems, open systems, and isolated systems. A closed system is a system for which only energy (but not matter) can cross the boundaries. An open system can experience both matter and energy transfer across its boundaries. An isolated system can experience neither matter nor energy transfer across its boundaries. A control volume is a fixed region in space that is defined and studied as a thermodynamic system. Often the control volume is used to help in the analysis of open systems. Steady state refers to a condition where the properties at any given point within the thermodynamic system are constant over time. Neither mass nor energy accumulate (or deplete) in a steady state system.

thermometer. An instrument or device for measuring temperature.

thermonuclear. Pertaining to nuclear reactions in which very high temperatures (millions of kelvins) are needed to bring about the fusion (joining) of light nuclei, such as deuterium (D) and tritium (T), with the accompanying release of energy.

Thousand Astronomical Unit (TAU) mission. A conceptual future NASA mission involving an advanced-technology robot spacecraft that travels on a 50-year journey into very deep space more than 1,000 astronomical units (about 160 billion kilometers) away from Earth. The *TAU* spacecraft would feature an advanced multimegawatt nuclear reactor, ion propulsion, and a laser (optical) communications system.

throttling process. In thermodynamics, an adiabatic process in which the enthalpy of a working fluid remains constant and no work is done. This process, usually involves a mechanical throttling device that restricts fluid flow and leads to a decrease in fluid pressure.

thrust (symbol: T). The forward force provided by a reaction motor or device.

time-sharing. A computer environment in which multiple users can employ the computer essentially simultaneously by means of a program that time-allocates the use of computer resources among the users in a near-optimum manner.

tolerance. The allowable variation in measurements within which the dimensions of an item are judged acceptable.

tolerance stackup. Additive effects of all the allowable manufacturing tolerances on the final dimensions of the assembly; also called tolerance buildup.

ton (symbol: T or t). A unit of mass in both the SI and American standard system of units. In SI units, one ton (sometimes spelled tonne) is defined as 1,000 kilograms. Also called the *metric ton*. In the American standard system of units, one (short) ton is defined as 2,000 pounds-mass (lbm). In the United Kingdom, an imperial (or long) ton contains approximately 2,240 pounds-mass (lbm).

top-down approach. An approach to problem-solving in AI that is goal-directed or expectation guided based on models or other knowledge. Also referred to as *"hypothesize and test" approach*.

top-down logic. A problem-solving approach employed in production systems, in which production rules are used to find a solution path by chaining backward from the goal.

torque (symbol: t). In physics, the moment of a force about an axis; the product of a force and the distance of its line of action from the axis.

torr. A unit of pressure named in honor of the Italian physicist Evangelista Torricelli (1608–1647). One standard atmosphere (on Earth) is equal to 760 torr; or 1 torr = 1 millimeter of mercury = 133.32 pascals.

torsion. The state of being twisted.

total impulse (symbol: I_T). The thrust force (T) integrated over an interval of time (*t*).

touchdown. The precise moment when a robot spacecraft lands on the surface of a planet or moon, surviving the process so it can pursue its mission objectives. If the robot spacecraft does not touchdown properly, it most likely gets crunched and the planned safe landing becomes a crash.

toughness. In engineering and materials science, the ability of a material (especially a metal) to absorb energy and deform plastically before fracturing.

trajectory. The path traced by any object or body moving as a result of an externally applied force, considered in three dimensions.

transceiver. A combination of transmitter and receiver in a single housing, with some components being used by both units.

transducer. General engineering term for any device that converts one form of energy (usually in some type of signal) to another form of energy. For example, a microphone is an electroacoustic transducer in which sound waves (acoustic signals) are converted into corresponding electrical signals that then can be amplified, recorded, or transmitted to a remote location. The photocell and thermocouple are also transducers, converting light and heat (respectively) into electrical signals.

transient. The condition of a physical system (such as a robot) in which the engineering parameters of importance (for example, temperature, pressure, vibration levels, velocity, attitude, etc.) vary significantly with time.

translation. Movement in a straight line without rotation.

transmission grating. A diffraction grating in which incoming signal energy is resolved into spectral components upon transmission through the grating.

transmitter. A device for the generation of signals of any type and form that are to be transmitted. For example, in radio and radar, a transmitter includes electronic circuits designed to generate, amplify, and shape the radio frequency (RF) energy, which is delivered to the antenna from where it is then radiated out into space.

transpiration cooling. A form of mass transfer cooling, which involves controlled injection of a fluid mass through a porous surface. This process basically is limited by the maximum rate at which the coolant material can be pumped through the surface.

transponder. A combined receiver and transmitter whose function is to transmit signals automatically when triggered by an appropriate interrogating signal.

tree structure. A graph in which one node, called the root, has no predecessor node, and all other nodes have exactly one predecessor. For a state space representation, the tree starts with a root node, which represents the initial problem situation. Each of the new states, which can be produced from this initial state by application of a single operator, is represented by a successor node of the root node. Each successor node branches in a similar way until no further states can be generated or a solution is reached. Operators are represented by the directed arcs from the nodes to their successor nodes.

tribology. The branch of engineering science that deals with friction, lubrication, and the behavior of lubricants.

truncation error. The error resulting from the use of only a finite number of terms of an infinite series or from the approximation of operations in the infinitesimal calculus by operations in the calculus of finite differences.

truth-maintenance. A method of keeping track of beliefs (and their justifications) developed during problem-solving, so that if contradictions occur, the incorrect beliefs or faulty lines of reasoning and all conclusions resulting from them, can be retracted.

truth value. One of the two possible values—TRUE or FALSE—associated with a proposition in logic.

tumble. In general, to rotate end over end; with respect to a gyro, to process suddenly and to an extreme extent as a result of exceeding its operating limits of pitch.

turbine. A machine that converts the energy of a fluid stream into mechanical energy of rotation. The working fluid used to drive a turbine can be gaseous or liquid. For example, a highly compressed gas drives an expansion turbine, hot gas drives a gas turbine, steam (or other vapor) drives a steam (or vapor) turbine, water drives a hydraulic turbine, and wind spins a wind turbine (or windmill).

two-phase flow. Simultaneous flow of gases and solid particles (e.g., condensed metal oxides); or the simultaneous flow of liquids and gases (vapors).

ultra-. A prefix meaning beyond or surpassing a specified limit, range, or scope.

ultrahigh frequency (UHF). A radio frequency in the range 0.3 gigahertz to 3.0 gigahertz.

ultrasonic. Of or pertaining to frequencies above those that affect the human ear, that is, acoustic waves at frequencies greater than approximately 20,000 hertz; for example, an ultrasonic vibrator.

ultraviolet (UV) radiation. That portion of the electromagnetic spectrum that lies beyond visible (violet) light and is longer in wavelength than X-rays. Generally taken as electromagnetic radiation with wavelengths between 400 nanometers (just past violet light in the visible spectrum) and about 10 nanometers (the extreme ultraviolet cutoff and the beginning of X-rays).

unit. The unit defines a measurement of a physical quantity, such as length, mass, or time. There are two common unit systems in use in aerospace applications today, the International System of Units (SI), which is based on the meter-kilogram-second (mks) set of fundamental units, and the American standard system of units, which is based on the foot-pound-mass-second (fps) set of fundamental units. In the American standard system of units, 1 pound-mass (lbm) is defined as equaling 1 poundforce (lbf) on the surface of Earth at sea level. Derived units, such as energy, power, and force, are based on combinations of the fundamental units in accordance with physical laws, such as Sir Isaac Newton's laws of motion.

universal constructor (UC). As proposed by John von Neumann (1903–1957) in the late 1940s and early 1950s, a self-replicating system that operates in a cellular automata environment. One important characteristic of von Neumann's self-replicating machine is that it features open-ended evolution, meaning the universal constructor (a cellular automaton) remains separate from its own description (in von Neumann's day, a compilation of instructions and data punched on a long paper tape). When errors (mutations) happen in making copies of this descriptive tape, functioning variants of the original automaton occur, that is, machine offspring experiences some degree of change and (possibly) evolution. Following a machine version of "natural selection," some of the machine offspring will be inferior variants and not survive; while others will be superior variants. The superior machine progeny not only survive, but also proceed to replicate an improved version of the original machine.

universal time coordinated (UTC). The worldwide scientific standard of timekeeping, based on carefully maintained atomic clocks. It is kept accurate to within microseconds. The addition (or subtraction) of leap seconds as necessary at two opportunities every year keeps UTC in step with Earth's rotation. Its reference point is Greenwich, England. When it is midnight there on Earth's prime meridian, it is midnight (00:00:00.000000) UTC, often referred to as "all balls" in engineering jargon.

unmanned. Without human crew; unpersoned or uncrewed are more contemporary terms.

unmanned aerial vehicle (UAV). A robot aircraft flown and controlled through teleoperation by a distant human operator. *Also called a* **remotely piloted vehicle (RPV).**

uplink. The telemetry signal sent from a ground station to a UAV, robot spacecraft, or planetary probe.

uplink data. Information that is passed from a ground station on Earth to a UAV, a robot spacecraft, a space probe, or a space platform.

vacuum. The absence of gas or a region in which there is a very low gas pressure. This is a relative term. For example, a soft vacuum (or low vacuum) has a pressure of about 0.01 pascal (i.e., 10^{-2}pascal); a hard vacuum (or high vacuum) typically has a pressure between 10^{-2} and 10^{-7} pascal; while pressures below 10^{-7} pascal are referred to as an ultrahard (or ultrahigh) vacuum. Engineers designing robot systems to operate under hard vacuum conditions (such as outer space) must give special attention to component movement and lubrication issues, as well as potential outgassing and thermal control problems.

Valles Marineris. An extensive canyon system on Mars near the planet's equator, discovered in 1971 by NASA's *Mariner 9* robot spacecraft.

valve. Mechanical device by which the flow of fluid may be started, stopped, or regulated by a movable part that opens, closes, or partially obstructs a passageway in a containing structure, called the valve housing.

vapor. The gaseous phase of a substance; in thermodynamics, this term often is used interchangeably with gas.

vaporization. In thermodynamics, the transition of a material from the liquid phase to the gaseous (or vapor) phase, generally as a result of heating or pressure change.

vapor pressure. The pressure exerted by the atoms or molecules of a given vapor. For a pure substance confined within a container, it is the vapor's pressure on the walls of its vessel; for a vapor (or gas) mixed with other vapors (or gases), it is that particular vapor's contribution to the total pressure—its partial pressure. The total pressure is the sum of the partial pressures of all the component vapors (or gases) in a mixture or system.

vapor turbine. A turbine in which part of the thermal energy (heat) supplied by a vapor is converted into mechanical work of rotation. The steam turbine is a common type of vapor turbine. Sometimes called a *condensing* turbine.

variable. A quantity or function that can assume any given value or set of values.

vector. Any physical quantity, such as force, velocity, or acceleration, that has both magnitude and direction at each point in space, as opposed to a scalar, which has magnitude only.

vector steering. A steering method for rockets and robot spacecraft in which one or more thrust chambers are gimbaled so that the direction of the thrust (that is, the thrust vector) may be tilted in relation to the vehicle's center of gravity to produce a turning movement.

velocity. A vector quantity that describes the rate of change of position. Velocity has both magnitude (speed) and direction, and it is expressed in terms of units of length per unit of time (such as, meters per second).

velocity of light (symbol: c). *See* **speed of light**.

Venera. The family of Russian robot spacecraft (flybys, orbiters, probes, and landers) that successfully explored Venus, including its inferno like surface, between 1961 and 1984.

vent valve. Pressure-relieving valve that is operated on external command, as contrasted to a relief valve, which opens automatically when pressure reaches a given level.

vernier engine. A low-thrust rocket engine used primarily to obtain a fine adjustment in the velocity and trajectory or in the attitude of a robot spacecraft.

very large scale integration (VLSI). An integrated circuit containing more than 64,000 transistors. *See also* **integrated circuit**.

Viking Project. NASA's highly successful Mars exploration effort in the 1970s in which two orbiter and two lander spacecraft conducted the first detailed study of the Martian environment and the first (albeit inconclusive) scientific search for life on the Red Planet.

virtual reality (VR). A computer-generated artificial reality that captures and displays in varying degrees of detail the essence or effect of physical reality (that is, the "real-world" scene, event, or process) being modeled or studied. With the aid of a data glove, headphones, and/or head-mounted stereoscopic display, a person is projected into the three-dimensional world created by the computer.

viscosity. A measure of the internal friction or flow resistance of a fluid when it is subjected to shear stress. The *dynamic viscosity* is defined as the force that must be applied per unit area to permit adjacent layers of fluid to move with unit velocity relative to each other. The dynamic viscosity is sometimes expressed in poise (centimeter-gram-second unit system) or in pascal-seconds (SI unit system). One poise is equal to 0.1 newton-second per square meter (1 poise $= 10^{-1}$ N s m^{-2}). The *kinematic viscosity* is defined as the dynamic viscosity divided by the fluid's density. The kinematic viscosity can be expressed in stokes (centimeter-gram-second unit system) or in square meters per second (SI unit system). One stoke is equal to 10^{-4} m^2 s^{-1}. In general, the viscosity of a liquid usually decreases as the temperature is increased; the viscosity of a gas increases as the temperature increases.

viscous fluid. A fluid whose molecular viscosity is sufficiently large to make the viscous forces a significant part of the total force field in the fluid.

volt (symbol: V). The SI unit of electric potential difference and electromotive force. One volt is equal to the difference of electric potential between two points of a conductor carrying a constant current of one ampere when the power dissipated between these points equals one watt. This unit is named after the Italian scientist Count

Figure 8-16 A scientist at the Argonne National Laboratory (ANL) uses an automatic virtual reality environment to examine the way complex biological molecules link up and form strings. The scientist can stop the simulation at any point and quite literally move around inside the image to see changes and unique linkages from all angles. This application of virtual reality is a powerful starting point in biomedical research, pharmaceutical development projects, and materials science. (Credit: Image courtesy of U.S. Department of Energy/Argonne National Laboratory.)

Alessandro Volta (1745–1827), who performed pioneering work involving electricity and electric cells.

volume (symbol: V). The space occupied by a solid object or a mass of fluid (liquid or confined gas).

von Neumann architecture. The commonly encountered computer architecture that uses sequential processing. Initially proposed by the mathematician John von Neumann (1903–1957).

Voyager. NASA's twin robot spacecraft that explored the outer regions of the solar system, visiting all the Jovian planets. *Voyager 1* encountered Jupiter (1979) and Saturn (1980) before departing on an interstellar trajectory. *Voyager 2* performed the historic "Grand Tour" by visiting Jupiter (1979), Saturn (1981), Uranus (1986), and Neptune (1989). Both RTG-powered spacecraft are now involved in the Voyager Interstellar Mission (VIM) and each carries a special recording ("Sounds of Earth")—a digital message for any intelligent species that finds them drifting between the stars millennia from now.

watt (symbol: W). The SI unit of power (that is, work per unit time). One watt is defined as 1 joule (J) per second. In electrical engineering, 1 watt corresponds to the product of 1 ampere (A) times 1 volt (V). This represents the rate of electric energy dissipation in a circuit in which a current of 1 ampere is flowing through a voltage difference of 1 volt. This unit is named in honor of James Watt (1736–1819), the Scottish engineer who developed the steam engine.

wave. A periodic disturbance that is propagated in a medium in such a manner that at any point in the medium, the quantity serving as a measure of the disturbance is a function of time, while at any instant the displacement at a point is a function of the position of the point. At each spatial point there is an oscillation. The number of oscillations that occur per unit time is the frequency (symbol: ν). The distance between one wave crest to the next wave crest (or one trough to the next trough) is called the wavelength (symbol: λ).

wavelength (symbol: λ). In general, the mean distance between maxima (or minima) of a periodic pattern. Specifically, the least distance between particles moving in the same phase of oscillation in a wave disturbance. The wavelength is measured along the direction of propagation of the wave, usually from the midpoint of a crest (or trough) to the midpoint of the next crest (or trough). The wavelength (λ) is related to the frequency (ν) and phase speed (c) (that is, speed of propagation of the wave disturbance) by the simple formula: $\lambda = c/\nu$. The reciprocal of the wavelength is called the wave number.

weight (symbol: w). Generally, the force with which a body is attracted toward Earth by gravity. Within the context of physics and engineering, the product of the mass (m) of a body and the gravitational acceleration (g) acting on the body, namely $w = mg$. For example, a robot with a mass of 100 kilograms on the surface of Earth would experience a downward force or weight of approximately 980 newtons. While the mass of an object remains the same throughout the solar system (and the universe), the object's weight varies on other planets in accordance with the local value of the acceleration of gravity. For example, a 100-kilogram-mass robot on the lunar surface has a weight of about 163 newtons.

welding. The process of joining two or more pieces of metal by applying thermal energy (heat), pressure, or both, with or without filler material to produce a localized union through fusion or recrystallization across the interface.

white room. A clean, dust-free room that is used for the assembly, calibration, and (if necessary) repair of delicate robot components and microelectronic devices, such as tiny gyros and microsensor systems. Many space robots are assembled and tested in white rooms. So-named because all the surfaces are colored white to make it easy to detect and remove dirt, dust, and tiny machine components. Also known as a *clean room*.

wick. A group or braid of thin fibers that transports (or quite literally "sucks up") a liquid, if the adhesive force between the fiber and the liquid is greater than the liquid's cohesive force. The wick uses capillary force to move a fluid without the action of a mechanical pump. Wicks are found in many engineered devices that require self-lubrication or thermal control.

work (symbol: *W*). In physics and engineering, work (*W*) is defined as the energy (*E*) expended by a force (*F*) acting though a distance (*d*). The science of thermodynamics defines mechanical work as the organized (reversible) flow of energy into or out of a system; in contrast to heat, which is the disorganized (irreversible) flow of energy into or out of a system. The SI unit system, work is expressed in joules (*J*). Specifically, when a force of 1 newton (*N*) moves through a distance of one meter, one joule of work is performed.

work envelope. The area (or volume) that an industrial robot can sweep across or touch with the end of its arm. A manipulator robot with a versatile articulated arm may use different tools or end effectors for different industrial operations. As a result, the work envelope could be different in each application and the dimensions of the work envelope would depend on the specific tool or end effector in use.

working fluid. A fluid (gas or liquid) used as the medium for the transfer of energy from one part of a system to another part. Working fluids play a major role in heat engines, pneumatic systems, hydraulic systems, and jet/rocket propulsion systems.

world knowledge. In AI, knowledge about the world or the domain of interest.

world model. A representation of the current situation.

wrist. A mechanical unit mounted on the end of an industrial robot's arm. For the robot to perform a specific task, a certain tool, gripper, or end effector is attached to the wrist.

X-ray. A penetrating form of electromagnetic radiation of very short wavelength (approximately 0.01 to 10 nanometers or 0.1 to 100 angstroms) and high photon energy (approximately 100 electron volts to some 100 keV). X-rays are emitted when either the inner orbital electrons of an excited atom return to their normal energy states (these photons are called *characteristic X-rays*) or when a fast-moving charged particle (generally an electron) loses energy in the form of photons upon being accelerated and deflected by the electric field surrounding the nucleus of a high atomic number element (this process is called *bremsstrahlung*, or "braking radiation"). Unlike gamma rays, X-rays are nonnuclear in origin.

yaw. The rotation or oscillation of an object or system about its vertical axis so as to cause the object's longitudinal axis to deviate from the direction of motion or heading in its horizontal plane. *See also* **pitch; roll.**

Yohkoh. A Japanese robot spacecraft launched in 1991. The main objective of this scientific satellite was to study the high-energy radiations from solar flares, as well as presolar flare conditions. *Yohkoh*, (sunbeam in Japanese) was a three-axis stabilized observatory-type satellite in a nearly circular Earth orbit. It carried four instruments—two imagers and two spectrometers.

Zond. A family of early Russian robot spacecraft that explored the Moon, Mars, Venus, and interplanetary space in the 1960s.

9

Associations

This chapter presents a selected collection of interesting organizations and companies that are involved in developing, applying, or promoting the technology of robotics. Some of the entries included here are major, government-sponsored agencies whose raison d'être is the timely performance of specific missions that are enhanced or enabled by the use of robots or remote manipulators systems. For example, missions involving the application of nuclear energy or the exploration of outer space use a wide variety of robot systems. Entries listed here also include a selected number of commercial companies that focus on development and application of robotic technologies within the defense, scientific, commercial, or public-services sector. Other entries exist to promote the application of industrial or military robots. Still other entries involve associations and societies that represent specific scientific or engineering disciplines that directly or indirectly support robotic technology. While this chapter's collection of organizations is not totally inclusive, it does provide an important sampling of the many associations, facilities, companies, and organizations in the United States and around the world that actively contribute to the further development and use of robot systems in this century. Please recognize, however, that any commercial entities mentioned here are listed for information purposes only and do not necessarily imply a specific endorsement by either the author or the publisher.

American Nuclear Society (ANS)

555 North Kensington Avenue
LaGrange Park, Illinois 60526 USA
1-708-352-6611
1-708-352-0499 (Fax)
http://www.ans.org/

The American Nuclear Society is an international scientific and educational organization that promotes and unifies professional activities within the diverse fields of nuclear science and nuclear technology. Founded in December 1954, the society has a current membership of approximately 11,000 engineers, scientists, administrators, and educators. The mission of the ANS is to assist its members in their professional efforts to develop and safely apply nuclear science and technology for public benefit. The society uses knowledge exchange, professional development, and enhanced public understanding to fulfill this mission. Within the ANS, the Robotics and Remote Systems Division (RRSD) promotes the use of robotics and remote systems technologies in hazardous environments. The RRSD traces its roots back to 1960, when the division was the first professional division of the ANS. In 1965, the original name of the division, which was the Hot Laboratories Division, was changed to Remote Systems Technology Division. Then, in 1992, the ANS adopted the division's present name to be more representative of contemporary activities. Professional members of the RRDS continue to make significant contributions to all fields involving the use of robotics and remote technologies, including the field of nuclear energy. Every two years the ANS presents the Raymond C. Goertz Award to honor a member of the society's RRSD who has made outstanding contributions to the field of remote technology. The biennial award honors the Late Ray Goertz, who conceived and developed the master–slave manipulator in the late 1940s for use in the American nuclear industry.

American Society of Mechanical Engineers (ASME)

ASME International (Headquarters)
Three Park Avenue
New York, New York 10016-5990 USA
1-800-843-2763
http://www.asme.org/

Founded in 1880 as the American Society of Mechanical Engineers, ASME is now a worldwide professional organization that focuses on technical, educational, and research issues of the engineering and technology communities. ASME conducts one of the world's largest technical publishing operations, sponsors numerous technical conferences around the world each year, and offers a wide variety of professional development courses. ASME also sets internationally recognized industrial and manufacturing codes. The professional society's official monthly publication, *Mechanical Engineer*, covers the development and application of the wide-ranging technologies and tools of the modern engineering profession, including robotics and nanotechnology. The ASME Mechanisms and Robotics Committee operates under the auspices of the ASME Design Engineering Division. The purpose of this committee is to promote advances in research and education in the theory and application of mechanisms and machine systems, including robotic systems, micromachines, and nanoscale machines.

Ames Research Center (ARC)

Moffett Field, California 94035 USA
1-650-604-5000
http://www.nasa.gov/centers/ames/

The Ames Research Center (ARC) is one of 10 NASA field installations. ARC is located in the heart of California's Silicon Valley at the core of the research cluster of high-tech companies, universities, and laboratories that define the region's character. Ames plays a critical role in virtually all NASA missions. As a leader in information technology research with a focus on supercomputing, networking, and intelligent systems, Ames conducts the critical research and development and pursues the enabling information technologies that make NASA missions possible. For example, the Computational Sciences Division at ARC has a long history and extensive experience in field robotics and human/robot field-testing. Ames has been running robotic field experiments in planetary analogue sites since 1993, and has the staff and expertise to design and build robotic test platforms and embedded control systems. The Intelligent Robotics Group (IRG) at ARC has personnel experienced in mechanical design and fabrication, electronics, instrumentation, embedded control, computer vision, robotic navigation, state estimation, diagnosis, networking, educational robotics outreach, and user interface design. Recognizing that nanoscale devices and sensors have the potential to enable revolutionary advances in onboard data processing, communication, and sensing, ARC also serves the NASA mission as a leader in nanotechnology. For example, researchers at Ames are pursuing computer-aided design (CAD) of nanoscale devices and sensors as a cost efficient way of infusing emerging nanoelectronics technologies into the onboard information processing systems of future space exploration missions. One promising research area involves chemical and electromechanical sensors based on carbon nanotubes and nanowires.

Air Force Research Laboratory (AFRL)

Headquarters Air Force Research Laboratory-Public Affairs Office (HQ AFRL/PA)
1864 Fourth Street
Wright-Patterson Air Force Base, Ohio 45433-7132 USA
1-937-904-9851
1-937-255-4073 (Fax)
http://www.afrl.af.mil/

The overall mission of the Air Force Research Laboratory (AFRL) is to promote the discovery, development, and integration of affordable warfighting technologies to support the air and space forces of the United States. In addition to its central (headquarters) staff, AFRL contains nine technology directorates located throughout the United States and the Air Force Office of Scientific Research (AFOSR). AFRL is responsible for planning and executing the entire science and technology budget for the United States Air Force (about $1.5 billion in 2006). AFRL's customers include the major commands of the U.S. Air Force that operate

Figure 9-1 The K-9 (background) and Gromit (foreground) space robots in action at the outdoor "Marscape" (a planetary analogue research site simulating the surface of Mars) at the NASA Ames Research Center, located in California's Silicon Valley. The K-9 and Gromit space robots are smart enough to make decisions about how to achieve objectives on a planet or moon without receiving detailed instructions from human beings. Researchers at Ames are also investigating "mobile agent " software that may someday help robots and human beings communicate effectively with each other whether the human/mobile robot system teams are operating somewhere on Earth, on the Moon, or on Mars. (Credit: Photograph courtesy of NASA/Ames Research Center.)

Figure 9-2 During Operation Iraqi Freedom (OIF) robots from the Air Force Research Laboratory (AFRL) helped to counter bomb threats. This photograph shows an all-purpose remote transport system (ARTS) clearing an area at Balad Air Base, Iraq, of unexploded ordnance (UXO). Airmen responsible for explosive ordnance disposal (EOD) activities use such remotely operated systems to safely and effectively clear important ranges, while staying out of harm's way. (Credit: Photograph courtesy of U.S. Air Force.)

and maintain the full spectrum of air force weapon systems. Robotic systems, artificial intelligence, microelectromechanical systems (MEMS), and nanotechnology are all experiencing growing roles in AFRL's overall science and technology program.

The Robotics Research and Development Group within AFRL's Materials and Manufacturing Directorate (AFRL/ML) conducts research and development of advanced robotic technologies and unmanned ground systems to protect, support, and augment the American warfighter in the accomplishment of dirty, dull, dangerous, and impossible (for human beings) missions. For example this group is working on the All-Purpose Remote Transport System (ARTS) to provide a remote standoff solution to operational needs to locate, remove, and neutralize unexploded ordnance (UXO) and improvised explosive devices (IED). The group's Advanced Robotic Modules and Systems Program focuses on the development and validation of state-of-the-art navigation, detection, control, and communication subsystems to a multitude of existing and future platforms. The effort also emphasizes modularity and interchangeability of components, as well as cooperation with other government agencies and military services, industry, and academia. Other groups within AFRL are investigating the use of emerging nanotechnologies.

Argonne National Laboratory (ANL)

9700 South Cass Avenue
Argonne, Illinois 60439 USA
1-630-252-2000
1-630-252-5274 (Fax, Office of Public Affairs)
http://www.anl.gov/

Argonne National Laboratory (ANL) is one of the largest research centers of the U.S. Department of Energy. ANL is a direct descendant of the University of Chicago's Metallurgical Laboratory (Met Lab)—part of the American Manhattan Project to build an atomic bomb during World War II. It was at the Met Lab on December 2, 1942, that Enrico Fermi and his team of about 50 colleagues created the world's first controlled nuclear chain reaction in a squash court at the University of Chicago. In 1946, the Met Lab team received a charter to create first national laboratory of the United States. As part of its charter, Argonne received the important mission of developing nuclear reactors for peaceful purposes. However, ANL never functioned as a nuclear weapons laboratory. Following World War II, ANL scientists and engineers supported expanding nuclear research and development activities involving highly radioactive materials by applying the principles of cybernetics to manipulator design and constructed the first electric master–slave manipulator system. The device represented a major milestone in teleoperation and robotics. Today, the University of Chicago operates ANL for the U.S. Department of Energy. Over the years, ANL's research activities have expanded to include many other areas of science, engineering, and technology. These diverse areas include nanotechnology research and the innovative application of robotic systems in life sciences research.

Army Research Laboratory (ARL)

Public Affairs Office (AMSRD-ARL-O-PA)
2800 Powder Mill Road
Adelphia, Maryland 20783-1197 USA
1-301-394-3590
http://www.arl.army.mil/

The Army Research Laboratory (ARL) is the "corporate" basic and applied research laboratory of the United States Army. ARL's mission is to provide innovative science, technology, and analysis to enable full-spectrum operations. ARL consists of the Army Research Office (ARO) and six directorates: the Weapons and Materials Research Directorate (WMRD), the Sensors and Electron Devices Directorate (SEDD), the Human Research and Engineering Directorate (HRED), the Computational and Information Sciences Directorate (CISD), the Vehicle Technology Directorate (VTD), and the Survivability and Lethality Directorate (SLAD). The ARO pursues scientific and far-reaching technological discoveries in extramural organizations (such as educational institutions, nonprofit organizations, and private industry). The U.S. Army relies on ARL for scientific discoveries, technical advances, and analyses to provide modern warfighters with the advanced capabilities needed to succeed on the battlefield.

Figure 9-3 Argonne National Laboratory (ANL) scientists work with the first robot of its type in the United States to automate protein purification. The robot, which is housed in a refrigerator, is an integral part of the Argonne-based Midwest Structural Genomics Center's plan to automate the protein crystallography process. The ANL robotic system can purify six proteins per day versus previous procedures that required two to three days per protein. (July 2004) (Credit: Photograph courtesy of the U.S. DOE and Argonne National Laboratory.)

A number of projects and programs being undertaken by ARL involve robotic systems or the application of emerging developments in nanotechnology. For example, the HRED is conducting scientific research and technology directed toward optimizing soldier performance and soldier–machine interactions to maximize battlefield effectiveness, and to ensure that soldier performance requirements are adequately considered in the design and development of new

Figure 9-4 A U.S. Army soldier of Company A, 101st Military Intelligence Battalion, 3rd Brigade Combat Team, 1st Infantry Division prepares a Shadow 200 unmanned aerial vehicle (UAV) for launch at Forward Operating Base Warhorse in Iraq (September 24, 2004). Researchers at ARL are studying how to improve human-robot interactions to optimize the performance of robot-soldier teams in future combat environments. (Credit: Photograph courtesy of the U.S. Army.)

battlefield technology and military robot systems. One important effort within the basic human-robot interaction (HRI) research area is entitled: "Technology for Human-Robotic Interaction in Soldier-Robot Teaming." The project seeks to reduce the workload and improve combat performance for the soldier–robot team in future battlefield environments. The U.S. Army anticipates that this ARL-sponsored research will result in a better understanding of how improvements in human–robot interactions will translate into more efficient soldier-robot teams on the battlefield.

Association for Unmanned Vehicle Systems International (AUVSI)

2700 S. Quincy Street
Suite 400
Arlington, Virginia 22206 USA
1-703-845-9671
1-703-845-9679 (Fax)
http://www.auvsi.org

The Association for Unmanned Vehicle Systems International (AUSVI) is the world's largest nonprofit organization devoted exclusively to promoting and advancing the unmanned systems community. With members from government organizations, industry, and academia, AUVSI fosters and helps in development of unmanned systems and related technologies. AUVSI's bimonthly publication, called *Unmanned Systems*, highlights current developments and unveils new technologies in air, ground, maritime, precision strike, and space unmanned systems. This magazine covers systems and developments of interest to both civilian and military organizations.

Defense Advanced Research Projects Agency (DARPA)

3701 North Fairfax Drive
Arlington, Virginia 22203-1714 USA
1-703-526-6630
http://www.darpa.mil/

The Defense Advanced Research Projects Agency (DARPA) plays a unique role with the U.S. Department of Defense (DOD). Since DARPA is not tied to a specific operational mission, the agency supplies technological options for the entire defense department. By pursuing leading edge, often high risk–high payoff research and development projects are designed to serve as the "technological engine" capable of driving and transforming the DOD to meet capabilities a military commander might want in the future. Through a variety of technical demonstration projects, DARPA provides leadership and funding that often accelerates the arrival of innovative capabilities. These "futuristic" capabilities not only provide options to future commanders, but often change minds of senior military leadership about what is technologically possible today.

DARPA was born at the beginning of the space age in response to the technology surprise brought about when the former Soviet Union launched the first Earth-orbiting satellite, *Sputnik 1*. Since its founding DARPA has stimulated the development of many important technologies, including those involving robotics, micromachines, and nanotechnology. For example, the Global Hawk and Predator unmanned aerial vehicles (UAVs) have played prominent roles in Operation Enduring Freedom (OEF) in Afghanistan and Operation Iraqi Freedom (OIF). DARPA started the concept of a high-altitude, long-range, extended loiter unmanned military aerial robot system in the 1970s with the TEAL RAIN program. After a number of significant technical breakthroughs, the Global Hawk high-altitude endurance UAV transitioned from DARPA to the U.S. Air Force in 1998. The Tier 2 Predator medium-altitude endurance UAV evolved directly from DARPA's AMBER and Gnat 750-45 designs. In April 1996, the secretary of defense selected the U.S. Air Force as the operating service for the Predator.

DARPA is working with the U.S. Army, Navy, and Air Force toward a vision of a strategic and tactical battlespace filled with networked manned and unmanned systems. The goal is not simply to replace people with machines, but to team people with autonomous platforms to create a more capable, agile, and cost-effective military force capable of achieving its mission with significantly

Figure 9-5 On April 28, 2006, DARPA and the U.S. Army unveiled the Crusher unmanned ground combat vehicle. Crusher is a six-wheeled, all-wheel drive, hybrid electric, skid-steered unmanned ground combat vehicle (UGCV). The robot vehicle is being equipped with state-of-the-art perception capabilities, and will be used to validate the key technologies necessary for future UGCVs to perform military missions autonomously. The National Robotics Engineering Center (NREC) at Carnegie Mellon University in Pittsburgh, Pennsylvania, is the prime contractor for Crusher. (Credit: Photograph courtesy of the U.S. Army.)

lower risk of American casualties. The successful use of UAVs in Afghanistan and Iraq is regarded as an important initial step in demonstrating the transformational potential of this concept. In late April 2006, DARPA and the U.S. Army unveiled the Crusher, an advanced unmanned ground combat vehicle (UGCV). With its highly mobile vehicle design and innovative autonomous control system Crusher represents the state of the art in autonomous unmanned ground vehicles.

DARPA also sponsors major competitions for autonomous vehicles designed by teams from academia and industry. For example, DARPA's Third Grand Challenge is scheduled to take place on November 3, 2007, and will feature autonomous ground vehicles executing simulated military supply missions safely and effectively in a mock urban area. DARPA will award prizes for the top three autonomous ground vehicles (AGVs) that compete in a final event where the AGVs must safely complete a 100-kilometer (60-mile) urban area course in fewer than six hours. First prize in DARPA's Urban Challenge is two million dollars.

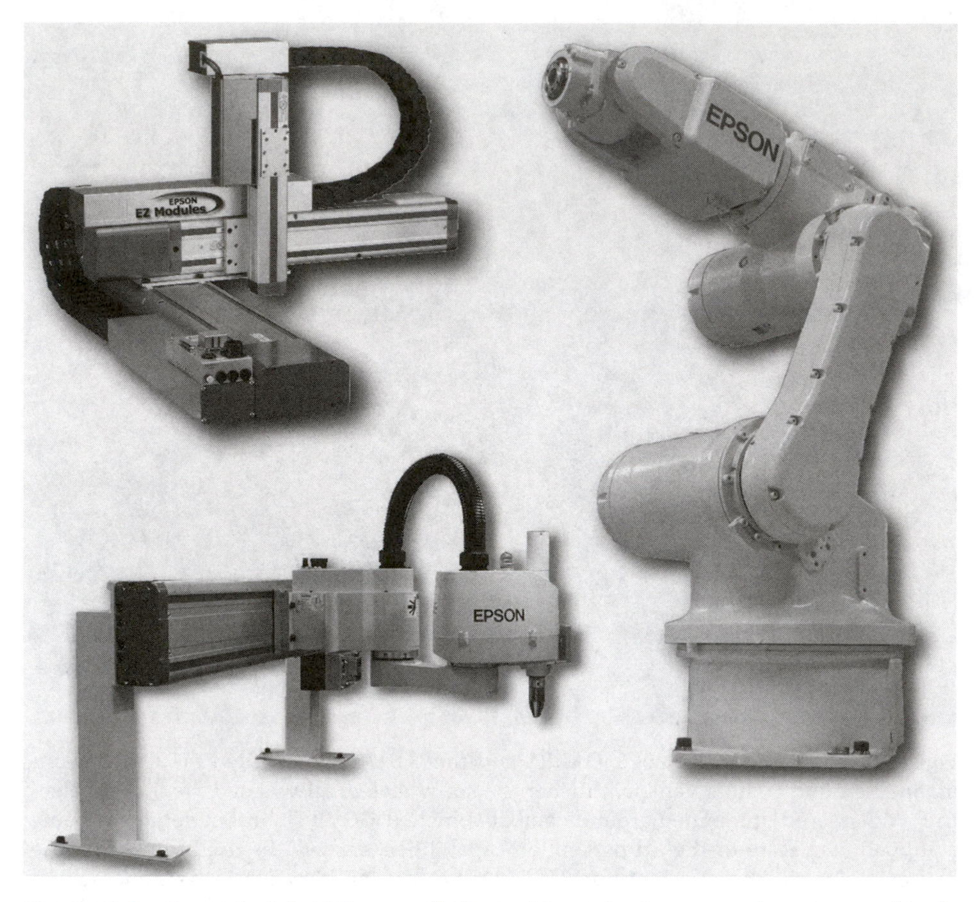

Figure 9-6 Several of the PC-controlled, precision robotic automation systems developed and marketed by EPSON Robots. Theses types of modern industrial robots are typically designed for small parts assembly, dispensing, laboratory automation, machine tending, material handling, medical device manufacturing, packaging, food handling, and a variety of other applications that require speed, precision, and smooth operation. (Credit: Photograph courtesy of EPSON Robots.)

EPSON Robots

18300 Central Avenue
Carson, California 90746
1-562-290-5910
1-562-290-5999 (Fax)
http://www.robots.epson.com/

For more than two decades, EPSON Robots has served customers around the world as a major supplier of industrial robots that feature easy to use PC-based controls, high speed, and high precision. The company, one of many major product divisions within the multi-billion-dollar Seiko Epson Corporation, also supplies related factory automation equipment.

FANUC Robotics America, Inc.

3900 West Hamlin Road
Rochester Hills, Michigan 48309-3253 USA
1-248-377-7000

FANUC Robotics is an international robotics technology company that has over 155,000 of its industrial robots installed worldwide. Among its many products, the company makes a variety of assembly robots, material handling robots, welding/laser robots, material removal robots, and painting and dispensing robots, robots for the aerospace and defense industries, automotive industry robots, medical device robots, food and beverage industry robots, and electronics and clean room robots.

Food and Drug Administration (FDA)

5600 Fishers Lane
Rockville, Maryland 20857 USA
1-888-463-6332
http://www.fda.gov/

The United States Food and Drug Administration (FDA) is one of the nation's oldest and most respected consumer protection agencies. FDA's mission is: to promote and protect the public health by helping safe and effective products reach the market in a timely manner; to monitor products for continued after they are in use; and to help the public get the accurate science-based information needed to improve health. Since the public trusts FDA to drugs and medical devices that are safe and effective, new pharmaceutical products that involve the use of nanotechnology and new medical devices that involve the use of robotic systems fall within the regulatory responsibility of this administration. While FDA does not regulate "technologies," the administration regulates new products based on emerging technologies on a product-by-product basis. For example, in the field of molecular medicine and "nanosized" drugs, FDA will test, evaluate, and approve the use of each such nanotechnology-based medical product—including miniaturized microelectromechanical system (MEMS) medical devices that involve the release of nanoparticles in human beings or animals. The FDA also evaluates, regulates, and approves robot-assisted surgical systems intended for use by physicians within the United States.

Foster-Miller, Inc.

350 Second Avenue
Waltham, Massachusetts 02451-1196 USA
1-781-684-4000
http://www.foster-miller.com/

Foster-Miller, Inc. is a diversified engineering, development, and manufacturing company with its headquarters located in the suburban Boston area. In November 2004, the company became an independent, wholly owned subsidiary of QinetiQ. Of particular interest here is the fact that Foster-Miller produces a line of mobile robots for the Department of Defense—most notably the lightweight, man-portable Talon™ robot, which is serving many military roles in Iraq and Afghanistan, including reconnaissance, armed reconnaissance, and bomb disposal operations. In the late 1980s, Foster-Miller developed its Cecil® robot under contract with the Electric Power Research Institute (EPRI) to provide inspection and cleaning capability within the tube bundle of steam generators used in nuclear power plants. Cecil® robots are now in use in nuclear plants in Japan, Korea, France, Canada, and the United States.

Hafmynd–Gavia Ltd.

Fiskislod 73
101 Reykjavik, Iceland
+354-5112990
+354-5112999 (Fax)
http://gavia.is/

Hafmynd is a developer of innovative underwater technologies. The company's Gavia (the Great Northern Diver) autonomous underwater vehicle (AUV) provides solutions to customers involved in underwater research, surveying, and monitoring tasks. The Gavia AUV is a fully modular, man-portable robot vehicle, which is available in production vehicle depth ratings in excess of 1,000 meters and in customized vehicle versions to depth ratings of 2,000 meters or more. There are a wide variety of sensors available for this modular, commercial AUV. The name of this AUV comes from the genus name of the Great Northern Diver, a diving bird found in Iceland, Greenland, parts of the United States, and Canada. As the company's literature suggests the Gavia is a small 2,000-meter depth rated AUV that represents "an ideal tool for any (underwater) research, monitoring or surveillance task where autonomy, cost, and ease of deployment matter." In April 2006, Hafmynd demonstrated the Gavia AUV in a port security application for the Icelandic Coast Guard, special police and Reykjavik harbor authorities. As part of this exercise, dummy mines were placed within the tight confines of Reykjavik harbor and surveillance sensor-equipped Gavia AUV was then tasked to navigate on its own through the tricky harbor environment and detect the dummy mines. Released reports indicate the Gavia performed the underwater survey of the harbor effectively.

Honda Motor Company, Ltd.

2-1-1 Minami Aoyama
Minato-ku Tokyo
107-8556, Japan
+81-(0)3-3423-1111
http://world.honda.com/

Since being founded in 1948, the Honda Motor Company has grown to become the world's largest motorcycle manufacturer and one the world's leading automobile makers. Honda uses a wide variety of modern industrial robots to manufacture products ranging from small general-purpose engines and

Figure 9-7 With a weapons platform mounted on a Talon™ robot, the SWORDS system allows American soldiers to fire small arms by remote control from as far away as 1,000 meters. (Credit: Photograph courtesy of U.S. Army.)

scooters to special sports cars. In 1986, a team of Honda engineers took on the challenging task of developing a people-friendly, autonomous bipedal (two-legged) humanoid robot. In 2000, the company presented the eleventh in a line of two-legged prototypes, a humanoid robot called ASIMO. Although officially named the Advanced Step in Innovative Mobility robot, the popular humanoid robot became known around the world by the acronym, ASIMO. In December 2005, Honda presented the new ASIMO, a significantly improved version of original ASIMO. The new ASIMO features the ability to perform important tasks in a real-life office or home environment and is also more agile than its mechanical predecessor. For example, the new ASIMO can run at a pace of about six kilometers per hour and twist and turn while running.

Idaho National Laboratory (INL)

1765 North Yellowstone Highway
Idaho Falls, ID 83415 USA
1-208-526-0111
http://www.inl.gov/

The Idaho National Laboratory (INL) is a multiprogram national laboratory operated for the U.S. Department of Energy (DOE) by the Battelle Energy Alliance (BEA). Having been formally designated as the nation's center for advanced civilian nuclear technology research and development, the laboratory is undergoing a major transformation. As with the other major national laboratories, INL still performs work in support of DOE's four strategic goal areas: energy, defense, environment, and science. However, INL's major emphasis is to serve as the nation's lead laboratory for nuclear energy research and development. In support of the DOE's science strategic goal, INL researchers also perform work in intelligent automation and remote systems. For example, INL is developing tools for synergistic interaction between autonomous robots and human operators. As part of the continuing cleanup of cold war nuclear sites, INL staff members have developed numerous robotic systems to assist in environmental restoration work—nuclear robots and remotely operated equipment that saves human workers from unnecessary exposures to ionizing radiations or toxic materials.

IEEE Robotics and Automation Society

IEEE Corporate Office
3 Park Avenue, 17th Floor
New York, New York 10016-5997 USA
1-212-419-7900
1-212-752-4929 (Fax)
http://www.ieee.org/

The IEEE (Institute of Electrical and Electronics Engineers, Inc.) is a nonprofit organization and based on its current global membership (more than 365,000 members in 2006) represents the world's leading professional association for the advancement of technology. The IEEE serves as a leading authority on such areas as aerospace systems, computers, and telecommunications, biomedical engineering, electric power, and consumer electronics among others. The *IEEE Spectrum Magazine* is the organization's flagship professional publication, although the IEEE currently publishes a total of 128 transactions, journals and magazines and hosts more than 300 conferences worldwide each year.

The Robotics and Automation Society (RAS) with IEEE addresses both applied and theoretical issues in robotics and automation. The Society considers robotics to include intelligent machines and systems used, for example, in space exploration, human services, or manufacturing. Similarly, the Society's definition of automation includes the use of automated methods in various applications—as, for example, factory, office, home, laboratory automation, or transportation systems to improve performance and productivity. As interpreted and understood by RAS members, robotics and automation involves designing and implementing intelligent machines that can perform tasks considered too dirty, too dangerous, too tedious, or too precise for human workers. RAS members often have professional interests in robotics and automation that push the boundary on the level of intelligence and technical capability for many forms of autonomous, semiautonomous, and teleoperated machines. As frequently discussed in the Society's sponsored publications, intelligent machines have applications in medicine, defense, space and underwater exploration, service industries, disaster relief, manufacturing and assembly, and entertainment.

The IEEE-RAS is the sole sponsor of three IEEE publications: *IEEE Transactions on Robotics*; *IEEE Transactions on Automation Science and Engineering*; and *IEEE Robotics and Automation Magazine*. The Society is also a cosponsor of several other IEEE publications, including: *IEEE/ASME Transactions on Mechatronics*; *IEEE Transactions on Nanotechnology*; *IEEE Sensors Journal*; and *IEEE/ASME Journal of Micro-Electrical-Mechanical Systems (MEMS)*. Finally, for the past two decades, RAS has sponsored the IEEE International Conference on Robotics and Automation. The Society also sponsors several smaller, more narrowly focused conferences and workshops, such as the IEEE-RAS International Conference on Humanoid Robots.

Intuitive Surgical, Inc.

950 Kifer Road
Sunnyvale, California 94086
1-408-523-2100
1-408-523-1390 (Fax)

Intuitive Surgical, Inc. manufactures robotics surgical systems, with the da Vinci™ Surgical System serving as the company's flagship product. The da Vinci system has three main components: the surgical cart, a computerized vision system, and a surgeon's console. The surgical cart, stationed adjacent to the operating table, has three robot arms—one for the surgeon's right hand, one for the surgeon's left hand, and a middle mechanical arm to hold the laparoscope that the surgeon uses to "see" inside the patient's body. Effectively, the medical robot becomes the mechanical hands and eyes of the surgeon, who is seated just a few meters away at the surgeon's console. The computerized vision system transforms the images captured by the tiny camera inside the patient into three-dimensional (3-D), real-time images that the physician views at the surgical console. Robotic controls allow the surgeon to make natural hand movements—in contrast to the "counterintuitive" instrument movements that are characteristic of nonrobotic, standard laparoscopic surgery. Counterintuitive instrument movement involves an operating condition similar to the surgeon working on the patient while looking in a mirror. As carefully designed and engineered, the da Vinci™ Surgical System provides the surgeon with nearly all-natural movements of the human wrist, making the robot-assisted minimally invasive surgery feel more like open surgery. The robotic system also eliminates natural hand tremor and improves dexterity thereby allowing the surgeon to perform ever-finer surgery in a more controlled manner. Selected hospitals and medical centers around the world use the da Vinci™ Surgical System to perform various minimally invasive surgical procedures.

iRobot Corporation

63 South Avenue
Burlington, Massachusetts 01803
1-781-345-0200
1-781-345-0201 (Fax)
http://www.irobot.com/

Founded in 1990 by roboticists from the Massahusetts Institute of Technology (MIT), iRobot Corporation specializes in behavior-based robots that help human beings complete tasks in a better way—whether the task is boring and dirty like cleaning floor or extremely dangerous like defusing bombs or other improvised explosive devices (IEDs). Powered by iRobot's proprietary AWARE™ robot intelligence systems, the company's robots can navigate in complex and dynamic real-world situations. In 1998, DARPA awarded iRobot a contract under the agency's Tactical Mobile robot program. This contract led to the iRobot PackBot®. Today, more than 300 PackBots® (of various designs) support American troops in Afghanistan and in Iraq. The iRobot PackBot® Scout Tactical Mobile Robot is a rugged, lightweight reconnaissance robot used daily in Afghanistan and Iraq to search buildings and caves for hostile forces. The iRobot PackBot® EOD Tactical Mobile Robot is a bomb-disposal robot used by American troops in Iraq and Afghanistan to disarm roadside bombs or other IEDs. The company's PackBot® Explorer Tactical Robot allows soldiers to stay at safe standoff distances, while the mobile military robot relays real-time video, audio, and sensor readings. The iRobot Corporation is also developing several more advanced military robots, such as the Wayfarer—a PackBot® type of mobile robot with fully autonomous urban reconnaissance capabilities.

The iRobot Corporation also makes commercial and industrial robots. In September 2002, the company introduced the Roomba® Vacuuming Robot—a domestic robotic system that can clean hardwood floors and carpets, detect dirt under furniture, avoid stairs, and then return to its docking station for recharge. To date, the company has sold more than 1.5 million Roomba® Vacuuming Robots worldwide. In May 2005, iRobot introduced the Scooba™ Floor Washing Robot—the world's first floor washing robot available for home use.

Figure 9-8 This photograph shows a U.S. Army explosive ordnance disposal (EOD) robot—here an iRobot PackBot® Tactical Mobile Robot—carrying a stick of C4 plastic explosive down the street in Samarra, Iraq, to the site of a suspected improvised explosive device (IED). The suspicious device was found by Iraqi policemen on November 3, 2004. Controlled by a team of American soldiers assigned to the 731st Ordnance Company, the mobile military robot used its mechanical arm and two on-board cameras to safely survey the suspected IED, while the bomb disposal troops remained at a safe distance. As a result of the EOD robot's careful inspection, the soldiers determined that the suspicious device was actually a decoy—most likely set up by terrorists or insurgents who planned to launch a rocket-propelled grenade against American troops whenever they passed by on patrol. (Credit: Photograph courtesy of the U.S. Navy/Journalist 1st Class Jeremy L. Woods photographer.)

Jet Propulsion Laboratory (JPL)

California Institute of Technology
4800 Oak Grove Drive
Pasadena, California 91109 USA
1-818-354-0112 (Public Services Office)
1-818-393-4641 (Fax)
http://www.jpl.nasa.gov/

The American space age began on January 31, 1958, with the launch of the first U.S. satellite, *Explorer 1*—an Earth-orbiting robot spacecraft built and controlled by the Jet Propulsion Laboratory (JPL). For almost five decades since then, JPL has led the world in exploring the solar system with robot spacecraft. The JPL is a federally funded research and development facility managed by the California Institute of Technology for the National Aeronautics and Space Administration (NASA). The laboratory is located in Pasadena, California,

Figure 9-9 This high-technical fidelity artist's rendering of NASA's *Mars Reconnaissance Orbiter* features the spacecraft's main bus facing down, toward the Red Planet. The large circular feature above the robot spacecraft's bus is the high-gain antenna—the spacecraft's primary means of communicating with scientists and engineers on Earth. (Credit: Artist's rendering courtesy of NASA/JPL.)

approximately 20 miles (32 kilometers) northeast of Los Angeles. In addition to the Pasadena site, JPL operates the worldwide Deep Space Network (DSN), including a DSN station, at Goldstone, California. JPL performs leading-edge activities associated with deep space automated scientific missions, such as subsystem engineering, instrument development, and data reduction and analysis required by deep space flight. The *Cassini* spacecraft's exploration of Saturn and the *Mars Reconnaissance Orbiter* (MRO) are current JPL space robot missions.

Los Alamos National Laboratory (LANL)

(Mailing Address)
Los Alamos National Laboratory (LANL)
Public Affairs Office
P.O. Box 1663
Los Alamos, NM 87545 USA
http://www.lanl.gov/

Founded by Robert Oppenheimer in 1942, Los Alamos National Laboratory (LANL) served the United States as the country's first nuclear weapons laboratory—a high-security research complex devoted exclusively to the design, development, and testing of nuclear explosive devices. Called the Los Alamos Scientific Laboratory (LASL) during the Manhattan Project, Oppenheimer's technical team delivered two successful nuclear weapon designs in 1945: a gun-assembled design, called Little Boy, that used highly enriched uranium-235 as the nuclear material and an implosion device, called Fat Man, which squeezed a ball of plutonium-239 into a supercritical mass. During the first decade of the cold war, Los Alamos scientists pioneered many other innovations in the design of fission bombs and then successfully developed the world's first hydrogen bomb. The University of California operates the laboratory for the National Nuclear Security Administration (NNSA) of the U.S. Department of Energy (DOE). Although LANL scientists contribute to many other areas of science and technology, the central mission of the laboratory is supporting and enhancing the national security of the United States. Most Los Alamos employees are working to help ensure the safety and reliability of the nuclear weapons in the American stockpile. Many other employees work to prevent the spread of weapons of mass destruction and to protect the American homeland from terrorist attack. Consequently, the stewardship and management of the American nuclear stockpile remains a major responsibility for laboratory personnel. Carefully selected civilian research and development programs, often in partnership with universities and industry, complement this mission by allowing LANL personnel to maintain a solid foundation in science and state-of-the-art technology.

At LANL, robots and other automated devices have been used for years, primarily to transport, store, and handle hazardous materials. More recently, robotic systems have helped characterize and clean contaminated equipment and perform chemical analyses. The laboratory's expertise in robotics technology is being integrated into major projects such as ARIES—a system to dismantle nuclear weapons safely. Other researchers at LANL are investigating the use of minimal autonomous robots that work without computers or human supervision, basing these devices on novel, fundamental principles of machine control. These robots "learn" how to do their jobs in relatively unstructured environments, rather than being programmed for specific tasks. Potential uses for these autonomous devices include detecting and destroying landmines, acting as tactical "scouts" in battlefield situations, and cleaning areas of hazardous waste material. LANL researchers envision that in the future, such autonomous robots will gain problem-solving abilities by being linked with computer-based neural networks, giving them animal-like skills to tackle real-world problems.

National Aeronautics and Space Administration (NASA)

Headquarters Information Center
Washington, DC 20546-0001 USA
1-202-358-0000
1-202-358-3251 (Fax)
http://www.nasa.gov/

The National Aeronautics and Space Administration (NASA) is the civilian space agency of the United States government and was created in 1958 by an act of Congress. NASA's overall mission is to plan, direct, and conduct American civilian (including scientific) aeronautical and space activities for peaceful purposes. Since its founding, NASA has promoted the development of a progressively more sophisticated family of robot space vehicles and probes—including flyby, orbiter, lander, and surface rover spacecraft, as well as automated orbiting observatories—with which to explore the solar system and beyond. Today, NASA, especially through robot development programs at the Ames Research Center, the Goddard Space Flight Center, the Johnson Space Center, and the Jet Propulsion Laboratory continues to serve as a major stimulus in the emergence of progressively more autonomous robotic systems and spacecraft, as well as humanoid robots that can cooperatively function and interact with human space explorers.

National Institute of Standards and Technology (NIST)

100 Bureau Drive
Gaithersburg, Maryland 20899-1070 USA
1-301-975-6478 (Public Inquires Unit)
http://www.nist.gov/
NIST Boulder Laboratories
325 Broadway
Boulder, Colorado 80305-3328 USA
1-303-497-5507 (Public Inquiries)

Founded in 1901, the National Institute of Standards and Technology (NIST) is a nonregulatory federal agency within the U.S. Commerce Department's Technology Administration. NIST's mission is to promote American innovation and industrial competitiveness by advancing measurement science, standards, and technology in ways that enhance economic security and improve the quality of life. The NIST has two major operating locations: a campus in Gaithersburg, Maryland (which also serves as the agency's headquarters), and a campus in Boulder, Colorado. The NIST laboratories are located in both Maryland and Colorado and conduct research in a wide variety of physical and engineering sciences, in order to advance technology infrastructure of the United States. For example, the NIST's Manufacturing Engineering Laboratory (MEL) develops measurement methods, standards, and technologies to improve American manufacturing capabilities. MEL also maintains the basic units for measuring mass and length in the United States.

The Intelligent Systems Division within NIST's MEL is developing scientific and engineering foundations for metrics and standards of intelligent (robotic) systems. This work includes interaction with first responders, technology developers, and mobile robot manufacturers to examine and develop performance standards for mobile autonomous robot vehicles to be used in search and rescue operations. Therefore, NIST staff members have developed and have created the Reference Test Arenas for Autonomous Mobile Robots to focus research efforts, provide direction, and accelerate the advancement of mobile robot capabilities. These arenas, modeled from buildings in various stages of collapse, allow objective performance evaluation of robots as they perform a variety of urban search and rescue (USAR) tasks in both physical and virtual domains.

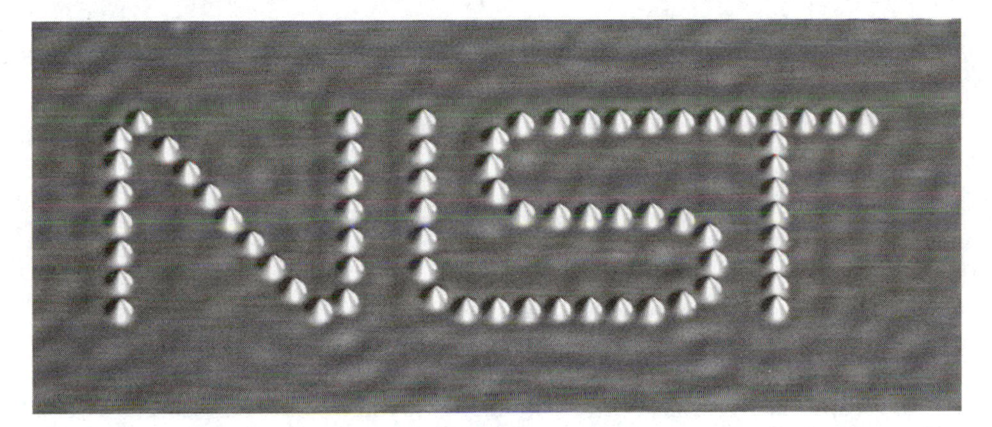

Figure 9-10 A 40-nanometer-wide National Institute of Standards and Technology (NIST) logo made with cobalt atoms on a copper surface. The ripples in the background are made by electrons, which create a fluid-like layer at the copper surface. Each atom on the surface acts like a pebble dropped in a pond. (Credit: Photograph courtesy of NIST.)

First responders often face daunting challenges during search and rescue tasks in dangerous environments. As such, the concept of including robots as a part of the responders' tool cache is being accepted, as robots have the potential of helping take responders out of harm's way and augmenting their capabilities. The Department of Homeland Security (DHS) Science and Technology (S&T) Directorate started an effort with NIST to develop comprehensive standards related to the development, testing, and certification of effective technologies for Urban Search and Rescue (USAR) robotics. These USAR robotic performance standards cover sensing, mobility, navigation, planning, integration, and operator control in order to ensure that the robots can meet operational requirements under the extremely challenging conditions that rescuers are faced with, including long endurance missions.

NIST's multidisciplinary Center for Nanoscale Science and Technology (CNST) is providing science and industry essential measurement methods, instrumentation, and standards to support all phases of nanotechnology development—from discovery to production. The CNST consists of both a research arm and a nanofabrication facility. The goal of this center is to partner with industrial, academic, and government organizations to solve the nanoscale measurement problems that now impede or prevent the more rapid implementation of nanotechnology.

National Institutes of Health (NIH)

9000 Rockville Pike
Bethesda, Maryland 20892 USA
1-301-496-4000
http://www.nih.gov/

The National Institutes of Health (NIH), a part of the U.S. Department of Health and Human Services, is the primary federal agency for conducting and supporting medical research. Composed of 27 institutes and centers, NIH provides leadership and financial support to researchers in every state and throughout the world. NIH is the steward of medical and behavioral research in the United States. The agency's mission is science in pursuit of fundamental knowledge about the nature and behavior of living systems and the application of that knowledge to extend healthy life and reduce the burdens of illness and disability. Pioneering research sponsored (in whole or in part) by NIH include robots for use in physical rehabilitation and robots for use in minimally invasive surgery (MIS). Nanomedicine is another exciting new research area being pursued within the NIH. Nanomedicine, an offshoot of nanotechnology, refers to highly specific medical intervention at the molecular scale for curing disease or repairing damaged tissues, such as bones, muscle, or nerve. The NIH's National Center for the Design of Biomimetric nanoconductors is tackling the broad challenge of capturing the capabilities of biological membranes in nanoscale devices. Biological membranes generate electrical and chemical signals, generate electrical power, perform osmotic pumping, and transform energy from one form to another. The key to their ability to accomplish these functions is biochemically directed self assembly that creates arrays of specific and regulated ion conductors embedded in lipid bilayers. Nanomedicine researchers working in this NIH effort seek to employ synthetically directed self-assembly on arrays of nanopores in silicon to create nano-engineered ion-conducting membranes to desired functional specifications. In essence, this particular center within the NIH is using interdisciplinary thinking in an effort to understand biology, design devices, and develop therapies at the nanoscale.

National Reconnaissance Office (NRO)

Office of Corporate Communications
14675 Lee Road
Chantilly, Virginia 20151-1715 USA
1-703-808-1198
1-703-808-1171 (Fax)
http://www.nro.gov/

Headquartered in Chantilly, Virginia, the National Reconnaissance Office (NRO) develops and operates unique and innovative space reconnaissance systems and conducts intelligence-related activities essential for the national security of the United States. The mission of the NRO is to ensure that the United States has the technology and space-borne assets needed to acquire intelligence worldwide. This mission is accomplished through research, development, acquisition, and operation of the nation's intelligence satellites. Reconnaissance satellites are complex and technically sophisticated space robots that are used to collect high-resolution images of Earth or to gather electronic signals for use in the practice of signals intelligence (SIGINT). The NRO's assets collect intelligence to support such functions as intelligence and warning (I&W), monitoring of arms control agreements, military operations and exercises, and monitoring of natural disasters and environmental issues.

NRO was officially established in September 1961 as a classified agency in the Department of Defense. The existence of the NRO and its mission of overhead reconnaissance were declassified by the U.S. government in September 1992. CORONA was the first American imagery satellite system. It was launched in 1960 and remained operational from 1960 to 1972. In 1995, the U.S. government declassified the CORONA program, along with the ARGON and LANYARD reconnaissance satellites. CORONA collected over 800,000 images. The intentional and careful diffusion (within U.S. government agencies) of certain spacecraft technologies from this pioneering photo reconnaissance satellite system greatly contributed to the development of more sophisticated robot spacecraft. From a technical perspective, the CORONA system was the first space program to recover an object from orbit and the first to deliver photoreconnaissance information from a satellite. CORONA would go on to be the first program to use multiple reentry vehicles, pass the 100-mission mark, and produce stereoscopic space imagery. Its most remarkable technological advance, however, was the improvement in its ground resolution from an initial 7.6- to 12.2-meter capability to an ultimate 1.82-meter resolution. GRAB, the first American SIGINT satellite was launched in 1960 and remained operational until 1962. This robot spy satellite program was declassified in 1998. Today, NRO partners with the National Geospatial Intelligence Agency (NGA) and the National Security Agency (NSA) to provide signals intelligence and near real-time imagery, global communications, and critical information to customers (military and civilian) within the U.S. government. Specific NRO satellite capabilities, numbers, and names are classified. The NRO Office of Corporate Communications is responsible for handling all public inquiries about the NRO.

National Robotic Engineering Center (NREC)

Ten 40th Street
Pittsburgh, Pennsylvania 15201 USA
1-412-681-6900
1-412-681-6961 (Fax)
http://www.rec.ri.cmu.edu/

The National Robotics Engineering Center (NREC) is a technology transfer organization that designs, develops, and tests robotic systems and vehicles for government and industrial clients. NREC is an operating unit within Carnegie Mellon University's Robotics Institute (RI) and maintains close ties with the faculty and staff at the campus-based Robotics Institute. Opened in July 1996, NREC staff members frequently adapt and refine robotic technology developed at the Robotics Institute in order to suit the needs of specific government or industrial clients. For example, in April 2006, the DARPA and the U.S. Army unveiled Crusher—an UGCV that has NREC as the prime contractor. This project is part of a major initiative within the Department of Defense to provide numerous operational unmanned ground combat vehicles to American military forces over the next decade. For every unmanned vehicle deployed in combat, American military personnel are removed from harm's way and have a reduced risk of become casualties. NREC is also pursuing autonomous vehicle applications in space exploration and industrial activities (such as automated turf care equipment).

National Science Foundation (NSF)

4201 Wilson Boulevard
Arlington, Virginia 22230 USA
1-703-292-5111
http://www.nsf.gov/

The National Science Foundation (NSF) is an independent federal agency created by the U.S. Congress in 1950 to (in the words of the congressional mandate) "promote the progress of science, to advance the national health, prosperity, and welfare; and the secure the national defense." Each year, the NSF serves as a funding source for about 20 percent of all federally supported basic research conducted by American institutions of higher learning. In many fields, such as mathematics and computer science, NSF is the major source of federal sponsorship. NSF supports fundamental research in robotics and nanotechnology. Recent NSF-sponsored research in robotics places emphasis on systems operating in unstructured environments with a high level of uncertainty, on the interaction and cooperation of humans and robots, and on advanced sensory systems, particularly computer vision. Research topics include theoretical, algorithmic, experimental, and hardware issues in robotics (including those on macro-, micro-, and nanoscale); robotics for unstructured environments (including issues of robustness, fault tolerance, and reconfigurability); personal robots, with an emphasis on human-centered end use; novel and advanced approaches to sensing, perception, and actuation (including embedded and highly distributed systems); understanding and processing of visual data; representation, reasoning, and planning for complex physical tasks involving temporal and spatial relationships; robots to extend human capabilities into unknown and hazardous environments; communication and task sharing between humans and machines, and among machines; intelligent cooperation among multiple robots; other research topics in robotic and computer vision applications, such as systems for surgery, undersea, space, search-and-rescue, and agriculture. For example, NSF funded research work on a robotic scrub technician that anticipates a (human) surgeon's request for an instrument during surgery, while also sponsoring the development of various-sized soccer robots that competed in international "RoboCup" championship matches. NSF is also sponsoring basic research in nanoscience and nanotechnology—recognizing that the agency's public investment in this new area of science, engineering, and technology will help create important knowledge, which could lead to a variety of exciting future applications. Nanotechnology is a bustling enterprise at many universities and colleges that spans the sciences, from physics to robotics to medicine. Some of the most interesting NSF-sponsored research projects involve the application of nanotechnology to the practice of medicine, such as microcapsules that can deliver precise amounts of a drug to where it will do the most good in the body of a patient.

Oak Ridge National Laboratory (ORNL)

One Bethel Valley Road
Oak Ridge, TN 37831 USA
1-865-574-1000 (Main number)

1-865-574-4160 (External Relations Office)
http://www.ornl.gov/

Oak Ridge National Laboratory (ORNL) is a multiprogram science and technology laboratory managed for the U.S. DOE by a partnership between the University of Tennessee and Battelle. Oak Ridge was established in 1943 as part of the secret Manhattan Project to pioneer a method for producing and separating plutonium. In the 1970s, ORNL's research program expanded beyond the study of nuclear energy into other areas, such as energy production, transmission, and conservation. Today, the laboratory's six major mission roles include neutron science, energy, high-performance computing, systems biology, materials science at the nanoscale, and national security.

Mechanical manipulators have long been used in nuclear science and chemistry in special facilities called hot cells to protect workers from radioactive materials. Starting in the late 1970s, ORNL researchers devised remotely controlled dexterous servomanipulators whose functions and performance could be viewed on closed-circuit television. Such teleoperation techniques allowed work to be accomplished in radioactive zones that were too hazardous for human beings. Staff members at ORNL extended earlier teleoperation techniques developed at ANL, and soon advanced remote manipulation technologies developed at ORNL found applications throughout the DOE complex in such important activities as nuclear fuel reprocessing, military field munitions handling, and environmental cleanup operations. The study of human-amplifying machines, such as exoskeletons, represents a special area of robotics and automation at ORNL. The effort focuses on systems that work in smooth synergy with humans in augmenting their physical strength abilities while maintaining complete task awareness through feedback to the human worker. In the late 1970s and early 1980s, ORNL staff members explored the development and use of mobile robot technologies. Some of these pioneering activities led to spin-off commercial companies (such as Remotec—now a division of Northrop-Grumman), which manufacture mobile robots for military, law enforcement, environmental cleanup, and scientific research applications. Today, major areas of robotic systems research at ORNL include mobile robots, advanced manipulators, and combined mobility manipulation systems.

The Spallation Neutron Source makes ORNL the world's foremost center for neutron science research and supports a variety of leading-edge projects at the laboratory's Center for Nanophase Materials Sciences (CNMS).

Occupational Safety and Health Administration (OSHA)

U.S. Department of Labor
200 Constitution Avenue, NW
Washington, District of Columbia 20210 USA
1-800-321-6742 (General information and questions)
http://www.osha.gov/

The Occupational Safety and Health Administration (OSHA) is a federal organization within the U.S. Department of Labor. OSHA has the mission of ensuring worker safety and health in the United States by working with employers and employees to create better working environments. Since its inception in 1971, OSHA has helped to reduce workplace fatalities by more than 60 percent and occupational injury and illness rates by 40 percent. At the same time, employment within the United States has doubled from 56 million workers at 3.5 million worksites to more than 115 million workers at 7.2 million sites (as of January 2005). Under its national charter to promote worker safety, OSHA is responsible for and has issued guidelines for robotic systems in the workplace. These guidelines describe some of the elements of good safety practices and techniques that should be used in the selection and installation of robots, especially industrial robots, and companion robot safety systems, control devices, robot programming activities, and employee training. OSHA currently recommends that the selection of an effective robotics safety system should be based on hazard analysis of the operation involving a particular (industrial) robot. Among the factors to be considered are the task an industrial robot is programmed to perform, the startup and programming procedures, the location of the robot and any pertinent environmental conditions, requirements for corrective tasks to sustain normal operations, human errors, and possible malfunctions of the robot system. Modern industrial robots are programmable multifunction mechanical devices designed to move material, parts, tools, or specialized devices through variable programmed motions in order to perform a variety of tasks. Studies in Japan and Sweden have indicated that many industrial robot accidents, which involved injury or death to human beings, did not take place under normal operating conditions, but rather during programming, adjustment, testing, cleaning, inspection, and repair periods.

Office of Naval Research (ONR)

One Liberty Center
875 North Randolph Street, Suite 1425
Arlington, Virginia 22203-1995 USA
1-703-696-5031 (Public Affairs Office)
1-703-696-5940 (Fax: Public Affairs Office)
http://www.onr.navy.mil/

The Office of Naval Research (ONR) coordinates, promotes, and conducts the science and technology programs of the United States Navy and Marine Corps through academic institutions, government laboratories, and nonprofit and for-profit organizations. This office also provides advice to the Chief of Naval Operations and the Secretary of the Navy and works with industry to improve manufacturing technology, especially as related to future military systems and naval ships. ONR supports programs in robotic systems (including AUVs and UAVs), as well as projects involving nanotechnology and the use of virtual reality systems. For example, in the late 1960s, ONR funding helped SRI (formerly called the Stanford Research Institute) create Shakey the mobile robot. Shakey had television (TV) eyes, tacticle sensors, an optical range finder, and an elementary navigation system, making this pioneering robot the technical ancestor to the modern, autonomous mobile robot. SRI roboticists designed Shakey so it could plan and execute simple tasks, such as finding objects and manipulating them, while

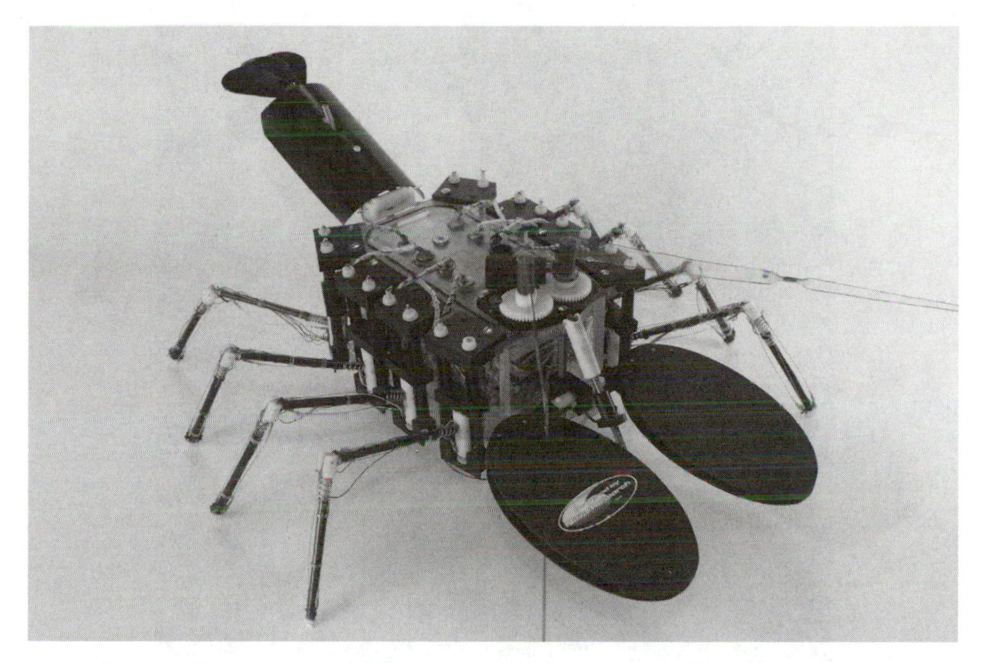

Figure 9-11 The biomimetic underwater robot, called Robolobster, at Northeastern University's Marine Science Center in Nahant, Maine. (Credit: Photograph courtesy of the U.S. Navy/ONR.)

avoiding obstacles. More recently, ONR was instrumental in the development of the biomimetic underwater robot, called Robolobster. Biomimetic robots are, at least in principle, relatively small, agile, and generally inexpensive. Such systems rely on electronic nervous systems, sensors, and novel actuators to deal (much like living animals) with real-world environments.

Pacific Northwest National Laboratory (PNNL)

902 Battelle Boulevard
Richland, WA 99352 USA
1-509-375-2121
1-888-375-7665 (Toll Free)
http://www.pnl.gov/

The Pacific Northwest National Laboratory (PNNL) is one of nine U.S. Department of Energy (DOE) multiprogram national laboratories. PNNL is operated by Battelle for the U.S. DOE. Founded in 1965, PNNL's current mission is to deliver science-based solutions to DOE's major challenges in expanding energy, ensuring national security, and cleaning up and protecting the environment. The PNNL scientific staff is well recognized for an ability to successfully integrate the chemical, physical, and biological sciences in the solution of complex problems. Robotics and advanced controls research, development, and applications form a major thrust area at PNNL. Activities at this national laboratory include mobile

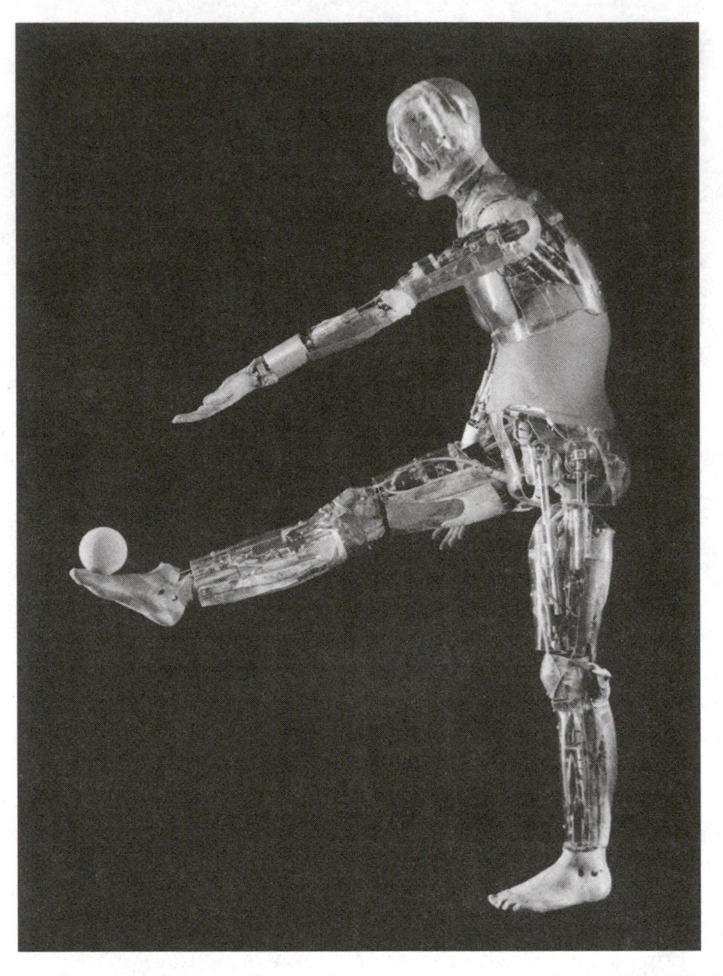

Figure 9-12 This is Manny—a life-sized robotic mannequin built by PNNL for the U.S. Army in the late 1980s to test protective military clothing. Manny is an example of an early anthropomorphic robot. The robot could reproduce human motions, such as sitting, walking, bending, and flexing. It could also simulate breathing and sweating, and the temperature of its "skin" surface could also be controlled and adjusted. (Credit: Photograph courtesy of U.S.DOE/PNNL.)

robotic platforms, teleoperated work systems, robot autonomy and control system design, and human emulation—an activity involving the development of robotic systems that emulate human motion and physiological functions for developing and testing human worker protective products.

RedZone Robotics Incorporated

484 West 7th Avenue
Homestead, Pennsylvania 15120 USA
1-412-476-8980
1-412-476-8981 (Fax)
http://redzone.com/

Founded in 1987 as a spin-off of the Robotics Institute of Carnegie Mellon University, RedZone Robotics designs, manufactures, and distributes products that clean, inspect, and rehabilitate pipes and tanks using robotic and remotely controlled mobile equipment. As the company's name "red zone" implies, product emphasis is on robotic systems that can successfully operate in harsh and hazardous environments. The company's government customers include the U.S. Department of Energy (DOE), NASA, and several of the DOE national laboratories. RedZone's Pioneer robot was originally designed and built for DOE and NASA sponsored research project to demonstrate the feasibility of using a teleoperated mobile robot system for performing structural analysis within the sarcophagus that had been constructed over the damaged Unit 4 reactor at Chernobyl, Ukraine. Another example of RedZone's hazardous environment robots is the company's Houdini robot—a bulldozer-like mobile robot that folds to fit through a 60-centimeter (24-inch) diameter opening. The development of this robot was sponsored by the DOE to help remediate nuclear waste storage tanks at the ORNL and other locations within the DOE nuclear weapons complex. Similarly, RedZone's Rosie robot, a heavy-payload, long-reach remote work vehicle for decontamination and decommissioning applications, was used at the ANL to assist in the dismantlement of the Chicago Pile Five (CP-5) nuclear reactor. The company has developed many other hazardous environment robots. In addition to hazardous environment robots, RedZone also produces the RedScore Soccer Robot—a new omni directional mobile base developed in collaboration with Carnegie Mellon University (CMU) "small-size" robot soccer team for use in RoboCup competition. This soccer robot features a kicker, a powered dribbler, and omni directional roller wheels.

Remotec

353 JD Yarnell Parkway
Clinton, Tennessee 37716 USA
1-865-483-0228
1-865-483-1426 (Fax)
http://www.es.northropgrumman.com/remotec/

Remotec manufactures mobile robot systems for hazardous-duty operations. For example, Remotec robots are used by the American military during combat operations and explosive ordnance disposal (EOD) missions, by local, state, and federal law enforcement authorities during hostage situations and suspicious package investigations, and by first responders and environmental cleanup teams during hazardous materials (HAZMAT) operations. Remotec is a subsidiary of Northrop Grumman Land Combat Systems within the parent company's Electric Sensor and Systems Sector (which is headquartered in Baltimore, Maryland). Remotec was founded in Oak Ridge, Tennessee, in 1980 to provide remote handling consultation to the nuclear industry. Remotec staff members soon identified the opportunity to expand and apply the fledgling company's technical competencies to meet mobile robot system requirements outside the nuclear field. This led to the company's decision in 1986 to purchase ANDROS (mobile robot) technology. Since acquiring the ANDROS technology, Remotec

Figure 9-13 This is the Pioneer robot that was designed and constructed in the late 1990s to demonstrate the feasibility of performing structural analysis and other activities within the huge shelter (sarcophagus) erected over the severely damaged Unit 4 reactor at Chernobyl, Ukraine. The robot is a track-driven machine that is electrically powered and teleoperated by means of a 100-meter (300-foot) umbilical cable. The project was sponsored by the U.S. Department of Energy and NASA in collaboration with academic and industrial partners, including the National Robotics Engineering Consortium at Carnegie Mellon University and RedZone Robotics. (Credit: Photograph courtesy of the U.S. Department of Energy and the Lawrence Livermore National Laboratory.)

has remained at the technical forefront in developing and supplying a large family of hazardous duty, mobile robots to a variety of customers.

Robot Hall of Fame®

225 Smith Hall
Carnegie Mellon University
Pittsburgh, Pennsylvania 15213 USA
1-412-268-9656

1-412-268-5497 (Fax)
http://www.robothalloffame.org/

In 2003, the School of Computer Science at Carnegie Mellon University established the Robot Hall of Fame® to identify and honor landmark achievements in robotics technology and the increasing contributions that robots are making to human activities and the planet's global civilization. The Robot Hall of Fame recognizes and celebrates the highest accomplishments of robots in both science and science fiction. Accordingly, the annual inductees (both real world and fictional) are selected by a jury of scholars, researchers, roboticists, designers, and entrepreneurs and then honored at an induction ceremony conducted by Carnegie Mellon University. To be selected, the robots from science must have demonstrated unique machine skills in accomplishing the purpose for which they were created or else have served useful or potentially useful functions. Autonomous entertainment robots qualify for consideration. Fictional robots must have achieved worldwide notoriety, inspiring human beings to create real robots that are productive, entertaining, or helpful. Real-world robots that have been selected for induction into the Robot Hall of Fame include NASA's Mars Pathfinder mission minirover (also known as *Sojourner*) and Shakey, the pioneering mobile robot that was developed at SRI in the late 1960s. Inductees from science fiction include Robby the Robot from the 1956 motion picture, *Forbidden Planet*, and R2-D2 the dutiful and efficient "droid" from the *Star Wars* motion picture saga.

Robotic Industries Association (RIA)

900 Victors Way
P.O. Box 3724
Ann Arbor, Michigan 48106
1-734-994-6088
1-734-994-3338 (Fax)
http://www.roboticsonline.com/

Founded in 1974, the Robotic Industries Association (RIA) is a trade group organized in North America specifically to serve the robotics industry. Member companies include leading robot manufacturers, the users of robot systems, robotic system integrators, component suppliers, consulting firms, and research groups. RIA sponsors *Robotics Online*, an Internet-based source of information that should prove quite helpful to engineers, managers, and executives who are considering the use of robotic and flexible automation systems.

Robotics Institute

Carnegie Mellon University
5000 Forbes Avenue
Pittsburgh, Pennsylvania 15213-3890 USA
1-412-268-3818
1-412-268-6436 (Fax)
http://www.ri.cmu.edu/

The Robotics Institute (RI) at Carnegie Mellon University was established in 1979 to perform basic and applied research in robotics and related technologies. Faculty, students, and staff members at the Robotics Institute give special attention to robotics technologies for industrial applications and the performance of societal tasks. The Field Robotics Center (FRC) is an embedded research organization within the Robotics Institute. This center has and continues to develop a wide variety of remotely operated and autonomous field robots capable of functioning on land, at sea, in the air, or in outer space. Some of the contemporary research projects at FRC include: the lunar rover initiative, which seeks to land a robot rover on the Moon to explore regions of high hydrogen (suspected water ice) concentration; unmanned security robots, which involves the development of autonomous all-terrain vehicles to secure borders and facility perimeters; and an emergency response robotics effort, which addresses multiagent robotic technologies that can assist first responders under emergency situations.

Sandia National Laboratories (SNL)

(New Mexico)
Sandia National Laboratories (Albuquerque)
1515 Eubank Blvd., SE
Albuquerque, NM 87123 USA
(mail address for Media Relations Office)
Sandia National Laboratories, New Mexico
P.O. Box 5800
Albuquerque, NM 87185-0165 USA
1-505-845-0011
1-505-844-8066 (Media Relations Office)
http://www.sandia.gov/

(California)
Sandia National Laboratories (Livermore)
7011 East Avenue
Livermore, CA 94550 USA
(mail address for Media Relations Office)
Sandia National Laboratories, California
P.O. Box 969
Livermore, CA 94551-9111 USA
1-925-294-2447 (Media Relations Office)
http://www.sandia.gov/

Sandia National Laboratories (SNL) is a government-owned/contractor operated (GOCO) facility. Lockheed Martin currently manages SNL for the National Nuclear Security Administration (NNSA) of the U.S. DOE. Sandia's technical roots extend back to the Manhattan Project and the development of the first American nuclear weapons. The original technical emphasis involved nuclear ordnance engineering, that is, the transformation of the nuclear physics packages produce by the Los Alamos National Laboratory and (later) the Lawrence Livermore National Laboratory into deployable nuclear weapons. Today, Sandia has expanded its role in supporting national security by pursuing the continued safety and reliability of stockpiled nuclear weapons, nuclear nonproliferation initiatives within the DOE, the development of innovative nuclear treaty monitoring technologies, and performing studies that help protect national energy and other critical infrastructures against international terrorism.

Sandia's Intelligent Systems and Robotics Center (ISRC) is one of the world's leading organizations in creating miniature to macro-sized teleoperated to autonomous vehicles for military and industrial applications. From environmental cleanup to the battlefield, the ISRC has the expertise to develop unique intelligent mobile systems. Current and future efforts focus on extreme mobility (such as multimode locomotion to permit a single vehicle to maneuver on combinations of land, water, and air) and dexterous mobile manipulation (such as using "arms" and "fingers" on miniature to macro-sized vehicles for the purpose of interacting with physical objects. For example, the ISRC is assisting the DOE's Accident Response Group (ARG) in the development of the Accident Response Mobile Manipulator System (ARMMS). This effort is providing the U.S. government with a sophisticated response unit that has both vehicular and robotic characteristics, which can support salvage and recovery operations involving a nuclear weapon accident or other hazardous material spill. Another example is SandDragon—a man-portable ground robot developed for the U.S. Marine Corps Warfighting Laboratory (MCWL). This military robot's mission is to conduct networked surveillance, reconnaissance, target acquisition, and response (lethal and nonlethal) in coordination with other sensors and robots (aerial and ground). Researchers at Sandia Laboratories are also involved in the design, development, and application of microelectronic machines (MEMs).

Society of American Military Engineers (SAME)

607 Prince Street
Alexandria, Virginia 22314-3117 USA
1-703-549-3800 (Main Telephone Number)
1-703-684-0231 (Fax Executive Office)
http://www.same.org/

The mission of the Society of Military Engineers (SAME) is to promote and facilitate engineering support for national security by developing and enhancing relationships and competencies among uniformed services, public and private sector engineers, and related professionals. SAME is the only nonprofit professional engineering education organization that actively promotes the advancement of both individual technical knowledge and the collective engineering capabilities of governments (local, state, and federal), the uniformed services, and private industry. The society's hallmark publication is the professional magazine *The Military Engineer*. Leaders in the uniformed services now recognize the hazards faced by many of the military and civilian engineers in day to day reconstruction operations in hostile areas such as Iraq, where threats due to unconventional warfare and terrorist activities have all but replaced previous, traditional military conflict scenarios. There are often no so-called "front lines" in contemporary conflict zones, especially regions involving terrorism and local insurgencies. So the uniformed services—each in its own distinct ways—are

Figure 9-14 "Robugs" are miniaturized robots that can carry arrays of microsensors and communications systems. Shown here is the prototype of the mini-robot called MARV (mobile autonomous robot vehicle) developed at the Sandia National Laboratories to help engineers understand and overcome some of the problems of building tiny autonomous mini-robots. The penny in the foreground provides a convenient dimensional scale. (Credit: Photograph courtesy of U.S. DOE/Sandia National Laboratories.)

actively exploring the use of remotely controlled heavy equipment and other robotic systems to keep military engineers out of harm's way under these hostile circumstances.

Sony Corporation

6-7 35 Kitashinagawa
Shinagawa-ku, Tokyo,
141-0001 Japan
+81-3-5448-2111
http://www.sony.net/

The Sony Corporation is a multinational Japanese company that focuses on consumer electronics, games (such as the Sony PlayStation), and entertainment (such as music and motion pictures). In 1999, Sony introduced its first generation of four-legged entertainment robots, called AIBO. The initial ERS-110 model could not only provide its human owner with a range of performances but AIBO could also express emotions and even "learn." Over the next several years Sony engineers expanded and improved upon the AIBO entertainment robot and also introduced (in 2000) a prototype humanoid robot, a 50-centimeter tall robotic system called SDR-3X. This prototype humanoid robot was capable of bipedal (two-legged) motion and could even balance on one foot and kick a ball. In 2003, Sony introduced an improved version of its line of humanoid robots, a prototype robotic system named QRIO. Sony engineers successfully integrated walking, jumping, and running movements into the people-friendly QRIO robot. Corporate literature also suggests that QRIO is the world's first running humanoid robot. That same year, Sony also introduced the ERS-7 version of its popular AIBO entertainment robot, an advanced robotic system loaded with a wide variety of sensors and capable of providing a range of versatile expressions.

Space and Naval Warfare Systems Center, San Diego (SSC San Diego)

Public Affairs Office
San Diego, California 92152-5001 USA
1-619-553-2717
http://www.spawar.navy.mil/robots/

The Space and Naval Warfare Systems Center in San Diego abbreviates its command title as *SSC San Diego* to avoid confusion with its parent organization, the Space and Naval Warfare Systems Command (SPAWAR), which is one of the U.S. Navy's major acquisition commands. Since the early 1960s, SSC San Diego and its predecessor organizations have been involved in various aspects of robotics for military applications. SSC San Diego is the U.S. Navy's research and development, test and evaluation, engineering, and fleet support center for command, control, communications systems, and ocean surveillance. The Advanced Systems Division of SSC San Diego conducts research and development on land and aerial robots, while the center's Ocean System Division deals with underwater robots. Present and past SSC San Diego robotics activities include: autonomous land robots (such as ROBART I, II, III), UGVs, UAVs, unmanned surface vehicles (USGs), and unmanned underwater vehicles (UUVs) (such as the family of Snoopy underwater robots). The center also conducts research in telepresence and virtual reality. The center has engineering development projects concerning the use of marsupial robots, tandem robots, nonlethal weapon pods on robots, and a railway intrusion detection system. SSC San Diego has also performed studies concerning the use of robotic systems for security and law enforcement.

Stäubli Robotics

201 Parkway West
Duncan, South Carolina 29334 USA
1-800-257-8235
http://www.stabuli.com

In 1982, the Stäubli Corporation, with its century long tradition and experience in engineered devices with mechanical motions, made a strategic decision to specialize in robotics. The corporate diversification into robotics began through an affiliation with Unimation (at the time a division of Westinghouse USA). In 1988, Stäubli acquired Unimation and then proceeded to expand its line of industrial robots by introducing the company's RX line of robots in 1992, followed by the TX line of robots in 2004. Today, the company offers an extended product range of industrial robots from a variety of SCARA robots to heavy payload six-axis robots, many of which can be run using a PC-based control platform.

UCI Center for Minimally Invasive Surgery (MIMS)

UCI Medical Center
101 The City Drive South
Orange, California 92868
1-714-456-7890
http://www.ucihealth.com/

Surgeons at the UCI Center for Minimally Invasive Surgery (MIMS) use a 7-foot (2-meter) machine assistant with three arms, known as the da Vinci™ Surgical System, to perform delicate surgeries that extend beyond the limits of the human hand. The da Vinci system gives doctors unprecedented control over the tiny instruments they use during minimally invasive surgery, also known as keyhole surgery. Furthermore the use of the medical robot system permits a more detailed view of the surgical site than unaided human eyes allow. The first robot-assisted surgery was performed at the UCI Medical Center on May 3, 2002. The surgeon used the center's da Vinci™ Surgical System to successfully perform a laparoscopic, or minimally invasive, procedure to treat the patient's gastroesophageal reflux disease. The robotic surgical system provided the surgeon with three-dimensional (3-D) imaging of the operating field and intuitive hand movement—two major improvements over standard laparoscopic surgery, which is characterized by two-dimensional (2-D) imaging and "counterintuitive" instrument movement. Counterintuitive instrument movement involves an operating condition similar to the surgeon working on the patient while looking in a mirror. In contrast to the conditions experienced during traditional laparoscopic surgery, the UCI Medical Center's da Vinci™ Surgical System provided the surgeon with nearly all-natural movements of the human wrist. The system also eliminated natural hand tremor and improved dexterity to enable the surgeon to perform ever-finer surgery in a more controlled manner.

United States Department of Energy (DOE)

(Headquarters Mailing Address)
U.S. Department of Energy (DOE)
Forrestal Building
1000 Independence Avenue, SW
Washington, DC 20585 USA

1-800-dial-DOE (Toll Free Information Gateway) (1-800-3425-363)
1-202-586-5000 (Main Number)
1-202-586-4403 (Fax, Secretary of Energy's Office)
http://www.energy.gov/

The U.S. Department of Energy traces its origins to the Manhattan Project and the race to develop an American atomic bomb during World War II. The Atomic Energy Act of 1946 established civilian control over nuclear energy applications when this legislation placed the newly created U.S. Atomic Energy Commission (USAEC) in charge of all atomic research and development by the federal government. During the early years of the cold war, the USAEC focused its efforts on the design and production of nuclear weapons and on the development of nuclear reactors for the propulsion of naval ships. Responding to President Dwight Eisenhower's "Atoms for Peace" initiative, the Atomic Energy Act of 1954 ended exclusive government control over nuclear energy and encouraged the USAEC to promote civilian nuclear technology applications—especially the growth of a commercial nuclear power industry within the United States. Responding to changing political and social needs, the Energy Reorganization Act of 1974 abolished the USAEC and replaced it with two new agencies the Nuclear Regulatory Commission (NRC) to regulate the nuclear power industry, and the Energy Research and Development Administration (ERDA) to manage the nuclear weapon, naval reactor, energy development programs. The energy crisis of the 1970s suggested the need for a more unified federal energy program and so the U.S. Department of Energy came into being in October 1977. This new organization combined the responsibilities ERDA and parts and programs of several other federal agencies. Today, the Department of Energy is responsible for enhancing the security of the United States through four major programmatic efforts. First, the National Nuclear Security Administration (NNSA), an organization embedded within the DOE, supports national security by insuring the integrity and safety of American nuclear weapons, promoting nuclear nonproliferation initiatives, and continuing to provide safe and efficient nuclear power plants for the United States Navy. Second, the DOE energy program focuses on increasing domestic energy production, encouraging energy conservation and efficiency, and promoting the development of renewable and alternative energy sources. Third, the DOE environmental program is responsible for the remediation of the environmental legacy from the cold war nuclear weapons program and the permanent and safe disposal of radioactive wastes generated as a result of the both civilian and military application of nuclear technology. Finally, the DOE science program sponsors cutting-edge research and development efforts intended to revolutionize the way the United States finds, generates, and delivers energy in this century. As part of these programmatic thrusts, the DOE (primarily through its national laboratories) sponsors the development and use of a wide variety of robotic systems, ranging from experimental swarms of microbots to heavy-duty remotely operated equipment for decontamination and decommissioning projects at nuclear weapons production sites. The DOE also provides support for a variety of leading edge research and development projects in nanotechnology. Finally, the DOE also supplies NASA the plutonium-238 fuel for the radioisotope thermoelectric generators (RTGs) that provide a long-lasting supply of electricity to robot spacecraft operating in deep space or on planetary bodies with hostile environments.

Woods Hole Oceanographic Institution (WHOI)

Fenno House MS #40 (Public Relations Office)
Woods Hole, Massachusetts 02543-1050 USA
1-508-289-3340
1-508-457-2180 (Fax)
http://www.whoi.edu/

Founded in 1930 as a permanent independent research laboratory to conduct a worldwide program of oceanographic research, the Woods Hole Oceanographic Institution (WHOI) grew substantially during World War II to support defense-related research, and later began a steady growth in staff and research fleet. Today, WHOI is a private, independent, not-for-profit corporation dedicated to research and higher education at the frontiers of ocean science. The institution's primary mission is to develop and effectively communicate a fundamental understanding of the processes and characteristics that govern how the world's oceans function and how they interact with Earth as a whole. To accomplish this mission, WHOI promotes the development and use of advanced instrumentation and systems. REMUS is an acronym that stands for remote environmental monitoring units, a family of AUVs, which resemble torpedoes and can operate underwater without human assistance and without cable connecting them to research vessels at the sea surface. REMUS was invented and continues to be developed at WHOI; a commercial company, Hydroid, Inc. of East Falmouth, Massachusetts, manufactures the vehicles. REMUS AUVs are used by ocean scientists, U.S. Navy personnel, and underwater archaeologists to autonomously explore, measure, and survey conditions beneath the ocean's surface. After entering the water, REMUS uses acoustic navigation to independently survey the area while sensors inside the robotic vehicle sample and record data.

10

Demonstration Sites

This chapter provides a selective listing of facilities, technical exhibits, science and technology museums, and several unusual experience sites around the world that are related to the development or application of robotic systems or their fundamental technologies. At these selected demonstration sites, a person can learn about the scientific discoveries, engineering breakthroughs, or special events associated with the development of modern robot technology, or perhaps some of the basic scientific principles and technologies necessary to bring about the modern robot. Several of the facilities listed (like the "live" stage appearance of ASIMO—the world's most advanced humanoid robot—in the Honda ASIMO Theater in *Innoventions* located inside Disneyland® Resort's Tomorrowland) are part of a major attraction that host millions of guests each year. Other demonstration sites (like the Titan Missile Museum south of Tucson, Arizona, and Museum of Art and History in Neuchâtel, Switzerland) are much more specialized in their content and size. Still other demonstration sites (like the Robot Hall of Fame at Carnegie-Mellon University) are robotics technology-related experiences that are best enjoyed by means of the Internet.

The selected sites encompass a wide variety of content and presentation formats. This chapter gives special emphasis to representative science and technology museums at strategic locations around the globe. Many modern science and technology museums now demonstrate scientific principles and engineering applications through visitor interactive "hands-on" exhibits and displays. Such facilities often provide visitors of all ages an excellent pathway to experiencing and understanding the fundamental physical laws and engineering approaches that make modern robot systems possible. Other science and technology museums focus their efforts on the process of preserving and displaying important technical artifacts. Some science and technology museums use both approaches. Each demonstration site listed in this chapter plays a distinctive role in helping to tell the overall story of modern robots and their machine ancestors, which helped shape and form the modern industrial world. To foster science and engineering education, many of the sites have hosted local, regional, or national-level robotics competitions. So, it is a good idea to visit the Web sites of nearby facilities on a regular basis to see what is new in the way of traveling technical exhibits or upcoming robot competitions.

As with planning any type of successful travel, it is wise to inquire ahead (preferably by telephone or via the Internet) to make sure that the particular site you wish to visit will actually be accessible during the time period desired. Generally, the accompanying Web site provides a great deal of useful information about the demonstration site—including hours of operation, admission prices (if any), specific location, and travel directions. These are an excellent source of current information. Many of the Web sites even provide a preview (virtual tour) of the exhibits available. Please recognize, however, that any commercial entities discussed in this chapter are representative and do not necessarily imply a specific endorsement by either the author or the publisher.

Some science and technology museums in the United States and around the world have hosted excellent, specialized exhibits (permanent or traveling) about robots or technologies supporting the emergence of robotics (such as the development of electricity). But in the most science and technology museums, robot technology generally represents only a small portion of the many fine exhibits and displays at such technology museums or science centers. Again, because of the dynamic nature of today's science-museum industry, the visitor is strongly encouraged to inquire ahead, as to whether a robot-technology exhibit is currently available at the particular science experience center or technology museum. This advice applies to bout traveling exhibits, as well as to so-called permanent exhibits, because the latter might be closed for maintenance, repair, or refurbishment. Finally, as described in this chapter as well as in Chapter 11, the Internet provides a continuously expanding opportunity to learn more about robotics from the comfort of your home or school computer.

American Museum of Science and Energy (ASME)

300 South Tulane Avenue
Oak Ridge, TN 37830 USA
1-865-576-3200
http://www.amse.org/

This museum opened in 1949 in an old World War II cafeteria. The facility was originally called the *American Museum of Atomic Energy* and guided tours took visitors through all the peaceful applications of nuclear energy. The present facility opened in 1975 and has expanded its focus to provide the general public with information about all forms of energy. In keeping with the changing exhibit emphasis, the museum changed its name in 1978 to the American Museum of Science and Energy. Present day exhibits at ASME include: *The Story of Oak Ridge*, which provides a panorama of historical photographs, documents and artifacts explaining the Manhattan Project and the construction of Oak Ridge to support the American atomic bomb effort; *World of the Atom*, which discusses pioneering atomic scientists, the natural radiation environment, nuclear fusion, and the use of nuclear energy in space exploration and which contains a cross section model of a nuclear reactor and a simulated underground nuclear waste storage area; *Y-12 and National Defense*, which features models of nuclear weapons and how protective clothing and tools used in working with radiation sources at the Y-12 plant in Oak Ridge; *Earth's Energy Resources*, which explores the various energy sources (including coal, oil, geothermal, hydropower, and natural gas) found here on Earth; and *Energy—The American Experience*, which provides an interesting historical display of labor-saving devices found in a typical preelectricity home. ASME also has a display about the new Spallation Neutron Source being constructed at the nearby Oak Ridge National Laboratory and a *Robot Zoo* that should appeal to both young and old visitors alike. As a special assistance to teachers, the museum's exhibits and programs allow students to investigate several themes within each of the following basic educational goals: *process of science* (themes: observing, explaining, and communicating); *concepts of science* (themes: interactions, explaining, conservation); *habits of mind* (themes: historical and cultural perspective, science and technology, creative enterprise); and *science in society* (themes: attitudes, career goals, and politics). There is a modest charge for admission.

Argonne Information Center

Argonne National Laboratory (ANL)
9700 South Cass Avenue
Argonne, Illinois 60439 USA
1-630-252-2000
1-630-252-5562 (Community Relations Office)
1-630-252-5274 (Fax, Office of Public Affairs)
http://www.anl.gov/

The Argonne Information Center features interactive computerized exhibits, displays, and historical artifacts from more than 50 years of scientific research at the first national (nuclear) laboratory in the United States. The center features a state-of-the-art learning laboratory where teachers and students can take advantage of the many educational opportunities of the Internet. Current exhibits include the following: a user-controllable, table-top electron accelerator that allows a visitor to use magnets to bend and control a charged particle beam's path; an interactive tour of Argonne's Advanced Photon Source—the most brilliant X-ray source for research in the United States; an interactive video tour of Argonne's Fuel Conditioning Facility, where experimental nuclear fuel is treated for safe disposal; and demonstrations of research that is helping scientists develop better global climate models by accurately measuring how much solar energy falls on Earth. There are also many interesting displays, including one describing how X-ray research fights cancers and viruses, a model of an inherently safe nuclear reactor that recycles its own nuclear waste, and information about environmental restoration and toxic waste cleanup.

The Argonne Information Center is located at the laboratory site's main gate. Admission is free and the center is open to anyone interested in the laboratory's programs or science, technology, and nature in general. Children of any age are welcome at the center, if accompanied by an adult. The use of the center is also available (upon request) to teachers, student groups, civic groups, and other organizations. No registration is required to visit the Argonne Information Center, except for Internet and modern technology training in the center's learning lab. Please contact the ANL Community Relations Office (1-630-252-5562) to reserve the Argonne Information Center for a special group or event or to obtain additional information about the center or the laboratory.

Guided tours of Argonne National Laboratory are also available. However, visitors wishing to take a laboratory tour must make reservations in advance through ANL's Community Relations Office and be at least 16 years old. Most tours are conducted on Saturday mornings or afternoons.

Arizona Science Center

600 East Washington Street
Phoenix, Arizona 85004 USA
1-602-716-2000
1-602-716-2099 (Fax)
http://www.azscience.org/

The Arizona Science Center is an example of a contemporary science center with an overall mission to serve the science and technology experience needs of guests of all ages, who live in the surrounding region. The metropolitan Phoenix area is one of the most rapidly growing in the United States and enjoys a regional economy powered by many high-technology companies, great and small. The center uses more than 300 interactive displays and hands-on exhibits (conveniently divided into five themed galleries) to introduce its guests to both the basic principles of science and also the exciting modern technologies (including electronics and robotics), which are shaping civilization on a regional, national, and global scale. As commonly practiced by regional science centers, core displays that explain scientific principles and are complemented by a variety of expanded (though often temporary) featured exhibits. The center also maintains (on a permanent basis) several regionally specific customized exhibits. The center's Fab Lab allows visitors of all ages to perform basic science experiments in an open-ended fashion. Each guest can explore the forces of gravity, friction, and magnetism at their own pace. Visitors are also invited to observe scientific demonstrations by visiting experts, who to live and work in the Phoenix area.

Boston Museum of Science (MOS)

Science Park
Boston, MA 02114 USA
1-617-723-2500 (General Information)
1-617-589-0250 (Media Relations)
1-617-589-454 (Fax, Media Relations)
http://www.mos.org/

The mission of the Museum of Science (MOS) in Boston is to stimulate interest in and further understanding of science and technology and their importance for individuals and for society. Of particular interest here is the Lightning Presentation in the Thomson Theater of Electricity. Each day, museum guests can safely experience a high-voltage demonstration of lightning as created by the world's largest air-insulated Van de Graaff generator. Following its service as a research tool and teaching device, in the early 1950s the Massachusetts Institute of Technology (MIT) donated the machine to the museum. Dr. Robert J. Van de Graaff, a professor at MIT, designed and built this generator in the 1930s for use in early "atom smashing" and high-energy X-ray research. Then, as more power particle accelerators came into being, the generator became an instructional device. Today, it supports science education by vividly demonstrating electricity and lightning to public and school audiences. There is a charge for admission.

Bradbury Science Museum

1350 Central Avenue, MS C330
Los Alamos, NM 87544 USA
1-505-667-4444
1-505-665-6932
http://www.lanl.gov/museum/

The Bradbury Science Museum is rich in nuclear technology history and a uniquely rewarding nuclear tourism destination. Founded in 1963, the museum's name honors Norris E. Bradbury, who served from 1945 to 1970 as the second director of the Los Alamos National Laboratory (LANL). The modern facility is actually a publicly accessible component of the laboratory, a multipurpose national security-oriented laboratory operated for the U.S. Department of Energy (DOE) by the University of California.

In April 1993, the Bradbury Science Museum moved to its present location in the heart of downtown Los Alamos, New Mexico, at the corner of 15th Street and Central Avenue. The modern museum serves the following mission: to interpret LANL research, activities, and history to official visitors, the general public, and laboratory employees; to promote greater public understanding of the laboratory's role in the security of the United States; to contribute to a visitor's knowledge of science and technology; and to improve the quality of mathematics and science education in northern New Mexico. Admission to the facility is free. At LANL, robots and other automated devices have been used for years, primarily to transport, store, and handle hazardous materials. More recently, robots have helped characterize and clean contaminated equipment and soil. Scientists at Los Alamos have also been involved in the development of a family of insect-like robots, called BEAM (biology, electronics, aesthetics, and mechanics) robots.

Since this museum serves as a window to LANL, the visitor will encounter over 40 high-technology interactive exhibits within five galleries that explain the laboratory's defense, technology, and basic research projects, as well as the history of the Manhattan Project. In the History Gallery for example, life-sized statues of Dr. Robert Oppenheimer and General Leslie Groves greet each visitor. They are perhaps the two most famous personalities that led the development of the world's first atomic bomb at Los Alamos during World War II. Many of the exhibits throughout the museum incorporate "hands-on" displays and multimedia experience activities, such as computer programs, interactive learning devices, and videos. The National Security Gallery contains replicas of the bomb casings of the world's first nuclear weapons, Little Boy and Fat Man. The visitor will also encounter an air-launched (nuclear-weapon capable) cruise missile and a Vela nuclear test ban treaty monitoring satellite—both exhibits float above the museum floor and represent important contributions made by the LANL during the cold war. Museum visitors can also view a 20-minute film that describes the history of the LANL and the race to build an atomic bomb during World War II. This film is shown throughout the day. Museum guides are present to answer questions. The Bradbury Science Museum also maintains an archival collection of over 500 artifacts dating the Manhattan Project and representing most of the major scientific efforts made by the laboratory.

The museum has many other interesting exhibits, including those that deal with computers, environmental monitoring and restoration, and the biosciences (especially the Human Genome). During weekdays, science educators also give live, hands-on science demonstrations for visitors and school groups. The museum is open every day, except Thanksgiving, Christmas Day and New Year's Day.

Carnegie Science Center

One Allegheny Avenue
Pittsburgh, Pennsylvania 15212-5850 USA
1-412-237-3400
http://www.carnegiesciencecenter.org/

Having received the 2003 National Award for Museum Service, the Carnegie Science Center continues to inspire and entertain its guests by connecting science and technology with modern life and everyday experiences. The center also engages in educational outreach programs that serve the diverse population of the greater Pittsburgh metropolitan area.

The center's Exploration Station is a large exhibition area focused on provided fun-filled activities and science-related hand-on displays that encourage open-ended exploration and problem solving by visitors of all ages, but especially young people in their most formative, career-deciding years. In this exhibition area, guests discover at their own pace the how and why of important science concepts and physical processes. The center's Works Theater is a live showcase of exciting demonstrations that are presented in a one-of-a-kind industrial setting. Works Theater shows include: *Frankenscience: Mary Shelley's Prophecy*. Targeted at upper elementary and middle school-level audiences, this show brings the famous, fictional Doctor Victor Frankenstein's monster to life. A one million volt Tesla coil helps visitors trace the birth of science fiction, safely experience a zap of electricity, and explore the research of contemporary "Frankensteins."

The Carnegie Science Museum supports its overall mission, while keeping the exhibition areas refreshed with interesting new materials, by developing, showing, and then renting (to other science museums) special scientific exhibits. The three scientific exhibits currently available for rented use by other museums and science centers are: Robotics, Zing!, and Zap!. Robotics draws upon the extensive amount of robot-related industrial and academic activities that are taking place in the Pittsburgh area. The exhibit introduces the visitor to the design and operation robots and introduces different career opportunities available in the robotics industry. Guests learn about the basic science that underlies the design and operation of robot systems. They are also introduced to the various components that make up a robot and compare the processes by which robots and human beings complete different functions (such as sensing, thinking and performing tasks).

The exhibit, Zap! Surgery in the Cutting Edge introduces a series of scientific concepts and shows the visitor how different forms of energy are used in surgical procedures to treat human beings. The third exhibit, entitled Zing!, uses highly interactive displays to allow guests to experiment with simple machines, magnetism, sound, motion, balance, energy transfer, and other physical science principles.

Because the Carnegie Science Center is located on Pittsburgh's North Shore along the banks of the Ohio River in very close proximity to Heinz Field, the center is closed (due to traffic congestion) for the home games of the National Football League's Pittsburgh Steelers.

CERN–Microcosm Visitor Centre

Microcosm
CERN
CH-1211 Geneva
Switzerland
41-22-767-8484
41-22-767-8710 (Fax)
http://microcosm.web.cern.ch/Microcosm/

The European Organization for Nuclear Research (CERN) is the world's largest particle physics center. Located just outside Geneva, Switzerland, the large accelerators of the international research center actually straddle the Franco-Swiss border. CERN has a hands-on exhibition center for visitors, called Microcosm. Admission is free and Microcosm is open Monday to Saturday, from 09.00 to 17.30. Microcosm takes the visitor into the hidden corners of the universe. The exhibition contains models, videos, computer games, and original pieces of equipment. Each year, about 40,000 people visit CERN's Microcosm—many of these visitors are schoolchildren who participate in a stimulating science encounters. No advanced reservations are necessary to visit Microcosm.

In 1989, British computer scientist Sir Timothy Berners-Lee, while working at CERN, proposed a computer-based, global hypertext project that permitted people and organizations to work together more easily and to share information. His concept became known as the World Wide Web. In December of 1990, the Web became available within CERN and by the summer of the1991 became available on the Internet. So, in addition to sustaining the modern physics scientific revolution of the twentieth century at the subnuclear particle physics level, CERN directly contributed to the information revolution that now embraces the entire planet.

There is another, subnanoscale dimension to visit at CERN. Each year, more than 20,000 people come from all over Europe to get a behind-the-scene look at the world's largest particle physics laboratory. If you wish to enjoy a guided tour and peek behind the scenes of a world-class nuclear physics laboratory, you must make a request in advance through CERN's Visits Service. A typical guided tour lasts half a day. It starts with an introduction to CERN presented by one of the laboratory's guides. Following this opening presentation, a choice of itineraries allows the guests to experience an escorted visit to one of the experimental areas of this very large laboratory—containing giant particle accelerators that are many kilometers in circumference. Contact information for guided tours is provided below:

Visits Service
CERN
CH-1211 Geneva
Switzerland
41-22-767-8484
41-22-767-8710 (Fax)
http://visitsservice.web.cern.ch/VisitsService/

Computer History Museum

1401 North Shoreline Blvd.
Mountain View, California 94043 USA
1-650-810-1010
1-650-810-1055 (Fax)
http://www.computerhistory.org/

The Computer History Museum was established in California in 1996 and obtained its current location in Mountain View in 2002. The overall mission of this museum is to preserve and present the artifacts and stories of the information age. Companies and individuals from around the globe have collaborated with the Computer History Museum to assemble one of the world's largest collections of computer science and information technology artifacts. As of 2006, the museum's collection includes 4,000 artifacts, 10,000 images, 4,000-linear feet (1,220-meters) of cataloged documentation, and several gigabytes of software. The museum's collection of artifacts includes such interesting items as a Hollerith census machine, a Cray-3 supercomputer, and a German (World War II era) Enigma machine. The museum places emphasis on education and preservation and its unique collection serves as a valuable resource for researchers, historians, scientists, computer industry professionals, and students.

Every Wednesday, Friday, and Saturday the museum's Visible Storage exhibit area is open for docent-led tours. Self-guided tours are also available on Saturday afternoons. (It is advisable to inquire ahead of time concerning the specific hours currently being made available for such access.) Group tours are also welcome, but arrangements should be made in advance with museum personnel.

The museum serves as a forum for intellectual activities in the heart of Silicon Valley. Accordingly, the museum offers lectures, seminars, and workshops that provide scholarly historical perspectives on the computing industry and some of the innovative pioneers, who helped bring about the information revolution. The museum honors leaders and pioneers of the computer industry at an annual awards celebration. Finally, the museum hosts online exhibitions that deal with a variety of interesting topics associated with the history of computation.

Deutsches Museum

Museumsinsel 1
D-80538 München
Germany
49-89-2179-1
49-89-2179-324 (Fax)
http://www.deutsches-museum.de

The Deutsches Museum, located in Munich, Germany, is a major science and technology experience. Among its numerous artifacts and outstanding displays, this museum has several of robotics technology-related exhibits. There are several masterpiece artifacts that should prove especially interesting. These important artifacts include a Leibniz mechanical calculating machine (built circa 1700), an early nineteenth-century Jacquard loom, a nineteenth-century mechanical trumpet player (automaton) constructed by Friedrich Kaufmann (1785–1866), and a World War II era Enigma enciphering and deciphering machine. Well-designed exhibits and displays treat the fundamental aspects and historic development of many important areas in science and technology. The museum's exhibits of special interest here are physics, power machinery, computer science and engineering, machine tools, machine components (especially gears and power trains), hydraulic engineering, energy technology (including photovoltaics), telecommunications, electricity, and chronometry (measurement of time, including historic mechanical clocks). The museum also offers a special exhibit honoring German inventors, engineers, and entrepreneurs. There is a charge for admission to the museum.

Franklin Institute Science Museum

222 North 20th Street
Philadelphia, Pennsylvania 19103 USA
1-215-448-1200
http://sln.fi.edu/

The Franklin Institute Science Museum is a major science and technology learning experience in the northeastern part of the United States. Benjamin Franklin was not only a great patriot, who served his newly independent country during and after the Revolutionary War, but he was also a world-class scientist and inventor, who performed pioneering research in electricity. The Benjamin Franklin National Memorial is located in the rotunda of the museum. Dedicated by the U.S. Congress in 1976, the Memorial Hall features a 6-meter (20-foot) high marble statue of Franklin sitting on a pedestal of white marble. The Memorial Hall houses many of Franklin's original possessions, including several of his original publications. However, the electrostatic machine that he used to perform his pioneering experiments in electricity is on display in another part of the museum, called the Franklin Gallery.

Historic scientific artifacts and contemporary hands-on science learning displays abound throughout the museum. In the Joy and George Rathmann Hall of Science, for example, visitors will encounter the permanent exhibit: *Franklin . . . He's Electric*, as well as *The Wonderland of Science* section. Located in the Franklin Gallery, the *Franklin . . . He's Electric* exhibit celebrates the patriot's far-reaching scientific legacy and helps visitors discover his scientific genius, which ranged from meteorology and music, to electricity, optics, and aquatics. The exhibit features many artifacts of historical significance, including an original Franklin lightning rod. The museum's "The Wonderland of Science" section highlights the role that the Franklin Institute has played over its 175-year history in promoting major scientific breakthroughs.

In addition to well-displayed artifacts, the museum provides many hands-on science exhibits for the entertainment and education of its visitors. The museum's Mandell Center offers hands-on exhibits such as the Science Park, Newton's Dream, and Mechanics and Patterns—all of which help make understanding physics fun. For example, guests will discover the six simple machines that make up most of the machines used today (including robots).

In the fall of 1928, a truck delivered a somewhat mysterious ruined brass machine that had been donated to the Franklin Institute. A machinist at the institute began tinkering with the unusual device and soon got it to function. To everyone's surprise, the complex machine turned out to be a long-lost Henri Maillardet automaton, called the *Draughtsman-Writer*. Constructed by the Swiss watchmaker in about 1805, this automaton of a young boy has the largest mechanical memory of any such machine ever made. Maillardet's elaborate machine can make four different drawings and write three poems (two in French and one in English). The device's mechanical memory is contained in cams (brass disks) that are turned by a clockwork motor—all neatly tucked away in the base below the mechanical doll. Museum officials plan to display this automaton as a featured artifact in a revised exhibit involving amazing machines.

The Henry Ford Museum

20900 Oakwood Blvd.
Dearborn, Michigan 48124-4088 USA
1-313-982-6001
http://www.hfmgv.org/

The Henry Ford Museum represents a major history destination in the north central portion of the United States. Collectively, the complex offers exhibits, demonstrations, programs, and reenactments regarding past American traditions and modern innovations—including factory automation and the automobile assembly line. The Henry Ford encompasses five distinct attractions: the Henry Ford Museum, Greenfield Village, the Ford Rouge Factory Tour, an Imax Theater, and the Benson Ford Research Center. Without question, Henry Ford's innovative use of the assembly line to mass-produce affordable automobiles at the start of the twentieth century refashioned the American way of life. Today, automobile manufacturing around the world represents the largest single consumer of industrial robots.

The Henry Ford Museum showcases people and ideas that have changed modern life. Of specific interest are the museum's presentations about Henry Ford and Thomas Edison. The museum's automotive history showroom displays the 15-millionth Model T Ford, as well as many other interesting automobile-related exhibits. The Ford Rouge Factory tour provides guests a first-hand look at American automobile manufacturing practices: past and present. During a visit to the final assembly plant (production schedules permitting), visitors will see (from an elevated walkway) where the new Ford F-150 trucks are assembled in a modern, flexible manufacturing plant. There is a charge for admission to these attractions and visitors are reminded to call the general information number (listed above) for specific details, prices, and schedules. The Henry Ford Museum is an independent, nonprofit, educational institution that is not affiliated with the Ford Motor Company or the Ford Foundation.

Honda ASIMO Theater in Innovations

Disneyland® Resort's Tomorrowland
1313 S. Harbor Blvd.
Anaheim, California 92803-3232 USA
1-714-781-4565 (Resort Information)
http://secure.disney.go.com/disneyland/
http://asimo.honda.com/disneyland.asp (Special Honda Web site)

The Honda Motor Company's advanced humanoid robot, ASIMO, provides a live 15-minute show for audience daily at the Honda ASIMO Theater in Innovations, an entertainment attraction located within the Tomorrowland portion of the Disneyland® Resort in Anaheim, California. ASIMO (Advanced Step in Innovative Mobility) is a people-friendly, 1.3-meter tall humanoid robot that wears a stylistic white spacesuit. In addition to experiencing ASIMO's engaging and educating performance, guests to the Honda ASIMO Theater will also learn about the Honda's robotics program and the great effort Honda engineers made to develop a humanoid robot capable of bipedal (two-legged) motion that mimics the motion of a human being. (*Please note* that the Honda ASIMO Theater is located within and is part of the Disneyland® Resort theme park—a theme park for which all guests must pay an admission fee.)

Hong Kong Science Museum

2 Science Museum Road
Tsimshatsui East
Kowloon, Hong Kong
(*Note*: The government of Hong Kong is a Special Administrative Region of the People's Republic of China)
+852-2732-3232
+852-2311-2248 (Fax)
http://www.lcsd.gov.hk/CE/Museum/Science/

The Hong Kong Science Museum was opened in April 1991. The modern, four-story-high facility has approximately 500 exhibits in the permanent exhibition area. These exhibits are divided into 18 galleries and cover a wide range of science and technology topics, including motion, mechanics, electricity, magnetism, computer science, communications, energy, and robotics. A high percentage (about 70) of the permanent exhibits are visitor-interactive (or hands-on) displays, providing for an enhanced learning experience.

Invent Now®—National Inventors Hall of Fame™

221 South Broadway
Akron, Ohio 44308-1505 USA
1-330-762-4463
1-330-762-6313 (Fax)
http://www.invent.org/

Invent Now®—National Inventors Hall of Fame™ in Akron, Ohio, offers visitors an interesting insight into the creative spirit. Guests discover what inspired and motivated the more than 200 men and women honored at this facility, as they developed and patented the important inventions, which formed the basis of America's contemporary economy and society. Visitors to the museum will encounter exhibits and hands-on displays designed to stimulate the sense of innovation that lies deep inside each human being. Throughout history, inventors have worked hard to create labor-saving devices, capable of freeing human beings from tasks that were either boring, dangerous, or disgusting. Modern robot engineers are driven by similar goals.

Jet Propulsion Laboratory (JPL)

California Institute of Technology
4800 Oak Grove Drive
Pasadena, California 91109 USA
1-818-354-0112 (Public Services Office)
1-818-393-4641 (Fax)
http://www.jpl.nasa.gov/

The American space age began on January 31, 1958, with the launch of the first U.S. satellite, *Explorer 1*—an Earth-orbiting robot spacecraft built and controlled by the Jet Propulsion Laboratory (JPL). For almost five decades since then, JPL has led the world in exploring the solar system with robot spacecraft. The Jet Propulsion Laboratory (JPL) is a federally funded research and development facility managed by the California Institute of Technology for the National Aeronautics and Space Administration (NASA). The Laboratory is located in Pasadena, California, approximately 20miles (32km) northeast of Los Angeles. In addition to the Pasadena site, JPL operates the worldwide Deep Space Network (DSN), including a DSN station, at Goldstone, California.

The JPL Public Services Office offers tours free of charge for groups or individuals on an advance reservation basis. However, in order to reserve a tour, the visitor must speak with a Public Services Office representative. JPL does not permit reservations to be made via e-mail, voice mail, or facsimile (fax) transmissions. All tours range between two and nearly three hours in duration and usually include a multimedia presentation on JPL entitled "Spirit of Exploration," which provides an overview of the Laboratory's activities and accomplishments. Guests may also visit the von Karman Visitor Center, the Space Flight Operations Facility, and the In-Situ Instruments Laboratory. Tours for groups of 10 or more persons (with a maximum of 40 persons) are available throughout the week and are booked by an initial telephone call to make a tentative reservation. Morning and afternoon group tours are offered Monday through Friday. Groups must provide a confirmation letter within 10 working days of the initial phone call, as well as a roster of all participants at least one month before the day of the tour.

Several times per month, JPL offers Visitor Day tours for individuals and families (up to nine persons). These tours take place approximately once per week on Monday or Wednesday on an alternating basis. Visitor Day tours are generally starts at 1:00 P.M.

JPL requires that all U.S. citizens, 18 years of age or older, present official government-issued photo identification (driver's license or passport) before being allowed entry to the laboratory. All non-U.S. citizens, 18 years of age or older, must present a passport or resident visa (green) card) before being allowed entry. It is highly recommended that all persons planning to visit JPL contact the Public Services Office well before the day of the planned visit to inquire about any other security restrictions that may be in effect at the time.

JPL also hosts an Open House event (typically on a Saturday and Sunday) several times a year. The laboratory uses this popular activity to celebrate its accomplishments. There are exhibits and demonstrations about the laboratory's ongoing research and space exploration and the event is designed as a fun and educational experience for adults as well as children. During the Open House event, the laboratory has special hands-on activities for the younger visitors. Contact the JPL Public Services Office to obtain the date and details about the next scheduled Open House event.

Lawrence Hall of Science

Centennial Drive
University of California, Berkeley
5200
Berkeley, CA 94720-5200 USA
1-510-642-5132 (General Information)
http://www.lhs.berkeley.edu/

The mission Lawrence Hall of Science (LHS), at the University of California at Berkeley, is to develop model programs for teaching and learning science and mathematics, and to disseminate these to an ever-increasing audience. The Hall is a resource center for children, parents, educators, and policymakers seeking to improve the understanding and increase the enjoyment of science and mathematics. Established in 1968 in honor of Ernest O. Lawrence, the University of California's first Nobel laureate, the LHS is a singular resource center for preschool

Figure 10-1 An aerial view of the Jet Propulsion Laboratory (JPL) in Pasadena, California, the preeminent space robot "factory" in the United States. (Credit: Photograph courtesy of NASA/JPL.)

through high school science and mathematics education, and a public science center with many exciting hands-on experiences for visitors of all ages. Of special interest to here is an exhibit called the nanoZone. It allows visitors of all ages to discover the world of the ultrasmall. As part of the nanoZone experience there are daily live demonstrations at 12:00 noon and at 2:00 P.M. Developed at the LHS, this permanent exhibit is one of the first anywhere to explore cutting-edge developments in nanotechnology. For example, visitors can use a simulated scanning electron microscope to zoom in an ultra tiny view of the world. The LHS is open daily (except Labor Day, Thanksgiving Day, and Christmas Day). For current programs, directions to the facility, or admission information are available use the telephone number or the Web site provided above.

MIT Museum

265 Massachusetts Avenue N52-200
Cambridge, Massachusetts 02139 USA
1-617-253-4444 (Recorded Information Line)
1-617-258-9118 (Visitor Services)
1-617-253-8994 (Fax: Visitor Services)
http://web.mit.edu/museum/

Home to world famous collections in science and technology, holography, architecture, and nautical engineering, the Massachusetts Institute of Technology (MIT) Museum has both permanent and temporary exhibits, as well as a variety of public programs. The museum's collections support research, publication, restoration, education, and exhibitions. The cutting-edge Emerging Technologies Gallery provides an especially exciting experience for visitors interested in what technologies might shape the world of tomorrow.. The main building of the MIT Museum is located at 265 Massachusetts Avenue. There are also two satellite galleries managed by the museum, located within 77 Massachusetts Avenue.

The ongoing exhibition, entitled *Robots and Beyond: Exploring Artificial Intelligence @ MIT* provides a multimedia excursion into the world of artificial intelligence (AI). But persons should expect an unusual twist, when they enter this particular exhibit. The moment a guest enters *Robots and Beyond*, he or she is actually participating in research at MIT. The exhibit details the overall research strategy behind building intelligent robots empowered by advanced levels of AI. MIT researchers are searching for ways to make their intelligent robots interact better with the environment, especially in ways that mimic human-like ways. For example, the motivation behind Cog—MIT's groundbreaking humanoid robot (developed circa 1997–1998)—is the fundamental hypothesis that the creation of humanoid intelligence requires humanoid interactions with the world. Kismet (developed from 1993 to 2000) is another famous anthropomorphic robot from MIT. Researchers constructed this robot in such a way that the system's advanced level AI can communicate the robot's needs and wants to human beings using human-like facial expressions, body position, the direction of its gaze, and voice. Visitors to the museum can see a series of photographs, which depict the full range of Kismet's expressions. *Robots and Beyond* also features walking, hopping, and running robots. These particular mobile robots provide MIT scientists the data they need to develop better technical tools to assist people with mobility impairment.

Robots are often designed to travel along the ocean floor, a regime that it is difficult for human beings to personally explore and conduct research. Visitors to the exhibit entitled *Deep Frontiers: Ocean Engineering at MIT* will discover some of the latest advances in underwater robotics. This exhibit is found in MIT Museum's Hart Nautical Gallery.

Museum of Science and Industry

57th Street and Lake Shore Drive
Chicago, Illinois 60637-2093 USA
1-773-684-1414
http://www.msichicago.org

Chicago's Museum of Science and Industry is a world-renowned destination for anyone interested in exploring science and technology, presented in a highly entertaining and educational manner. This museum has something for everyone, including robot technology enthusiasts. Permanent exhibits and displays are continuously enhanced temporary exhibits that involve a wide range of important technical topics and issues. For those interested in the history of science, the frieze, located above the balcony that encircles the rotunda of the museum, contains the names of many of the most influential scientific thinkers throughout history. The names of many of the great scientists and engineers discussed in this book also appear in the museum's frieze. When you visit the museum, it might be fun to see how many of these names you can locate.

One of the most delightful and unusual exhibits at the museum is called *Robots Like Us*. This exhibit features the Robert Lesser collection of robots and space toys from the mid-twentieth century. The artifacts presented tell the story of a very interesting technology transition era—from the end of World War II (1945) until the *Apollo 11* lunar landing mission (1969). During this era, popular culture in the United States (and elsewhere) was greatly influenced by visions of the future, enhanced and inspired by science fiction fantasies. Robot and space toys became the hallmark of these future visions, representing the imagined tools of exploring the unknown. In an interesting historic perspective, the children who played with these fascinating toys have grown up—and so have robots in industry, defense, and space travel.

Another delightful exhibit at the museum is called *Toymaker 3000*. Guests experience a true adventure in automation, as they watch their own personalized toy (a Graviton top) being made by a collection of robots on a completely automated assembly line. This fascinating exhibit shows how computer-integrated manufacturing (CIM) combines the speed and efficiency of digital data with the real world automated machines and robots. While guests may first think that the CIM assembly complex is just one giant machine, they soon discovery that this automated assembly device is really a well-coordinated group of smaller machines and robots that function together in a very precise order. Each robot and machine has a specific task to perform in the assembly process. While the museum's CIM complex produces customized toys, industrial engineers use the same concepts to manufacture finished products as quickly, efficiently, and cost-effectively as possible in modern CIM complexes that contain a number of computer-controlled robots working in a precisely organized manner.

Guests who have an interest in mechanics and machine components will enjoy the exhibit, *Gears from the Century of Progress*. The collection of gears was originally prepared by the Borg-Warner Corporation for display in the Travel and Transportation Building at the Century of Progress Exposition held in Chicago in 1933. Younger guests will enjoy a visit to the Idea Factor, the centerpiece of the museum's Imagination Station exhibit. Basic scientific principles, the fundamentals of construction, and the operation of simple machines become tangible through a variety of hands-on displays and learn-through-play exhibits. The robotic technology experience for guests continues when they enter *Petroleum World* (one of the museum's newest permanent exhibits) and are greeted by a remotely operated (underwater) vehicle (ROV).

National Air and Space Museum (NASM)

Smithsonian Institution
National Mall Building
6th and Independence Avenue, SW
Washington, DC 20560 USA
1-202-633-1000 (General Visitor Information)
1-202-633-8982 (Fax)
http://www.nasm.edu

The National Air and Space Museum (NASM) of the Smithsonian Institution contains the largest collection of historic aircraft, missiles, and spacecraft in the world. The museum building that is located on the National Mall in Washington, DC, offers its millions of annual visitors hundreds of professionally displayed aerospace artifacts. Of particular interest here are the hundreds of artifacts related to robot spacecraft and intermediate range and intercontinental ballistic missiles. The ballistic missile displays include the World War II era German V-1 flying robot bomb, a German V-2 ballistic missile, a Tomahawk cruise missile, an American Pershing-II ballistic missile, and a Soviet SS-20 ballistic missile. The (now retired) Pershing-II was a U.S. Army nuclear weapon (5 to 50 kiloton yield) carrying intermediate range (approximately 500 to 5,500 kilometers) surface-to-surface tactical ballistic missile. The (now retired) SS-20 was a mobile, nuclear-armed (three independently targeted thermonuclear warheads, each with yield of 250 kilotons) missile of the Soviet Strategic Rocket Forces. There are also numerous artifacts on display that feature space robots, including NASA's Ranger, Surveyor, Viking (lander), and Voyager spacecraft. General admission to NASM (on the National Mall) is free but there are charges for participation in special events, shows, and programs. Use the NASM Web site to obtain the latest information about operating schedules, special exhibits, and any supplemental admission fees. This Web site also provides an excellent overview of the aerospace artifacts on exhibit and the history behind many of the displays.

National Atomic Museum

1905 Mountain Road NW
Albuquerque, NM 87104 USA
1-505-245-2137
1-505-242-4537 (Fax)
http://www.atomicmuseum.com

The National Atomic Museum is the only "atomic museum" in the United States chartered by Congress to preserve and communicate nuclear science heritage and history. Responding to this mission, the museum offers visitors a wide variety of exhibits and educational programs concerning the people, technologies, and events that shaped the nuclear age. A variety of permanent and changing exhibits and displays describe the diverse applications of nuclear energy and the men and women who became the great pioneers of nuclear science. Exhibits, artifacts, and authentic replicas document the Manhattan Project, the cold war era, and the history of nuclear arms control. Development of nuclear weapons encouraged development of sophisticated robotic devices, such as master–slave manipulators, to allow human workers to safely handle a variety of highly radioactive materials. The museum has replicas of several nuclear weapons, some of which (such as the Titan II missile's W53 multimegaton warhead and Mark 6 reentry vehicle) served as the business end of American intercontinental ballistic missiles (ICBMs). The nuclear weapon tipped ICBM is generally considered to be the most lethal robot weapon system every developed. The National Atomic Museum is scheduled to become the *National Museum of Nuclear Science and History*. Allow with the new title, this museum will have an expanded mission to provide an even great collection of hands-on displays and interactive exhibits, including the use of robotics in nuclear technology applications. The museum is located in the heart of Old Albuquerque, within the city's museum corridor that contains three additional facilities. There is a charge for admission to the National Atomic Museum and its Web site provides directions, operating hours, and updated information about any new exhibits and special programs.

National Museum of the United States Air Force

1100 Spaatz Street
Wright-Patterson Air Force Base, Ohio 45433-7102 USA
1-937-255-3284
http://www.wpafb.af.mil/museum/

The National Museum of the United States Air Force at Wright-Patterson Air Force Base near Dayton, Ohio, contains a well-preserved and displayed collection of over 300 aircraft and missiles, along with a large number of interesting robotic aerospace and military artifacts (including radio controlled aircraft, guided missiles, and unmanned aerial vehicles). For example, there is a JB-2 ("Loon") on display. The Loon is an American-made copy of the famous German V-1 unpiloted flying bomb—the first operational surface-to-surface cruise missile, which was used against targets in the United Kingdom starting in 1944. There are also several ballistic missiles on display, including the German V-2 rocket and the U.S. Air Force Minuteman III, ICBM. The museum's collection of military robots ranges from the early era of powered flight up to the present day. Specifically, robotic artifacts range from a World War I era flying weapon, called the Kettering Aerial Torpedo (or "Bug") to the RQ-1A Predator unmanned aerial vehicle (UAV), which flies military and counterterrorism surveillance missions over Iraq, Afghanistan, and other trouble spots. The museum is open to the public seven days a week from 9:00 A.M. to 5:00 P.M. (closed on Thanksgiving, Christmas, and New Year's Day). Contact the museum for additional information, directions, or to inquire about any enhanced security conditions that could influence public access.

National Museums of Scotland (NMS)

Royal Museum
Chambers Street
Edinburgh
Scotland, United Kingdom
44-(0)-131-225-7534
44-(0)-131-225-3848 (Fax)
http://www.nms.ac.uk/

The National Museums of Scotland (NMS) is Scotland's national museum service. This organization cares for many of Scotland's museum collections of national and international importance. NMS has four museums, which open to the public daily (except Christmas Day) and two museums (the Museum of Flight and Shambellie House Museum of Costume), which open April to October. Most closely complementing some the technical topics and science principles discussed in this book are the robotics and artificial intelligence exhibits found in Connect Gallery within the Royal Museum. The Royal Museum is a magnificent Victorian building, which houses international collections themed as follows: decorative arts, science and industry, archaeology, and the natural world. Together, the museum's exhibits and displays reflect the diversity of life on Earth and the ingenuity of humankind. In the robotics area of the Connect Gallery, visitors will get a chance to view Freddie, the world's first thinking robot. Constructed by researchers at the University of Edinburgh in the early 1970s, Freddie was designed to assemble a child's toy, such as a ship, from a pile of random parts. This robot was also taught to tidy up a set of scattered toys and put them away in a box. There is also Alphabot, a working robot arm designed for simple, repetitive tasks. A visitor can type his (her) name at the interactive station and then watch Alphabot select the right blocks to spell out their name correctly. Another interactive exhibit involves a model robot exploring the surface of an alien planet. In this part of the Royal Museum, visitors will also find out how robots are developing and what these machines might do for human beings in the future.

Neuchâtel Museum of Art and History (Musée d'Art et d'Historie)

Esplanade Léopold-Robert 1
Case postale
CH-2001 Neuchâtel

Figure 10-2 The General Atomics RQ-1A Predator, a modern unmanned aerial vehicle (UAV), as displayed at the National Museum of the United States Air Force. (Credit: Photograph courtesy of the U.S. Air Force.)

Switzerland
+41 (0) 32 717-79-20
+41 (0) 32 717-79-29 (Fax)
http://www.mahn.ch/

Starting in 1768 continuing until about 1774, the Swiss watchmaker, Pierre Jaquet-Droz (in collaboration with his son, Henri-Louis) constructed several elaborate automata that were very popular among members of high-class European society. The Museum of Art and History in Neuchâtel, Switzerland, displays three of his most popular automata. These mechanical doll masterpieces are: *The Writer*—a boy scribe, who dips his pen in an inkwell and writes a letter; *The Draughtsman* (or *Draftsman*)—a young boy has such intricate mechanical mechanisms that the automaton first draws and then blows off dust from the drawing paper; and *The Musician*, a young girl in an elegant blue-and-gold dress. The *Musician* plays tunes on an eighteenth-century harpsichord-like instrument, moves her eyes and head, and then rises and bows gracefully. Jaquet-Droz's automata are some of the most complex and elaborate mechanical systems ever constructed for entertainment.

Nobel Museum

Börshuset, Stortorget
Gamla Stan (Old Town), Stockholm
Sweden

+46 (0) 8 534 818 00
+46 (0) 8 23 25 07 (Fax)
http://nobelprize.org/nobel/nobelmuseum/

The Nobel Foundation in collaboration with the Swedish government, and the City of Stockholm established the Nobel Museum as a permanent way of communicating to visitors the contributions of the Nobel Prize laureates—viewed within their cultural and social contexts. This treatment provides an important perspective between science and culture. The concept of a permanent museum in Stockholm emerged in 2001 from the great success of the Nobel Prize Centennial exhibition, entitled "Cultures of Creativity." This very popular exhibition examined the question of creativity by presenting selected Nobel laureates and their cultural environments from the one hundred year history of the Nobel Prize. The exhibition also presented Alfred Nobel as the idealist, inventor, entrepreneur, and cosmopolitan, who donated his entire fortune to establish the Nobel Prize system.

Through its film room and artifact theater the Nobel Museum presents Nobel laureates and their creative work, as well as the environment that inspired their important efforts. Well-known Nobel Prizes are discussed against a background of historic (twentieth century) events. To enhance the guest experience, the museum has its own café (Kafé Satir), which was modeled on Café Museum in Vienna—one of the many cafés that served as informal meeting places for many of the great intellectuals who shaped and molded the twentieth century. Today, over a cup of coffee, guests at the Nobel Museum can enjoy some of the more humorous and satirical comments that appeared over the past century with respect to the Nobel Prizes. The Nobel Foundation has also taken steps to ensure that this unusual museum, dedicated to honoring human creativity and intellectual excellence, will never stagnate. Every year, as the dozen or so new Nobel laureates are nominated, temporary exhibitions at the museum will be used to explain their great achievements and to portray these accomplishments historically in their scientific and cultural context. New permanent exhibitions, focusing on individual Nobel Prize winners from the past (like Albert Einstein), will also be used to refresh the museum's contents on a regular basis. Companion traveling exhibitions will carry the same message to major science museums around the globe.

Pacific Science Center

200 Second Avenue North
Seattle, Washington 98109 USA
1-206-443-2001
1-206-443-3631
http://www.pacsci.org/

The Pacific Science Center is located near the famous Space Needle in Seattle, Washington. The center strives to inspire a lifelong interest in science, technology, and mathematics through interactive and innovative exhibits and programs. The structure that houses the Pacific Science Center originally served as the United States Science Pavilion during the 1962 Seattle World's Fair. When this world's fair came to an end, the Science Pavilion was given a new life and mission as the private, not-for-profit Pacific Science Center. This administrative transformation of the Science Pavilion allowed the Pacific Science Center to become the first museum in the United States founded as a science and technology center.

One of the center's most popular permanent exhibits is called *Dinosaurs: A Journey Through Time*. In this exhibit, guests can travel back in time to the Mesozoic Era, where they then encounter seven moving and roaring robotic dinosaurs displayed in a lifelike prehistoric environment. Visitors can also take control of the *Pneumoferrosaurus* (the imaginary robotic "Air and Iron Lizard") to see how animatronic dinosaurs actually work. In the center's Insect Village, guests encounter giant robotic insects, operate interactive exhibits, and can see live animal displays. All of these have been carefully prepared to give visitors a close-up look at the world of insects and other anthropods. For robot engineers and hobbyists interested in designing flying insect robots, a walk through the center's Tropical Butterfly House provides a very special opportunity to observe living, beautiful butterflies, as they fly about a specially created and maintained, tropical enclosure that guests can enter. How does a motor work? The center's Science Playground permits young, scientists-in-training to explore (at their own pace) the answer to this important technical question and many others. Finally, the center's technology exhibits include computers, robots, and virtual reality demonstrations. A visitor can challenge an industrial robot to a game of tic-tac-toe or defend his team's goal in VR Keeper—the full-body virtual reality soccer experience.

Powerhouse Museum

500 Harris Street Ultimo
PO Box K346 Haymarket
Sydney New South Wales 1238, Australia
+61-2-9217-0111
http://www.powerhousemuseum.com/

The Powerhouse Museum is Australia's largest and most popular museum. It is located in Darling Harbor, Sydney, and has a unique and diverse collection of over 385,000 objects, which involve history, science, industry, the decorative arts, transportation, music, and space exploration. The museum has 22 permanent exhibits and several temporary exhibits, which are complemented by more than 250 interactive displays. A visit to this popular museum and its everchanging program of exhibitions might include the use of touch screen computers, audio phones, science experiments, virtual reality presentations, as well as entertaining films and lectures.

The Museum of Applied Arts and Sciences (MAAS) is a public museum operated by the state government for the people of New South Wales. Established in 1879, MAAS comprises of the Powerhouse Museum and the Sydney Observatory. The mission of the Powerhouse Museum is to develop collections and present exhibits and programs that explore science, design, and history. In 1879, Sydney staged an international exhibition to showcase invention and industry.

Unfortunately, a fire swept through the exhibition building in 1882 and destroyed virtually all it contained. However, the basic idea of a technology museum remained and—after several relocations—the Powerhouse Museum emerged in 1988 in a new building constructed from the shell of an old power station. The Ultimo power station dates back to 1899, when the facility was built to provide power to Sydney's electric tram system, which ceased operations in 1963.

Among the numerous exhibits and artifacts to be enjoyed at the Powerhouse Museum, there are several which especially complement topics and technologies appearing in this book. These include: Cyberworlds–computers and connections; a Boulton and Watt steam engine; the Steam Revolution; and Experimentations. Cyberworlds–computers and connections is an interactive exhibition of the past, present, and future world of computers, including displays that relate to robotics and artificial intelligence and even a portion (artifact specimen) of Charles Babbage's Difference Engine, the general-purpose calculating machine that anticipated the principles behind and the basic structure of the modern computer. Babbage intended that his computer would be powered by steam. The Powerhouse Museum also has the world's oldest surviving rotative steam engine, a Boulton and Watt steam engine, which was originally installed in a London brewery in 1785 and then delivered to Sydney in 1888. With 12 working steam engines and a variety of hands-on displays and videos, the museum's steam revolution exhibition shows how steam power changed the world and touched millions of lives during the First Industrial Revolution. Steam power (in the form of modern Rankine cycle heat engines) remains the world's primary way of generating electricity, the enabling form of energy for the Second Industrial Revolution and today's information technology era revolution. Finally, visitors will explore the basic physical principles of motion, gravity, light, pressure, temperature, electricity, and magnetism in the museum's Experimentations exhibition. The Powerhouse Museum is open daily (except Christmas Day) and there is a modest fee for admission.

Robot Hall of Fame®

225 Smith Hall
Carnegie Mellon University
Pittsburgh, Pennsylvania 15213 USA
1-412-268-9656
1-412-268-5497 (Fax)
http://www.robothalloffame.org/
Note: Use the Internet to visit this demonstration site

In April 2003, Carnegie Mellon University (CMU) created the Robot Hall of Fame® to bring attention to the impact robots have and continue to make on the trajectory of human civilization and modern society. In preparation for each formal induction cycle, highlighted by a formal induction ceremony at CMU, the university assembles a panel of scholars, researchers, designers, entrepreneurs, and writers, who then serve as a jury and select real world (physical) robots as well as fictional robots for recognition and induction into the Robot Hall of Fame.

This panel of experts examines candidates from two basic categories of robotics: robots from science and robots from fiction.

According to the basic guidelines established by members of the School of Computer Science at Carnegie Mellon University, robots from science are real (physical world) robots that "have served useful or potentially useful functions and demonstrated unique skills in accomplishing the purpose for which they were created." Entertainment robots can be included in this basic category, as long as the candidate robot system can function autonomously. Similarly, robots from science fiction are fictional (imaginary) robots that have inspired people to "create real robots that are productive, helpful, and entertaining." Candidate fictional robots are those that have achieved worldwide recognition and by the fictional characteristics have encouraged human beings to form opinions about the important role, function, and value of real world robots—present and future.

The real world 2003 inductees to the Robot Hall of Fame were: NASA's Mars Pathfinder robot minirover (called *Sojourner*) and the Unimate industrial robot, developed by Joseph F. Engelberger and George C. Devol, Jr. The science fiction 2003 inductees to the Robot Hall of Fame were R2-D2 (the popular little "droid" from George Lucas's *Stars Wars* motion picture series), and HAL 9000 (the mischievous fictional computer/character in the film *2001 A Space Odyssey*). In this highly acclaimed 1968 motion picture from producer Stanley Kubrick and science fiction writer Sir Arthur C. Clarke, HAL 9000 (an acronym meaning heuristically programmed algorithmic computer) is the advanced onboard computer designed to essentially run the interplanetary ship *Discovery*, as it carries a team of human astronauts to the vicinity of Jupiter on a mysterious mission.

The real world 2004 inductees were SRI's Shakey (the pioneering mobile robot with artificial intelligence) and Honda's ASIMO (the advanced humanoid robot that is demonstrating people–robot interactions in both workplace and home environments). The science fiction 2004 candidates were: Robby the Robot (from the 1956 picture *Forbidden Planet*); C-3PO (the protocol humanoid robot from the *Star Wars* motion picture series); and Astro Boy (a popular fictional robot which originated in Japan in 1951).

The real world robots inducted in 2006 to the Robot Hall of Fame were Sony's AIBO (the dog-like, robot pet sold commercially from 1999 to 2006) and the SCARA industrial robot. Because its shoulder, elbow, and wrist joints provide motions that are well suited for the assembly of consumer products, the Selective Compliance Assembly Robot Arm (or SCARA) has become a widely used type of industrial robot. The science fiction robots inducted in 2006 were: Maria (the female robot in Fritz Lang's classic 1927 motion picture *Metropolis*); Gort, (the giant and powerful metallic robot from outer space who worked with a humanlike alien visitor to sponsor peace in the 1951 classic science fiction-fantasy thriller, *The Day The Earth Stood Still*); and David (the boy-like android in Steven Spielberg's 2001 science fiction-fantasy movie, *Artificial Intelligence: AI*).

The Science Museum

Exhibition Road
South Kensington,

London
SW7 DD
United Kingdom
+44 (0) 870-870-4771
http://www.nmsi.ac.uk/

The Science Museum in London traces its origins directly back to the nineteenth-century movement to improve science and technology education in the United Kingdom. At the time, Great Britain, a world power, served as economic and technical stimulus for the First Industrial Revolution. Prince Albert (Queen Victoria's husband) was a leading figure in this science education movement. His efforts were primarily responsible for the Great Exhibition of 1851—an exhibition that promoted achievements of science and technology. Many of the artifacts of this transitional era are now on display at the museum. In a visionary step, profits from the very successful 1851 Exhibition were used to purchase land in South Kensington to establish institutions that would promote and improve industrial technology. Following passage of the National Heritage Act of 1983, the Science Museum experienced a rapid period of expansion, including the addition of new interactive galleries and the use of supplementary (temporary) exhibitions to place science and technology education in a less-traditional, more entertaining contemporary context. Today, the museum consists of many skillfully prepared artifact exhibits from the museum's vast collections, as well as contemporary hands-on displays. Together, these exhibits and displays tell the story of how the modern world developed. Several examples are described briefly below. The museum provides an extensive online description of its many the exhibits (at the following Web site: www.sciencemuseum.org.uk/galleryguide/).

The power exhibit in the East Hall explains to visitors how the ingenuous use of steam to generate mechanical (rotary motion) power supported the industrialization of Great Britain. Another gallery, entitled *Making of the Modern World*, offers visitors a collection of 150 artifacts from the museum's collections (covering the period from 1750 to 2000). The artifacts displayed significantly assisted in the development of our contemporary global civilization. The artifacts are arranged in chronological order, so guests can visually appreciate the rapid rate of technical progress that occurred in little over two and one half centuries. On display at the museum, for example, is the paddle-wheel apparatus used by James Prescott Joule (circa 1847–1849) to develop a precise physical relationship between heat and mechanical work. Joule's pioneering work quantified the conservation of energy principle (or first law of thermodynamics) and opened the way for scientists and engineers to more efficiently use heat engines to power the Second Industrial Revolution.

The museum's computing and information-technology collection covers the devices, machines, and systems—ranging from an early mechanical calculator (circa 1623) to the present day. Exhibits depict electromechanical and electronic calculation, analog and digital computation, data management and processing, and cryptography. Visitors can see a part of the original Babbage Difference Engine 1 (circa 1830–1832). The portion on display at the museum consists of 2,000 parts, which are still in good working order. This exhibit shows the first successful automatic calculator and represents a fine example of precision engineering in the nineteenth century. By studying Charles Babbage's original papers and designs, the museum's engineers were able to complete the construction of a complete calculating machine in 1991. This exhibit, called the Babbage Difference Engine 2, has over 4,000 parts and weighs over three metric tons. To dramatize technical progress, museum visitors are invited to compare the capability of Babbage's Difference Engine Number Two to their home computer, or even a small hand-held calculator. If that message is not impressive enough, in the same gallery visitors will find Pilot ACE (automatic computing engine), one of Great Britain's earliest computers. The design of this computing machine (a forerunner of the electronic computer) was based on the ideas of Alan Turing—the brilliant mathematician and code-breaker during World War II.

Sony Wonder Technology Lab

550 Madison Avenue Annex
New York, New York 10022-3301 USA
1-212-833-8100
http://wondertechlab.sony.com/

Right in the heart of Manhattan, the Sony Wonder Technology Lab provides visitors four floors of hands-on interactive exhibits in technology and communications. There is no charge for admission to the Lab, but due to the popularity of the facility, advanced reservations are recommended. After picking up a timed-entry ticket in the museum lobby, visitors are greeted and welcomed by *b.b. Wonderbot*—a telepresence robot. Upon admission, guests receive a brief introduction to the Lab and are given swipe cards for use throughout the visit.

All visitors begin their interactive experience on the fourth floor by logging into the computer network. At one of the eight log-in exhibits, visitors personalize their bar-coded swipe cards. Individuals type in their name, take their picture, and can even record their voice. In the Communications Bridge exhibit, visitors travel over a series of ramps that serve as a visual and auditory showcase in which the history of communications technology and electronic entertainment are highlighted. Featured landmark inventions from the past 150 years include the camera, telephone, radio, television, and computer.

In the Technology Workshop portion of the Lab visitors can actively explore the basic elements and inner mechanics of communications technology by using touch screen monitors at three exhibit areas: Signal Viewers, Audio Lab, and Image Lab. Proceeding on to the Professional Studios portion of Sony's Wonder Technology Lab, visitors explore several exciting new fields related to communications technology and entertainment: robotic engineering, environmental crisis response, medical imaging, and the television production studio. Of special interest here is the Factory Automation exhibit in which visitors program a robot to perform a simple assembly line task. When the visitor completes the programming, the robot will run through all the commands at top speed. At the Remote Inspection exhibit, visitors can experience telepresence, as they let another robot serve as their eyes, while searching through a frame of interlocking steel for a hidden leak.

The environmental-crises management, medical imaging, and television-production studio interactive experiences are equally exciting and educational experiences for visitors. Sony's Wonder Technology Lab provides guests many other interesting interactive experiences at exhibits called: the High Definition Theater, the Wonders of Games, Music, and Digital Entertainment, the Wonder of Imagination, the Shadow Garden, and Sand Interactive. As the visitors emerge from the Wonder of Imagination they have completed their journey through the Lab. A series of six interactive electronic log-out stations allow the guests to swipe their cards one last time and receive a printed color certificate, which lists all of the activities they have participated during their visit.

The Tech Museum of Innovation

201 South Market Street
San Jose, California 95113 USA
1-408-294-8324
http://www.thetech.org/

The Tech Museum of Innovation is a hands-on science and technology museum with over 250 exhibits to engage people of all ages and backgrounds to explore and experience the technologies that are affecting their lives. Located in the heart of Silicon Valley, The Tech also strives to inspire young people to become innovators in the technologies of the future, including robotics. The vast majority of the interactive exhibits and displays at The Tech are original or exhibits custom-made expressly for use at the museum.

The Tech is focused at inspiring the spirit of innovation that abides in every human being. For some people innovation is a way of life, but for many other people, the pressures of daily life have created a rigid pattern of conformity that has all but crushed any desire to innovate. Visitors who wander through the museum's themed galleries and experience the hands-on and interactive exhibits found in the Imagination Playground should have no trouble awakening or enhancing their own spirit of innovation. To improve the visitor experience, many of the exhibits at The Tech are being renewed and updated, without interrupting ongoing operations. This transformation process should be completed in 2007 and result in the appearance of several new major exhibits including: IDEA House, What's New?, View from Space, and Green by Design. The Hackwork IMAX Dome Theater also helps make a visit to The Tech a memorable experience. Guests interested in robot rovers operating on other worlds will definitely enjoy the IMAX feature *Roving Mars*, which describes the adventures of NASA's *Spirit* and *Opportunity* robot rovers on Mars—an exciting surface exploration mission that started in 2004.

Robotic technology is featured in many of the exhibits at The Tech. For example, the Alphabot is a robot, nicknamed *Vanna* that spells out a guest's name by selecting and then correctly arranging the correct set of lettered blocks. Guests might wonder how the robot does this, since Alphabot does not have a machine vision system. Alphabot accomplishes the task by using its computer memory to carefully pick and place blocks from an inventory of blocks whose locations are precisely known. Thanks to its computer memory *Vanna* "knows" where all the appropriate lettered blocks can be found and then selects the lettered blocks that correspond to the guest's name. When it is finished, the robot dutifully places the blocks back into their proper storage locations and awaits the next visitor.

A person can also sit down and have his or her portrait made by a robot artist. In this robotic technology exhibit, after a television camera captures an image of the guest, a computer decides what the robot artist should draw. When the portrait is complete, the robot gives the picture to the visitor. Another exhibit introduces the guest to the use of remote robotics (teleoperation) in exploration. From the control panel of the teleoperated robot, a visitor can operate a distant camera-equipped robotic arm and follow the movements of other guests as they enter the gallery. The Tech also provides an excellent online exhibit (found at www.thetech.org/robotics/) called *Robotics: Sensing, Thinking, Acting.*

The museum also hosts the Robert N. Noyce Center for Learning (NCFL), which serves as a professional home for teachers while they are at The Tech. The center supports K-12 professional development opportunities for teachers with an emphasis on the role of creativity and innovation in science and mathematics education. Robert N. Noyce was the founder of Intel and a pioneer in the information revolution of the twentieth century.

Titan Missile Museum

1580 West Duval Mine Road
Sahuarita, AZ 85629 USA
1-520-625-7736
1-520-625-9845 (Fax)
http://www.pimaair.org/ (This is the Web site of the Arizona Aerospace Foundation and serves as a portal to the Titan Missile Museum)

A registered National Historic Landmark, this site is the sole remaining Titan II intercontinental ballistic missile (ICBM) complex of the 54 that were "on alert" between 1963 and 1987, during the cold war. Deactivated under the terms of the Strategic Arms Limitation Treaty (SALT), the complex has been converted into a unique museum. The facility is located about 40 kilometers (25 miles) south of Tucson, Arizona. Except for certain treaty-required deactivation modifications, the site is an authentic, walk-through example of the liquid-fueled ICBM launch facilities used by the Strategic Air Command. The Titan II missile's reentry vehicle carried the largest yield, single nuclear warhead (megaton range) used in the American land-based ICBM program. Built in response to the "missile gap" panic of the late 1950s and early 1960s, Titan II Missile Site 571-7 now provides a unique window into the design, construction, and operation of a weapon system designed to survive a Soviet first-strike nuclear attack and then be able to launch its retaliatory missile, if so ordered. The site has retained all of the above and below ground command and control facilities, as well as the missile silo itself, and a (deactivated) Titan II missile and reentry vehicle. In May 1986, the United States Air Force responded to requests from the people of Arizona and transferred this site for use as a public museum. Today, visitors go underground to see an actual Titan II missile in its silo and tour the launch control center. In April 1994, the U.S. Interior Department designated the missile site as a National Historic Landmark and in November 2003, the museum opened the Count Ferdinand von Galen Education and Research Center. This companion center houses an expanded Exhibits Gallery, a classroom for educational activities, and various artifacts of the Titan II ICBM program. With its multimegaton W-53 nuclear warhead, the Titan II was one of the most powerful robot weapon systems to emerge during the nuclear arms race of the cold war. The museum is open daily (except Thanksgiving and Christmas Day) from November 1 to April 30; and Wednesday to Sunday from May 1 to October 31. There is a modest charge for admission.

Tokyo Science Museum

2-1, Kitanomaru-koen
Chiyoda-ku
Tokyo 102-0091
Japan
+81-03-3212-8544
+81-03-3212-8540 (Fax)
http://www.jsf.or.jp/eng/ (English language Web site)

The Science Museum of Tokyo was established in April 1964 for the purpose of spreading scientific knowledge to the general public. Founded and managed by the Japan Science Foundation, this modern five-story-tall science and technology museum is located near the Imperial Palace and is surrounded by beautiful Kitanomaru Park. The outer image of the five-story-high building was designed in the image of scattered stars in space. Most of the exhibits are interactive displays and hands-on devices, which provide guests entertainment, as well as information about basic scientific principles and the latest technological advances. Of particular interest is the extensive mechanics exhibition area (on the fifth floor), which contains an entertaining and educational collection of exhibits that explain how machine works and explain the principles of operation of gears, screws, levers, and springs—the fundamental mechanical components of many machines, including robots. One interesting mechanics exhibit allows visitors to lift a heavy metal ball by properly using simple machines, such as the lever, pulley, wheel, slope (inclined plane), and screw. Another exhibit, called the mega-wheel, consists of a handle attached to a complex train of gears. Visitors are invited to give the handle a few turns, but it will take 25.2 million turns (or about 10 years), to turn the big gear in this exhibit just once.

The Tokyo Science Museum offers visitors equally interesting interactive exhibits about motors, electricity and magnetism, the world of iron (and steel)—the essential metal of modern civilization, computers, the generation and distribution of electric power, and many other technological topics. The staff is constantly updating the museum's interactive exhibits and adding new ones.

United States Patent and Trademark Office (USPTO) Museum

600 Dulany Street
Alexandria, Virginia 22314-5782 USA
571-272-8400 (USPTO Public Affairs)
http://www.uspto.gov/

Established in 1995, the U.S. Patent and Trademark Office (USPTO) Museum strives to educate the American public about the patent and trademark systems, and the important role intellectual property protection plays in the Nation's social and economic health. The museum is operated for the USPTO by the National Inventors Hall of Fame (of Akron, Ohio) and houses both permanent and changing exhibits that feature inventors, inventions, and trademarks. Through interactive exhibits, artifacts, videos, and touch-screen technology, visitors learn how intellectual property protection, patents, and trademarks impact and improve the daily lives of millions of people. Visitors also discover how the spirit of innovation drove famous and not-so-well-known inventors to create the important devices, which form and shape the modern world. Located in the Atrium of the Madison Building at 600 Dulany Street in Alexandria, Virginia, the museum is open to the public on Monday to Friday from 9:00 A.M. to 5:00 P.M. and on Saturday from noon until 5:00 P.M. However, the museum is closed on Sundays and on federal holidays. School and group tours are welcome, but visits should be coordinated several weeks in advance with USPTO personnel in the Office of Public Affairs.

11

Sources of Information

This chapter describes additional sources of information about robot technology. The list of more traditional sources (such as selected books, publications, and educational resource centers) is complemented by a special collection of cyberspace resources. The information and the exponential growth of the Internet have produced an explosion in electronically distributed materials. Unfortunately, unlike a professionally managed library or a well-stocked bookstore within which you can confidently locate desired reference materials, the Internet is a vast digitally formatted information reservoir that is overflowing with both high-quality, technically accurate materials and inaccurate, highly questionable interpretations of history, technology, or the established scientific method.

To help you make the most efficient use of your travels through cyberspace in pursuit of additional information about robot technology, this chapter provides a selected list of Internet addresses (that is, Web sites), which can conveniently serve as your starting point. Many of the Web sites suggested here contain links to other interesting Internet locations that contain complementary, often more specific information. With some care and reasoning, you should be able to rapidly branch out and customize the particular robot technology information search. Using the contents of this book and especially this chapter as a guide, you can effectively harness the power of the modern global information network. One important tool in conducting productive searches in cyberspace is to use appropriate keywords and phrases.

The following key words and phrases should prove quite useful in starting your customized Internet searches: *actuator, adaptive control system, aerobot, android, artificial intelligence (AI), assembly robot, automaton, autonomous robot, autonomous underwater vehicle (UAV), bang-bang robot, biomimetic system, Cartesian robot, central processing unit (CPU), charge coupled device (CCD), computerized robot, computer vision, continuous path robot, controller, cryobot, cylindrical coordinate robot, cyborg, degrees of freedom (DOF), digital computer, drone, electric robot, end effector, end-point robot, entertainment robot, expert system, explosive ordinance disposal (EOD) robot, extravehicular activity (EVA) robot, field of view (FOV), field robot, gripper, humanoid robot, hydraulic robot, industrial robot, insect robot, intelligent robot, intravehicular activity (IVA) robot, knowledge engineering, limited sequence robot, limit switch, machine, machine intelligence (MI), machine vision; manipulator, Mars surface rovers, marsupial robot, master/slave manipulator, microelectronics, microrobotics, military robot, mobile robot, mother-spacecraft, nanorover, nanotechnology, nonservo robot, pick-and-place robot, pneumatic robot, point-to-point robot, programmable robot, remote control, remotely operated vehicle (ROV), remotely piloted vehicle (RPV), remote manipulator system (RMS), remote sensing, Robonaut, robot, robot-assisted surgery, robotics, robot rover, robot spacecraft, rover, Santa Claus machine, self-replicating system (SRS), sensory robot, sentry robot, serial robot, servo robot, smart robot, space robot, spherical coordinate robot, surface rover spacecraft, teleoperation, telemedicine, telepresence, telescience, universal constructor (UC), unmanned aerial vehicle (UAV), unmanned ground system (UGS), unmanned underwater system (UUV)*, and *virtual reality (VR)*.

Also, as found within this book, the proper names of robot-technology pioneers (such as Norbert Weiner), and programs or projects (such as the NASA's *Surveyor* Project and the U.S. Air Force's Predator unmanned aerial vehicle program) will prove helpful in initiating other specialized information searches on the Internet.

SELECTED BOOKS

Angelo, Joseph A., Jr. *Space Technology*. Westport, CT: Greenwood Press, 2003.

———. *Nuclear Technology*. Westport, CT: Greenwood Press, 2004.

Ballantyne, Garth H., Jaques Marescaux, and Pier Cristoforo Giulianotti, eds. *Primer of Robotic and Telerobotic Surgery*. Philadelphia, PA: Lippincott Williams & Wilkins, 2004.

Bekey, George A. *Autonomous Robots: From Biological Inspiration to Implementation and Control*. Cambridge, MA: The MIT Press, 2005.

Bergren, Charles. *The Anatomy of a Robot*. New York: McGraw-Hill Companies, 2003.

Berns, Karsten, and Rudiger Dillmann, eds. *Climbing and Walking Robots: From Biology to Industrial Applications*. Hoboken, NJ: John Wiley & Sons Inc., 2001.

Berube, David M. *Nano-Hype: The Truth Behind the Nanotechnology Buzz*. Amherst, NY: Prometheus Books, 2006.

Breazeal, Cynthia L. *Designing Sociable Robots*. Cambridge, MA: The MIT Press, 2004.

Capek, Karel. *R.U.R. (Rossum's Universal Robots)*. New York: Doubleday, Page and Company, 1923. (Translated by Paul Selver.)

Colestock, Harry. *Industrial Robotics: Selection, Design, and Maintenance*. New York: McGraw-Hill Companies, 2004.

Conrad, James M., and Jonathan W. Mills. *STIQUITO for Beginners: An Introduction to Robotics*. Hoboken, NJ: John Wiley & Sons Inc., 2000. (Six-legged robot kit included with book.)

Constable, George, and Bob Somerville. *A Century of Innovation: Twenty Engineering Achievements that Transformed our Lives*. Washington, DC: National Academies Press (Joseph Henry Press Imprint), 2003.

Davidson, J.K., and K. H. Hunt. *Robots and Screw Theory: Applications of Kinematics and Statistics to Robotics*. New York: Oxford University Press, 2004.

Drexler, K. Eric. *Engines of Creation: The Coming Era of Nanotechnology*. New York: Doubleday, 1986.

Drexler, K. Eric, Chris Peterson, and Gayle Pergamit. *Unbounding the Future: The Nanotechnology Revolution*. New York: William Morrow, 1991.

Engelberger, Joseph F. *Robotics in Service*. Cambridge, MA: The MIT Press, 1989.

Ferrari, Mario, Giulio Ferraii, Syngress Publishing Staff, and Ralph Hempel (editor). *Building Robots with LEGO Mindstorms*. Sebastopol, CA: Syngress Publishing (distributed in United States by OReilly Media, Inc.), 2001.

Gilbert, Horace D., ed. *Miniaturization*. New York: Reinhold, 1961. (Contains Richard P. Feynman's lecture: "There's plenty of room at the bottom.")

Gurstelle, William. *Building Bots: Designing and Building Warrior Robots*. Chicago: Chicago Review Press, 2002. (Intended for robot hobbyists and battlebot builders.)

Hally, Mike. *Electronic Brains: Stories from the Dawn of the Computer Age*. Washington, DC: National Academies Press (Joseph Henry Press Imprint), 2005.

Hannold, Chris. *Combat Robot Weapons*. New York: McGraw-Hill Companies, 2003. (Intended for robot hobbyists and battlebot builders.)

Hrynkiw, David, and Mark W. Tilden. *Junkbots, Bugbots, and Bots on Wheels: Building Simple Robots with Beam Technology*. New York: McGraw-Hill Companies, 2002. (Intended for robot hobbyists.)

Imahara, Grant. *Kickin' Bot: An Illustrated Guide to Building Combat Robots*. Hoboken, NJ: John Wiley & Sons Inc., 2003. (Intended for robot hobbyists and battlebot builders.)

Iovine, John. *Robots, Androids and Animatrons*. 2nd edn. New York: McGraw-Hill Professional, 2001.

Jones, Joseph L., Anita M. Flynn, and Bruce A. Seiger. *Mobile Robots: Inspiration to Implementation*. 2nd edn. Wellesley, Mass.: A. K. Peters Ltd, 1999.

Martin, Fred G. *Robotic Explorations: A Hands-on Introduction to Engineering*. Upper Saddle River, NJ: Prentice Hall, 2000.

Meadhra, Michael, and Peter J. Stouffer. *Lego Mindstorms: for Dummies*. Hoboken, NJ: John Wiley & Sons Inc., 2001.

Mulhall, Douglas. *Our Molecular Future: How Nanotechnology, Robotics, Genetics, and Artificial Intelligence Will Transform Our World*. Amherst, NY: Prometheus Books, 2002.

Murphy, Robin R. *An Introduction to AI Robotics*. Cambridge, MA: The MIT Press, 2000.

National Research Council. *Interfaces for Ground and Air Military Robots: Workshop Summary*. Washington, DC: National Academies Press, 2005.

Naval Studies Board. *Autonomous Vehicles in Support of Naval Operations*. Washington, DC: National Academies Press (Joseph Henry Press Imprint), 2005.

Niku, Saeed. *Introduction to Robotics: Analysis, Systems, Applications*. Upper Saddle River, NJ: Prentice Hall, 2001.

Noble, David F. *Forces of Production—A Social History of Industrial Automation*. New York: Alfred A. Knopf, 1984.

Nolfi, Stefano, and Dario Floreano. *Evolutionary Robotics: The Biology, Intelligence, and Technology of Self-Organizing Machines*. Cambridge, MA: The MIT Press, 2004.

Perkowitz, Sidney. *Digital People: From Bionic Humans to Androids*. Washington, DC: National Academies Press (Joseph Henry Press Imprint), 2004.

Predko, Michael. *123 Robotics Experiments for the Evil Genius*. New York: McGraw-Hill Companies, 2003. (Printed circuit board included with book.)

Reintjes, J. Francis. *Numerical Control: Making a New Technology*. New York: Oxford University Press, 1991.

Stone, Brad. *Gearheads: The Turbulent Rise of Robotic Sports*. New York: Simon & Schuster (Adult Publishing Group), 2003.

Thrun, Sebastian, Wolfram Burgard, and Dieter Fox. *Probabilistic Robotics*. Cambridge, MA: The MIT Press, 2005.

Webb, Barbara, and Thomas R. Consi, ed. *Biorobotics*. Cambridge, MA: The MIT Press, 2001.

Wiener, Norbert. *Cybernetics: Or the Control and Communication in the Animal and the Machine*. Cambridge, MA: The MIT Press, 1961 (paperback edition).

Wise, Edwin. *Robotics Demystified*. New York: McGraw-Hill Companies, 2004.

SELECTED PUBLICATIONS AND PERIODICALS

Artificial Life. Quarterly journal of the International Society of Artificial Life (ISAL). http://www.mitpressjournals.org/loi/artl

Journal of Biomechanical Engineering. Professional technical journal from ASME, dealing with artificial organs and prostheses and bioinstrumentation. http://scitation.aip.org/ASMEJournals/Biomechanical/

Journal of Field Robotics. Scholarly journal published by Wiley InterScience, dealing with fundamentals of robotics in unstructured and dynamic environments. http://www3.interscience.wiley.com/cgi-bin/jabout/111090262/ProductInformation.html

Journal of Microelectromechanical Systems. Professional engineering journal published under joint sponsorship by IEEE and ASME. (Internet Portal) http://www.ieee.org/portal/pages/pubs/transactions/jms.html

Journal of Robotic Systems. Scholarly journal published by Wiley InterScience, dealing with all aspects of robotic systems. http://www3.interscience.wiley.com/cgi-bin/jabout/35876/ProductInformation.html

Mechanical Engineering. Monthly professional magazine of the American Society of Mechanical Engineers (ASME). http://www.memagazine.org/

Planetary Report. Bimonthly space exploration magazine of the Planetary Society. http://www.planetary.org/

Presence: Teleoperators and Virtual Environments. Bimonthly technical journal for serious investigators of teleoperators and virtual environments. http://www.mitpressjournals.org/loi/pres

The Futurist. Bimonthy publication of the World Future Society. http://www.wfs.org/futurist.htm

The International Journal of Robotics Research (IJRR). Professional international journal for scientists and engineers produced by Sage Publications, dealing with robotics research. http://wwwijrr.org/

The Military Engineer. Bi-monthly publication of the Society of American Military Engineers. http://www.same.org/

EDUCATIONAL RESOURCES—WEB SITES

The following Internet sites offer useful educational materials concerning robotic systems and their applications. Educational robot competitions are also included.

NASA's Robotics Alliance Project. http://robotics.nasa.gov/

NASA's Robotics Curriculum Clearinghouse (RCC) http://robotics.nasa.gov/rcc/

NASA's Space Place provides a wide variety of space-related educational resources (including robotic systems and projects) for educators and students. http://spaceplace.nasa.gov/

ImagiBotics© is a multilingual robotics program developed for children in the Pre-K through eighth grades. http://imagiverse.org/

PreK-12 Engineering is a website that provides resources to educators who wish to integrate engineering concepts and activities (including robotics) into preK through twelfth grade curricula and classroom activities. http://www.prek-12engineering.org/

ROBOTICS COMPETITIONS

Battlebots—an educational and commercial program centered around building and competing robot system in four combat robot classes: lightweight, middleweight, heavyweight, and super heavyweight. http://www.battlebots.com/

Best Robotics, Inc.—a nonprofit, volunteer organization whose mission is to inspire students to pursue careers in engineering, science, and technology through participation in sports-like, science and engineering-based robotic system competition. http://www.bestinc.org/

Botball—a hands-on learning experience in robotics that stimulates interest in learning more about engineering, science, technology, and mathematics. http://www.botball.org/

FIRST (For Inspiration and Recognition of Science and Technology)—a multinational, nonprofit organization with the mission to make mathematics and science enjoyable by using interest in robotic systems and competitions. http://www.usfirst.org/

FIRST LEGO League—an educational program that uses the LEGO® MINDSTORMS™ Robotics Invention System™ technology to stimulate younger students (ages 9 through 14) in science and technology through the construction and experimentation with robots. http://www.usfirst.org/jrobtcs/fllego.htm

CYBERSPACE SOURCES: A COLLECTION OF SELECTED ROBOT TECHNOLOGY-RELATED INTERNET SITES

Selected Organizational Home Pages These organizations and agencies sponsor the research, development, testing, and the application of robotic systems for a variety of missions and goals, such as national defense, space exploration, underwater exploration, or environmental restoration and monitoring.

The Department of Energy (DOE) is responsible for maintaining the U.S. nuclear stockpile, promoting civilian energy programs, cleaning up the environmental legacy of the cold war's nuclear weapons program, developing a nuclear waste repository, and supporting fundamental science programs. http://energy.gov/

The European Space Agency (ESA) is an international organization whose task is to provide for and promote, exclusively for peaceful purposes, cooperation among European states in space research, technology, and applications. http://www.esrin.esa.it/

The National Aeronautics and Space Administration (NASA) is the civilian space agency of the United States government, which plans, directs, and conducts the American civilian space activities for peaceful purposes. http://www.nasa.gov/

The National Oceanic and Atmospheric Administration (NOAA) was established in 1970 as an agency within the U.S. Department of Commerce to ensure the safety of the general public from atmospheric phenomena and to provide the public with an understanding of Earth's environment and resources. http://noaa.gov/

The National Science Foundation (NSF) an independent federal agency that promotes progress in science, including fundamental research in robotics and nanotechnology. http://www.nsf.gov/

SELECTED ORGANIZATIONS WITHIN THE U.S. DEPARTMENT OF DEFENSE INVOLVED IN THE RESEARCH, DEVELOPMENT, CONSTRUCTION, AND OPERATION OF ROBOTIC SYSTEMS

Air Force Research Laboratory (AFRL). http://www.afrl.af.mil/
Army Research Laboratory (ARL). http://www.arl.army.mil/
Defense Research Project Agency (DARPA). http://www.darpa.mil/
The National Reconnaissance Office (NRO). http://www.nro.gov
Office of Naval Research (ONR). http://www.onr.navy.mil/
Space and Naval Warfare Systems Center, San Diego (SSC San Diego). http://www.spawar.navy.mil/robots/
The United States Air Force (USAF). http://www.af.mil/
The United States Army (USA). http://www.army.mil/
The United States Navy (USN) http://www.navy.mil/

SELECTED LIST OF OTHER U.S. GOVERNMENT-SPONSORED FEDERAL LABORATORIES AND CENTERS INVOLVED IN ROBOTIC SYSTEM RESEARCH, DEVELOPMENT, TESTING, OR OPERATION

DOC's National Institute of Standards and Technology (NIST). http://www.nist.gov/
DOE's Argonne National Laboratory (ANL). http://www.anl.gov/
DOE's Idaho National Laboratory (INL). http://www.inl.gov/
DOE's Los Alamos National Laboratory (LANL). http://www.lanl.gov/
DOE's Oak Ridge National Laboratory (ORNL). http://www.ornl.gov/
DOE's Pacific Northwest National Laboratory (PNNL). http://www.pnl.gov/
DOE's Sandia National Laboratories (SNL). http://www.sandia.gov/
DOH's National Institutes of Health (NIH). http://www.nih.gov/
DOL's Occupational Safety and Health Administration (OSHA). http://www.osha.gov/
NASA's Ames Research Center (ARC). http://www.nasa.gov/centers/ames
NASA's Jet Propulsion Laboratory (JPL). http://www.jpl.nasa.gov/

PROFESSIONAL SOCIETIES, ORGANIZATIONS, AND ACADEMIC INSTITUTIONS INVOLVED IN ROBOTIC SYSTEM DEVELOPMENT

Aerospace Robotics Laboratory, Stanford University. http://sun-valley.stanford.edu/home.html
American Nuclear Society (ANS). http://www.ans.org/
American Society of Mechanical Engineers (ASME). http://www.asme.org/
Association for Unmanned Vehicle Systems International (AUVSI) http://www.auvsi.org/
Field Robotics Center, Carnegie Mellon University. http://www.frc.ri.cmu.edu/
IEEE Robotics and Automation Society. http://www.ieee.org/
National Robotic Engineering Center (NREC). http://www.rec.ri.cmu.edu/
Robotic Industries Association (RIA). http://www.roboticsonline.com/
Robotics Institute (RI). http://www.ri.cmu.edu/
Society of American Military Engineers (SAME). http://www.same.org/
Space Systems Lab, University of Maryland. http://www.ssl.umd.edu/
Wisconsin Center for Space Automation and Robotics (WCSAR). http://wcsar.engr.wisc.edu/
Woods Hole Oceanographic Institution (WHOI). http://www.whoi.edu/

ROBOT CLUBS, GROUPS, AND INFORMAL ORGANIZATIONS

Over the past decade there has been an exponential increase in the number of robot system hobbyists, who enjoy designing and building warrior robots (battlebots) for machine competitions, as well as a wide variety of robots for fun, education, and various competitive robot sports, such as soccer. Often, these "amateur" robot enthusiasts form informal groups and organizations to their share experiences and participate in competitions. Here is a short (geographically dispersed) list of some of the informal robot clubs and organizations in the United States and Canada.

Arts and Robots Group (ARG), Toronto, Canada. http://interaccess.org/arg/
Atlanta Hobby Robot Club (AHRC), Atlanta, Georgia. http://www.botlanta.org/
Carnegie Mellon Robotics Club, Pittsburgh, Pennsylvania. http://www.roboticsclub.org/
Central Illinois Robotics Club (CIRC), Peoria, Illinois. http://circ.mtco.com/
ChiBots-Chicago Area Robotics Group, Chicago, Illinois. http://www.chibots.org/index.php
Connecticut Robotics Society (CRS), Hartford, Connecticut. http://www.ctrobots.org/
Dallas Personal Robotics Group (DPRGF), Dallas, Texas. http://www.dprg.org/
HomeBrew Robotics Club (HBRC), San Jose, California. http://www.hbrobotics.org/index.html
Nashua Robot Club, Nashua, New Hampshire. http://nashuarobotbuilders.org/
Phoenix Area Robotics Experimenters (PAREX), Phoenix, Arizona. http://www.parex.org/
Portland Area Robotics Society, Portland, Oregon. http://www.portlandrobotics.org/
Robomo-Missouri Area Robotics Society, St. Louis, Missouri. http://robo.com/
Robotics Society of Southern California (RSSC), Fullerton, California. http://www.rssc.org/

Sacramento Area Robotics Group, Sacramento, California. http://www.sacrobotics.org/

San Francisco Robotics Society of America (SFRSA), San Francisco, California. http://www.robots.org/

Seattle Robotics Society (SRS), Seattle, Washington State. http://www.seattlerobotics.org/

The Robot Group, Austin, Texas. http://www.robotgroup.org/navigation.html

Triangular Amateur Robotics, Raleigh, North Carolina. http://www.triangleamateurrobotics.org/

Twin Cities Robotics Group (TCRG), St. Paul, Minnesota. http://www.tcrobots.org/

Vancouver Island Robotics Club, Vancouver, British Columbia, Canada. http://vancouverroboticsclub.org/

Western Canadian Robotics Society, Calgary, Alberta, Canada. http://www.robotgames.net/

Index

Accelerated life test, 258
Acceleration, 258
Accelerometer, 258
Acceptance test, 258
Accumulator, 259
Acronym, 259
Active control, 259
Active homing guidance, 259
Active sensor, 259
Actuator, 109, 259
Adaptive control system, 260
Aeolipile, 60. *See also* Hero of Alexandria
Aerobot, 260
AI. *See* artificial intelligence
AIBO entertainment robot, 17, 51, 362–63
Air Force Research Laboratory (AFRL), 330–32
Alien life form, 260
Alphabot, 393
Alphanumeric, 260
American Museum of Science and Energy (ASME), 368–69
American Nuclear Society (ANS), 328–29
American Society of Mechanical Engineers (ASME), 329
Ames Research Center (ARC), 330–31
Ampère, André-Marie, 4, 32, 74–76
Ampere (unit), 76, 260
Amplitude modulation (AM), 260. *See also* communications
Analog computer, 35, 261; Vannevar Bush's differential analyzer, 35
Analog-to-digital converter, 261
Analytical Engine, 32, 79–80. *See also* Charles Babbage
Androgynous interface, 261
Android, 20, 261
Angular momentum, 261
Antenna, 262; dipole antenna, 275; directional antenna, 275; diplexer, 275; duplexer, 276
Antisatellite (ASAT) spacecraft, 262
Aqua spacecraft, 262
Archimedes of Syracuse, 3, 28, 56–58
Argonne Information Center, 369–70
Argonne National Laboratory (ANL), 333–34
Arizona Science Center, 370
Army Research Laboratory (ARL), 333–35
ARPANET, 262. *See also* Defense Advanced Research Projects Agency (DARPA)
Artificial intelligence (AI) 22–26, 262; backward chaining, 264; blackboard approach, 265; blind search, 265; bottom-up control structure, 265; breadth-first search, 266; cognition, 268; computer vision, 268; connectives, 269; constraint propagation, 269; depth-first search, 274; difference reduction, 274; event-driven, 278; expert system, 278;

Artificial intelligence (AI) (*cont.*)
first order predicate logic, 281; generate and test, 283; goal regression, 283–84; heuristic search technique, 284; image understanding (IU), 287; knowledge base, 290; knowledge engineering, 290; least commitment, 291; John McCarthy, 41; means-ends analysis, 294; Marvin Minsky, 41; MIT Museum, 381–82; model-driven, 296; origin of term artificial intelligence, 41; personal AI computer, 300; problem-solving, 302; production rule, 303; property list, 303; propositional logic, 303; pseudo-reduction, 303; recursive operations, 306; relaxation approach, 306; smart robot, 235–45, 313–14; speech recognition, 316; speech understanding, 316; theorem-proving, 319; top-down approach, 321; truth-maintenance, 322; world knowledge, 327; world model, 327

Artificial life, 1, 22, 53, 66, 80–83, 226–28; *Frankenstein: The Modern Prometheus*, 1, 22, 53, 81–82; Golem, 1, 22; machine consciousness, 66, 226–28

ASIMO (humanoid robot), 17–18, 46–47, 51, 52, 342, 377; Honda ASIMO Theater in Innovations, 377

Asimov, Isaac, 18, 25–26, 102–4; three laws of robotics, 18, 25–26, 54, 103; zeroth law of robotics, 26, 103–4

Assembler, 263. *See also* Santa Claus machine

Assembly line 4, 31, 35; The Henry Ford Museum, 376–77; factory automation exhibit, 392–93; Eli Whitney, 31

Assembly robot, 263. *See also* industrial robot

Association for Unmanned Vehicle Systems International (AUVSI), 335–36

Astronaut, 44, 263; robots versus astronauts in space exploration, 213–14

Astronomical unit (AU), 263

Atmospheric probe, 263. *See also* robot spacecraft

Atomic clock, 263

Atomic force microscope (AFM), 11

Attitude, 264; pitch, 301; roll, 310; yaw, 32

Attitude control system, 135–36, 264; pitch, 301; roll, 310; spin stabilization, 316; tumble, 322; yaw, 327

Aura spacecraft, 264

Automated loom, 30, 31, 91, 375; Deutsches Museum, 375. *See also* Joseph-Marie Jacquard

Automatic pilot, 264

Automation, 110; hard automation, 110; flexible automation, 110

Automaton 3, 28, 30, 31, 264, 375–76; Deutsches Museum, 375; Franklin Institute Science Museum, 375–76; Pierre Jaquet Droz automatons, 3, 30, 386; Henri Maillardet, 31; mechanical duck, 30, 264; Jacques de Vaucanson, 3, 30, 264;

Autonomous ground vehicle, 337–38; DARPA Grand Challenge robot races, 337–38

Autonomous underwater vehicle (AUV) 110, 193–95, 340; Gavia AUV, 340; Space and Naval Warfare Systems Center, 363; REMUS, 195, 366; Woods Hole Oceanographic Institution (WHOI), 195, 366

Babbage, Charles, 32, 78–80; Analytical Engine, 32, 79–80; Difference Engine, 32, 79; Powerhouse Museum, 388–89; The Science Museum, 390–91

Backward chaining, 264. *See also* artificial intelligence

Bang-bang robot. *See* pick-and-place robot.

Battery, 31, 265; Count Alessandro Volta, 31, 71–73

Baud, 33, 265

Bel, 33, 265

Binning, Gerd, 10–11

Biomimetic system, 265. *See also* insect robot.

Biomimetic underwater robot, 355

Blackboard approach, 265. *See also* artificial intelligence

Black box, 265

Blind search, 265. *See also* artificial intelligence

Boole, George, 33

Boston Museum of Science (MOS), 370–71

Bottom-up control structure, 265. *See also* artificial intelligence

Bradbury Science Museum, 371–72

Breadboard, 266

Breadth-first search, 266. *See also* artificial intelligence

Čapek, Karel, 1, 18, 35, 92–94
Carnegie Mellon University, 337, 351, 358–59, 389–90; National Robotics Engineering Center (NREC), 337, 351; Robot Hall of Fame®, 358–59, 389–90; Robotics Institute, 359–60
Carnegie Science Center, 372–73
Cartesian coordinate system, 64, 116, 266
Cartesian robot, 116, 266
Central processing unit (CPU), 267
CERN–Microcosm Visitor Centre, 373–74
Chandra X-ray Observatory (CXO), 49, 187–89, 267
Clean room, 267; white room, 326
Clepsydra, 59
Closed loop, 267. *See also* feedback
Cog (humanoid robot), 381–82
Communications, 33–34; amplitude modulation, 260; antenna, 264; baud, 264; bel, 264; bent-pipe communications, 264; demodulation, 274; digital transmission, 275; direct readout, 275; electromagnetic (EM) communications, 33–34, 277; frequency modulation (FM), 281; Internet, 289; link, 291; modulation, 296; neper, 298; one-way communications (OWC), 141, 299; phase modulation (PM), 300; pulse code modulation (PCM), 303; readout station, 306; robot spacecraft telecommunications, 141–46; signal, 313; signal-to-noise ratio (SNR), 313; telecommunications, 141–46; 318; telemetry, 318; telephone, 33; transceiver, 321; transmitter, 321; transponder, 321; ultrahigh frequency (UHF), 322; uplink, 323
Compton Gamma Ray Observatory (CGRO), 187–89, 268
Computer History Museum, 374–75
Computer-integrated manufacturing (CIM), 382–83; Museum of Science and Industry, 382–83
Computerized robot, 268
Computer vision, 268; image understanding (IU), 287
Console, 269
Continuous path robot, 117, 270
Controller, 270
Coulomb, Charles-Augustin de, 4, 31, 70–71
Coulomb (unit), 71, 270
Cruise missile, 146–47, 270–71; Tomahawk, 147
Crusher (UGCV), 337
Cryobot, 271–72
Ctesibius of Alexandria, 3, 28, 58–59; clepsydra, 59
Cybernetics, 39, 94–96, 114–16. *See also* Norbert Wiener
Cyborg, 20–22
Cylindrical coordinate robot, 272
C-3PO, 20. *See also* robots and smart computers in the cinema

da Vinci™ Surgical System, 157, 343–44, 364; Intuitive Surgical, Inc., 343–44; UCI Center for Minimally Invasive Surgery, 364
Deep Space Network (DSN), 141–46, 273
Defense Advanced Research Projects Agency (DARPA), 7–8, 336–37; ARPANET, 262; Grand Challenge robot races, 337–38; Shakey, 7–8
Defense Support Program (DSP), 171–72, 273; early warning satellite, 276
Degrees of freedom (DOF), 108–9, 116, 274
Descartes, René, 25, 29, 55, 63–66, 226–27; Cartesian coordinate system, 64; dualism, 65–66
Deutsches Museum, 375
Devol, George C., Jr., 5, 7, 39, 40, 41–42, 46–47, 98–100
Difference Engine, 32, 79, 389, 391. *See also* Charles Babbage
Digital Revolution, 5–6, 35, 36, 37–38, 40, 45, 49; first electronic computers, 36–40; Internet, 49; transistor, 39
Discourse on Method (Descartes), 29, 55, 63–66
Docking mechanism, 275
Domestic robot, 344; Roomba® vacuuming robot, 344; Scooba™ floor washing robot, 344
Drone, 39, 275; nuclear cloud sampling drone, 39
Dyne (unit), 276

Edison, Thomas Alva, 33–34, 84–86; conflict with Nikola Tesla and George Westinghouse, 33–34, 83–84, 87–88; direct current electricity, 85; electric lighting, 33–34, 85–86; talking doll, 85–86
Educational robot, 192–95, 204–5
Electric generator, 77
Electricity, 31, 33–34, 276; ampere (unit), 32, 76, 260; capacitor, 266; electric potential, 276; electrode, 276; farad, 280; impedance, 287; rectifier, 306; resistance, 32, 307; siemens (unit), 313; solar cell, 314; solar photovoltaic conversion, 314; solenoid, 314; volt, 324
Electric motor, 32, 77–78
Electric robot, 276
Electric tabulating machine, 91–92. *See also* Herman Hollerith
Electromagnetic (EM) communications, 277. *See also* communications
Electromagnetic radiation (EMR), 187–89, 277; gamma rays, 283; infrared radiation (IR), 288; microwave radiation, 295; multispectral sensing, 296; radio frequency (RF) radiation, 305; remote sensing, 187–89, 307; ultraviolet (UV) radiation, 31, 322; X-ray, 34, 327
Electron, 34, 277
Electronics, 34–36, 39–40, 277; solid-state device, 39–40, 314; very large scale integration (VLST), 40, 324
Electron volt (eV), 277
End effector, 110, 277–78
Engelberger, Joseph F., 5, 7, 41–42, 46–47, 104–5, 112–13; PUMA, 46–47, 104; 112–13; Unimation, 104
ENIAC, 38–39, 97. *See also* John von Neumann
Entertainment robot, 17–22, 85–86, 278, 362–63; AIBO, 17, 362–63; ASIMO, 17–18, 342; Edison's talking doll, 85–86; Honda ASIMO Theater in Innovations, 377; QRIO, 363; robotic dinosaurs, 388; robotic pirates, 22
EPSON Robots, 108, 338
Explorer 1 spacecraft, 279
Explosive ordnance disposal (EOD) robot, 7, 155–56, 171–77, 206, 211, 279, 344–45; iRobot PackBot®, 344–45
Extraterrestrial contamination, 219–26, 279
EVA robot, 280

FANUC Robotics America, Inc., 339
Farad (unit), 280
Faraday, Michael, 4, 31, 32, 76–78; electric generator, 77
Feedback, 94–95, 114–16, 281; closed loop, 267; open loop, 299
Feynman, Richard P., 9, 41
Field robot, 183, 281–82
Fluid mechanics, 281; hydrostatics, 67; ideal gas, 285–86; incompressible fluid, 287; inviscid fluid, 289–90; perfect fluid, 300; perfect gas, 300; poise, 301; pressure, 302; pump, 303; regulator, 306; Osborne Reynolds, 33; sealant, 311; slip flow, 313; surface tension, 318; tribiology, 322; two-phase flow, 322; valve, 323; vapor, 323; vent valve, 324; viscosity, 324; viscous fluid, 324; working fluid
Food and Drug Administration (FDA), 339
Force, 30, 281; impulse, 287; inertia, 288; kinetic energy, 290; line of force, 291; mass, 293; moment of inertia, 296; momentum, 296; pressure, 302; reaction engine, 306; tap, 318; thrust, 320; torque, 321; total impulse, 321; work, 327. *See also* Sir Isaac Newton
Ford, Henry, 4, 35, 162; The Henry Ford Museum, 376–77
Foster-Miller, Inc., 155–56, 339–41
Frankenstein: The Modern Prometheus, 1, 22, 53, 81–82, 372; Carnegie Science Center 372–73. *See also* Mary Wollstonecraft Shelley
Franklin, Benjamin, 4, 30, 375–76
Franklin Institute Science Museum, 375–76
Freddie the robot, 385
Frequency, 281; Heinrich Rudolf Hertz, 89–90, 284
Frequency modulation (FM), 281. *See also* communications
Fuel cell, 283

Galilei, Galileo, 3, 29, 60–62, 186–87
Galileo Project, 283
Galvani, Luigi, 4, 72
Gavia AUV, 340

Geographic information system (GIS), 283
Giotto spacecraft, 283
Global Hawk (UAV), 51, 153–55, 171–75
Global Positioning System (GPS), 283
Goertz, Raymond C., 36–41, 175–80. *See also* nuclear robot
Gravity assist, 284
Guided missile, 36–37, 146–49, 166–69; proliferation of missile technology, 210–12; revolution in strategic warfare, 166–69; V-2 rocket, 36–37
Gyroscope, 284

Hafmynd–Gavia, Ltd., 340; Gavia AUV, 340
HAL 9000, 19–20, 44, 196. *See also* robots and smart computers in the cinema
Hard landing, 284
Heat engine, 32, 284
Henry (unit), 284
Henry, Joseph, 4, 32, 77–78, 284; electric motor, 77–78
The Henry Ford Museum, 376–77
Hero of Alexandria, 3, 59–60; aeolipile, 60
Heron. *See* Hero of Alexandria
Hertz (unit), 284
Hertz, Heinrich Rudolf, 33, 89–90, 284; radio waves, 90
High Energy Astronomy Observatory (HEAO), 285
Hollerith, Herman, 90–92
Honda Motor Company, Ltd., 47–47, 51–52, 340–42; Honda ASIMO Theater in Innovations, 377
Hong Kong Science Museum, 377
Hubble Space Telescope (HST), 48, 132–34, 285
Humanoid robot, 8, 46–47, 51–52, 285–86, 342; ASIMO, 46–47, 51–52, 342; Cog, 381–82; Honda ASIMO Theater in Innovations, 377; Kismet, 381; MIT Museum, 381–82; QRIO, 363
Huygens probe, 16, 125, 128, 285
Hydraulic robot, 117, 285
Hydrostatics, 67

Idaho National Laboratory (INL), 342
IEEE Robotics and Automation Society, 342–43
Industrial Revolution, 3–6, 31, 162; First Industrial Revolution, 3–4, 31, 162; Luddites, 197–99; Second Industrial Revolution, 4–5, 162; social issues with industrialization, 196–98; steam engine, 31; superindustrialization, 200
Industrial robot 41–42, 110–18, 287–88; assembly robot, 264; Carnegie Science Center, 372–73; classifications, 112–18; components, 119; EPSON Robots, 108, 338; FANUC Robotics America, Inc., 339; first industrial robot, 41; hydraulic robot, 117, 285; job displacement caused by robots, 199–200; Museum of Science and Industry, 382–83; paths generated by industrial robots, 118; pick-and-place robot, 301; pitch, 301; pneumatic robot, 301; point-to-point robot, 117, 301; programmable robot, 303–4; psychological impact of robots in workplace, 200–202; Robotic Industries Association (RIA), 359; SCARA configuration, 112, 116–18; sensory robot, 118, 311; serial robot, 311; servo robot, 312; spherical coordinate robot, 118, 316; Stäubli Robotics, 48, 112, 363–64; teaching industrial robots, 120–21; Unimation, 41–42, 104; work envelope, 327; wrist, 46, 327
Infrared radiation (IR), 288
Insect robot, 49, 265, 288–89
Intercontinental ballistic missile (ICBM), 6, 42–43, 146–49, 166–69; 385; Minuteman (ICBM), 148–49, 166–69, 385; National Museum of the United States Air Force, 385; proliferation of missile technology, 210–12; revolution in strategic warfare, 166–69; Titan Missile Museum, 394–95
Internet, 41, 48, 289
Interstellar probe, 288; star probe, 316
Intervehicular activity (IVA) robot, 289
Intuitive Surgical, Inc., 49, 343–44
Invent Now®–National Inventors Hall of Fame™, 378
iRobot Corporation, 344–45
iRobot PackBot®, 344–45

Jacquard, Joseph-Marie, 31, 73–74, 91; automated loom, 73–74, 91, 375
Jaquet-Droz, Pierre, 3, 30, 386; automatons in museum, 386
Jansky (unit), 290

Jet Propulsion Laboratory (JPL), 345–46, 378–80; Deep Space Network (DSN), 273; *Explorer 1* spacecraft, 279
Joule (unit), 32, 290

Kelvin (unit), 290
Kettering Aerial Torpedo, 35, 385; National Museum of the United States Air Force, 385
Kilogram (unit), 290. *See also* SI units
Kinetic energy, 290
Kismet (humanoid robot), 381–82
Knowledge base, 290
Knowledge engineering, 290. *See also* artificial intelligence

Lambert (unit), 291
Lander spacecraft, 291. *See also* robot spacecraft
Lawrence Hall of Science, 380–81
Leibniz, Gottfried, 3, 69
Light-year (ly), 291
Limited sequence robot. *See* pick-and-place robot
Limit switch, 291
Line of sight (LOS), 291
Longitudinal axis, 292
Los Alamos National Laboratory (LANL), 346–47; BEAM robot, 49; Bradbury Science Museum, 371–72
Lovelace, Lady Ada, 32, 80
Luddites, 197–99
Lumen (unit), 292
Lumped mass, 292
Luna, 41, 44, 45, 292
Lunar orbiter, 44, 292
Lunar Prospector, 292.
Lunar rover, 292. *See also* robot spacecraft
Lunokhod, 45, 292
Lux (unit), 292

Machine, 2–3, 107, 292; heat engine, 284; mechanical efficiency, 94; microelectromechanical systems (MEMS), 11, 21, 42, 255–56; reaction engine, 306; simple machines, 2–3; turbine, 322; vapor turbine, 323; work, 327; working fluid, 327
Machine consciousness, 22–26, 66, 226–28
Machine intelligence. *See* artificial intelligence
Machine vision. *See* computer vision
MANIAC, 40, 98. *See also* John von Neumann
Manipulator, 293. *See also* robot
Man-machine interface, 293
Manny, 356. *See also* military robot
Manufacturing, 31, 110, 161–66, 264, 293; arrival of industrial robots, 41–47, 164–66; automation, 110; batch manufacturing, 264; breadboard, 266; circuit board, 267; computer-integrated manufacturing (CIM), 111; flexible manufacturing, 165–66; Henry Ford, 4, 35, 162, 164; The Henry Ford Museum, 376–77; manufacturing cell, 293; mass production, 31; mock-up, 296; Museum of Science and Industry, 382–83; nondestructive testing, 298; prototype, 303; reliability, 306; Sony Wonder Technology Lab, 392–93; superindustrialization, 200; tolerance stackup, 321; Eli Whitney, 31
Marconi, Guglielmo, 34, 90; Nobel Museum, 386–87
Mariner, 43, 44, 293
Mars Exploration Rover (MER) 14–16, 123, 132–33
Mars Global Surveyor (MGS), 293
Mars Odyssey, 16, 293
Mars Pathfinder, 12–13, 49, 293
Mars Reconnaissance Orbiter (MRO) spacecraft, 346
Mars surface rover, 132–34, 293
Marsupial robot, 293. *See also* military robot
Mass, 293
Master/slave manipulator, 36, 38, 39, 41, 175–80, 294; Raymond C. Goertz, 36, 38, 39, 41, 175–80
Maxwell, James Clerk, 4, 33, 78
Mechanical efficiency, 294. *See also* machine
Medical robot, 46, 48–49, 157, 339, 343–44; da Vinci™ Surgical System, 157, 343–44; first robot-aided surgery, 46; Food and Drug Administration (FDA), 339; Intuitive Surgical, Inc., 343–44; National Institutes of Health (NIH), 349–50; pharmacy robot, 163; UCI Center for Minimally Invasive Surgery, 364
Meter (unit), 295. *See also* SI units
Metric system, 295

Microelectromechanical systems (MEMS), 11, 12, 42, 255–56
Micrometer, 295
Micron, 295
Microorganism, 295
Microwave (radiation), 295
Milestone, 295
Military robot, 5–8, 16–17, 146–56, 166–75, 295, 330–37; Air Force Research Laboratory (AFRL), 330–32; Army Research Laboratory (ARL), 333–35; Association for Unmanned Vehicle Systems International (AUVSI), 335–36; cruise missile, 146–47; Crusher, 337; Defense Advanced Research Projects Agency (DARPA), 336–37; Defense Support Program (DSP), 171–72, 273; explosive ordnance disposal (EOD) robot, 7, 155–56, 171–77, 206, 211, 279, 344–45; Foster-Miller, Inc., 155–56, 339–41; future autonomous mobile military robot systems, 252–55; Global Hawk (UAV), 51, 153–55; Global Positioning System (GPS), 283; guided missile, 36–37, 146–49, 166–69; impact of mobile military robots, 171–75; intercontinental ballistic missile (ICBM), 6, 42–43, 146–49, 166–69; iRobot Corporation, 344–45; iRobot PackBot®, 344–45; Manny, 356; marsupial robot, 293; military satellite (MILSAT), 149–50, 169–71, 295; National Reconnaissance Office (NRO), 169–70, 350–51; Office of Naval Research (ONR), 354–55; pitch, 301; Predator (UAV), 5, 50, 52, 151–53, 385; proliferation of nuclear missile technology, 210–12; readout station, 306; reconnaissance satellite, 169–70, 306; remote control, 306; remotely piloted vehicle (RPV), 307; ROBART sentry robot, 363; roll, 310; Sandia National Laboratories (SNL), 360–62; sentry robot, 363, 311–12; Shadow 200 (UAV), 335; Society of American Military Engineers (SAME), 361–62; Space and Naval Warfare Systems Center, 363; surveillance satellite, 171–72, 318; Talon™, 155–56, 177, 340–41; telemetry, 318; teleoperation 318–19; Tomahawk, 146–47; unmanned aerial vehicle (UAV), 150–55, 252–55, 323, 335, 336–37; unmanned ground combat vehicle (UGCV), 252–55, 336–37; unmanned ground vehicle (UGV) 155–56; Vela spacecraft, 44, 150; yaw, 327
Military satellite (MILSAT), 149–50, 169–72, 295; information revolution in national security, 169–72
Miniaturized robot, 362
Minuteman (ICBM), 148–49, 166–69, 385
MIT Museum, 381–82
Mobile robot, 5–8, 16–17; impact of mobile military robots, 171–75; pipe inspection mobile nuclear robot, 182; Shakey the first mobile robot, 7–8, 351. *See also* robot
Modulation, 296
Mole (unit), 296. *See also* SI units
Molecule, 296. *See also* nanotechnology
Moment of inertia, 296
Momentum, 296
Mother spacecraft, 296
Multispectral sensing, 296. *See also* electromagnetic radiation
Museum of Science and Industry, 111, 165, 382–83; *Toymaker 3000*, 111, 165, 382

Nanometer (unit), 296
Nanorover, 296
Nanotechnology, 8–10, 11, 41, 46–48, 255–57, 263, 297; assembler, 263; atomic force microscope (AFM), 11; Gerd Binning, 10–11; CERN–Microcosm Visitor Centre, 373–74; Richard Curl, 11, 46, 46; Richard Feynman, 9, 41; Food and Drug Administration (FDA), 339; fullerenes, 11–12, 46, 48; Sir Harold Kroto, 11, 46, 48; National Institute of Standards and Technology (NIST), 348–49; promise of nanotechnology, 255–57; Heinrich Rohrer, 10; Sandia National Laboratories (SNL), 360–62; scanning tunneling microscope (STM), 10; Richard Smalley, 11, 46, 48
National Aeronautics and Space Administration (NASA), 187–89, 297, 347–48; Ames Research Center (ARC), 330–31; Great Observatories Program, 187–89; Jet Propulsion Laboratory (JPL), 345–46
National Institute of Standards and Technology (NIST), 348–49

National Institutes of Health (NIH), 349–50
National Museums of Scotland (NMS), 385; Freddie the robot, 385
National Reconnaissance Office (NRO), 169–70, 350–51; reconnaissance satellite, 306
National Robotics Engineering Center (NREC), 337, 351; Crusher (UGCV), 337; Pioneer robot, 358
National Science Foundation (NSF), 352
Neolithic Revolution, 2
Neper (unit), 298
Neuchâtel Museum of Art and History, 385–86; Pierre Jaquet Droz automatons, 385
Neumann, John von, 38–39, 40, 96–98; ENIAC, 38–39, 97; MANIAC, 40, 90; self-replicating system, 98, 229–33, 311; universal constructor (UC), 323
Newton, Sir Isaac, 3, 30, 68–70; force, 281; law of gravitation, 30, 298; laws of motion, 30, 298
Newton (unit), 298. *See also* SI units
Nobel Museum, 386–87
Nondestructive testing, 298
Nonservo robot, 109, 298. *See also* industrial robot
Nuclear-electric propulsion (NEP), 298
Nuclear radiation, 299
Nuclear robot, 36–41, 175–83, 340, 342; Cecil®, 181, 340; Foster-Miller, Inc., 339–41; Idaho National Laboratory (INL), 342; master/slave manipulator, 36–41, 175–80, 294; mobile nuclear robots, 180–83; nuclear weapons complex, 36–41, 175–80; Oak Ridge National Laboratory (ORNL), 183, 352–53; Pioneer robot, 358; RedZone Robotics Incorporated, 356–58; Remotec, 357–58; Sandia National Laboratories (SNL), 360–61; teleoperation 175–80, 318–19; remote control, 306; United States Department of Energy (DOE), 364–65

Oak Ridge National Laboratory (ORNL), 183, 352–53
Occupational Safety and Health Administration (OSHA), 16, 353–54
Oersted, Hans Christian, 31, 75, 77
Oersted (unit), 299
Office of Naval Research (ONR), 7–8, 354–55; biomimetic underwater robot, 355;
Robolobster, 355; Shakey the mobile robot, 7–8, 351
Ohm (unit), 32, 299. *See also* electricity
One-way communications (OWC), 141, 299. *See also* communications
Open loop, 299
Orbiter spacecraft, 299
Orbiting Astronomical Observatory (OAO), 299
Orbiting Quarantine Facility (OQF), 299

Pacific Northwest National Laboratory (PNNL), 355–56
Pacific Science Center, 387–88; robotic dinosaurs, 388
Pascal, Blaise, 3, 29, 67–68; mechanical calculator, 29, 67
Pascal (unit), 300. *See also* SI units
Pascaline, 29, 67. *See also* Blaise Pascal
Passive sensor, 300. *See also* sensor
Phase modulation (PM), 300. *See also* communications
Photon, 187–89, 300
Pick-and-place robot, 301. *See also* industrial robot
Pioneer 10, 11 spacecraft, 130, 301
Pioneer Venus mission, 130, 301
Pitch, 301
Pneumatic robot, 301
Point-to-point robot, 117, 301. *See also* industrial robot
Polyphase AC motor, 33–34, 87. *See also* Nikola Tesla
Power, 302. *See also* thermodynamics
Powerhouse Museum, 388–89
Predator (UAV), 5, 50, 52, 151–53, 171–75, 385; National Museum of the United States Air Force, 385–86
Pressure, 302; vacuum, 323; vapor pressure, 323
Principia (Newton), 3, 30, 68–70. *See also* Sir Isaac Newton
Probe, 130, 302. *See also* robot spacecraft
Programmable robot, 46–47, 104, 112–13, 303–4; PUMA, 104
Programmable Universal Machine for Assembly. *See* PUMA
Pulse code modulation (PCM), 303. *See also* communications

PUMA, 46–47, 104, 112–13. *See also* industrial robot
Pump, 303. *See also* fluid mechanics

QRIO (humanoid robot), 363
Quantum, 303

Radian (unit), 305
Radio frequency (RF) radiation, 305. *See also* electromagnetic radiation
Radio waves, 33–34, 90, 305; jansky (unit), 290; ultrahigh frequency (UHF), 322
Radioisotope thermoelectric generator (RTG), 135, 305; aerospace nuclear safety, 214–19
Ranger Project, 44, 122, 305
Reconnaissance satellite, 169–70, 306. *See also* military robot
RedZone Robotics Incorporated, 356–58. *See also* nuclear robot
Remotec, 357–58. *See also* nuclear robot
Remote control, 306
Remotely operated vehicle (ROV), 110, 190–91; Super Scorpio ROV, 191
Remotely piloted vehicle (RPV), 307. *See also* unmanned aerial vehicle
Remote manipulator system (RMS), 307–8. *See also* robot spacecraft
Remote sensing, 307; active sensor, 259; electromagnetic radiation, 277; passive sensor, 300; resolution, 307; spectroscopy, 316
REMUS (AUV), 191–93, 366
Rescue robot, 157–59
Research robot, 334, 356, 385; Freddie the robot, 385; Manny, 356
ROBART sentry robot, 363. *See also* military robot
Robby the Robot, 19, 40. *See also* robots and smart computers in the cinema
Robolobster, 355. *See also* military robot
Robonaut, 9, 262, 286, 309
Robot: 7–9, 54, 92–94, 309–10; agent of social change, 161–66; Alphabot, 393; android, 261, 262; assembly robot, 263; autonomous robot; 263; Čapek, Karel, 1, 54, 92–94; Cartesian robot, 116, 266; computerized robot, 268; continuous path robot, 117, 270; cryobot, 271–72; cylindrical coordinate robot, 272; Defense Advanced Research Projects Agency (DARPA), 336–37; educational robot, 192–95, 204–6; electric robot, 276; end effector, 110, 277–78; entertainment robot, 17–22, 51–52, 85–86, 278, 362–63; explosive ordnance disposal (EOD) robot, 7, 155–56, 279, 344–45; EVA robot, 280; fear of robots, 196–98; field robot, 281–82; humanoid robot, 8, 46–47, 51–52, 285–86, 342; hydraulic robot, 117, 285; industrial robot, 41–42, 110–18, 287–88; insect robot, 49, 265, 288–89; IVA robot, 289; job displacement caused by robots, 199–200; jointed arm, 280; manipulator, 293; marsupial robot, 293; medical robot, 46, 48–49, 157, 339, 343–44; military robot, 5–8, 16–17, 146–56, 171–75, 295, 330–37; nanorover, 296; nonservo robot, 108, 298; nuclear robot, 175–83, 340; origin of word robot, 92–94; pick-and-place robot, 301; pitch, 301; pneumatic robot, 301; point-to-point robot, 117, 301; programmable robot, 46–47, 104, 112–13, 303–4; QRIO, 363; probe, 130, 302; psychological impact of robots in workplace, 200–202; PUMA, 46–47, 104, 112–13; reliability, 306; remote control, 306; remotely piloted vehicle (RPV), 307; rescue robot, 157–59; robot spacecraft, 12–16, 42–46, 48, 50–52, 121–27, 182–90, 260, 262–64, 310, 314–15; *Rossum's Universal Robots (R.U.R.)*, 1, 35, 54, 92–94; safety issues, 206–10; SCARA configuration, 116–18; sensory robot, 118, 311; sentry robot, 363, 311–12; serial robot, 311; service and rescue robots, 16–17, 157–59; servo robot, 107, 108, 312; smart robot, 235–55, 313–14; spherical coordinate robot, 118, 316; steel-collar worker, 114; teleoperation 318–19; telepresence robot, 393–93; test bed, 319; work envelope, 327; wrist, 327
Robot clubs, 403–4
Robot Hall of Fame®, 51–52, 358–59, 389–90. *See also* robots and smart computers in the cinema
Robotic dinosaurs, 388
Robotic Industries Association (RIA), 359
Robotic mannequin, 356; Manny, 356
Robotics, 105–10, 310
Robotics competitions, 192–95, 401

Robotics Institute, 359–60
Robot safety, 206–10, 353–54; fatal accident case history, 206–7; Occupational Safety and Health Administration (OSHA), 353–54; potential hazards involving use of robots, 207–10; redline, 306; safety analysis, 310; safety device, 310; work envelope, 327
Robots and smart computers in the cinema: *Artificial Intelligence: AI* (android named David), 22; *Colossus: The Forbin Project* (malevolent super computer), 20; *The Day The Earth Stood Still* (powerful alien robot named Gort), 19, 39–40; *Fantastic Voyage* (miniaturized submarine), 19; *Forbidden Planet* (Robby the Robot), 19, 40; *Metropolis* (exotic dancer robot named Maria), 18, 35; *Star Trek: The Next Generation* (android named Data), 20–21; *Star Wars* series (R2-D2 and C-3PO) 20, 45, 51–52; *Terminator* (powerful cyborgs and killer robots), 1, 22, 196; *Westworld* (gunfighter android), 209; *2001: A Space Odyssey* (mischievous supercomputer named HAL 9000), 19–20, 44, 196
Robot spacecraft, 12–16, 42–46, 48, 50–52, 121–27, 182–90, 260, 262, 263, 264, 310, 314–15; aerobot, 130, 260; Ames Research Center (ARC), 330–31; antisatellite spacecraft; *Aqua* spacecraft; atmospheric probe, 263; attitude control system, 135–36, 264; *Aura* spacecraft, 264; automated approach, 264; *Cassini* spacecraft, 16, 125, 128, 184, 215; *Chandra X-ray Observatory* (CXO), 49, 133, 187–89, 267; classes of scientific spacecraft, 128–33; components of typical robot spacecraft, 133–38; *Compton Gamma Ray Observatory* (CGRO), 132–33, 187–89, 268; *Copernicus*, 270; *Cosmic Background Explorer* (COBE), 48; cryobot, 271–72; data handling, 139; *Deep Impact*, 52; Deep Space Network (DSN), 141–46, 273; *Deep Space One*, 126; Defense Support Program (DSP), 171–72, 273; docking, 275; docking mechanism, 275; *Explorer 1* spacecraft, 279; EVA robot, 280; extraterrestrial contamination 219–26, 279; flyby, 130, 281; future very smart space robots, 235–48; Galileo Project, 130, 186–87, 283; *Giotto* spacecraft, 283; Global Positioning System (GPS), 283; gravity assist, 284; hard landing, 284; High Energy Astronomy Observatory (HEAO), 285; *Hubble Space Telescope* (HST), 48, 132–34, 187–89, 285; *Huygens* probe, 16, 125, 128, 184, 285; interstellar probe, 245–47, 288; intervehicular activity (IVA) robot, 289; issue of robots versus astronauts in space exploration, 213–14; Jet Propulsion Laboratory (JPL), 345–46, 378–80; lander spacecraft, 131, 291; longitudinal axis, 292; Luna, 41, 44, 45, 292; lunar orbiter, 44, 130, 292; *Lunar Prospector*, 292; lunar rover, 45, 292; Lunokhod, 45, 292; *Magellan* spacecraft, 126–27; Mariner, 43, 293; *Mars Exploration Rover* (MER), 14–16, 123, 132–33; *Mars Global Surveyor* (MGS), 293; *Mars Odyssey*, 16, 293; *Mars Pathfinder*, 12–13, 49, 293; *Mars Reconnaissance Orbiter* (MRO) spacecraft, 346; Mars surface rover, 132, 293; military satellite (MILSAT), 295; mother spacecraft, 296; Museum of Science and Industry, 382–83; nanorover, 296; National Reconnaissance Office (NRO), 169–70, 350–51; navigation, 140; *New Horizons* spacecraft, 51–52; observatory spacecraft, 132–34; *Opportunity*, 14–16, 123, 132–33, orbiter spacecraft, 130, 299; Orbiting Astronomical Observatory (OAO), 299; Orbiting Quarantine Facility (OQF), 299; payload, 300; *Pioneer* 10, 11 spacecraft, 130, 185, 301; Pioneer Venus mission, 130, 301; pitch, 301; probe, 130, 302; Progress, 303; radioisotope thermoelectric generator (RTG), 31–32, 135, 214–19, 305; Ranger Project, 44, 122, 305; readout station, 306; reconnaissance satellite, 169–70, 306; remote control, 306; remote manipulator system (RMS), 307–8; resilience, 307; roll, 310; rover, 310; science payload, 310; self-replicating system (SRS), 98, 229–33, 248–52, 311;

soft landing, 314; solar cell, 135, 314; spacecraft clock, 139; spin stabilization, 316; *Spirit*, 14–16, 123, 132–33; *Spitzer Space Telescope*, 132–33, 187–89; star probe, 316; static testing, 316; stationkeeping, 317; surface penetrator spacecraft, 131–32, 317; surface rover spacecraft, 132, 317; surveillance satellite, 169–72, 318; Surveyor Project, 44, 129, 131, 318; telemetry, 141–46, 318; teleoperation 318–19; telepresence, 319; telescience, 319; thermal control, 137–38, 319; Thousand Astronomical Unit (TAU) mission, 320; touchdown, 321; trajectory, 321; tumble, 322; vector steering, 324; Venera, 45, 324; Viking Project, 12–13, 45–46, 128, 131, 185, 324; Voyager, 45, 48, 130, 185, 326; yaw, 327; *Yohkoh*, 327; Zond, 327
Robots versus astronauts in space exploration, 213–14
Robot weapon, 42–43, 146–56, 166–69, 383–84; cruise missile, 146–47, 270–71; intercontinental ballistic missile (ICBM), 6, 42–43, 146–49, 166–69, 385; Minuteman (ICBM), 148–49, 166–69, 385; National Air and Space Museum (NASM) 383–84; National Museum of the United States Air Force, 385–86; proliferation of nuclear missile technology, 210–12, remotely piloted vehicle (RPV), 307; Titan Missile Museum, 394–95
Rohrer, Heinrich, 10
Roll, 310
Roomba® vacuuming robot, 344
Rossum's Universal Robots (R.U.R.) (Čapek), 1, 35, 54, 92–94
Rover, 132–33, 310
R2-D2, 20, 45, 51–52. *See also* robots and smart computers in the cinema

Sandia National Laboratories (SNL), 360–61
Santa Claus machine, 19, 41, 310
Scanning tunneling microscope (STM), 10. *See also* nanotechnology
The Science Museum, 390–91
Scientific notation, 310
Scientific Revolution, 3, 28–29, 60–63, 68–70; Nicholas Copernicus, 3, 28–29
Scooba™ floor washing robot, 344
Self-replicating system (SRS), 98, 229–33, 248–52, 311; control of SRS units, 229–33; universal constructor (UC), 323
Sensor, 311; active sensor, 259; passive sensor, 300; remote sensing, 307; resolution, 307
Sensory robot, 108, 311
Sentry robot, 363, 311–12
Serial robot, 311
Service and rescue robots, 16–17, 157–59
Servo, 312
Servomechanism. *See* servo
Servo robot, 107, 108, 312. *See also* industrial robot
Shadow 200 (UAV), 335
Shakey the mobile robot, 7–8, 351
Shelley, Mary Wollstonecraft, 1, 80–84; *Frankenstein: The Modern Prometheus*, 1, 22, 53, 81–82, 372
Siemens (unit), 313
Simulation, 313; *See also* virtual reality
SI units, 313
Smart robot, 236–48, 252–55, 313–14; autonomous future military mobile robots, 252–55; future very smart space robots, 236–48. *See also* artificial intelligence
Solar cell, 135, 314
Solid angle, 314
Solid-state device, 314
Sony Corporation, 17, 51, 362–63; AIBO entertainment robot, 17, 51, 362–63; Sony Wonder Technology Lab, 392–93
Space and Naval Warfare Systems Center, 363. *See also* military robot
Spacecraft, 314–15. *See also* robot spacecraft
Spacecraft clock, 139, 315
Space robot, 121–27, 315. *See also* robot spacecraft
Speed of light, 316
Speed of sound, 316
Spherical coordinate robot, 118, 316. *See also* industrial robot
Spin stabilization, 135–36, 316
Spirit, 14–16, 123, 132–33
Star probe, 245–48, 316
Starship, 316
State of the art (SOA), 316
Static testing, 316

Stäubli Robotics, 48, 112, 363–64
Steel-collar worker, 114
Surveillance satellite, 169–72, 318
Surveyor Project, 44, 129, 131, 318
Synthetic aperture radar (SAR), 318

Talon™, 155–56, 177, 340–41. *See also* military robot
Tap (unit), 318
TEAL RAIN program, 336
The Tech Museum of Innovation, 393–94; Alphabot, 393
Technophobia, 1, 196–99
Telecommunications, 141–46, 318. *See also* communications
Telemetry, 318
Teleoperation 318–19; remote control, 306
Telepresence, 239, 319
Telepresence robot, 239, 392–93
Telescience, 319
Terraforming, 319
Tesla, Nikola, 33–34, 83–84, 86–89; conflict with Thomas Edison, 33–34, 87–88; interaction with George Westinghouse, 33–34, 83–84, 88
Tesla (unit), 319. *See also* SI units
Test bed, 319
Theory of Self-Replicating Automata. See John von Neumann
Thermocouple, 31, 320
Thermodynamics, 320; closed system, 267; energy, 278; entropy, 278; force, 281; heat, 32, 284; heat engine, 32, 284; ideal gas, 285–86; joule (unit), 32, 290; kelvin (unit), 290; kinetic energy, 290; open system, 299; photovoltaic conversion, 301; power, 302; pressure, 302; pump, 303; radiant heat transfer, 305; regenerator, 306; solar cell, 314; solar photovoltaic conversion, 314; specific volume, 316; steam engine, 30, 31; sublimation, 317; temperature, 319; thermal conductivity, 319; thermal equilibrium, 319–20; thermometer, 320; throttling process, 320; thrust, 320; transpiration cooling, 322; turbine, 322; vapor pressure, 323; vapor turbine, 323; volume, 324; watt, 326; work, 327; working fluid, 327
Thousand Astronomical Unit (TAU) mission, 320
Tokyo Science Museum, 395
Ton (unit), 321
Torr (unit), 321
Torricelli, Evangelista, 29, 67; torr, 321
Transducer, 321
Turing, Alan Mathison, 55, 100–102, 228; Turing test, 55, 228

UCI Center for Minimally Invasive Surgery, 364
Underwater robots, 190–93
Ultrahigh frequency (UHF), 322
Ultraviolet (UV) radiation, 322. *See also* electromagnetic radiation
Unimation, 41–42, 104; Stäubli Robotics, 363–64. *See also* Joseph F. Engelberger;
Unit, 322. *See also* SI units
United States Department of Energy (DOE), 175–82, 364–65; Argonne National Laboratory (ANL), 333–34; Idaho National Laboratory (INL), 342; Los Alamos National Laboratory (LANL), 346–47; nuclear robot, 36–41, 175–82, 340, 342; Oak Ridge National Laboratory (ORNL), 183, 352–53; Pacific Northwest National Laboratory (PNNL), 355–56; Sandia National Laboratories (SNL), 360–61; teleoperators, 175–80
United States Patent and Trademark Office (USPTO) Museum, 395–96
Universal constructor (UC), 323. *See also* self-replicating system
Universal time coordinated (UTC), 323
Unmanned aerial vehicle (UAV), 50–52, 150–55, 171–75, 252–55, 323, 335, 336–37; Association for Unmanned Vehicle Systems International (AUVSI), 335–36; Defense Advanced Research Projects Agency (DARPA), 7–8, 336–37; drone, 39, 275; Global Hawk, 52, 153–55, 171–75; National Museum of the United States Air Force, 385–86; Predator, 5, 50, 52, 151–53, 171–75, 385; readout station, 306; remote control, 306; remotely piloted vehicle (RPV), 307; roll, 310; Shadow 200, 335; Space and Naval Warfare Systems Center, 363; TEAL RAIN program, 336. *See also* military robot
Unmanned ground combat vehicle (UGCV), 252–55, 336–37; rover, 310

Unmanned ground vehicle (UGV), 5, 7–8, 155–56, 171–175. *See also* military robot.

Vacuum, 323
Vapor turbine, 323. *See also* thermodynamics
Vaucanson, Jacques de, 3, 30, 264
Viking Project, 12–13, 45–46, 128, 131, 324
Virtual reality (VR), 324–25
Volt (unit), 324
Volta, Count Alessandro, 4, 31, 71–73; volt, 324
Voyager, 45, 48, 130, 326

Water clock, 59
Watt, James, 30, 31, 389
Watt (unit), 326. *See also* SI units
Westinghouse, George, 33–34, 83–84; interaction with Nikola Tesla, 33–34, 83–84, 88; conflict with Thomas Edison, 33–34, 83–84, 88; support for alternating current, 33–34, 83–84
Wiener, Norbert, 39, 94–96, 114–16
Work, 327. *See also* thermodynamics
Work envelope, 327
Working envelope. *See* work envelope
Woods Hole Oceanographic Institution (WHOI), 366; REMUS AUV, 366

X-ray, 34, 327; Wilhelm Konrad Roentgen, 34. *See also* electromagnetic radiation

Yaw, 327
Yohkoh, 327;

Zond, 327

About the Author

JOSEPH A. ANGELO, JR., a retired U.S. Air Force officer (lieutenant colonel), is currently a consulting futurist and technical writer. He has a Ph.D. in nuclear engineering from the University of Arizona and served as a nuclear research officer in the U.S. Air Force (1967–1987) in a variety development and nuclear treaty monitoring. He is also an adjunct professor in the College of Engineering at Florida Tech, specializing in nuclear radiation protection and waste management. Dr. Angelo is the author of 15 other technical books, including *Space Technology* and *Nuclear Technology*.